THERMAL SCIENCES 16

THERMAL SCIENCES 16

Proceedings of the 16th Southeastern Seminar

VOLUME 2

Edited by

T. Nejat Veziroğlu
Clean Energy Research Institute, University of Miami

Springer-Verlag Berlin Heidelberg GmbH

Proceedings of the 16th Southeastern Seminar on Thermal Sciences, held in Miami, Florida, U.S.A., April 19–21, 1982, and presented by the Clean Energy Research Institute, School of Engineering and Architecture, University of Miami, Coral Gables, Florida, U.S.A.; in cooperation with the International Association for Hydrogen Energy, Oak Ridge Associated Universities, and the Department of Mechanical Engineering, University of Miami, Coral Gables, Florida, U.S.A.

EDITOR
T. Nejat Veziroğlu
Clean Energy Research Institute
University of Miami
Coral Gables, Florida 33124, U.S.A.

EDITORIAL BOARD
Halim Gürgenci
Clean Energy Research Institute
University of Miami
Coral Gables, Florida 33124, U.S.A.

M. Necati Özişik
North Carolina State University
Raleigh, North Carolina 27607
U.S.A.

Herbert W. Hoffman
Oak Ridge National Laboratory
Oak Ridge, Tennessee 37830, U.S.A.

John W. Sheffield
University of Missouri-Rolla
Rolla, Missouri 65401, U.S.A.

MANUSCRIPT EDITOR
Sheila M. Puryear
Clean Energy Research Institute
University of Miami
Coral Gables, Florida 33124, U.S.A.

THERMAL SCIENCES 16: Proceedings of the 16th Southeastern Seminar

Copyright © 1983 Springer-Verlag Berlin Heidelberg
Originally published by Hemisphere Publishing Corporation in 1983
Softcover reprint of the hardcover 1st edition 1983

1 2 3 4 5 6 7 8 9 0   B C B C   8 9 8 7 6 5 4 3

Library of Congress Cataloging in Publication Data

Southeastern Seminar on Thermal Sciences (16th : 1982 : Miami, Fla.)
    Thermal sciences 16.

    Proceedings of the 16th Southeastern Seminar on Thermal Sciences, held in Miami, Fla., Apr. 19–21, 1982, and presented by the Clean Energy Research Institute, School of Engineering and Architecture, University of Miami, Coral Gables, Fla.
    Bibliography: p.
    Includes index.
    1. Thermodynamics—Congresses. 2. Heat engineering—Congresses. I. Veziroğlu, T. Nejat. II. University of Miami, Clean Energy Research Institute. III. Title.
TJ265.S68   1982   621.402   83-10705
ISBN 0-89116-319-0   Hemisphere Publishing Corporation

DISTRIBUTION OUTSIDE NORTH AMERICA:
ISBN 978-3-662-13206-7   ISBN 978-3-662-13204-3 (eBook)
DOI 10.1007/978-3-662-13204-3

Contents

MELTING AND SOLIDIFYING

VOLUME 2

TWO-PHASE FLOWS AND HEAT TRANSFER

TWO-PHASE FLOW INSTABILITIES

MASS TRANSPORT

COMBUSTION

SOLAR ENERGY COLLECTION AND STORAGE

SOLAR ENERGY APPLICATIONS

COOLING AND DEHUMIDIFYING

THERMAL CURING

HYBRID ENERGY SYSTEMS

Preface

Continuity of research in basic sciences is indispensable for obtaining tangible results in applied fields. Sometimes it is a new technology introduced by the results of an apparently modest basic research project leading to novel methods of manufacturing; sometimes it is more than that, and new fields of human endeavor are opened before us. Basic scientific research creates the tricks to be released later by the magic hand of technology. These volumes include many papers of this nature, relating the research done to widen the limits of human knowledge. They also include papers dealing with advances in technological applications. As we are living in a world transformed by science every week, sometimes it is difficult to distinguish between the basic and the applied science fields. The time lags associated with the transition of a particular concept from one stage to another have been decreasing day by day.

Thermal sciences is one field in which basic sciences and engineering technologies should go hand in hand. Because of the large number of topics related to thermal sciences, the papers collected in the present volumes come from contributors in various disciplines of engineering sciences. A considerable number of papers cover topics in convection, which still constitutes one of the most important heat transfer research areas. With the new power plant technologies employing higher and higher design heat fluxes and with the introduction of heat transfer applications at very high or very low temperature differences with original system geometries, we need to have a better understanding of basic mechanisms of convection heat transfer.

The subject of multiphase flow and heat transfer is growing in importance. Multiphase flow applications are found in a wide range of engineering systems, such as boiling water nuclear reactors, pressurized water nuclear reactors, conventional steam power plants, evaporators of refrigeration systems, and a wide variety of evaporative and condensive heat exchangers in chemical and petroleum industries. The increasing importance of the field is reflected in the large number of papers contributed under the general title.

As the world's fossil fuel resources are being depleted irrevocably, the whole energy structure based on the premise of cheap oil is becoming obsolete. A number of new energy sources have been proposed as substitutes for oil. With the advent of new energy technologies, new applications of thermal sciences have come to the fore. Various energy-related aspects of thermal sciences are covered in the various parts. These volumes also include papers on other prominent thermal science fields such as radiation heat transfer, turbulence, thermophysical properties, applications, and measurement techniques.

The Southeastern Seminar on Thermal Sciences was initiated in 1965, with the first meeting held on the campus of the Georgia Institute of Technology. Since then, the seminar has convened on a yearly basis—with the exception of a few lapses—and has grown into an important annual event on thermal sciences. The 16th Southeastern Seminar on Thermal Sciences became a truly international event, attracting researchers not only from the United States but from 35 countries around the world.

The papers recommended by the session chairpersons and cochairpersons, together with the keynote address, have been divided by subject into 18 parts in two volumes. The reader should be advised that some papers overlapped our subject matter categories and were difficult to classify specifically. In such cases, we tried to make the best possible choice.

With one half having come from countries outside the United States, these volumes present the latest developments around the world in the ever-expanding field of thermal sciences.

T. Nejat Veziroğlu

Acknowledgments

The Organizing Committee gratefully acknowledges the assistance and cooperation of the Oak Ridge National Laboratory, Oak Ridge Associated Universities, International Association for Hydrogen Energy, and the Department of Mechanical Engineering, University of Miami.

We also wish to extend sincere appreciation to the keynote speaker, William H. Thielbahr, Energy Conservation Branch, Idaho Operations Office, U.S. Department of Energy, and to the banquet speaker, Arthur E. Bergles, Department of Mechanical Engineering, Iowa State University, Ames.

Special thanks are due our authors and lecturers, who provided the substance of the seminar as published in the present volumes.

And last, but not least, our debt of gratitude is owed to the session chairpersons and session cochairpersons in organizing and executing the technical sessions. In acknowledgment, we list these session officials on the following pages.

The Organizing Committee
16th Southeastern Seminar on Thermal Sciences

Conference Committee and Staff

ORGANIZING COMMITTEE

Herbert W. Hoffman
Oak Ridge National Laboratory

Sadik Kakaç
University of Miami

M. Necati Özişik
North Carolina State University

John W. Sheffield
University of Missouri-Rolla

T. Nejat Veziroğlu (Chairperson)
University of Miami

STAFF

Coordinators
G. D. Mayfield
Lucille Walter

Manuscript Editor
Sheila M. Puryear

Special Assistant
Ann Raffle

Graduate Assistants
M. M. Blanco
H. Gürgenci
I. Gürkan
A. Menteş
O. T. Yildirim

Undergraduate Assistants
Mark Drews
Anne Oberdieck
Cristina Robu

Session Officials

Chairpersons and Cochairpersons

K. M. Akyüzlü
University of New Orleans
New Orleans, Louisiana 70148
U.S.A.

W. Aung
National Science Foundation
Washington, D.C. 20550
U.S.A.

Y. Bayazitoğlu
Rice University
Houston, Texas 77001
U.S.A.

B. L. Bhatt
Oakland University
Rochester, Michigan 48063
U.S.A.

J. Z. Catz
University of Miami
Department of Mechanical
   Engineering
Coral Gables, Florida 33124
U.S.A.

N. G. Einspruch
University of Miami
School of Engineering
   and Architecture
Coral Gables, Florida 33124
U.S.A.

B. Farouk
Drexel University
Philadelphia, Pennsylvania 19104
U.S.A.

H. Gürgenci
University of Miami
Coral Gables, Florida 33124
U.S.A.

I. Gürkan
University of Miami
Coral Gables, Florida 33124
U.S.A.

H. W. Hoffman
Oak Ridge National Laboratory
Oak Ridge, Tennessee 37830
U.S.A.

S. Kakaç
University of Miami
Coral Gables, Florida 33134
U.S.A.

A. T. Kirkpatrick
Colorado State University
Fort Collins, Colorado 80523
U.S.A.

Z.-H. Lin
Xian Jiaotong University
Xian, Shaanxi Province
The People's Republic of China

P.-C. Lu
University of Nebraska
Lincoln, Nebraska 68588
U.S.A.

A. Menteş
University of Miami
Coral Gables, Florida 33124
U.S.A.

M. N. Özişik
North Carolina State University
Raleigh, North Carolina 27650
U.S.A.

I. A. Tag
Oak Ridge National Laboratory
Oak Ridge, Tennessee 37830
U.S.A.

W. M. Worek
Illinois Institute of Technology
Chicago, Illinois 60616
U.S.A.

D. W. Yarbrough
Tennessee Technological University
Cookeville, Tennessee 38501
U.S.A.

Y. Yener
University of Delaware
Newark, Delaware 19711
U.S.A.

O. T. Yildirim
University of Miami
Coral Gables, Florida 33124
U.S.A.

TWO-PHASE FLOWS AND HEAT TRANSFER

Heat Transfer to Gas-Solid Suspensions Flowing Turbulently in a Vertical Pipe

TULAY A. ÖZBELGE and TARIK G. SOMER
Department of Chemical Engineering
Middle East Technical University
Ankara, Turkey

ABSTRACT

Considering the importance of convective heat transfer to gas-solid systems in engineering applications, the rate of heat transfer to a gas stream containing solids has been studied. The solution by theoretical means has been achieved by applying Galerkin's [1] and Newman's [2] methods. The results have been compared with the experimental results of Depew[3].

Nusselts numbers of suspensions at different solid loading ratios, at varying gas Reynolds numbers and particle sizes have been calculated.

The results indicate that Newman's numerical method requires less computer time compared to the Galerkin's method to achieve equal accuracy.

1. INTRODUCTION

The effect of solid particles on the rate of heat transfer to gas streams has been observed in numerous studies on fluidized beds. In recent years, the subject has been treated in broader scope to determine the effect of solid particles present in the flowing gas stream. This effect appears to be the introduction of additional turbulence both to the turbulent core and the boundary layer in the conduit, thus enhancing the rate of heat transfer[3 - 9].

Although several correlations for the wall heat transfer coefficient have been proposed[7], they are for limited ranges of parameters only. The main objective is, of course, to obtain a general solution for heat transfer to flowing suspensions at uniform wall heat flux (U.H.F.).

This problem has been theoretically treated first by Tien[8], later by Depew and Farbar [9] with certain simplifications. It should be noted, however, that, these workers have not considered the radial heat transfer due to eddy mixing of solids to be significant. In this study, radial heat transfer of solids is also included.

2. HEAT TRANSFER TO SUSPENSIONS

A heat transfer balance in pipe flow yields the coupled equations which are:

$$C_g \rho_f u_f \frac{\partial T_f}{\partial x} + n_p A_p h_p (T_f - T_s) = \frac{C_g \rho_f}{r} \frac{\partial}{\partial r} \left[r \left(\frac{\nu_g}{Pr} + \epsilon_{H,f} \right) \right] \frac{\partial T_f}{\partial r} \tag{1}$$

for the fluid phase, and

$$C_s \rho_s u_s \frac{\partial T_s}{\partial x} + n_p A_p h_p (T_s - T_f) = \frac{C_s \rho_s}{r} \frac{\partial}{\partial r} \left[r \epsilon_{H,s} \right] \frac{\partial T_s}{\partial r} \tag{2}$$

for the solids; the boundary conditions are:

at $x = 0$, $T_f = T_s = T_o$ (inlet temperature) \qquad (3)

at $r = 0$, $\dfrac{\partial T_f}{\partial r} = \dfrac{\partial T_s}{\partial r} = 0$ \qquad (4)

at $r = R$, $k_g \dfrac{\partial T_f}{\partial r} = q$, $\dfrac{\partial T_s}{\partial r} = 0$ \qquad (5)

In equations (1) and (2), the first term on the left hand side is the heat convected, the second term is the heat transferred between the phases, and the term on the right hand is the radial heat transfer caused by the radial mixing of phases.

The main assumptions involved in the above equations are:

(a) the solid particles are sufficiently small and numerous to be described as a continuum,

(b) viscous dissipation is negligible,

(c) axial conduction is negligible; this is reasonable when Péclét number exceeds 100,

(d) the system is at steady state,

(e) temperature gradients do not exist in individual particles due to high solid thermal conductivity and due to small and uniform particle sizes,

(f) the suspension is very dilute; therefore solid particles do not affect the gas velocity profile which has been obtained from Wasan - Tien - Wilke correlation [10] for a turbulent flow field,

(g) developed velocity profile at the onset of heating,

(h) plug flow for solids velocity,

(i) gas - particle heat transfer coefficient is obtained from Ranz-Marshall correlation [11] ,

(j) the solid particles are uniformly distributed throughout the pipe,

(k) as a heat transfer mechanism, it is assumed that heat is transferred

first from the pipe wall to the gas and later from the gas to the particle.
Direct heat transfer from the pipe wall to the solid particle is assumed to
be negligible due to the small contact area and the small residence time
during collisions between particles and the pipe wall. Therefore the solids
phase temperature will be lagging behind the fluid phase temperature.

Heat transfer is coupled with the hydrodynamics of the system which has been
analysed previously [12].Equations (1-5) are given in dimensionless form as:

$$u_f^+ \frac{\partial \theta_f}{\partial x^+} + \beta_f(\theta_f - \theta_s) = \frac{1}{r^+} \frac{\partial}{\partial r^+}(r^+ \gamma_f \frac{\partial \theta_f}{\partial r^+}) \tag{6}$$

for the fluid phase, and

$$u_s^+ \frac{\partial \theta_s}{\partial x^+} + \beta_s(\theta_s - \theta_f) = \frac{1}{r^+} \frac{\partial}{\partial r^+}(r^+ \gamma_s \frac{\partial \theta_s}{\partial r^+}) \tag{7}$$

for the solids; the boundary conditions are:

at $x^+ = 0$ $\qquad\qquad$ $\theta_f = \theta_s = 0$ $\qquad\qquad\qquad\qquad\qquad\qquad$ (8)

at $r^+ = 0$ $\qquad\qquad$ $\dfrac{\partial \theta_f}{\partial r^+} = \dfrac{\partial \theta_s}{\partial r^+} = 0$ $\qquad\qquad\qquad\qquad$ (9)

at $r^+ = r_o^+$ $\qquad\qquad$ $\dfrac{\partial \theta_f}{\partial r^+} = 1, \quad \dfrac{\partial \theta_s}{\partial r^+} = 0$ $\qquad\qquad$ (10)

where

$$\beta_f = \frac{3h_p \rho_s \nu_g}{\rho_f C_g R_p \overset{*}{u}_f^2 \rho_p} \qquad \text{and} \tag{11}$$

$$\beta_s = \frac{3h_p \nu_g}{C_s R_p \overset{*}{u}_f^2 \rho_p} \tag{12}$$

Eddy diffusivity functions for the fluid, $\varepsilon_{H,f}(r^+)$, and the solids phases,
$\varepsilon_{H,s}(r^+)$ depend on the radial position in the pipe. The eddy diffusivity of
heat for the fluid phase is obtained from Wasan – Tien – Wilke correlation[10],
which is:

$$\frac{\varepsilon_{H,f}}{\nu_g} = (4A_1 y_p^3 - 5A_2 y_p^4)/(1 - 4A_1 y_p^3 + 5A_2 y_p^4) \tag{13}$$

for $y_p = r_o^+ - r^+ \leq 20$ (viscous sublayer and buffer region)

where $A_1 = 1.097 \times 10^{-4}$, $A_2 = 3.295 \times 10^{-6}$ and

$$\frac{\epsilon_{H,f}}{\nu_g} = \frac{y_p}{2.5} - 1 \tag{14}$$

for $y_p > 20$ (turbulent core region)

The relationship between the eddy diffusivities of fluid and solids has been decided from the literature [13].

Turbulent velocity profile of the fluid phase is given, in terms of dimensionless quantities, as [10]:

$$u_f^+ = y_p - A_1 y_p^4 + A_2 y_p^5 \qquad \text{if} \qquad y_p \leq 20 \tag{15}$$

and

$$u_f^+ = 2.5 \, \ell n \, y_p + 5.5 \qquad \text{if} \qquad y_p > 20 \tag{16}$$

Solutions to equations (6) and (7) with the given boundary conditions (8), (9) and (10) have been obtained by the use of the numerical method of Newman and the approximate method of Galerkin.

3.   NUMERICAL METHOD OF SOLUTION

After the fluid is far downstream from the beginning of the heated section, one can assume that the temperatures of both of the phases will increase linearly with x^+ for the constant heat flux. Hence the following dimensionless equations seem reasonable for large x^+ :

$$\theta_f = \frac{4x^+}{Pe_m} + \psi_f(r^+) + C_1 \tag{17}$$

$$\theta_s = \frac{4x^+}{Pe_m} + \psi_s(r^+) + C_1 \tag{18}$$

where Pe_m is the Péclét number of the suspension defined as [3]

$$Pe_m = (Re)(Pr) \left(1 + \frac{C_s W_s}{C_g W_f}\right) \tag{19}$$

The following set of ordinary differential equations and the boundary conditions are obtained in the dimensionless form by applying equations (17) and (18) to equations (6 – 10) :

$$u_f^+ \frac{4}{Pe_m} + \beta_f(\psi_f - \psi_s) = \frac{1}{r^+} \frac{d}{dr^+}(r^+ \gamma_f \frac{d\psi_f}{dr^+}) \tag{20}$$

$$u_s^+ \frac{4}{Pe_m} + \beta_s(\psi_s - \psi_f) = \frac{1}{r^+} \frac{d}{dr^+} (r^+ \gamma_s \frac{d\psi_s}{dr^+}) \tag{21}$$

$$\text{at } r^+ = 0 \qquad \frac{d\psi_f}{dr^+} = \frac{d\psi_s}{dr^+} = 0 \tag{22}$$

$$\text{at } r^+ = r_o^+ \qquad \frac{d\psi_f}{dr^+} = 1, \quad \frac{d\psi_s}{dr^+} = 0 \tag{23}$$

It is apparent that equation (8) cannot be satisfied for the solution at large x^+. Hence it is replaced by

$$2\pi Rxq = 2\pi \int_0^R \rho_f u_f C_g (T_f - T_o) r \, dr + 2\pi \int_0^R \rho_s u_s C_s (T_s - T_o) r dr \tag{24}$$

or in the dimensionless form, equation(24) is equivalent to

$$r_o^+ x^+ = \int_0^{r_o^+} \frac{\rho_f C_g \nu_g}{k_g} \theta_f u_f^+ r^+ dr^+ + \int_0^{r_o^+} \frac{\rho_s C_s \nu_g}{k_g} \theta_s u_s^+ r^+ dr^+ \tag{25}$$

Equations(17) and (18) are substituted in equation (25), which is simplified afterwards to get constant C_1 as:

$$C_1 = - \frac{\left[\rho_f C_g \int_0^{r_o^+} u_f^+ \psi_f(r^+) r^+ dr^+ + \rho_s C_s \int_0^{r_o^+} u_s^+ \psi_s(r^+) r^+ dr^+ \right]}{\left[\rho_f C_g \int_0^{r_o^+} u_f^+ r^+ dr^+ + \rho_s C_s \int_0^{r_o^+} u_s^+ r^+ dr^+ \right]} \tag{26}$$

3.1 Newman's Numerical Method

This method gives the solution of coupled, linear, difference equations [2]. Ordinary differential equations have to be linearized about a trial solution if they are not linear and then put into finite difference form.

At each point j, except $j = 1$ and $j = j_{max}$, there are "n" equations of the form

$$\sum_{k=1}^{n} A_{i,k}(j) C_k(j-1) + B_{i,k}(j) C_k(j) + D_{i,k}(j) C_k(j+1) = G_i(j) \tag{27}$$

The unknowns are C_k. The subscript i denotes the equation number, and each of the equations can involve all of the unknowns C_k, through the sum. $A_{i,k}$, $B_{i,k}$, and $D_{i,k}$ are coefficients of the unknowns at the mesh points $j-1$, j, and $j+1$, and G_i contains all terms independent of the unknowns C_k.

At $j = 1$ the equations are

$$\sum_{k=1}^{n} B_{i,k}(j)C_k(j) + D_{i,k}(j)C_k(j+1) + X_{i,k}C_k(j+2) = G_i(j) \tag{28}$$

At $j = j_{max}$ the equations are

$$\sum_{k=1}^{n} Y_{i,k}C_k(j-2) + A_{i,k}(j)C_k(j-1) + B_{i,k}(j) C_k(j) = G_i(j) \tag{29}$$

The coefficients $X_{i,k}$ and $Y_{i,k}$ in equations (28) and (29) respectively, allow the introduction of complex boundary conditions. In this study, they are not necessary.

According to the method, equations (20) and (21) are written in finite difference formulae [2] for the first and second derivatives. The unknowns here are :

$$C_1(j) = \psi_f(j) \tag{30}$$

$$C_2(j) = \psi_s(j) \tag{31}$$

Comparing the coefficients of the unknowns in the finite-difference forms of equations (20) and (21) with equation (27), the coefficients of the matrix to be solved can be written for the mesh points $1 < j < j_{max}$ as follows:

$$A_{1,1}(j) = \frac{\gamma_f(j)}{h^2} - \frac{1}{2h} \frac{d\gamma_f}{dr^+}(j) - \frac{\gamma_f(j)}{2h\, r^+(j)} \tag{32}$$

$$A_{2,2}(j) = \frac{\gamma_s(j)}{h^2} - \frac{1}{2h} \frac{d\gamma_s}{dr^+}(j) - \frac{\gamma_s(j)}{2h\, r^+(j)} \tag{33}$$

$$B_{1,1}(j) = -\beta_f(j) - \frac{2\gamma_f(j)}{h^2} \tag{34}$$

$$B_{1,2}(j) = \beta_f(j) \tag{35}$$

$$B_{2,1}(j) = \beta_s(j) \tag{36}$$

$$B_{2,2}(j) = -\beta_s(j) - \frac{2\gamma_s(j)}{h^2} \tag{37}$$

$$D_{1,1}(j) = \frac{\gamma_f(j)}{h^2} + \frac{1}{2h} \frac{d\gamma_f}{dr^+}(j) + \frac{\gamma_f(j)}{2h\, r^+(j)} \tag{38}$$

$$D_{2,2}(j) = \frac{\gamma_s(j)}{h^2} + \frac{1}{2h} \frac{d\gamma_s}{dr^+}(j) + \frac{\gamma_s(j)}{2hr^+(j)} \tag{39}$$

$$G_1(j) = \frac{4u_f^+(j)}{Pe_m} \tag{40}$$

$$G_2(j) = \frac{4u_s^+(j)}{Pe_m} \tag{41}$$

Other coefficients of the matrix are equal to zero.

The first boundary condition, $r^+ = 0$ is a singular point. Therefore, one has to expand ψ_f and ψ_s around this point according to the Taylor Series Expansion to satisfy the boundary condition at $r^+ = 0$ or at $j = 1$. Otherwise, a trivial solution is obtained. These expansions are:

$$\psi_f(r^+) = \psi_f(0) + \psi_f'(0)r^+ + \frac{1}{2!}\psi_f''(0)r^{+2} + \mathscr{O}(h^2) \tag{42}$$

$$\psi_s(r^+) = \psi_s(0) + \psi_s'(0)r^+ + \frac{1}{2!}\psi_s''(0)r^{+2} + \mathscr{O}(h^2) \tag{43}$$

where error of truncation for the series after the second derivative is in the order of h^2. From equation (22)

$$\psi_f'(0) = \psi_s'(0) = 0 \tag{44}$$

Substituting equations (42), (43) and (44) into equations (20) and (21), then setting r^+ equal to zero, the following results are obtained :

$$u_f^+(0)\frac{4}{Pe_m} + \beta_f(0)\left[\psi_f(0) - \psi_s(0)\right] = 2\gamma_f(0)\psi_f''(0) \tag{45}$$

$$u_s^+(0)\frac{4}{Pe_m} + \beta_s(0)\left[\psi_s(0) - \psi_f(0)\right] = 2\gamma_s(0)\psi_s''(0) \tag{46}$$

$\psi_f''(0)$ and $\psi_s''(0)$ are obtained from equations (45) and (46), then they are substituted back in equations (42) and (43) respectively to get

$$\psi_f(r^+) = \psi_f(0) + \frac{1}{2}\left\{\frac{2u_f^+(0)}{Pe_m\gamma_f(0)} + \frac{\beta_f(0)}{2\gamma_f(0)}\left[\psi_f(0) - \psi_s(0)\right]\right\}r^{+2} \tag{47}$$

$$\psi_s(r^+) = \psi_s(0) + \frac{1}{2}\left\{\frac{2u_s^+(0)}{Pe_m\gamma_s(0)} + \frac{\beta_s(0)}{2\gamma_s(0)}\left[\psi_s(0) - \psi_f(0)\right]\right\}r^{+2} \tag{48}$$

Now, by taking $r^+ = h$, equations (47) and (48) can be written in the finite-difference form:

$$\psi_f(j+1) - \left[1 + \frac{\beta_f(j)h^2}{4\gamma_f(j)}\right]\psi_f(j) + \frac{\beta_f(j)h^2}{4\gamma_f(j)}\psi_s(j) = \frac{u_f^+(j)h^2}{Pe_m\gamma_f(j)} \tag{49}$$

$$\psi_s(j+1) - \left[1 + \frac{\beta_s(j)h^2}{4\gamma_s(j)}\right]\psi_s(j) + \frac{\beta_s(j)h^2}{4\gamma_s(j)}\psi_f(j) = \frac{u_s^+(j)h^2}{Pe_m\gamma_s(j)} \tag{50}$$

Comparing equations (49) and (50) with equation (28) the coefficients of the unknowns are obtained as follows:

$$B_{1,1}(j) = -1 - \frac{\beta_f(j)h^2}{4\gamma_f(j)} \tag{51}$$

$$B_{1,2}(j) = \frac{\beta_f(j)h^2}{4\gamma_f(j)} \tag{52}$$

$$B_{2,1}(j) = \frac{\beta_s(j)h^2}{4\gamma_s(j)} \tag{53}$$

$$B_{2,2}(j) = -1 - \frac{\beta_s(j)h^2}{4\gamma_s(j)} \tag{54}$$

$$D_{1;1}(j) = 1 \tag{55}$$

$$D_{2,2}(j) = 1 \tag{56}$$

$$G_1(j) = \frac{u_f^+(j)h^2}{\gamma_f(j)Pe_m} \tag{57}$$

$$G_2(j) = \frac{u_s^+(j)h^2}{\gamma_s(j)Pe_m} \tag{58}$$

Other coefficients of the matrix created for $j = 1$ are zero.

Equation (23) is changed to a logically equivalent form in equation (59);

$$\text{at} \quad r^+ = r_o^+ \qquad \psi_f = 0, \quad \frac{d\psi_s}{dr^+} = 0 \tag{59}$$

If one of the unknowns is not initialized, the method does not give a regular solution, but a trivial one.

Similarly, $\psi_s(r^+)$ is expanded around the point r_o^+, in Taylor Series as follows:

$$\psi_s(r^+) = \psi_s(r_o^+) + \psi_s'(r_o^+)(r^+ - r_o^+) + \frac{1}{2!}\psi_s''(r_o^+)(r^+ - r_o^+)^2 + \mathcal{O}(h^2) \tag{60}$$

After substituting equations (59) and (60) in equation (21), then setting $r^+ = r_o^+$, the following equation is obtained:

$$u_s^+(r_o^+)\frac{4}{Pe_m} + \beta_s(r_o^+)\,\psi_s(r_o^+) = \gamma_s(r_o^+)\,\psi_s''(r_o^+) \tag{61}$$

$\psi_s''(r_o^+)$ from equation (61) is substituted back in equation (60) to get

$$\psi_s(r^+) = \psi_s(r_o^+) + \left[\frac{2\,u_s^+(r_o^+)}{\gamma_s(r_o^+)Pe_m} + \frac{\beta_s(r_o^+)\,\psi_s(r_o^+)}{2\gamma_s(r_o^+)}\right](r^+ - r_o^+)^2 \tag{62}$$

Considering the mesh interval as $h = r^+ - r_0^+$, the finite-difference form of equation (62) is:

$$\psi_s(j-1) - \left[1 + \frac{\beta_s(j)h^2}{2\gamma_s(j)} \right] \psi_s(j) = \frac{2 \, u_s^+(j)h^2}{\gamma_s(j)Pe_m} \tag{63}$$

The first equation for the first unknown, ψ_f, at $r^+ = r_0^+$ or $j = j_{max}$ is written from equation (59) as:

$$\psi_f(j) = 0 \tag{64}$$

Comparing equations (63) and (64) with equation (29), the following coefficients of the matrix at $j = j_{max}$ are obtained:

$$A_{2,2}(j) = 1 \tag{65}$$

$$B_{1,1}(j) = 1 \tag{66}$$

$$B_{2,2}(j) = -1 - \frac{\beta_s(j)h^2}{2\gamma_s(j)} \tag{67}$$

$$G_2(j) = \frac{2u_s^+(j)h^2}{\gamma_s(j)Pe_m} \tag{68}$$

The rest of the coefficients are zero.

The resulting coupled, tridiagonal matrices can be solved by computer. The arrays A,B,D and G are supplied by the main program for each value of j, and a special subroutine BAND(J) [2] is to be called for each value of j to obtain the values of the unknowns ψ_f and ψ_s at each mesh point.

4. APPROXIMATE METHOD OF SOLUTION

The Galerkin's method, which is generally used for approximate analysis of boundary value problems involving parabolic and elliptic type differential equations, has been applied for the solution of two-phase flow heat transfer problem in this study.

Defining a new dimensionless variable, ξ, and the dimensionless temperatures as

$$\frac{r^+}{r_0^+} = \xi \tag{69}$$

$$\theta_f = \frac{4x^+}{Pe_m} + \psi_f(\xi) \tag{70}$$

$$\theta_s = \frac{4x^+}{Pe_m} + \psi_s(\xi) \tag{71}$$

$$\phi_f = \psi_f(\xi) - \psi_w(x^+) \tag{72}$$

$$\phi_s = \psi_s(\xi) - \psi_w(x^+) \tag{73}$$

equations (20) and (21) can be written as follows:

$$u_f^+ r_o^{+2} \frac{4}{Pe_m} + \beta_f r_o^{+2}(\phi_f - \phi_s) = \frac{1}{\xi} \frac{d}{d\xi} (\xi\gamma_f \frac{d\phi_f}{d\xi}) \tag{74}$$

for the fluid phase and:

$$u_s^+ r_o^{+2} \frac{4}{Pe_m} + \beta_s r_o^{+2} (\phi_s - \phi_s) = \frac{1}{\xi} \frac{d}{d\xi} (\xi\gamma_s \frac{d\phi_s}{d\xi}) \tag{75}$$

for the solids phase. $\psi_w(x^+)$ is the dimensionless wall temperature at any dimensionless axial distance, x^+.

Equations (72) and (73) have been written to reduce the nonhomogeneous boundary condition given in equation (23) to the homogeneous one by assuming that

$$\psi_f = \psi_s = \psi_w(x^+) \qquad at \qquad \xi = 1 \tag{76}$$

Then the boundary conditions become

$$at \qquad \xi = 0 \qquad \frac{d\phi_f}{d\xi} = \frac{d\phi_s}{d\xi} = 0 \tag{77}$$

$$at \qquad \xi = 1 \qquad \phi_f = 0 \quad , \quad \phi_s = 0 \tag{78}$$

4.1 Galerkin's Approximate Method

The basis of the Galerkin's method is that of seeking solutions to the system in the following forms

$$\phi_{f_n} = \sum_{i=1}^{n} D_i\phi_i(\xi) \tag{79}$$

$$\phi_{s_n} = \sum_{i=1}^{n} E_i\phi_i(\xi) \tag{80}$$

where ϕ_{f_n} and ϕ_{s_n} are approximate temperature profiles of phases, i.e., n-term approximation; $\phi_i(\xi)$ is a set of known functions chosen at onset of problem such that they satisfy the given boundary conditions. D_i and E_i in equations (79) and (80) are the unknown coefficients to be determined by the Galerkin's method.

Bessel functions of first kind, zero order have been chosen as a set of known trial functions, $\phi_i(\xi)$, to satisfy the differential equations and the boundary conditions. For a two-term approximation, the approximate solutions are

$$\phi_f = D_1 J_o(\beta_1 \xi) + D_2 J_o(\beta_2 \xi) \tag{81}$$

$$\phi_s = E_1 J_o(\beta_1 \xi) + E_2 J_o(\beta_2 \xi) \tag{82}$$

where β_1, β_2 are the roots of the Bessel function J_o.

Equations (81) and (82) are substituted in the differential equations (74) and (75) to get

$$L \left[\sum_{i=1}^{2} D_i \phi_i(\xi) \right] = 0 \tag{83}$$

$$L \left[\sum_{i=1}^{2} E_i \phi_i(\xi) \right] = 0 \tag{84}$$

Equations (83) and (84) are multiplied by $\phi_j(\xi)$ and integrated over the region R, to yield 4 algebraic equations to determine the unknown coefficients D_1, D_2 and E_1, E_2 which are substituted afterwards in equations (81) and (82) to get two-term approximate solution of the problem.

The integrated forms of equations (83) and (84) according to the Galerkin's method are;

For $\phi_1(\xi) = J_o(\beta_1 \xi)$:

$$\{ r_o^{+2} \int_0^1 \beta_f(\xi) \left[J_o(\beta_1 \xi) \right]^2 \xi d\xi + \beta_1 \int_0^1 \frac{d\gamma_f}{d\xi} J_1(\beta_1 \xi) J_o(\beta_1 \xi) \xi d\xi$$

$$+ \beta_1^2 \int_0^1 \gamma_f(\xi) \left[J_o(\beta_1 \xi) \right]^2 \xi d\xi \} D_1 + \{ r_o^{+2} \int_0^1 \beta_f(\xi) J_o(\beta_2 \xi) J_o(\beta_1 \xi) \xi d\xi$$

$$+ \beta_2 \int_0^1 \frac{d\gamma_f}{d\xi} J_1(\beta_2 \xi) J_o(\beta_1 \xi) \xi \, d\xi + \beta_2^2 \int_0^1 \gamma_f(\xi) J_o(\beta_2 \xi) J_o(\beta_1 \xi) \xi \, d\xi \} D_2$$

$$- \{ r_o^{+2} \int_0^1 \beta_f(\xi) \left[J_o(\beta_1 \xi) \right]^2 \xi \, d\xi \} E_1 - \{ r_o^{+2} \int_0^1 \beta_f(\xi) J_o(\beta_2 \xi) J_o(\beta_1 \xi) \xi d\xi \} E_2$$

$$= - \frac{4}{Pe_m} r_o^{+2} \int_0^1 u_f^+ J_o(\beta_1 \xi) \xi \, d\xi \tag{85}$$

For $\phi_2(\xi) = J_o(\beta_2 \xi)$

$$\{ r_o^{+2} \int_o^1 \beta_f(\xi) J_o(\beta_1\xi) \, J_o(\beta_2\xi) \, \xi d\xi + \beta_1 \int_o^1 \frac{d\gamma_f}{d\xi} J_1(\beta_1\xi) J_o(\beta_2\xi) \, \xi d\xi$$

$$+ \beta_1^2 \int_o^1 \gamma_f(\xi) J_o(\beta_1\xi) J_o (\beta_2\xi) \, \xi \, d\xi \} D_1 + \{ r_o^{+2} \int_o^1 \beta_f(\xi) \left[J_o(\beta_2\xi) \right]^2 \xi d\xi$$

$$+ \beta_2 \int_o^1 \frac{d\gamma_f}{d\xi} J_1(\beta_2\xi) J_o(\beta_2\xi) \, \xi d\xi + \beta_2^2 \int_o^1 \gamma_f(\xi) \left[J_o(\beta_2\xi) \right]^2 \xi \, d\xi \} \, D_2$$

$$- \{ r_o^{+2} \int_o^1 \beta_f(\xi) J_o(\beta_1\xi) J_o(\beta_2\xi) \, \xi d\xi \} \, E_1 - \{ r_o^{+2} \int_o^1 \beta_f(\xi) \left[J_o(\beta_2\xi) \right]^2 \xi \, d\xi \} \, E_2$$

$$= - \frac{4}{Pe_m} r_o^{+2} \int_o^1 u_f^+ J_o(\beta_2\xi) \, \xi \, d\xi \tag{86}$$

From the differential equation of the solids phase for $\phi_1(\xi) = J_o(\beta_1\xi)$

$$- \{ r_o^{+2} \int_o^1 \beta_s(\xi) \left[J_o(\beta_1\xi) \right]^2 \xi \, d\xi \} \, D_1 - \{ r_o^{+2} \int_o^1 \beta_s(\xi) J_o(\beta_2\xi) J_o(\beta_1\xi) \xi \, d\xi \} \, D_2$$

$$+ \{ r_o^{+2} \int_o^1 \beta_s(\xi) \left[J_o(\beta_1\xi) \right]^2 \xi \, d\xi + \beta_1 \int_o^1 \frac{d\gamma_s}{d\xi} J_1(\beta_1\xi) J_o(\beta_1\xi) \, \xi \, d\xi$$

$$+ \beta_1^2 \int_o^1 \gamma_s(\xi) \left[J_o(\beta_1\xi) \right]^2 \xi d\xi \} \, E_1 + \{ r_o^{+2} \int_o^1 \beta_s J_o(\beta_2\xi) J_o(\beta_1\xi) \xi d\xi$$

$$+ \beta_2 \int_o^1 \frac{d\gamma_s}{d\xi} J_1(\beta_2\xi) J_o(\beta_1\xi) \, \xi \, d\xi + \beta_2^2 \int_o^1 \gamma_s(\xi) J_o(\beta_2\xi) J_o(\beta_1\xi) \, \xi \, d\xi \} \, E_2$$

$$= - \frac{4}{Pe_m} r_o^{+2} u_s^+ \int_o^1 J_o(\beta_1\xi) \, \xi \, d\xi \tag{87}$$

For $\phi_2(\xi) = J_o(\beta_2\xi)$

$$- \{ r_o^{+2} \int_o^1 \beta_s(\xi) J_o(\beta_1\xi) J_o(\beta_2\xi) \, \xi \, d\xi \} \, D_1 - \{ r_o^{+2} \int_o^1 \beta_s(\xi) \left[J_o(\beta_2\xi) \right]^2 \xi \, d\xi \} \, D_2$$

$$+ \{r_o^{+2} \int_o^1 \beta_s(\xi) J_o(\beta_1\xi) J_o(\beta_2\xi)\xi d\xi + \beta_1 \int_o^1 \frac{d\gamma_s}{d\xi} J_1(\beta_1\xi)J_o(\beta_2\xi)\xi \, d\xi$$

$$+ \beta_1^2 \int_o^1 \gamma_s(\xi) J_o(\beta_1\xi)J_o(\beta_2\xi)\xi d\xi\} \, E_1 + \{r_o^{+2} \int_o^1 \beta_s(\xi) \left[J_o(\beta_2\xi) \right]^2 \xi \, d\xi$$

$$+\beta_2 \int_o^1 \frac{d\gamma_s}{d\xi} J_1(\beta_2\xi)J_o(\beta_2\xi) \, \xi d\xi + \beta_2^2 \int_o^1 \gamma_s(\xi) \left[J_o \, \beta_2\xi \right]^2 \xi \, d\xi\} \, E_2$$

$$= - \frac{4}{Pe_m} r_o^{+2} u_s^+ \int_o^1 J_o(\beta_2\xi) \, \xi \, d\xi \qquad\qquad (88)$$

5.   DIMENSIONLESS PIPE - WALL TEMPERATURE

In the numerical   method, the dimensionless wall temperature is calculated by equations (17), (26) and (59); their combination gives the following equation,

$$\theta_w = \frac{4x^+}{Pe_m} + C_1 \qquad\qquad (89)$$

As it has been previously explained, C_1 is derived from equation (25). The latter equation is used for the derivation of the dimensionless wall temperature, $\psi_w(x^+)$, in the approximate method also. Substitution of equations (69-73) in equation (25) gives:

$$\psi_w(x^+) = \{x^+ - G_c r_o^+ \left[\frac{4x^+}{Pe_m} \int_o^1 u_f^+\xi d\xi + \int_o^1 \phi_f u_f^+ \, \xi \, d\xi \right]$$

$$- S_c \, r_o^+ u_s^+ \left[\frac{2x^+}{Pe_m} + \int_o^1 \phi_s \xi \, d\xi \right]\} / \left[G_c r_o^+ \int_o^1 u_f^+ \xi d\xi + S_c u_s^+ r_o^+/2 \right] \qquad (90)$$

where

$$G_c = \frac{\rho_f \, \nu_g C_g}{k_g} \qquad\qquad (91)$$

$$S_c = \frac{\rho_s \nu_g C_s}{k_g} \qquad\qquad (92)$$

6. NUSSELT NUMBER OF SUSPENSION

The definition of Nusselt number is

$$Nu_{s,N} = \frac{\left. \dfrac{dT_f}{dr} \right|_{wall} D}{T_w - T_{f,mm}} \tag{93}$$

In terms of the dimensionless variables used in the numerical solution, Nusselt number is expressed as:

$$Nu_{s,N} = \frac{\left. \dfrac{d\,\theta_f}{dr^+} \right|_{r^+ = r_o^+} (2r_o^+)}{\theta_{f,w} - \theta_{f,mm}} \tag{94}$$

where θ_f' is calculated from backward difference formula of numerical differentiation.

$$\left. \theta_f' \right|_j = (3\theta_f \left|_j - 4\theta_f \right|_{j-1} + \theta_f \left|_{j-2}\right.)/2h \tag{95}$$

Equation (95) is used where j is equal to j_{max} (at the wall).

Depew [3] derives the mixed-mean temperature from the total energy equation as

$$T_{mm} = \frac{W_f C_g}{W_f C_g + W_s C_s} \; T_{f,m} + \frac{W_s C_s}{W_f C_g + W_s C_s} \; T_{s,m} \tag{96}$$

Equation (96) is equivalent to

$$\theta_{f,mm} = \frac{\kappa}{1+\kappa} \left\{ \frac{\displaystyle\int_0^{r_o^+} \theta_f \, u_f^+ \, r^+ dr^+}{\displaystyle\int_0^{r_o^+} u_f^+ \, r^+ dr^+} \right\} + \frac{1}{1+\kappa} \left\{ \frac{\displaystyle\int_0^{r_o^+} \theta_f \, u_s^+ \, r^+ dr^+}{\displaystyle\int_0^{r_o^+} u_s^+ \, r^+ dr^+} \right\} \tag{97}$$

where

$$\kappa = \frac{W_f C_g}{W_s C_s} \tag{98}$$

After the temperatures of the phases are obtained at each mesh point, as a result of the numerical solution, Nusselt numbers are calculated from equations (89), (94),(95), and (97).

In the approximate solution, Nusselt numbers have been calculated from the following equations

$$Nu_{s,G} = \frac{2 \left. \dfrac{d\phi_f}{d\xi} \right|_{\xi=1}}{\phi_{f,w} - \phi_{f,mm}} \tag{99}$$

$$\phi_{f,w} = 0 \tag{100}$$

$$\phi_{f,mm} = \frac{\kappa}{1+\kappa} \left\{ \frac{\displaystyle\int_0^1 \phi_f u_f^+ \xi d\xi}{\displaystyle\int_0^1 u_f^+ \xi d\xi} \right\} + \frac{1}{1+\kappa} \left\{ \frac{\displaystyle\int_0^1 \phi_s u_s^+ \xi d\xi}{\displaystyle\int_0^1 u_s^+ \xi d\xi} \right\} . \tag{101}$$

7. RESULTS AND DISCUSSION

The dependence of Nusselt number to solids loading ratio is established in this study by numerical analysis. The results had to be compared with experimental data to see how closely these two checked. The experimental data was taken from Depew's study[3]. The Nusselt numbers of this study had to be recalculated by correcting for the wall temperatures as outlined below:

In Depew's study[3], Nusselt number is expressed as

$$Nu_H = \frac{h_i D_i}{k_g} = \frac{qD}{(T_w - T_{mm})k_g} \tag{102}$$

where T_w, wall temperature, was measured by a thermocouple. Noting that at high velocities the thermocouple readings would actually be the stagnation temperatures, thus including the kinetic energy term, the readings had to be corrected for the true gas-solid stream temperature. Having made these corrections by subtracting the kinetic energy term, Nusselt numbers were recalculated by means of equation (102).

The dependence of Nusselt numbers on solids loading ratio is shown in Figures 1-4. In these figures, the Reynolds numbers are 13500 and 27700; particle diameters 30 and 200µm. The results of this study by numerical analysis and those of Depew by experimental measurements are plotted together on the same figures.

As it can be seen, the two sets of data check each other very closely, indicating that the method of solution by numerical analysis is sound.

For small particles (30µm) Nusselt number increases by increasing solids loading ratio. For large particles (200µm), however, Nusselt number appears to be independent of solids loading ratio.

In an attempt to correlate the results by a relationship of the Dittus-Boelter equation type, Prandtl number being constant, it is established that $Nu_{s,N}/Re^{0.6}$ ratio is independent of the Reynolds number for the same particle size. This relationship, which could be expressed by

$$\frac{Nu_{s,N}}{Re^{0.6}} = f(W_s/W_f) \tag{103}$$

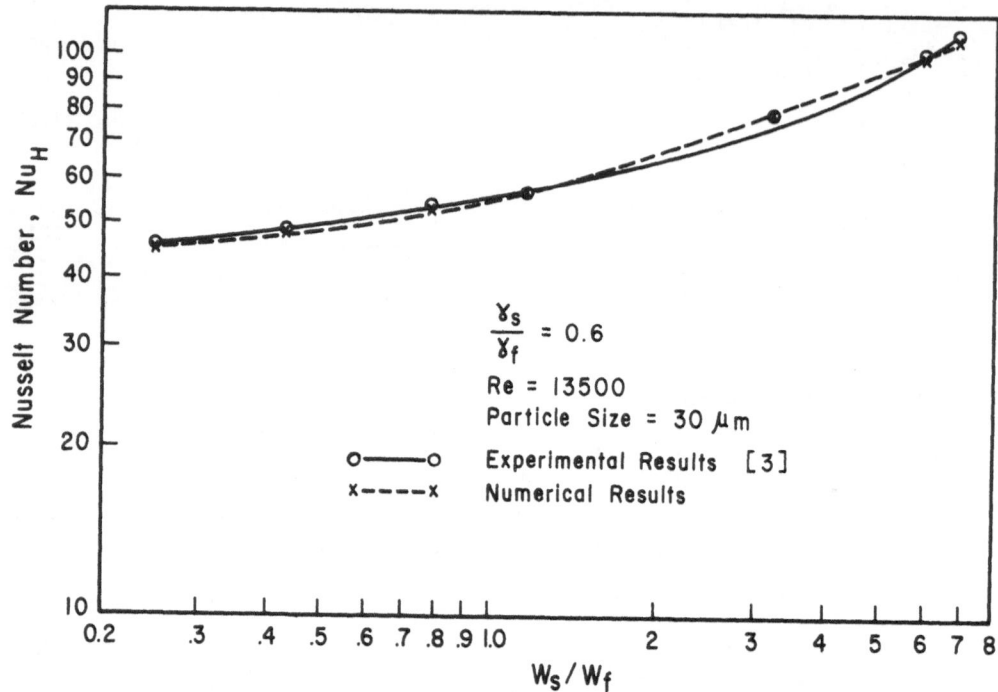

Figure 1.   Comparison of Numerical Results with Experimental Data

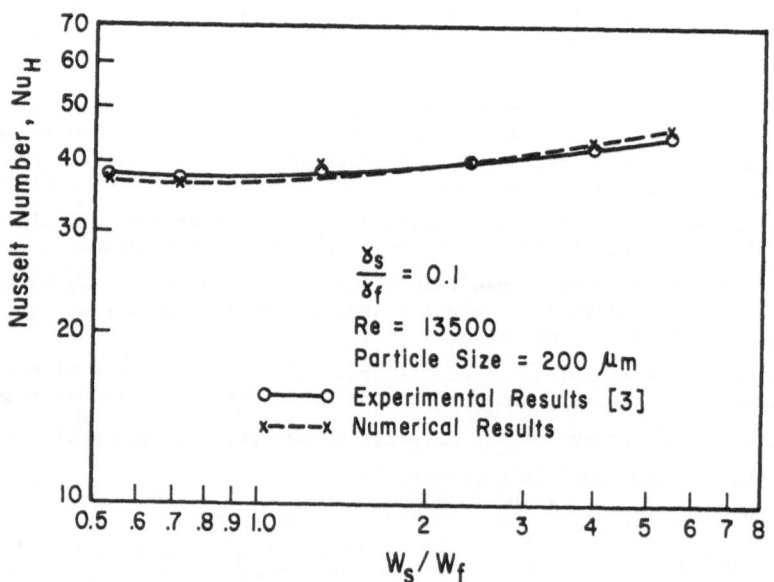

Figure 2.   Comparison of Numerical Results with Experimental Data

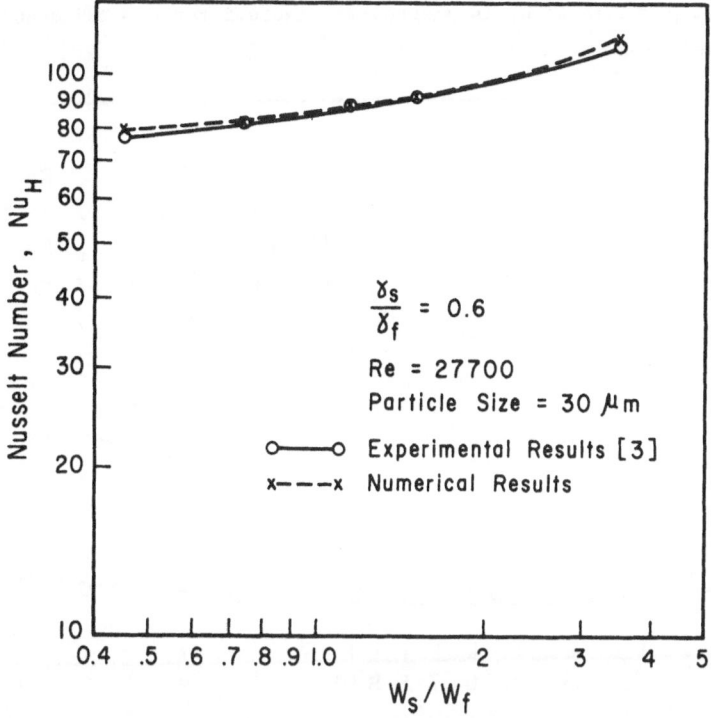

Figure 3. Comparison of Numerical Results with Experimental Data

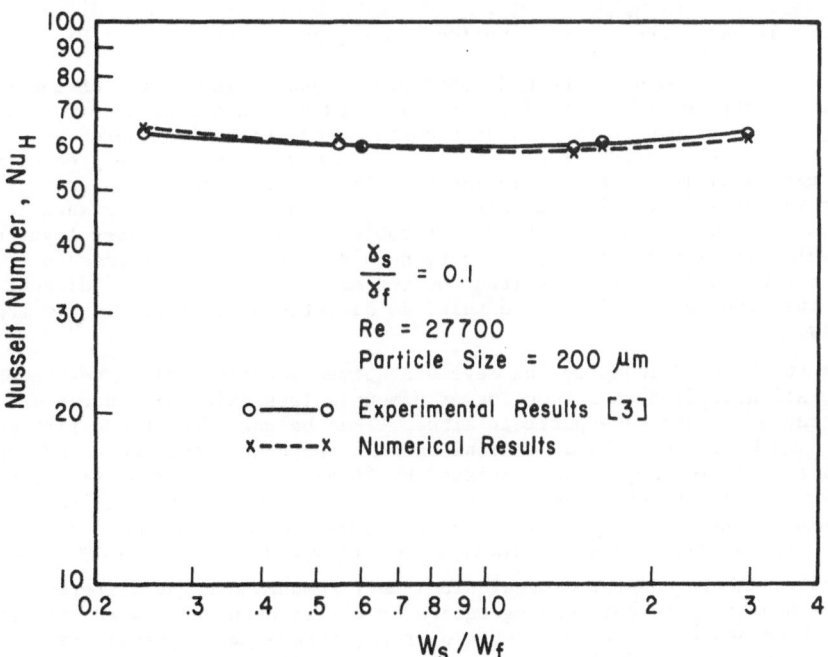

Figure 4. Comparison of Numerical Results with Experimental Data

for constant particle size, is plotted in Figure.5 for $D_p = 200\mu m$ and $D_p = 30\mu m$. The exact solution

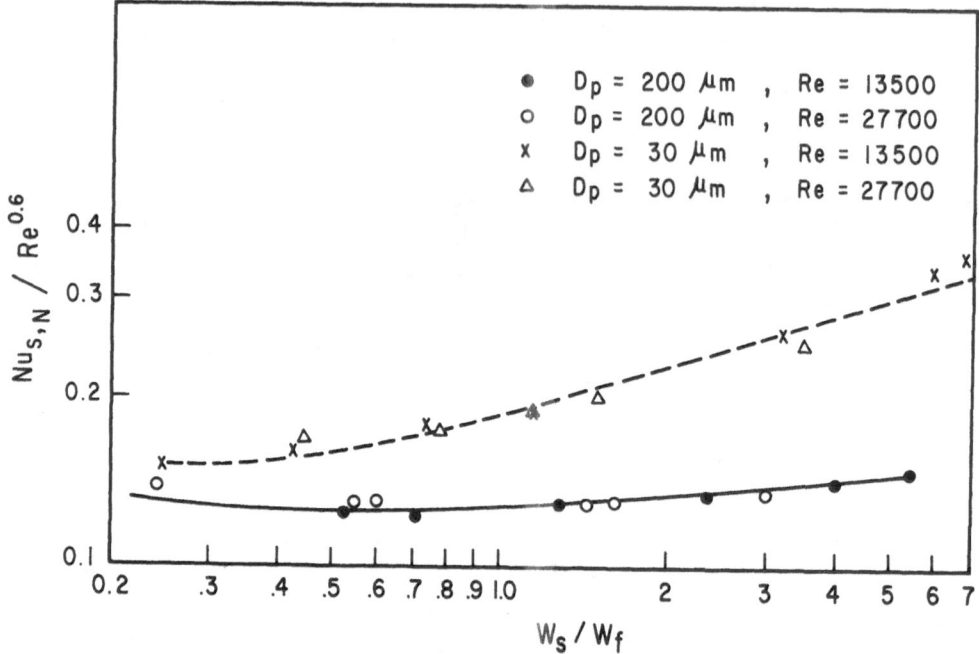

Figure 5. $Nu_{s,N}/Re^{0.6}$ vs.   Solids Loading Ratio for Constant $Pr = 0.72$

of equation (103) could not be made because of insufficient data; at least 4 or 5 different particle sizes should be experimented for an adequate correlation. From Figure 5, it appears that, heat transfer enhancement is favored by small particles, rather than by large particles. This may be due to the greater momentum gain by big particles in the direction of motion, preventing these particles to move (or vibrate) in the lateral direction. The small magnitude of such lateral motions would prevent the induction of turbulence into the boundary layer close to the wall. Smaller particles, however, should possess greater freedom to move towards the walls since the resultant of two momentums, one in the direction of motion, the other perpendicular to this, is directed by a greater angle towards the walls.

Figures 6 and 7 indicate the levels of dimensionless wall temperatures at uniform wall heat flux as a function of dimensionless axial distance and solids loading ratio for the same particle size. It can be seen from these figures that, for the same heat flux, the dimensionless wall temperature decreases by increasing solids loading ratio and increases by Reynolds number. This is another indication of heat transfer enhancement by solids loading ratio and reverse effect of high Reynolds numbers due to greater momentum acquisition to solid particles in the direction of gas velocity, thus preventing their lateral motions.

Nusselt numbers and the dimensionless wall temperatures at different axial distances and at different loading ratios have also been calculated with the Galerkin's method [1] of one-term, two-term, three-term approximations. The results indicate that, Nusselt numbers do not significantly change with the solids loading ratio and Nusselt numbers are very low compared with the experimental

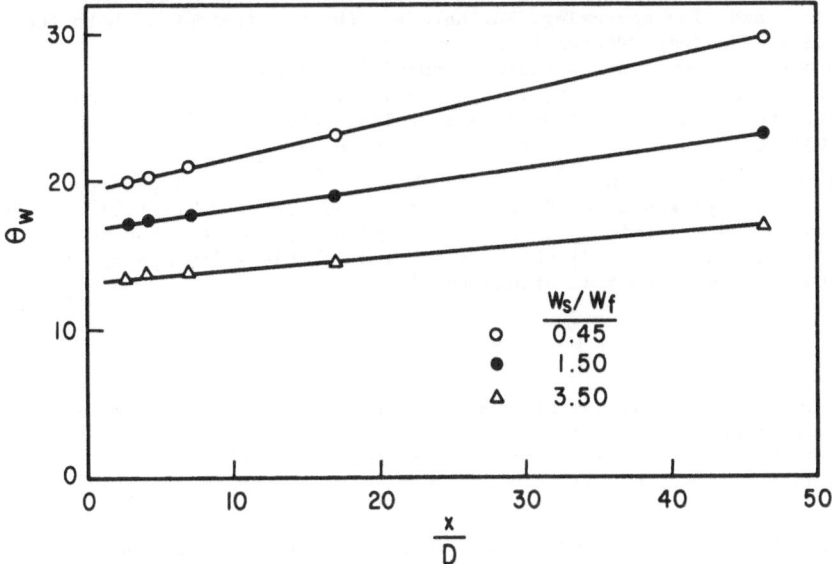

Figure 6. Dimensionless Wall Temperature vs. Dimensionless Axial Distance for D_p = 30µm, Re = 27700 and Various Solids Loading Ratios

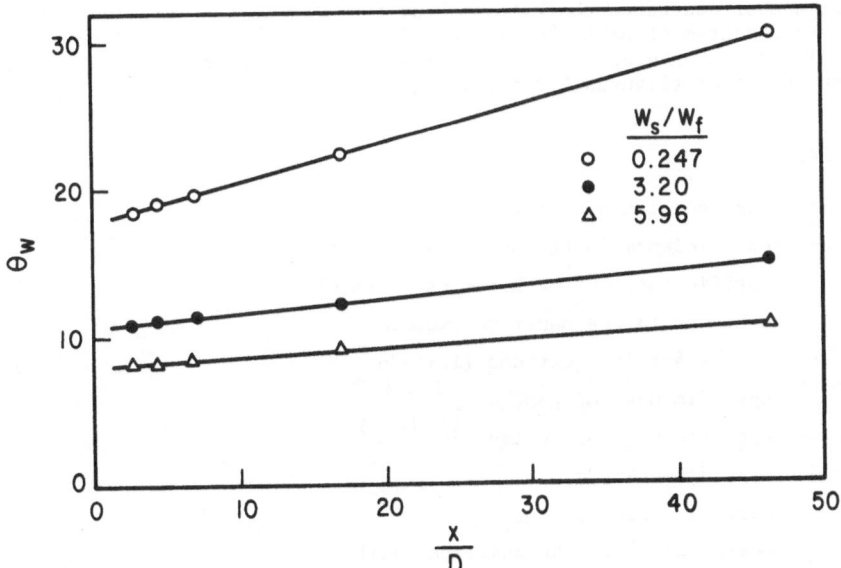

Figure 7. Dimensionless Wall Temperature vs. Dimensionless Axial Distance for D_p = 30µm, Re = 13500 and Various Solids Loading Ratios

values. As the number of terms in the approximate temperature profile of each phase increases, the value of Nusselt number increases as well, but the solution becomes tedious and time consuming. Furthermore, the requirement of both the fluid and the solids temperatures to be equal to the wall temperature at the wall in order to obtain a homogeneous boundary condition which makes the application of the Galerkin's method possible, is not physically realistic. It is more logical to think that the solids phase temperatures follow behind the fluid temperatures due to the absence of non-slip condition for the solids.

Having obtained the results for up to three terms, and having compared them with the experimental results of Depew[3], it was concluded that still more terms had to be considered to achieve reasonably close results between the two. Thus, the numerical solution method[2] which is applied in this study appears to be a better method compared to that of Galerkin's.

8.   CONCLUSIONS

In the determination of Nusselt number at the wall, at constant heat flux, to a gas-solid stream in turbulent flow, the numerical method applied in this study gives almost the same values as the experimental results (Figures 1-4). The experimental readings of wall temperatures must be corrected, however, for the kinetic energy term of the stream.

The Nusselt number of the suspension increases slightly with increasing solids loading ratio. This increase is less pronounced for larger particle sizes. It is also established that, if $Nu_{s,N} / Re^{0.6}$ ratio is plotted against solids loading ratio, the same curve is obtained at different Reynolds numbers for the same particle size (Figure.5).

Enhancement of heat transfer by solids in the stream is certain. This effect is more pronounced in the case of loading by smaller particles. An increase in the Reynolds number decreases the rate of heat transfer in the turbulent region due to greater momentum of particles in the direction of flow, thus preventing their lateral motions or vibrations to decrease the resistance of the boundary layer to heat transfer (Figures 6 and 7).

9.   NOMENCLATURE

A_p	surface area of a particle, m^2.
$A_{i,k}$	coefficients in the numerical method
$B_{i,k}$	coefficients in the numerical method
C_k	unknowns in the numerical method
C_1	a constant in equations (17) and (18)
C_g	specific heat of gas, $J\ kg^{-1}\ K^{-1}$
C_s	specific heat of solids, $Jkg^{-1}\ K^{-1}$
D	pipe diameter, m
D_p	particle diameter, μm
$D_{i,k}$	coefficients in the numerical method
D_1, D_2	coefficients in the Galerkin's method
E_1, E_2	coefficients in the Galerkin's method
f	function

G_i	coefficients in the numerical method
G_c	a constant defined in equation (91)
h_p	particle – gas heat transfer coefficient, $W/m^2 \cdot K$
k_g	thermal conductivity of gas, $Wm^{-1} K^{-1}$
L	Operator for the Galerkin's method
n_p	number of particles per unit volume
Nu_H	Local Nusselt number at U.H.F. (generally)
$Nu_{s,N}$	Nusselt number of suspension found by numerical method
$Nu_{s,G}$	Nusselt number of suspension obtained by Galerkin's method
Pe_m	Péclét number of mixture
Pr	Prandtl number
Re	Reynolds number
$r^+ = \dfrac{ru_f^*}{\nu_g}$	dimensionless radial variable
$r_o^+ = \dfrac{Ru_f^*}{\nu_g}$	dimensionless radius
R	pipe radius, m
S_c	a constant defined in equation (92)
T	temperature $^\circ K$, $^\circ R$
$<u>$	average velocity, m/s
u	point velocity, m/s
$u_f^* = <u_f>\sqrt{\dfrac{fw}{2}}$	friction velocity, m/s
$u^+ = \dfrac{u}{u_f^*}$	dimensionless velocity
W	mass flow rate, kg/s
x	axial distance, m
$x^+ = \dfrac{xu_f^*}{\nu_g}$	dimensionless axial distance

Greek Letters

β_f	a function of radial variable defined in equation (11)
β_s	a function of radial variable defined in equation (12)
ε_H	eddy diffusivity of heat, m^2/s
$\gamma_f = 1/Pr + \varepsilon_{H,f}/\nu_g$	dimensionless eddy diffusivity of fluid phase
$\xi = \dfrac{r^+}{r_o^+}$	another dimensionless radial variable used in Galerkin's method
$\theta = \dfrac{(T-T_o) k_g r_o^+}{qR}$	dimensionless temperature.

$\gamma_s = \varepsilon_{H,s}/\nu_g$ dimensionless eddy diffusivity of solids phase

ν kinematic viscosity, m^2/s

ρ density, kg/m^3

ϕ dimensionless temperature defined in equations (72) and (73)

ψ dimensionless temperature defined in equations (17),(18),(72),(73)

Subscripts

f   fluid phase

g   gas

H   at uniform wall heat flux condition

p   solid particle

s   solid phase

w   wall condition

mm   mixed mean condition

REFERENCES

1. Özışık, M.N. 1968. Boundary value problems of heat conduction. International Textbook Company. pp. 338 – 345.

2. Newman, J. 1967. Numerical solution of coupled, ordinary differential equations. University of California, Berkeley. Report UCRL – 17739.

3. Depew, C.A. 1960. Heat transfer to flowing gas – solids mixtures in a vertical circular duct, Ph.D. Thesis. University of California, Berkeley.

4. Danziger, W.J. 1963. Heat Transfer to fluidized gas – solids mixtures in vertical transport. I&EC Process Design and Development, Vol.2, No.4 pp. 269 – 276.

5. Depew, C.A. and Kramer, T.J. 1972. Heat transfer to flowing gas – solid mixtures. Advances in Heat Transfer, Vol.9, Academic Press, pp.113–180.

6. Shrayber, A.A. 1976. Turbulent heat transfer in pipe flows of gas – conveyed solids. Heat Transfer-Soviet Research, Vol.8, No.3, pp.60 – 67.

7. Boothroyd, R.G. 1971. Flowing gas – solids suspensions, Chapman and Hall LTD, London.

8. Tien, C.L., 1961. Heat transfer by a turbulently flowing fluid – solids mixture in a pipe. Trans.ASME. ser. C, Vol.83, No.2, pp.183 – 188.

9. Depew, C.A. and Farbar L. 1963. Heat transfer to pneumatically conveyed glass particles of fixed size. Trans. ASME, ser. C, Vol.85, No.2, pp.164–172.

10. Wasan, D.T., Tien. C.L. and Wilke, C.R. 1963. Theoretical correlation of velocity and eddy viscosity for flow close to a pipe wall. J. AIChE, Vol.9, No.4, pp. 567 – 569.

11. Bird, R.B., Stewart, W.E. and Lightfoot, E.N. 1960. Transport Phenomena. John Wiley&Sons, Inc.

12. Özbelge,T.,Ph.D. Thesis (to be published)

13. Rouhiainen, P.O. and Stachiewicz, J.W. 1970. On the deposition of small particles from turbulent streams. Trans. ASME, ser. C, Vol.92, No.1, pp.169–177.

The Characterization of a Gravity Assisted Heat Pipe with Internal Two-Phase Parallel Flow Throughout

VIC A. CUNDY and L.N. HA
Department of Mechanical Engineering
Louisiana State University
Baton Rouge, Louisiana 70803, USA

ABSTRACT

The paper describes the performance of a new, closed loop, gravity assisted heat pipe (term the unidirectional heat pipe) characterized by the parallel flow of the vapor and returning condensate at the condenser section. The device is completely passive achieving the desired flow configuration by geometrical means rather than the use of non-condensable gases or other mechanical means. In this respect, the device retains all of the favorable characteristics of conventional heat pipes sometimes termed thermosyphons. The unidirectional heat pipes were observed to completely alleviate the flooding limitation found in conventional two-phase gravity assisted heat pipes. Also, film thinning resulting as a consequence of parallel flow at the condenser section of the unidirectional heat pipe increased the inner condensing convective heat transfer coefficients on the order of 10 percent. A complete description of the new heat pipe along with its operational characteristics is presented in detail in the paper.

1. INTRODUCTION

The standard two-phase thermosyphon (gravity operated heat pipe) is a widely used terrestrial device exhibiting geometric flexibility, simplicity of construction, high internal thermal conducance in most applications, and extreme reliability. It has been successfully applied to the fields of heat recovery, solar energy utilization, permafrost stabilization, and highway deicing to mention only a few. The versatility of this device has led to an extensive amount of research directed toward providing fundamental physical phenomena which occurs during thermosyphon operation.

The closed, two-phase thermosyphon consists of a sealed chamber containing a condensable working fluid which transfers thermal energy in the form of latent heat of vaporization from one part of the arrangement to another part by means of vaporization of the liquid, transport and condensation of the vapor, and the subsequent return of the condensate from the condenser section to the evaporator section. A schematic of this device is presented in Figure 1.

A characteristic of standard two-phase thermosyphon is the counter flow of vapor and returning condensate. In many applications this flow does not adversely affect the performance of the device; however situations have been shown to exist in which this flow configuration significantly alters the operation characteristics of the thermosyphon.

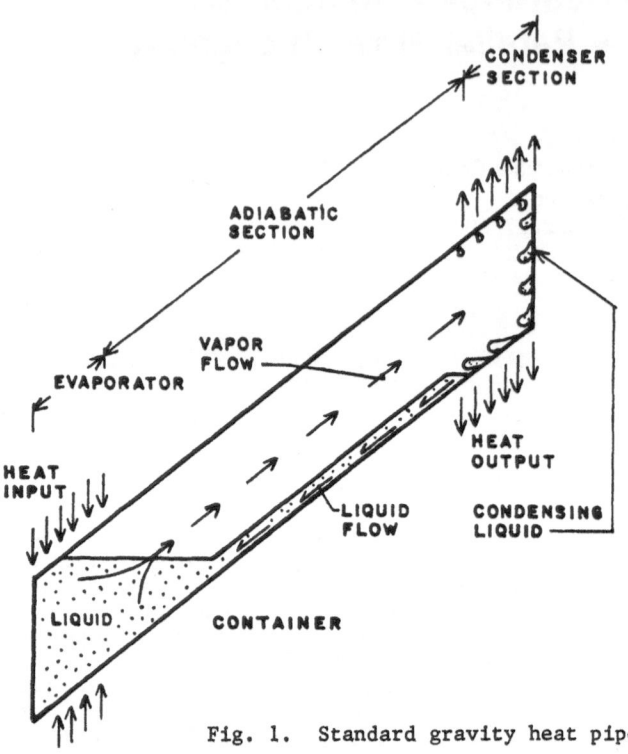

Fig. 1. Standard gravity heat pipe.

An example of such a situation involves thermosyphon configurations consisting of high axial heat fluxes with relatively low radial evaporator heat
fluxes. These conditions readily lead to the flooding phenomena or entrainment
limitation [1-5]. The flooding phenomena is caused by the collection of entrained
liquid droplets at the condenser end of the device. Only after the weight of
the collected liquid overcomes the vapor flow forces will the suspended liquid
pool collapse allowing condensate flow back to the evoporator section of the
thermosyphon. Operation of the thermosyphon at or near the flooding point leads
to severe stability problems manifested by eratic evaporator temperature fluctuations. Unless otherwise designed for, operation at the entrainment limiting
point can lead to significantly lower heat flux rates, resulting in possible
overall system failure at elevated heat transfer levels.

Another result of the counter current flow occurring in thermosyphons is
associated with the condensate film thickening which occurs in the condenser
end of the device. This phenomena is related to the flooding phenomena previously
described but can occur at significantly lower axial heat fluxes up to the final
entrainment limitation. Ultimately, the energy that is transported by heat must
pass through a thickness of condensate film, through the heat pipe container and,
finally through a thermal resistance to the heat sink. Optimum heat transfer
would occur with drop wise condensation; however in reality it has been shown
that film condensation more realistically represents the actual physical situation [6]. Moreover, Lee and Mital [6] have demonstrated that the inner condenser
film coefficient is very sensitive to the operating pressure and condensation
area. Bezrodnyi and Beloivan [7] similarly found that for low heat fluxes the
condensate return was annular in nature and relatively stable. As the axial flux
increased however, even with increasing pressure, the liquid film became unstable
forming waves ultimately resulting in a thickened condensate film.

As a result of these adverse effects, a number of modified thermosyphons have been developed which seperate the flow of the returning condensate from the rising vapor [5,7,8]. The device of Bezrodnyi and Beloivan [7] is not described in detail; however, this research did report a significant increase in maximum permissible heat flux. The thermosyphon of Steponchuk and Strelitsov [8] operates by capillary action and although favorably increasing the heat transfer characteristics and alleviating the flooding problem, this device was limited because of boiling which could occur in the syphon return line effectively stopping condensate return. The thermosyphon of Oshima and Masumoto [5], proved to be a versatile device separating the flow of returning condensate and rising vapor; but, has the disadvantage of utilizing noncondensable gases to achieve this result.

The purpose of this paper is to report on the design, fabrication, and testing of a closed, two-phase thermosyphon with separate channels for returning condensate and rising vapor. The device which has been developed is completely passive, requiring no additions of any external physical parameters. Parallel flow throughout has been obtained entirely with geometric modifications.

2. THE UNIDIRECTIONAL HEAT PIPE

A schematic of the unidirectional gravity operated heat pipe is presented in Figure 2. Originally, the device was constructed with the use of a one-way check value; however, subsequent testing clearly demonstrated that this value was not required. This is highly desireable since the use of such values may result in leakage and eventual value failure after long term operation. A requisite of the unidirectional heat pipe was to retain all of the favorable characteristics of the standard gravity operated heat pipe. The use of mechanical check values certainly did not meet this requirement.

As shown in Figure 2, the unidirectional, gravity operated heat pipe consists of a closed loop configuration with the evaporator section located at the bottom of the loop and the condenser section at the side of the loop. Between these sections is the adiabatic sections similar to that found in standard gravity operated heat pipes. The U-shaped tub connecting the evaporator and the condenser sections of the heat pipe serves as a vapor barrier blocking the flow of vapor generated in the evaporator section and thus forcing this vapor to travel to the adiabatic section. Upon reaching the condenser section, this vapor is condensed, gives up its latent heat of vaporization and then flows down the inner wall of the device to the U-shaped section. The accumulation of consensate in one side of the U-shaped tube causes a hydrostatic inbalance of the fluid in the tube. As a result, a pressure difference is generated causing the liquid in the condenser side to flow to the evaporator section. As long as the heat pipe is in operation, the U-shaped tube is always filled with liquid (acting as a vapor blockage) flowing from the condenser to the evaporator.

In direct opposition to the standard gravity operated heat pipe, the unidirectional gravity operated heat pipe exhibits film thinning in the condenser section rather than film thickening. The parallel direction of both the vapor and the returning condensate (both flowing downward) tends to enhance heat transfer by thinning the condensate film and removing it. Such action, as will be shown next, effectively raises the overall coefficient of heat transfer of the unidirectional heat pipe when compared with standard gravity operated heat pipes operating under the same conditions. Additionally, the likelihood of flooding occurring during the operation of the unidirectional heat pipe has been completely alleviated.

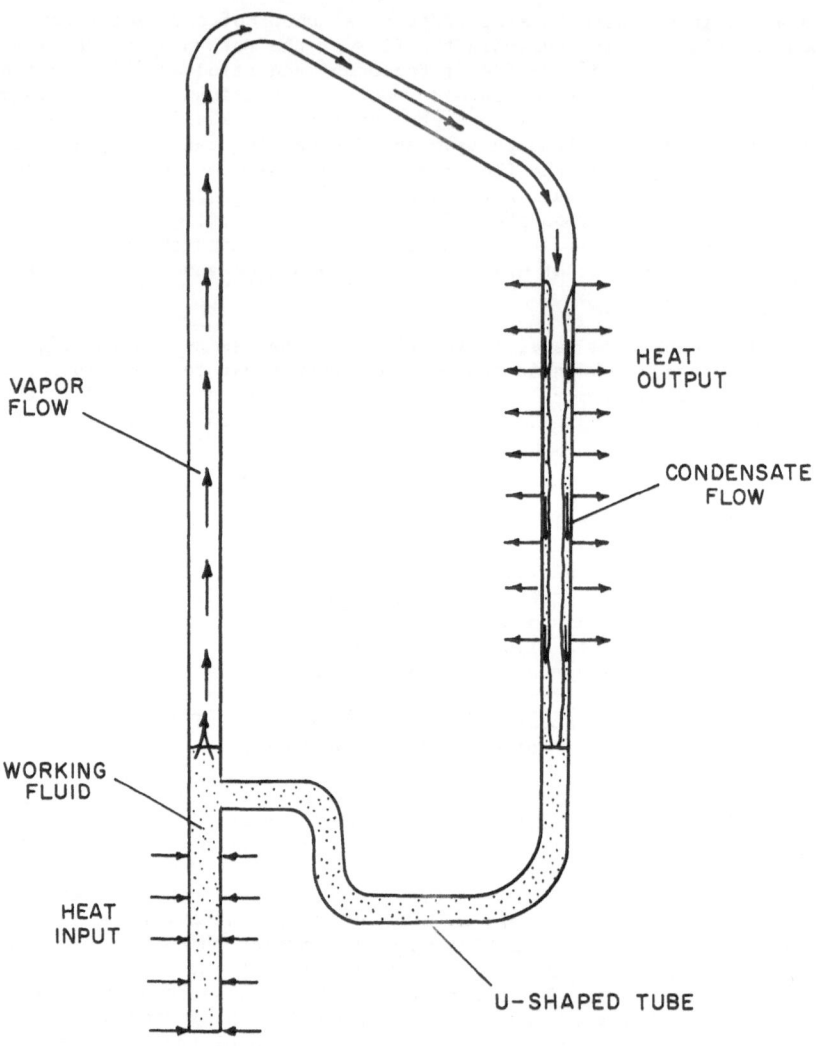

Fig. 2.   Unidirectional, gravity assisted heat pipe.

3.   EXPERIMENTAL APPARATUS AND CONFIGURATIONS

Figure 3 provides a representation of the unidirectional heat pipe which was fabricated for testing purposes. Actually, two heat pipes were constructed and tested, each with the configuration shown in Figure 3; the first made of 304 stainless steel and filled with ethanol as the working fluid and the second, made of copper and filled with water as the working fluid. It should be noted that compatability problems with ethanol and stainless steel previously reported in the literature [9, 10] did not occur primarily due to the short operational periods of the heat pipes. No long term tests were conducted. All tests conducted were of 4 to 6 hours in duration and the fluid within the heat pipes was routinely changed after 1 week of operation.

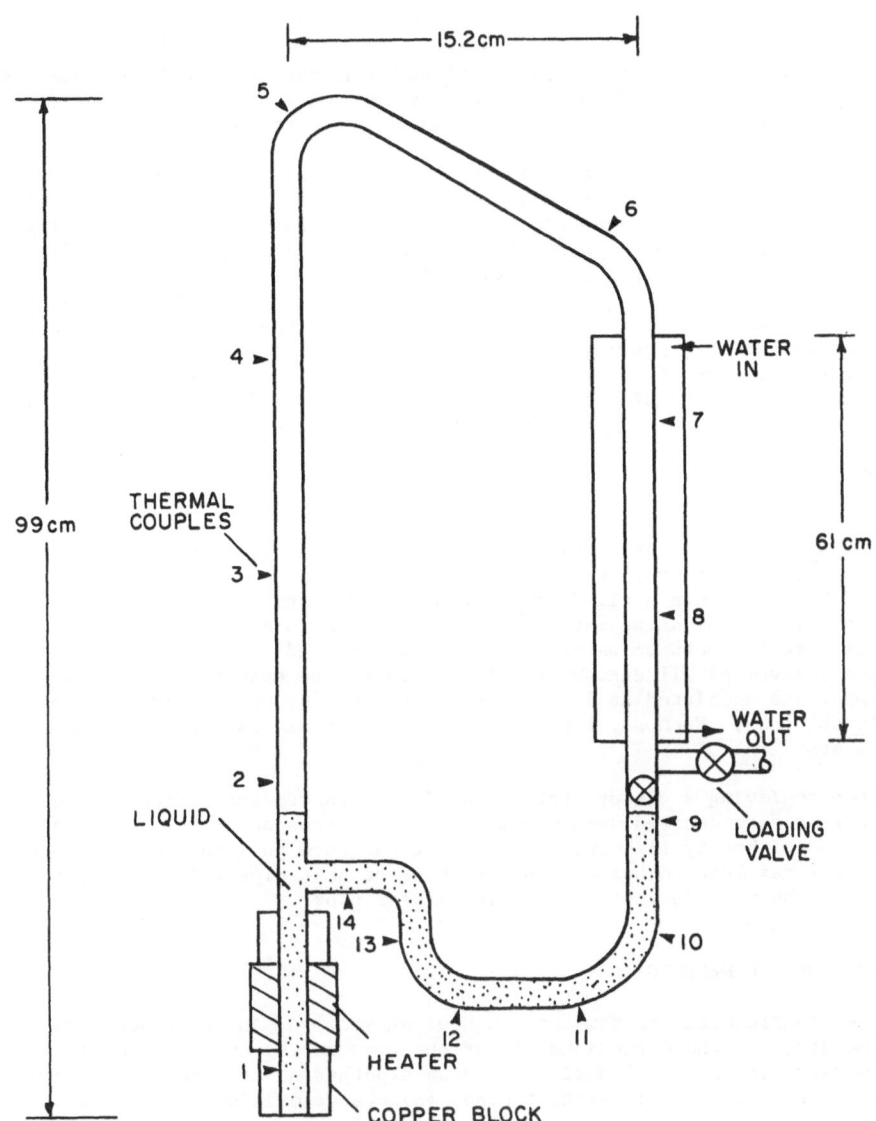

Fig. 3. Experimentally fabricated heat pipe.

Referring to Figure 3, the 6.35 mm tubing was bent into a loop shape with
the evaporator leg protruding at the lower end of the heat pipe. All fittings
and tube connections were either welded or made with vacuum tight fittings
(Swagelock) to insure the integrity of the device. A ring heater was mounted on
a split copper block clamped around the evaporator section of the heat pipe. The
heater, in turn, was connected to a reheostat and power meter. On the condenser
side, a flexible tube of 37.5 mm diameter surrounded the heat pipe for 61 cm.
During heat pipe operation, pressurized water at a constant temperature was
circulated through this tube to remove heat energy transferred to the condenser

section.

A high vacuum value was attached to the heat pipe near the condenser section. This value provided access to the vacuum system and working fluid loading reservoir. Fourteen Chromel-Alumel thermocouples were installed along the heat pipe wall to provide the axial temperature distribution along the heat pipe. Figure 3 shows the position of these thermocouples with a numbered reference for each senser. All thermocouple measurements were recorded with a Linseis 12-point printing recorder along with a Fluke 2200B data logger. A thick layer of fiberglass insulation was applied over the entire length of the heat pipe to insure negligible heat loss to the surroundings.

To compare performance of the unidirectional, gravity assisted heat pipe, two standard vertical gravity operated heat pipes were also constructed. These pipes were also made of 304 stainless steel and copper tubing of 6.4 mm diameter. The equivalent lengths of the evaporator, adiabatic, and condenser section were identical to those shown in Figure 3 for the unidirectional heat pipe.

3.1 Test Procedures

A typical experiment consisted of initiating cooling water flow to the condenser section of the heat pipe being tested, whether conventional or unidirectional. Heat was then applied to the evaporator section. In all experiments, the water flow rate was adjusted so that no appreciable temperature difference could be detected between water inlet and outlet conditions. This insured the complete removal of all energy transferred in at the evaporator section. The heat input was initiated at low levels so that steady state conditions might be readily obtained. Even so, exhibited start up transients occurred from 5 minutes to 1½ hours.

Upon achieving a steady state condition, power to the heater was incrementally increased to desired values which usually corresponded to the burnout condition caused either by flooding or achieving nucleate pool boiling. Either of these cases may lead to burnout in the standard heat pipe while only the latter could cause burnout in the unidirectional heat pipes.

4. EXPERIMENTAL RESULTS

The experimental program consisted of primarily four distinct phases: (1) demonstrating the directionality of the flow in the unidirectional heat pipe, (2) demonstrating the liquid film thinning hypothesis occurring in the condenser section of the unidirectional heat pipe, (3) characterizing the burnout limitations and (4) observing the start up behavior of the heat pipes. Each of these areas will be addressed separately.

4.1 Unidirectional Flow Verification

Before complete operational characterization of the unidirectional heat pipe was possible, verification of the one-way flow within the heat pipe was required. Confirming one-way flow could be achieved indirectly with temperature measurements, or directly by actually measuring the condensate flow in the condenser section. Both measurements were performed in order to assure the unidirectionality of the flow within the heat pipe. The first method to be described will be that of the indirect temperature measurements.

At various heat energy inputs below the classical burnout point, temperatures along the unidirectional heat pipe were observed. Table I and Table II present the typical observed temperature variations of the unidirectional heat pipe operating in the steady state mode as a function of various heat inputs. Refer to Figure 3 for the thermocouple locations. Obviously the temperature was a maximum at the evaporator section for all heat inputs and then gradually decreased along the adiabatic section. At the condenser section, the temperature dropped to essentially the temperature of the cooling water. At thermocouple No. 9 which was located midway between the condenser section and the U-shaped tube, the temperature of the working fluid was the same as the temperature of the cooling water in the condenser section. The temperature along the U-shaped tube gradually increased as it came closer to the evaporator section. Since the temperature above the condenser section (thermocouple No. 6) was much greater than below the condenser section (thermocouple No. 9), the pressure at the former must be greater than that at the latter. As a result, the working fluid must flow from the upper to the lower part of the condenser section. Moreover, since the temperature in the adiabatic U-shaped tube section below the condenser section was as low as the cooling water temperature, the fluid could not be vaporized in that section. Vaporization of the fluid occurred only in the evaporator section. As a result, the flow must be unidirectional from the evaporator section through the adiabatic, to the condenser section, through the adiabatic U-shaped tube and then back to the evaporator section.

The second method to confirm the one-way flow was achieved directly by measuring the condensate flow in the condenser section. A valve was inserted in the lower part of the condenser section. The location of this valve is shown in Figure 3. The purpose of the valve was to stop the condensate in the condenser section from flowing down the U-shaped tube and eventually returning to the evaporator section. If unidirectional flow was achieved, the amount of condensate built up above the closed valve in the condenser section must be equal to the amount of vaporized working fluid as determined by the power applied to the evaporator section.

As power was applied to the evaporator, the valve was shut off, allowing the condensate to accumulate in the lower part of the condenser section. The theoretical mass flow rate was calculated by dividing the power input to the fluid's latent heat of vaporization. By observing a distinct time period of power application, the amount of working fluid flowing to the condenser section could be determined theoretically. If the flow was actually one-way, the fluid accumulated in the lower part of the condenser must be equal to the fluid vaporized in the evaporator section as a result of the applied power. Actual weighing the amount of fluid accumulated in the condenser section was carried out. With 12 identical runs at power level settings on the ethanol-stainless steel heat pipe from 5 to 35 watts, the mean difference between the weight of the actual and theoretical amount of fluid was only three percent. This three percent difference could be credited to the heat loss to the atmosphere through the adiabatic sections of the heat pipe.

With direct and indirect verifications, results agreed on the unidirectionality of flow within the heat pipe. The flow within the heat pipe must be one-way around the heat pipe loop from the evaporator to the condenser section and back to the evaporator section through the adiabatic U-shaped tube.

4.2 Liquid Film Thinning Verification

One of the main objectives of this research has been to alleviate the problem of condensate liquid film thickening occurring in standard gravity

Table I. Temperature distribution in the ethanol–stainless steel
unidirectional heat pipe

Power Input W	Flux MW/m²	Cooling Water Flow Rate ml/sec	Temperature, °C Thermocouple Number													
			1	2	3	4	5	6	7	8	9	10	11	12	13	14
16.8	1.0	90	105	100	99	98	99	98	30	30	30	35	48	60	91	100
28.8	1.8	90	114	104	102	101	102	101	30	30	30	37	51	64	98	111
35.5	2.2	90	120	106	105	105	105	104	30	30	30	41	56	68	100	113
41.5	2.6	90	255	111	111	110	109	108	30	30	30	43	59	69	110	146
45.1	2.8	90	160	112	111	110	110	109	30	30	30	45	63	72	123	150

Table II. Temperature distribution in the water-copper unidirectional heat pipe.

Power Input W	Flux MW/m²	Cooling Water Flow Rate ml/sec	Temperature, °C Thermocouple Number													
			1	2	3	4	5	6	7	8	9	10	11	12	13	14
16.8	1.0	90	72	65	65	64	64	64	30	30	30	32	35	42	50	56
28.8	1.8	90	76	68	68	68	68	66	30	30	30	33	37	45	51	58
35.5	2.2	90	82	70	70	70	70	69	30	30	30	34	38	47	53	59
41.5	2.6	90	84	72	72	71	72	70	30	30	30	35	38	49	54	61
45.1	2.8	90	85	73	72	72	72	71	30	30	30	35	38	49	55	63
75.0	4.6	90	100	78	78	78	77	77	30	30	30	37	39	53	59	79

assisted heat pipes through the use of the proposed unidirectional, gravity assisted heat pipe. Verification of thinner condensate films will be based upon an elementary heat transfer analysis applied to both the standard heat pipe and the unidirectional heat pipe. An overall coefficient of heat transfer for both heat pipe configurations is defined by:

$$Q = A_p \, U_{hp} (T_{p,e} - T_{p,c})$$

where, Q is the heat transfer rate, $T_{p,e}$ is the temperature of the pipe at the outer surface of the evaporator, $T_{p,c}$ is the temperature of the pipe at the outer surface of the condenser, and U_{hp} is the overall heat transfer coefficient based on the cross sectional area of the heat pipe A_p. For the same power input Q, a smaller value of the temperature difference $(T_{p,e} - T_{p,c})$ would indicate a higher value of the overall heat transfer coefficient U_{hp} since A_p remains constant.

Also, the following relationship holds:

$$U_{hp} = \frac{1}{R_{p,e} + R_{e,i} + R_v + R_{c,i} + R_{p,c}}$$

With identical heat pipe conditions and equal amounts of working fluid, all values of the heat pipe thermal resistances must be the same for both the unidirectional heat pipe and the standard heat pipe except the value of $R_{c,i}$ which is the thermal resistance due to the liquid film inside the condenser section. The values of $R_{p,e}$ and $R_{p,c}$ which are the thermal resistances due to conduction through the evaporator section and the condenser section, respectively, must be identical since both heat pipes were made of the same tubing materials, either copper or stainless steel, and the same thickness. The thermal resistances R_v's due to the axial convective transport of latent heat by vapor from the evaporator section to the condenser section must be the same in both heat pipes. These resistances were dependent only on the vapor flows which were governed by the amount of heat input. The same power inputs resulted in the same vapor flow rates and consequently the same thermal resistances, R_v's. The thermal resistance $R_{e,i}$ due to the boiling heat transfer in the evaporator section was a function of the power input, the properties of the fluid, and the amount of fluid obtained in each heat pipe. Since all of these variables were the same for both heat pipes configurations, these values of $R_{e,i}$'s must be identical. Only the condensation thermal resistances $R_{c,i}$ were different since they were dependent on the condenser section liquid film thicknesses, which varied from the unidirectional to the standard heat pipes. A thicker film thickness results in a higher value of the thermal resistance $R_{c,i}$.

Figure 4 presents a typical set of data which was obtained to aid in the verification of the film thinning phenomena. This figure represents a comparison of the standard and unidirectional stainless steel heat pipes. The heater was adjusted for a axial flux of 1.8 MW/m^2 based upon the cross sectional area of the heat pipes. Steady state temperatures have been plotted as a function of distance from the evaporator. The results presented in this figure clearly show the decrease in $(T_{p,e} - T_{p,c})$ for the unidirectional heat pipe. As a result, the overall coefficient must increase and hence the thermal resistance due to film thickening must decrease.

The results of all tests (varying the axial heat flux from 1.0 to 2.8 MW/m^2) for both types of heat pipes (stainless steel-ethanol and copper-water) confirmed the tendency presented in Figure 4. In fact, both configurations (stainless steel-ethanol and copper-water) exhibited nearly a 10 percent reduction in the inner convection condensation heat transfer coefficient between the conventional gravity aided heat pipe and the new unidirectional heat pipe.

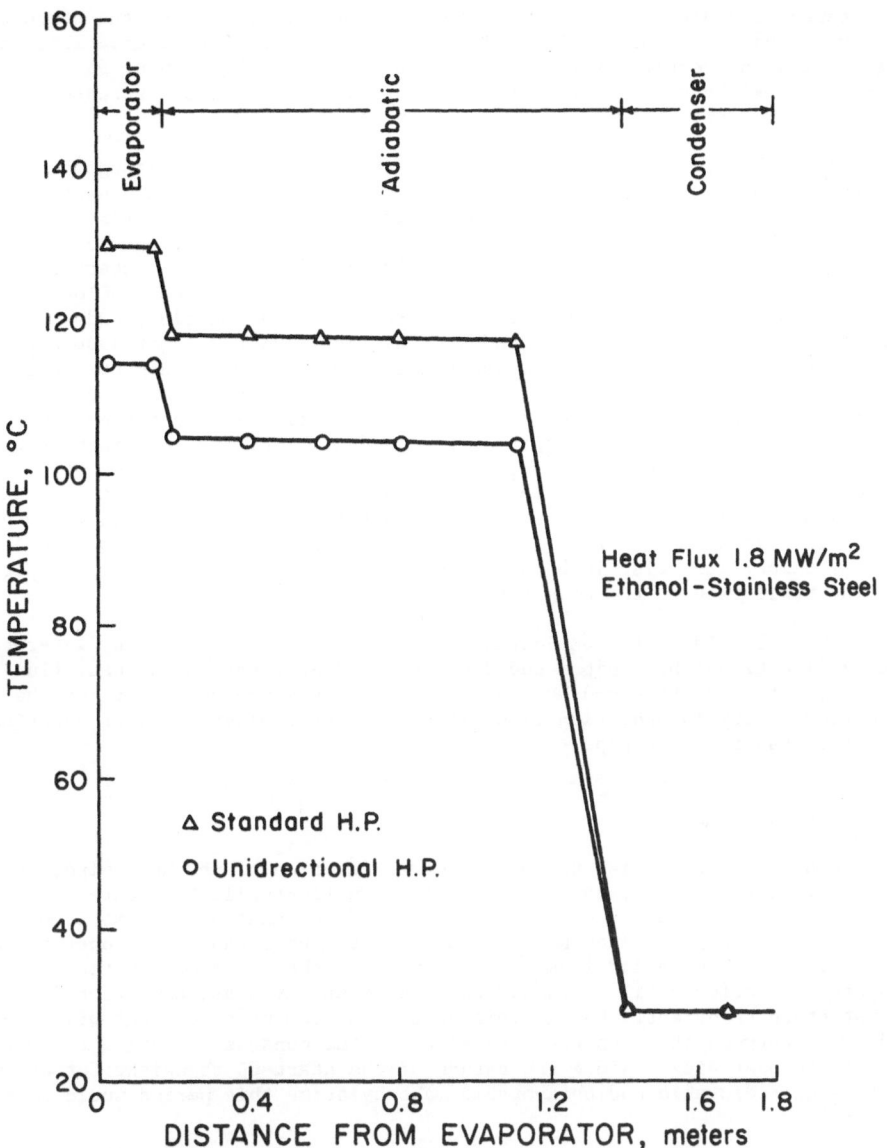

Fig 4. Typical comparison of heat pipe performance.

4.3 Comparison of Burnout Limitations

Classical burnout in a gravity operated heat pipe is characterized by a sudden increase in temperature of the evaporator section together with a decrease in temperature of both the condenser section and the adiabatic section. In this experiment, the burnout points in the unidirectional heat pipe and in the standard heat pipe were determined by the procedures described previously. With the ethanol-stainless steel combination, the burnout point for the standard heat pipe was found to occur at an axial heat flux level of 2.6 MW/m^2. This value agreed with the result given by Bezodnyi et al [8]. The burnout point for the unidirectional heat pipe was found to occur at 2.8 MW/m^2, an increase of almost 8 percent over the standard heat pipe burnout point.

With the water-copper combination, a rather complicated situation was observed. The classical burnout point was found in the standard heat pipe to be at 4.60 MW/m^2. Again, this value agreed with the data in Reference 8. In the unidirectional heat pipe, however, the classical burnout was not reached. A limitation totally different from the classical burnout was observed. First, a sudden increase in temperature of the evaporator section was followed with a small gradual increase in temperature of the adiabatic section, and a small increase in temperature of the condenser section. This was not a burnout point since both temperatures of the adiabatic section and the condenser section were increasing instead of decreasing as in the case of classical burnout. One explanation for this situation is due to the insufficient heat transfer area at the condenser section (the length of the condenser section was fixed at 0.61 m). Stated simply, the condenser could not transmit the high power density input at the evaporator section. Heat applied to the evaporator section at an axial heat flux of 5.40 MW/m^2 could not be transferred totally to the cooling water at the condenser section due to the lack of heat transfer area. This resulted in the gradual temperature rise in both the condenser section and the adiabatic section and the rapid rise in the evaporator section.

Obviously, for both combinations (ethanol-stainless steel and water-copper) the unidirectional heat pipes could be operated at higher axial heat flux levels than those for the standard heat pipes. This gives the unidirectional heat pipe more flexibility in operation since it can accept a wider range of power input than the standard heat pipe.

4.4 Start-Up Behavior

When power is applied to the evaporator section of the heat pipe, it will obviously heat this section up. A temperature front will be created and will gradually move toward the condenser section of the heat pipe. The temperatures along the heat pipe will eventually build up to the steady state operational temperatures. The period from the beginning of the power application to the evaporator section until the achievement of a steady state condition is called the start-up transient. During this period the temperature front will move back and forth between the evaporator section and the condenser section to heat up the entire heat pipe. Since the nature of the start-up transient for the most part is unpredictable and unfavorable, abbreviating this period would be highly desirable.

In all experiments conducted during this investigation it was observed that the unidirectional heat pipe exhibited both a longer and much more erratic start-up transient period than the conventional gravity assisted heat pipe. The temperatures along the unidirectional heat pipe build up to the steady state temperatures in an unpredictable and fluctuating manner. Again, this is unfortunate since many heat pipe applications are cyclic in nature and long, unpredictable transient conditions simply cannot be tolerated. However, the complete characterization of this phenomenon was not obtained and is presently an area of further study.

5. CONCLUSIONS

The results presented in this paper lead to the following conclusions:
1. A completely passive, unidirectionsl heat pipe is possible exhibiting parallel flow throughout.
2. Such a device completely alleviates the flooding phenonema sometimes found to occur in standard gravity operated heat pipes (thermosyphons).
3. Such a device enhances heat transfer through film thinning at the condenser section.
4. Preliminary results indicate that such a device also increases the boiling limitation over conventional heat pipes (thermosyphons).
5. Preliminary results indicate that such devices will exhibit more erratic and longer transient start up periods than conventional heat pipes (thermosyphons).

ACKNOWLEDGEMENTS

Initiation of this research was made possible through a Summer Faculty Research Award provided by Louisiana State University. The authors also wish to extend appreciation to Mr. J. Delhom for his contributions with fabrication of the heat pipes.

REFERENCES

1. Lee, Y., and Mital, U., "A Two-Phase Closed Thermosyphon", J. Heat Mass Transfer, 15, Pergammon Press, 1972, pp. 1695-1707.

2. Chi, N.H., Groll, M., "The Influence of Wall Roughness on the Maximum Performance of Closed Two-Phase Thermosyphons", Presented at the AIAA 15th Thermophysics Conference, Snowmass, Colorado, July, 1980, (AIAA paper No. 80-1503).

3. Groll, M., Chi, N.H., and Krahling, H., "Reflux Heat Pipes as Components in Heat Exchanges for Efficient Heat Recovery", Commission of the European Communities International Seminar "New Ways to Save Energy", Brussels, 1979, pp. 481-490.

4. Groll, M., Chi, N.H., and Krahling, H., "Warmeruckgewinnungsanlagen mit Reflux-Warmerohren als Bauelemente", Commission of the European Communities, Contract No. 379-77-10 EED, Final Report, IKE 5TF-363-80, 1980.

5. Chi, N.H., and Abhat, A., "Performance Evaluation of Gravity-Assisted Copper-Water Heat Pipes", 3rd International Heat Pipe Conference, Palo Alto, CA, 1978 and: AIAA Journal, Vol. 17, No. 9, 1979, pp. 1003-1011.

6. Oshima, K., and Masumoto, H., "A Heat Pipe with the Separated Liquid Return Passage", Presented at the 15th AIAA Thermophysics Conference, Snowmass, Colorado, July, 1980, (AIAA paper No. 80-1481).

7. Bezrodnyi, M.K. and Beloivan, A.I., "Investigation of the Maximum Heat Transfer Capacity of Closed Two-Phase Thermosyphons", Teplofizika Vysokikh Temperatur, Vol. 15, No. 2, 1977, pp. 370-376.

8. Steponchuk, V.F., and Strelitrov, A.I., "The Performance of a U-Shaped Evaporating Thermosyphon", Heat Transfer Soviet Research, Vol. 7, No. 2, 1975, pp. 129-134.

9. Chi, S.W., <u>Heat Pipe Theory and Practice</u>, McGraw-Hill Book Co., New York, 1976.

10. Dunn, P.D., and Reay, D.A., <u>Heat Pipes</u>, Pergamon Press, New York, 1978.

Using the Energy Spectrum Method to Calculate the Two-Phase Frictional Resistance

Z.Z. LIN
Jinan University
Guangzhou, The People's Republic of China

Z.H. LIN
Xian Jiao-Tong University
Xian, The People's Republic of China

ABSTRACT

The energy spectrum method is one of the statistical theories for single-phase turbulent flow, and has been used successfully in meteorology.

This paper uses the energy spectrum method to predict the two-phase frictional pressure drop in a tube. Analytical predictions are compared with experimental data. Good agreements have been obtained.

INTRODUCTION

Two-phase flow problems are often found in a wide range of engineering systems, such as boiling water reactors, breeder reactors, conventional steam boilers, evaporators of refrigeration systems, and evaporative and condensive heat exchangers in chemical and petroleum industries. Among these problems, one of the most essential and important issues is to predict the frictional pressure drop in a tube.

Knowing the pressure drop in a two-phase flow system is of primary interest to the designer in order to establish the pumping load and prescribe the longitudinal variation in pressure necessary to compute the fluid properties along a channel.

Many predicting correlations of this topic have been published. Most of them were obtained empirically, such as the correlations of Armand (I), Kosterin (2), Chenoweth and Martin (3), Isbin (4), Hughmark (5), etc. Some of them were obtained by assuming a model first and then modifying the model with experimental data, such as correlations of Mc Adams (6), Lockhart & Martinelli (7), Owen (8), Lockshen & Shevarz (9), Thom (10), etc. Some were obtained by other methods, such as Dukler's corrleation (II), obtained from dimensional analysis, Levy's correlation (I2) obtained from mixing length theory etc.

The energy spectrum method is one of the statistical theories for single-phase turbulent flow and has been used successfully in meterology. Up until now, this theory has not yet been used to solve two-phase flow problems.

In this paper, the energy spectrum method, together with the mixing length theory, are used to predict the two-phase frictional pressure drop in a tube. By treating the two-phase flow as a continuous medium, where the turbulent exchanges of momentum and density are equal, a predicting method is developed.

THEORETICAL CONSIDERATION

Suppose an isotropic homogeneous vapor-liquid two-phase turbulent flow is flowing along a horizontal tube. The two-phase flow has a velocity w in the x-direction, a velocity v in the y-direction, which is perpendicular to the x direction, a density ρ and a dynamic viscosity μ. Assuming that w' is the fluctuating velocity component of w, v' is the fluctuating velocity component of v and ρ' is the fluctuating velocity component of ρ, we may obtain the frictional stress for two-phase as follows:

$$\tau = \rho\overline{w'v'} + w\overline{\rho'v'} + \mu\frac{dw}{dy} \qquad (1)$$

where ———— refers to statistically averaged values, y is the distance measured from the tube wall.

From the mixing length theory:

$$\rho' = \ell_\rho\frac{d\rho}{dy}$$

$$w' = v' = \ell_w\frac{dw}{dy}$$

where ℓ_w and ℓ_ρ are the mixing lengths for the velocity and density distribution, assume that the turbulent exchange of momentum and density are equal, then

$$\ell = \ell_w = \ell_\rho$$

According to Obukhov's spectral theory (13), (14), the relationship between fluctuating velocity components is

$$w'^2 = v'^2 = 2/3 \int_k^\infty \phi(p)\,dp \qquad (2)$$

where k is the outer scale of the turbulent flow, and $\phi(P)$ is the spectral density.

Heisenburg (15), (16) improved Obukhov's theory and obtained the following relationship for spectral density:

$$\phi(p) = \alpha^{-9/4}(\varepsilon\nu^5)^{\frac{1}{4}} \cdot 16 \cdot 9^{1-s/s}x^{-7} \qquad (3)$$

where S is an arbitrary value, α is a constant, ε is the total energy, ν is the kinematic viscosity,

$$x = k/k_o', \quad k_o' = \alpha^{-3/4}(\varepsilon/\nu^3)^{\frac{1}{4}}.$$

Substituting equation (3) into (2), equation (2) becomes:

$$w'^2 = v'^2 = 2/3 \; \alpha^{9/4}(\varepsilon v^5)^{\frac{1}{4}} \cdot 16 \cdot 9^{(1-s)/s} \int_k^\infty x^{-7} dx$$

$$= A/6 \cdot k^{-6} \tag{4}$$

where $\qquad A = 2/3 \; \alpha^{9/4} (\varepsilon v^5)^{\frac{1}{4}} \cdot 16 \cdot 9^{(1-s)/s}$

Substituting equation (4) into (1) and considering $\ell = \ell_w = \ell_\rho$ yields

$$\tau = \rho A k^{-6}/6 + (\ell^2 w \frac{d\rho}{dy} + \mu) \frac{dw}{dy} \tag{5}$$

The relationship between velocity W and density ρ can be expressed by Pai's relationsip (17):

$$\rho = A'\rho w + B' \tag{6}$$

where A' and B' are proportionality constants, nondimensional. Considering the following boundary conditions: $y = 0$ (on the tube wall), $\rho = \rho_L$, $W = 0$; $y = R$ (in the tube center) $\rho = \rho_c$, $W = W_c$, and substituting them into correlation (6), we obtain the required relationship between ρ and W:

$$\frac{\rho}{\rho_L} = \frac{\rho_c w_c}{\rho_L w + \rho_c (w_c - w)} \tag{7}$$

where subscript L indicates liquid, and subscript c indicates the center of the tube.

Let $\rho^+ = \rho/\rho_L$, equation (7) can be written as follows:

$$w = \frac{\rho_c w_c (1 - \rho^+)}{(\rho_L - \rho_c)\rho^+} \tag{8}$$

$$\frac{dw}{dy} = - \frac{\rho_c w_c}{\rho_L - \rho_c} \cdot \frac{d\rho^+}{dy} \cdot \frac{1}{\rho^{+2}} \tag{9}$$

Assuming that the mixing length for two-phase flow is the same as that for the single flow, we may use the Van Drist correlation (18) for the mixing length ℓ:

$$\ell = k_1 y \left[1 - \exp(1-y/k_2) \right] \tag{10}$$

where k_1 and k_2 are mixing length constants, nondimensional. Combination equations (5), (8), (9) and (10) yield equation (11):

$$\tau = (\frac{d\rho^+}{dy^+})^2 \left[1 - \exp(-\frac{y^+}{k_2^+}) \right]^2 \frac{1-\rho^+}{\rho^+} \frac{y^{+2}}{\rho^{+2}} - \frac{\rho_c - \rho_L}{k_1^2 \rho_c w_c} \sqrt{\tau/\rho_L} \frac{\mu^+}{\rho^{+2}} \frac{d\rho^+}{dy^+} \tag{11}$$

Let
$$a = - \frac{\rho_c - \rho_L}{k_1 \rho_c w_c} \sqrt{\tau/\rho_L}$$
equation (11) becomes:

$$(\rho A k^{-6}/6 - \tau) a^2 /\tau = (\frac{d\rho^+}{dy^+})^2 \left[1 - \exp(-y^+/k_2^+) \right]^2 \frac{y^{+2}}{\rho^{+2}} \frac{1-\rho^+}{\rho^+}$$
$$+ \frac{a\mu^+}{k_1 \rho^{+2}} \frac{d\rho^+}{dy^+} \tag{12}$$

where
$$y^+ = y \sqrt{\tau/\rho_L} / (\mu_L/\rho_L) \quad, \quad \mu^+ = \mu/\mu_L \tag{13}$$

$$k_2^+ = k_2 \sqrt{\tau/\rho_L} / (\mu_L/\rho_L) \quad, \quad w^+ = \frac{w}{\sqrt{\tau/\rho_L}}$$

By derivation, we may obtain

$$(\rho A k^{-6}/6 - \tau) /\tau = \rho^+ - 1$$

Then equation (12) becomes

$$(\rho^+ - 1) a^2 = (d\rho^+/dy^+)^2 \left[1 - \exp(-y^+/k_2^+) \right]^2 (y^{+2}/\rho^{+2}) \frac{1-\rho^+}{\rho^+}$$
$$+ \frac{a}{k_1} \frac{\mu^+}{\rho^{+2}} \frac{d\rho^+}{dy^+} \tag{14}$$

The initial boundary condition for equation (14) is $\rho^+ = 1$, $y^+ = 0$. With this condition, equation (14) can be integrated stepwise to obtain ρ^+ as a function of y^+ for given values of a, k_1, k_2^+ and μ^+.

For a circular pipe of radium R, the mean mixture density ρ_a can be obtained from

$$\rho_a = 2 \int_0^1 (1-y/R)\ \rho\ d(y/R) \qquad (15)$$

Equation (15) can be rewritten

$$\rho_a^+ = \frac{2}{R^{+2}} \int_0^{R^+} \rho^+(R^+-y^+)\ dy^+ \qquad (16)$$

where $\quad R^+ = \frac{1}{2}\ Re_L\ \dfrac{ak_1}{1-\rho_a^+} \qquad (17)$

Re_L here is the Reynolds number based upon total flow and liquid properties   .

$$Re_L = GD/\mu_L \qquad (18)$$

The mass velocity G is equal to

$$G = 2 \int_0^1 (1-y/R)\rho w\ d(y/R) \qquad (19)$$

The velocity distribution is

$$w^+ = (1-\rho^+)\ /\ (k_1 a \rho^+) \qquad (20)$$

Substitution of equations (13) and (20) into equation (19) gives

$$G = \frac{\sqrt{\tau/\rho_L}\ \rho_L}{k_1 a}(1-\rho_a^+) \qquad (21)$$

Equation (21) can be rewritten

$$\tau = (w_o^2 \rho_L/2)\ \frac{2k_1^2 a^2}{(1-\rho_a^+)^2} \qquad (22)$$

where w_o is the fluid velocity when total flow is liquid. When total flow is liquid, the frictional stress is

$$\tau_L = \frac{1}{4}\lambda_L \rho_L w_o^2/2 \qquad (23)$$

where λ_L is the frictional factor for single-phase liquid. The ratio of the frictional pressure drop for two-phase flow ΔP_{TP} to the frictional pressure drop for single phase flow ΔP_o is:

$$\frac{\Delta p_{TP}}{\Delta p_o} = \tau/\tau_L = \frac{8k_1^2 a^2}{\lambda_L} (1-\rho_a^+)^2 \tag{24}$$

or

$$\lambda_L \frac{\Delta p_{TP}}{\Delta p_o} = 8k_1^2 a^2 (1-\rho_a^+)^2 \tag{25}$$

Equations (14), (16), (17) and (25) specify the desired solution if the constants K_1, K_2^+ and μ are known. According to Levy's investigation (12), $K_1 = 0.4$, $k_2^+ = 26$, $\mu = 1$. Therefore these equations can be solved in the following method. A value of parameter a is first assumed and equation (14) is integrated stepwise to obtain ρ^+ as a function of y^+. At each step of the solution ρ^+ is calculated from equation (16), and the corresponding value of Re number is computed from equation (17). The value of $\frac{\Delta p_{TP}}{\Delta p_o}\lambda_L$ is simultaneously obtained from equation (25). The step-by-step integration is thus carried out until the desired value of Re number is reched. A new value of a is then assumed and all of the preceding calculations are repeated to cover the desired range of ρ_a^+. The solution was programmed on a computer, and calculations performed for various values of the controlling parameters.

The calculated results are shown in figure 1. In figure 1, the two-phase frictional factor $\frac{\Delta P_{TP}}{\Delta P_o}\lambda_L$ in the horizontal flow is plotted against the mean density of the two-phase mixture. Seven different curves at various Re numbers are shown.

PREDICTION OF TWO-PHASE PRESSURE DROP AND COMPARISONS

The two-phase frictional pressure drop can be predicted as follows. According to given pressure p mass velocity G and mean quality X, the two-phase mixture velocity W_m can be calculated

$$W_m = \frac{G}{\rho_L} \left[1 + x(\rho_L/\rho_G - 1)\right] \tag{26}$$

According to W_m, ρ and the vapor volumetric flow fraction β, proportional factor c can be determined from figure 2 (19). Then the mean void fraction α can be determined as follows:

$$\alpha = c\beta \tag{27}$$

The equation for calculating mean two-phase mixture density is

$$\rho_a = \rho_L(1-\alpha) + \rho_G\alpha \tag{28}$$

then ρ_a can be determined.

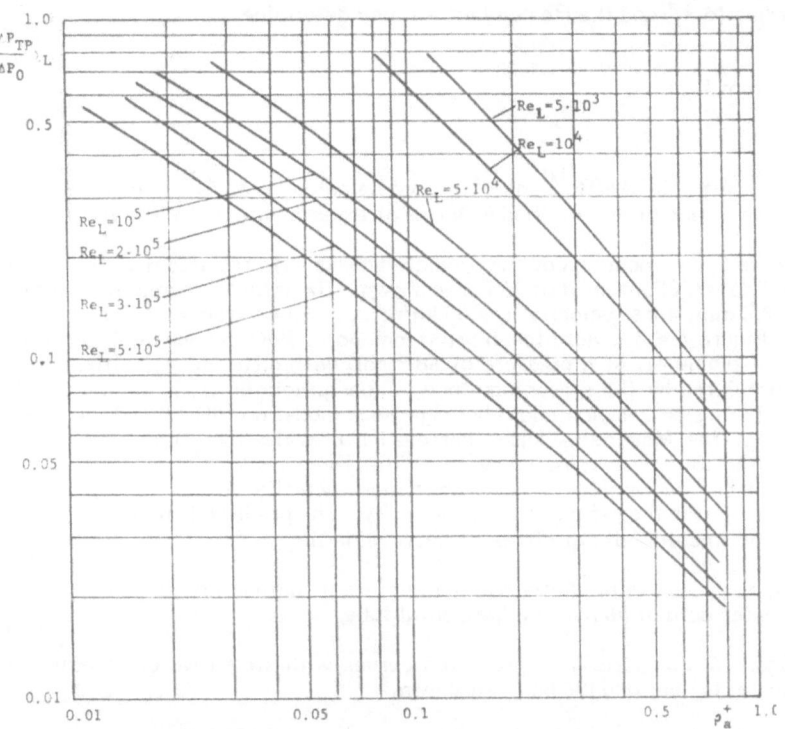

Figure 1 . The relationship among $\Delta P_{TP} \lambda_L / \Delta P_0$
Re_L and ρ_a^+

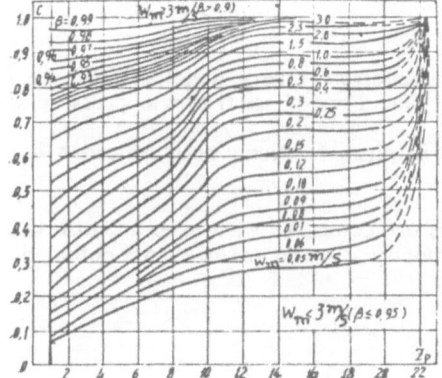

Figure 2. The proportional

factor C. $Z_p = \dfrac{p}{10}$ (unit of p is bar)

The Re number can be calculated from equation (18).

Thus, according to ρ_a^+ and the Re number, we may determine

$$\frac{\Delta P_{TP}}{\Delta P_o} \lambda_L \quad \text{from figure 1.}$$

Since $\lambda_L = 0.3164/\mathrm{Re}^{1/4}$ and ΔP_o can be calculated, ΔP_{TP} can therefore be predicted. Evaluation of this method in terms of experimental data is given.

In figure 3, a comparison of the predicted curve with the steam-water experimental data (20) obtained at 25 bar is shown. The internal diameter of the test tube is 2.6 mm, mass velocity is 5000 kg/m^2. S. Values from a homogeneous model are also shown. Figure 4 shows data taken for steam-water flow in a vertical round tube with 5 mm internal diameter, at a pressure of 68.9 bar, with different velocities (21). Besides the values predicted by the present method, values predicted from the Martinelli-Nelson correlation, Thom's correlation and a homogeneous model are also shown. From figure 4, it is clear that the Martinelli-Nelson correlation provides more accurate pressure drop estimates in the low mass velocity range. The Thom correlation and the homogeneous model give better agreement in the higher mass velocity range. The present method properly accounts for the effects of mass velocity. The predicted curves are in good agreement with experimental data of different mass velocities.

In figure 5, values predicted from the present method are compared with Moem's (22) steam water data at 96 bar in a horizontal tube.

In figure 6, comparison of predicted curves with steam-water experimental data (23), obtained at 147 bar and 196 bar, are shown.

From what has been compared above, we may see that the agreements between tests and analysis are generally good, especially at high and superhigh pressures.

CONCLUSION

By using the energy spectrum theory, together with the mixing length theory, a method for predicting vapor-liquid two-phase frictional pressure drop is developed.

In this method the two-phase frictional pressure drop ΔP_{TP} is function of $\mathrm{Re}_L, \lambda_L, P$ and X.

The predicted values are compared with experimental data of different pressures, flow directions and mass velocity. Good agreements are obtained.

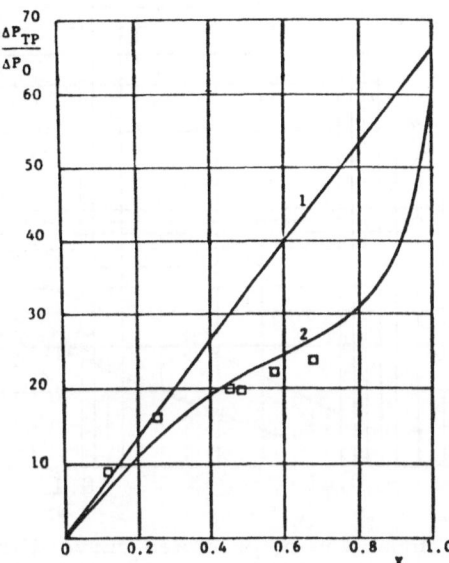

Figure 3. Comparison of predicted curve
with steam--water experimental data (20)
p=25bar. q=0 ρw=5000kg/m^{2}S
1---homogeneous model
2---present method

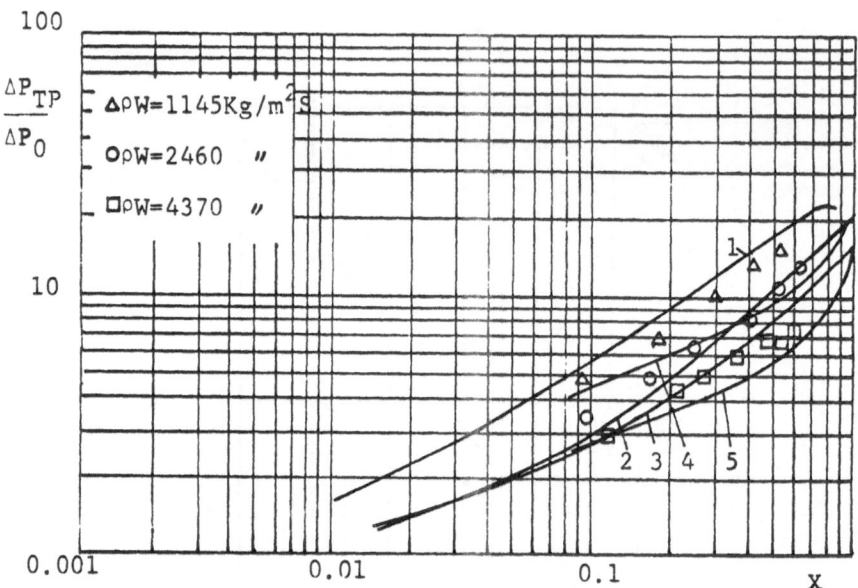

Figure 4. Comparison of predicted curve with steam-water
experimental data [21] p=68.9bar.

1---Martinelli and Nelson's correlation 2---Thom's correlation
3---homogeneous model , 4---present method($\rho w= 1145kg/m^2 S$)
5---present method $\rho w = 4370kg/mS$

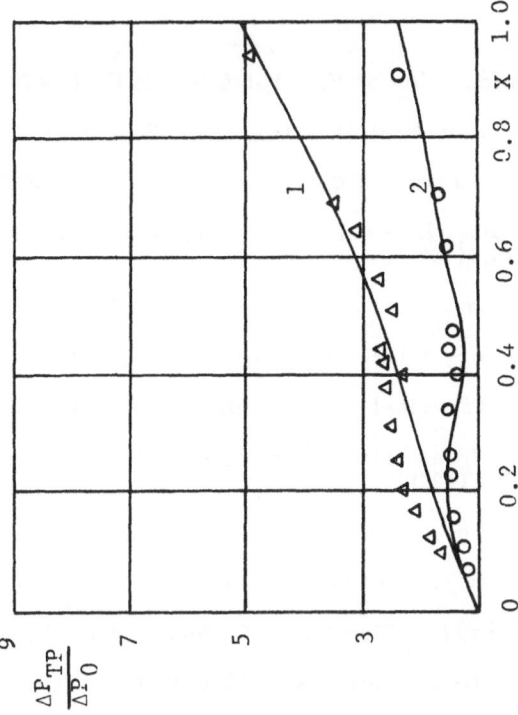

Figure 6. Comparison of predicted curve with steam–water experimental data(23).

1—p=147bar 2— p=196bar

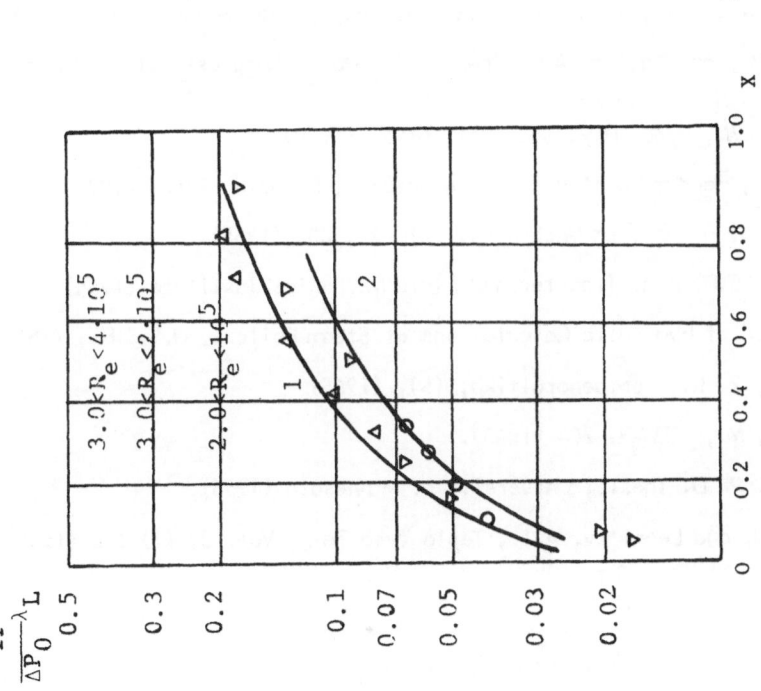

Figure 5. Comparison of predicted curve with steam–water experimental data(22)

p= 96bar

REFERENCES

1. Armand, A.A., Information of VTI, Vol 4, U.S.S.R. (1947).

2. Kostelin, S.I., Teploenergitika, (6), U.S.S.R. (1958).

3. Chenoweth, J.M., and Martin, M.W. , Pet. Ref., Vol. 34 (10),151 (1955).

4. Isbin, H.S., et al., Nuclear Eng. Part VI, Chem. Eng. Symp. Series, Vol. 55 (23), 75, (1959).

5. Hughmark, G.A., Chem. Eng. Sci., Vol. 20, 1007 (1965).

6. McAdams, W.H. et al., Trans. ASME, Vol. 64, 193 (1942).

7. Lockhart, R.W. and Martinelli, R.C., Chem. Eng. Prog., Vol. 45, 39 (1949).

8. Ownes, W.L., Jr., International Developments in Heat Transfer, Part II, ASME, 363 (1961)

9. Lockshen, V.A., et al., Teploenergitika, (8), U.S.S.R. (1959).

10. Thom, J.R.S., Inter. J. Heat and Mass Transfer, Vol. 7 (1964).

11. Dukler, A.E., Wicks, M., Cleveland R.C. AICHE, Vol. 10, 44 (1964).

12. Levy, S., J. Heat Trans., Vol. 85C, 137 (1963).

13. Obukhov, A.M., IZV., Acad. Sci. U.S.S.R. Geogr. and Geophys. Ser. (4,5) (1941).

14. Obukhov, A.M., and Yaglam, A.M. Trudy, III, Math. Congress, (1956), Acad. Press (1958).

15. Heisenberg, W., Proc. Roy. Soc. Vol. A195, (1948).

16. Panchev, S., Random Functions and Turbulence, Pergamon Press (1971).

17. Pai, S.I., J. of Applied Mechanics, Vol. 22, 604 (1955)

18. Van Driest, E.R. Heat Transfer and Fluid Mechanics Institute (1955).

19. The Standard of Hydraulic Calculations of Steam Boilers, U.S.S.R., (1973).

20. Miropolski, Z. L., Teploenergitika, (5), (1965).

21. Muscettola, M., AEEW-R. 284 (1963).

22. Moen, R.H., Ph.D. Thesis, University of Minnesota (1956).

23. Tarasova, N. and Leont'ev, A.I., Teplo Vyso Tem., Vol. 3, (1) 115 (1965).

Two-Phase Flow Measurements with Herschel Venturis

Z.H. LIN
Xian Jiao-Tong University
Xian, The People's Republic of China

ABSTRACT

Herschel venturi is a convenient and reliable device for quality measurements.In previous studies, the effective range of published correlations were in narrow limit. In this paper, on the
basis of a modified separated flow model and published experimental
data, a more general correlation is developed.

1. INTRODUCTION

Measurements of flow rate and quality of vapor liquid mixtures are
of interest in many fields of engineering such as power cycles, geo-
thermal, petroleum and control. As Herschel venturi is a convenient
and reliable device and has sufficient accuracy for these measurements, it has been receiving increasing attention in the recent two
decades, and a considerable number of papers on this topic have been
published. For instance, the use of Herschel venturi for quality and
flow rate metering is described by Ratnel (1), Smith (2), Bizon (3),
Collins (4) and Lavagno and Panella (5).

Table 1: SUMMARY OF EXPERIMENTAL RESULTS FOR HERSCHEL VENTURIS

AUTHOR	FLUID	PRESSURE (bar)	PRESSURE RATIO P/P_c	PIPE DIAMETER (mm)	DIAMETER RATIO	QUALITY X, %
Smith	steam-water	1.19	0.0052	12.7	0.5	0.88-7
Bizon	"	84.37	0.3745	26.64	0.57	5-50
Collins	"	68.90	0.3058	58.8-72.7	0.46-0.64	5-90
Lavagno	"	20.114	0.0888-0.5060	26 &13.88	0.394&0.55	up to 82
Ratnel	"	154-180	0.6837-0.7991	11	o.55	65-65

In previous studies, although the highest tested ratio of gas to liquid density has been ρ_g/ρ_l =0.1863 (Table 1), however, the effective
range of published correlations are in narrow limit. This paper in-
tends to develop a more general correlation. On the basis of a modified separated flow model, a simple and rational relationship is developed for the flow rate and quality by the introduction of a corrective coeeficient Θ, to be determined empirically. Θ is a function of

ρ_g / ρ_1 and is derived from the experimental data. The relationship is compared with other experimental data and proposed correlations.

2.   THEORETICAL CONSIDERATIONS

Reported correlations for two-phase flowmeters were mainly derived from two flow models: The Homogeneous flow model and the separated flow model. However, in reality, experiments showed that the
flow at the throat of a two-phase venturi was not homogeneous. Even
the flow pattern upstream of the throat was bubbly flow which is usually considered as a homogeneous flow. Three-dimensional suction
effects gave rise to an appreciable transverse pressure gradient at
the throat section. This gradient caused a localized increase in
void fraction near the wall (6). It was not a wholly separated
flow either. Therefore, this paper uses a modified separated flow
model to derive the correlation.

It is well-known that the equation for the single phase mass
flow rate in a venturi is:

$$W_{sp} = \frac{\psi C_d A}{\sqrt{1-\beta^4}} \sqrt{2\Delta P_{sp} \rho_{sp}} \tag{1}$$

In the case of two-phase flow, we assume the following: the vapor
and liquid phases flow separately through a venturi; the vapor
phase is imcompressible; the discharge coefficient C_d is the same
for both phases; the pressure drop for each phase is the same as
the pressure drop for the two-phase flow in device; there is no evaporation during the flow. Then, the mass flow rate of the vapor phase,
if flowing alone through a venturi would be:

$$W_g = \frac{\psi C_d A}{\sqrt{1-\beta^4}} \sqrt{2\Delta P_g \rho_g} \tag{2}$$

The mass flow rate of the liquid phase, if flowing alone through a
venturi, would be:

$$W_1 = \frac{\psi C_d A}{\sqrt{1-\beta^4}} \sqrt{2\Delta P_1 \rho_1} \tag{3}$$

The mass flow rate of the vapor phase when two phases flow together
is:

$$W_g = \frac{\psi C_d A_g}{\sqrt{1-\beta^4}} \sqrt{2\Delta P_{TP} \rho_g} \tag{4}$$

The mass flow rate of the liquid phase at that condition is:

$$W_1 = \frac{\psi C_d A_1}{\sqrt{1-\beta^4}} \quad \sqrt{2\Delta P_{TP}\rho_1} \tag{5}$$

The venturi flow area is:

$$A = A_g + A_1 \tag{6}$$

By using equations (2) through (6), one can get

$$\sqrt{\frac{\Delta P_{TP}}{\Delta P_g}} = \sqrt{\frac{\Delta P_1}{\Delta P_g}} + 1 \tag{7}$$

Equation (7) is the separated flow model correlation obtained under the above assumptions. It does not wholly correspond to the real case and should be modified by a corrective coefficient Θ. That is:

$$\sqrt{\frac{\Delta P_{TP}}{\Delta P_g}} = \Theta \sqrt{\frac{\Delta P_1}{\Delta P_g}} + 1 \tag{8}$$

The corrective coefficient, Θ, is to be determined by experimental data. The corrective coefficient Θ has certain physical meanings. Substituting equations (2), (3) and (4) into (8) and using the relationships $W_1 = W_{TP}(1-x)$. $W_g = W_{TP}x$, and $\alpha = Ag/_A$ gives

$$\alpha = \frac{1}{1 + \Theta\sqrt{\rho_1/\rho_g} \; [\frac{\rho_g}{\rho_1} (\frac{1-x}{x})]} \tag{9}$$

Where α is the void fraction. From two-phase flow theory, void fraction α can be expressed as follows:

$$\alpha = \frac{1}{1 + S[\frac{\rho_g}{\rho_1} (\frac{1-x}{x})]} \tag{10}$$

Where S is the slip ratio between the phase velocities. Comparing equation (9) to equation (10) one may find that the corrective coefficient θ is a function of slip ratio S and density ratio ρ_g/ρ_1. Therefore, Θ reflects the influence of slip ratio S and working pressure or density ratio ρ_g/ρ_1.

As a slip ratio S is also a function of ρ_g/ρ_1 (7), on the whole, should be a function of ρ_g/ρ_1.

Under given ρ_g/ρ_1, is a constant.

3. DETERMINATION OF CORRECTIVE COEFFICIENT

In this section, the corrective coefficient θ is to be determined by experimental data of different authors. First, Lavagno's data (5) obtained under the pressure of 20,40,70,100 bar, Bizon's data (3) obtained under the pressure of 84 bar, and Ratnel's data (1) obatined under the pressure of 160 bar are rearranged in the form of Martinelli's parameters $\sqrt{\Delta P_{TP}/\Delta P_g}$ and $\sqrt{\Delta P_1/\Delta P_g}$. The parameter $\sqrt{\Delta P_{TP}/\Delta P_g}$ is plotted against $\sqrt{\Delta P_1/\Delta P_g}$ at different pressure and density ratios. One of them is shown in figure 1. Experiments show $\sqrt{\Delta P_{TP}/\Delta P_g}$ varies approximately linearly with $\sqrt{\Delta P_1/\Delta P_g}$. From figure 1, according to equation (8), we may obtain $\theta=1.8$. By using the same method we may obtain different values for different experimental data obtained under different pressure. Thus we may obtain a set of values θ. In figure 2 the coefficient θ is plotted against ρ_g/ρ_1.

4. COMPARISON AND DISCUSSION

Equation (8) is not a convenient form for practical use when x=0, i.e., $\Delta P_g=0$, ΔP_{TP} does not converge to ΔP_1.

Multiplying equation (8) by $\sqrt{\Delta P_g/\Delta P_o}$ gives:

$$\sqrt{\Delta P_{TP}/\Delta P_o} = \theta \sqrt{\Delta P_1/\Delta P_o} + \sqrt{\Delta P_g/\Delta P_o} \tag{11}$$

Substituting equation (2), (3) and (13) into equation (11) and remembering $W_1=W_{TP}(1-x)$, $W_g=W_{TP}x$, produces:

$$\sqrt{\Delta P_{TP}/\Delta P_o} = \theta + x (\sqrt{\rho_1/\rho_g} - \theta) \tag{12}$$

In equation (12), ΔP_o is the pressure drop across venturi assuming total flow to be liquid, and is a function of total two-phase flow rate W_{TP} at the same time. $\sqrt{\Delta P_o}$ can be calculated from the following equation:

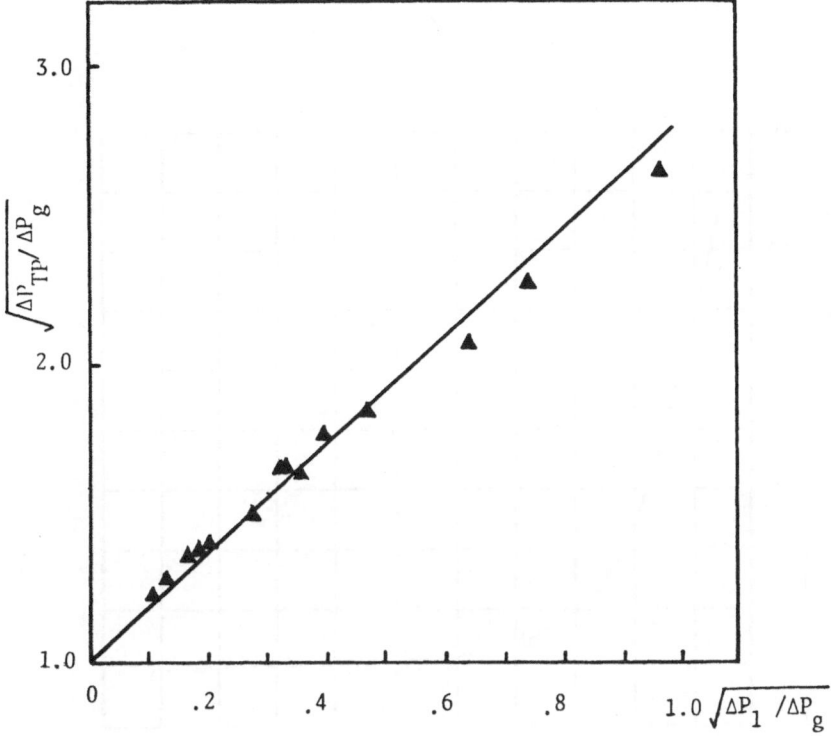

Figure 1. The relationship between $\sqrt{\Delta P_{TP}/\Delta P_g}$ and $\sqrt{\Delta P_l/\Delta P_g}$. Lavagno's data(5). P = 20 bar.

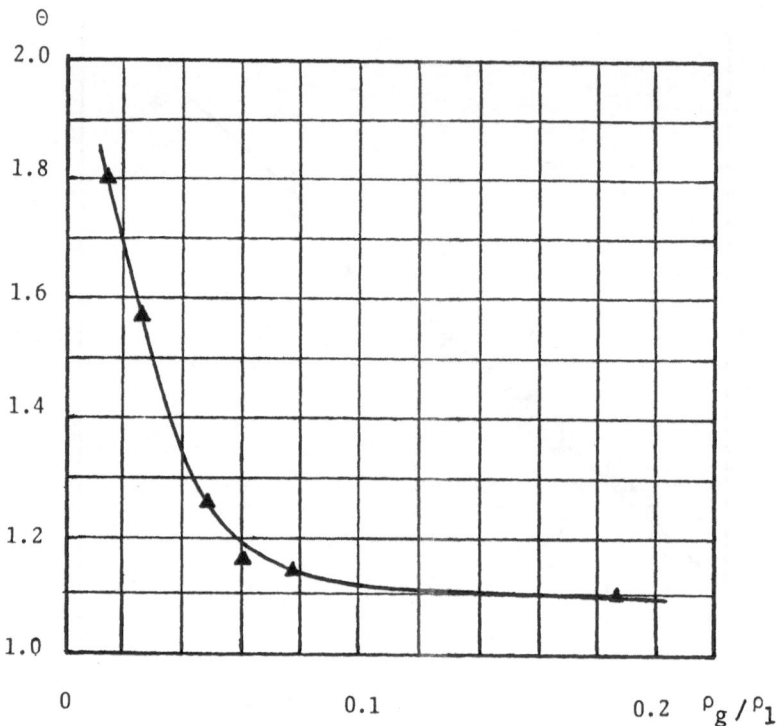

Figure 2 . Relationship between Θ and ρ_g / ρ_1.

$$\sqrt{\Delta P_O} = \frac{W_{TP}\sqrt{1-\beta^4}}{\Psi C_d A\sqrt{2\rho_1}} \tag{13}$$

So knowing ΔP_O, W_{TP} can be calculated at once.

Equation (12) is a more convenient form for application. Using figure 2 and equation (12) under measured pressure drop and working pressure, we may predict the value of ΔP_O, i.e., the total two-phase flow rate W_{TP}, for a given quality or vice versa. What is more, in equation (12) under given pressure, ρ_1/ρ_g and θ being constants, the relationship between x and parameter $\sqrt{\Delta P_{TP}/\Delta P_O}$ should be a straight line.

Equation (12) fits for $x \geqslant 0.1$. When $x < o.1$, $\sqrt{\Delta P_{TP}/\Delta P_O}$ can be calculated by using interpolation between the value of $\sqrt{P_{TP}/P_O}$ at x=o.1 and the value of $\sqrt{P_{TP}/P_O}$ at x=0. i.e., $\sqrt{P_{TP}/P_O} = 1.0$.

Figure 3 indicates that the present predictive equation a-grees very well with Collins' (4) steam water experimental data which have not been applied to derive the corrective coefficient θ curve. In the same figure, the pressure drop correlations for the homogeneous flow model (Collins (4) and Bizon (3)) are also given. The present predictive equation lies within the range of other predictive equations, and can predict Collins' data with R.M.S. errors smaller than 12.2%.

Because tested data do not express apparent influence of d/D and mass velocity on equation (12) within the experiment range, we may consider approximately that equation (12) can be used to calculate the flow rate or the quality of steam water mixture in the range 0.01157-0.18632 of the ρ_g/ρ_1 ratio and in pipe size ranging from 10-70 mm (β=o.4-0.6). The mean square root error of this method is about 12% when the quality x ranges from 5% to 90%.

When $\rho_g/\rho_1 = 1.0$, i.e. $P/P_c \ge 1.0$, then $\sqrt{P_{TP}/P_O}$ should equal 1.0. Substituting these values into equation (12), we may get $\theta = 1.0$. The curve shown in figure 2 expresses this tendency.

Substituting equation (2) and (3) into equation (8) and remembering $W_g = W_{TP}x$, $W_1 = W_{TP}(1-x)$, then gives:

$$W_{TP} = \frac{\psi C_d A \sqrt{2\Delta P_{TP}\rho_1}}{\sqrt{1-\beta^4}\ [\ (1-x)\theta + x\sqrt{\rho_1/\rho_g}\ ]} \tag{14}$$

Equation 14 is a direct equation for calculating W_{TP} from given x or vece versa, and can be used in the range of $x \geqslant 0.1$. When $x < 0.1$, as it should be calculated by the method of interpolation mentioned above, it is more simplified to use equation (12).

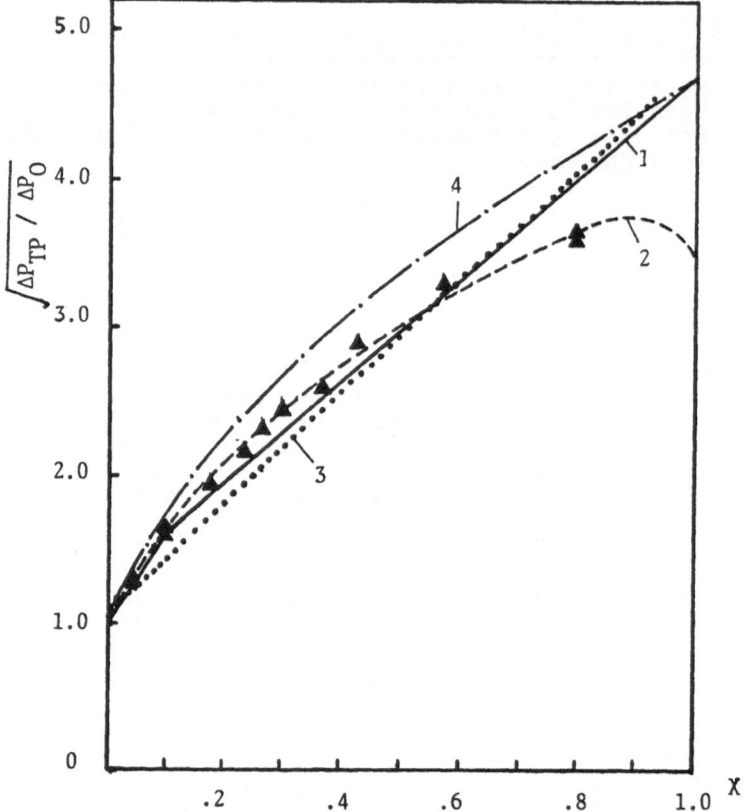

Figure 3 . Comparison among Collins'data (4)
(P= 68.9 bar) and different correlations.
1- eq. (12); 2-Collins'correlation;3-Bizon's
correlation; 4-homogeneous model.

Comparing equation (14) with equation (1), we may find the main difference between them is the two-phase factor $1/[(1-x)\Theta x \sqrt{\rho_1/\rho_g}]$. The two-phase factor is a function of Θ, x and ρ_1/ρ_g, i.e., a function of quality x and pressure P. When P is equal to P_C, and $\sqrt{\rho_g/\rho_1}$ should both be equal to 1.0. Thus, this factor is equal to 1.0 too. At that time, equation (14) is equal to equation (1). That is, equation (14) or (12) turns to the equation for single phase. When $x=1.0$, the two-phase factor is equal to $\sqrt{\rho_g/\rho_1}$, and equation (14) becomes the equation of vapor flow rate. When $x=0$, by usihg the proposed method of interpolation, we may also get from equation (12), that $\sqrt{\Delta P_{TP}/\Delta P_o}$ is equal to 1.0. From this result, by using equation (13), we may get the equation for liquid mass flow rate.

CONCLUSIONS

1. This paper presents a simple and practical relationship, equation (12) for calculating two-phase flow rate or quality whose mean square root error is less than 20% when the quality ranges from 5% to 90%. The comparison with experimental data shows that equation (12) can be used to calculate the quality or flow rate of low viscosity vapor liquid, such as steam-water mixture in the effective range)/01157-0.18632 of the ρ_g/ρ_1 ratio and in pipe size ranging from 10-70 mm ($\beta = 0.40$-0.60).

2. To further improve the proposed relationship, additional high and lower steam-water data are required.

NOMENCLATURE

A	venturi flow area	(m^2)
A_g	flow area occupied by gas or vapor phase at venturi	(m^2)
A_1	flow area occupied by liquid phase at venturi	(m^2)
C_d	venturi discharge coefficient	
d	venturi diameter	(m)
D	internal pipe diameter	(m)
P	working pressure	(Pa)
P_C	critical pressure	(Pa)
S	slip ratio	
x	quality	
W_g	mass flow rate of gas or vapor	(kg/s)
W_1	mass flow rate of liquid phase	(kg/s)
W_{SP}	mass flow rate of single phase fluid	(kg/s)
W_{TP}	mass flow rate of two-phase flow	(kg/s)

ΔP_g pressure drop across venturi for gas or vapor
phase alone flow   (Pa)

ΔP_1 pressure drop across venturi for gas or vapor phase
alone flow   (Pa)

ΔP_0 Pressure drop across venturi assuming total flow to be
liquid   (Pa)

ΔP_{SP} Pressure drop acrodd venturi for single phase fluid
flow   (Pa)

α void fraction, A_g/A

β Diameter ration, d/D

Θ corrective coefficient

ψ venturi thermal expansion factor

ΔP_{TP} pressure drop across venturi for two-phase flow   (Pa)

ρ_g gas or vapor density (kg/m^3)

ρ_1 liquid density (kg/m^3)

ρ_{sp} single phase density (kg/m^3)

REFERENCES

1. Ratnel A.V. 1958 The determination of the quality of high pressure steam. Teploenergetika, No. 5.

2. Smith, R. V. et al. 1962 The use of a venturi tube as a quality meter. J. of Basic Engi.

3. Bizon, E. 1965 Twophase flow measurements with sharp edged orifices and venturis. Atomic Energy of Canada LTD. Rep. No. AECL-2273.

4. Collins D.B. and Gacesa M. 1971. Measurements of steam quality in two-phase upflow with venturi meters and orifice plates. J. of Basic Eng., A.S.M.E.

5. Lavagno E. and Panella B 1979 Two-phase measurements with venturis: A correlation and a comparison with theoretical models. Multiphase Transport Function, Reactor Safety, Applications. Vol. 5. Edited by T. N. Veziroglu.

6. Thang N. T. and Davis M.R. 1979 The structure of bubbly flow through venturis. International J. of Multiphase Flow. Vol. 5, No. 1.

7. Thom J.R.S. 1964 Prediction of pressure drop during forced circulation boiling of water. International J. of Heat and Mass Transfer. Vol. 7.

A New Correlation for Heat Transfer during Flow Boiling

S.G. KANDLIKAR
Rochester Institute of Technology
Rochester, New York 14623, USA

B.K. THAKUR
Foster-Wheeler Energy Corporation
Dansville, New York 14437, USA

ABSTRACT

This paper is concerned with an investigation carried out to establish a
correlation for the prediction of the heat transfer coefficient during flow
boiling through tubes, vertical and horizontal flows, for water and some
organic fluids (including refrigerants). An additive mechanism of the
convective and the nucleate boiling heat transfer has been formulated to give
the net heat transfer during flow boiling. The correlation employs
three dimensionless parameters: the Convection number, the Boiling number and
the Froude number. It has been tested with the data (over 500) of five
investigators from twelve experimental cases with a mean deviation of 13
percent.

1.   INTRODUCTION

The heat transfer in boiling, i.e. during change of phase from liquid to
vapor, is a very efficient means of heat transfer, due to ability of the
boiling systems to attain large heat fluxes, while employing relatively small
temperature differences. This mode of heat transfer is applied in numerous
industrial systems, such as, boilers in steam power plants, equipments in
air-conditioning and refrigeration systems, and nuclear reactors.

When a liquid flows in a heated tube, heat transfer takes place by single
phase forced convection. As the bulk temperature reaches a level somewhat
below the saturation temperature, subcooled boiling may occur at the tube
wall. When the bulk fluid temperature reaches the saturation temperature,
saturation boiling commences and net vapor generation takes place in the flow
stream. Within the boiling region of the tube, several complex interacting
processes occur simultaneously. Heat is transferred from the tube wall into
the two-phase mixture, the net effect of this heat transfer being the
formation of vapor. Vapor is also formed by flashing because of the pressure
drop and the criterion of thermal equilibrium between the phases.

As briefly described above, flow boiling is very complex in nature;
several physical quantities and hydrodynamic conditions influence the flow
boiling in a confined channel. Many empirical correlations, based on
experimental results, have been suggested for different fluids and flow
conditions. However, none of these empirical correlations give satisfactory
results for all the fluids.

2.   EXISTING CORRELATIONS

Some of the more widely accepted correlations are discussed below:

Guerrieri and Talti [1] measured local heat transfer coefficients during vertical upflow of various organic fluids, including cyclohexane, through brass tubes and suggested the following correlation:

$$h_{TP} = 3.4 \left(\frac{1}{X_{tt}} \right)^{0.45} h_\ell \tag{1}$$

where

$$h_\ell = (1-x)^{0.8} h'_\ell \tag{2}$$

and h' is evaluated by the Dittus-Boelter equation,

$$h'_\ell = 0.023 \frac{k_\ell}{D} \left(\frac{DG}{\mu_\ell}\right)^{0.8} \left(Cp_\ell \frac{\mu_\ell}{k_\ell}\right)^{0.4} \tag{3}$$

The mean deviation with their own experimental data was 11.1%.

Dengler and Addoms [2] used water in an upflow system consisting of a 25.4 mm ID, 6096 mm long, vertical copper tube and recommended the following correlation to represent their data:

$$h_{TP} = 3.5 \frac{F_{DA}}{(X_{tt})^{0.5}} \times h'_\ell \tag{4}$$

In the above correlation, F_{DA} represents the correction factor to be applied for points where nucleate boiling exists. The mean deviation was 30.5%.

Bennet et al [3] made the measurements on vertical annulii upflow, the inner pipe being electrically heated. Vapor quality was generated outside the test section by mixing water with steam. They noted an effect of heat flux and proposed the following correlation

$$h_{TP} = 0.64 \left(\frac{1}{X_{tt}}\right)^{0.74} h_\ell \left(\frac{q}{A}\right)^{0.11} \tag{5}$$

The mean deviation was 11.9%.

Schrock and Grossman [4] used water in a vertical upward flow system. They used electrically heated test sections of 2.95, 6.02, and 10.96 mm ID tubes, 381 to 1016 mm in length. They introduced boiling number, Bo, as an additional variable and obtained the following correlation:

$$h_{TP} = 7400 \left[Bo + 1.5 \times 10^{-4} \left(\frac{1}{X_{tt}}\right)^{2/3} \right] h'_{\ell} \tag{6}$$

The mean deviation with their own experimental data was 35%.

Chen [5] proposed a correlation which is generally accepted as one of the best available. The correlation covers both the saturated nucleate boiling region and the two-phase forced convective region. He assumed that both the nucleation and the convection mechanisms occur to some degree over the entire range and that the contributions made by the two mechanisms are additive:

$$h_{TP} = h_{mac} + h_{mic} \tag{7}$$

where, h_{mac} is the contribution due to convection and h_{mic} is the contribution due to nucleate boiling, and are given by the following equations:

$$h_{mac} = 0.023 \left[\frac{G(1-x)D}{\mu_{\ell}}\right]^{0.8} Pr_{\ell}^{0.4} \left(\frac{k_{\ell}}{D}\right) F \tag{8}$$

and,

$$h_{mic} = 0.00122 \frac{k_{\ell}^{0.79} Cp_{\ell}^{0.45} \rho_{\ell}^{0.49}}{\sigma^{0.5} \mu_{\ell}^{0.29} h_{fg}^{0.24} \rho_{g}^{0.24}} \Delta Tsat^{0.24} \Delta Psat^{0.75} S \tag{9}$$

F, the two-phase correction factor, is a function of the Martinelli parameter X_{tt}, and S, is the suppression factor defined as the ratio of the mean superheat seen by the growing bubble, and the local wall superheat, $(T_{wall} - T_{sat})$.

The mean deviation for all (over 600) data points from ten experimental cases for vertical flow was 12%.

Jallouk [6] carried out an extensive study of two-phase flow characteristics of refrigerants in 19.9 mm ID and 3048 mm long vertical tubes. Jallouk observed that the Chen's correlation exhibited the least amount of scatter, although it consistently underpredicted the experimental results. He proposed a variable coefficient in the Forster Zuber's pool boiling correlation used in predicting the nucleate boiling part of heat transfer. The modified h_{mic} correlation for R-114 is as follows:

$$h_{mic} = 0.00270 \frac{k_{\ell}^{0.79} Cp_{\ell}^{0.45} \rho_{\ell}^{0.49}}{\sigma^{0.5} \mu_{\ell}^{0.29} h_{fg}^{0.24} \rho_{g}^{0.24}} \Delta Tsat^{0.24} \Delta Psat^{0.75} S \tag{10}$$

h_{mac} is evaluated by Eq. (9), as proposed earlier by Chen.

The majority of Jallouk's experimental values were correlated to within 30%.

Shah [7] has developed a chart correlation, a graphical method of calculation, which appears to be considerably superior to any other available predictive technique. It has been compared by Shah with about 800 data points from 18 different experimental studies. The chart correlation employs four dimensionless parameters:

The ratio $\psi = \dfrac{h_{TP}}{h_\ell}$

Convective number $Co = (\dfrac{1}{x} - 1)^{0.8} (\dfrac{\rho_g}{\rho_\ell})^{0.5}$

$$(11)$$

Boiling number $Bo = \dfrac{q}{Gh_{fg}}$

Froude number $Fr_\ell = \dfrac{G^2}{\rho_\ell^2 gD}$

The use of chart for calculation of h_{TP} is described in Shah's paper for various conditions. The mean deviation for all the data points is 14%.

3. DEVELOPMENT OF CORRELATION

It is seen from the literature survey that none of the existing correlations appear to be entirely satisfactory for general use. However, Chen's and Shah's correlations have had some success for a variety of liquids.

The correlation proposed by Chen has been derived for the following flow conditions:

   i)   saturated, two-phase fluid in convective flow
   ii) vertical, axial flow
   iii) stable flow
   iv) no slug flow
   v)   no liquid deficiency
   vi) heat flux less than critical flux.

These conditions usually occur with annular flow or annular mist flow in the quality range of approximately 1-70%.

The chart correlation proposed by Shah appears to be considerably superior to any other available predictive technique. However, it needs use of the chart for prediction of heat transfer coefficients after calculating the necessary parameters. The concept of additive contributions of convection and boiling heat transfer during flow boiling has been accepted by many investigators. Shah considers heat transfer coefficient to be a function of only Bo number (boiling contribution) in the nucleate boiling regime, and a function of only Co number, for low values of Co number corresponding to high vapor qualities or condition of pure convective boiling in which bubble nucleation has been assumed to be completely suppressed. The effect of Bo and Co numbers is significant only in the transition region.

3.1 Vertical Flow

In the present investigation [12] convective as well as boiling contributions have been considered in all the regimes of flow boiling, the convective part for the data analyzed is well represented by the following correlation:

$$h_c = \frac{0.533 \, h_\ell}{Co^{0.79}} \, , \qquad \text{For } Co > 0.65$$

$$= \frac{1.876 \, h_\ell}{Co^{0.79}} \, , \qquad \text{For } Co \leq 0.65$$

(12)

where, h_c is the two-phase heat transfer coefficient for convective contribution during flow boiling, and h_ℓ is the superficial heat transfer coefficient for liquid phase and is given by:

$$h_\ell = 0.023 \, \left[\frac{G(1-x)D}{\mu_\ell}\right]^{0.8} \, Pr^{0.4} \, \frac{k_\ell}{D}$$

(13)

For the nucleate boiling part of heat transfer the following correlation is recommended:

$$h_b = 2.3 \, h_\ell \, (Bo \times 10^4)^{0.5}, \text{ for } Co > 0.65$$

$$= 0.7 \, h_\ell \, (Bo \times 10^4)^{0.9}, \text{ for } Co \leq 0.65$$

(14)

where, h_b is the boiling contribution to the two-phase heat transfer coefficient and h_ℓ is given by Eq. (13).

The two-phase heat transfer coefficient, h_{TP}, is then obtained as the sum,

$$h_{TP} = h_\ell + h_b \tag{15}$$

The above equations in their final form for the case of vertical flow are given by:

$$h_{TP} = h_\ell \left[\frac{0.533}{Co^{0.79}} + 2.3 \left(Bo \times 10^4\right)^{0.5}\right], \text{ For } Co > 0.65 \tag{16}$$

and

$$h_{TP} = h_\ell \left[\frac{1.876}{Co^{0.79}} + 0.7 \left(Bo \times 10^4\right)^{0.9}\right], \text{ For } Co \leq 0.65 \tag{17}$$

3.2 Horizontal Flow

The flow patterns during generation of vapor in horizontal tubes are asymmetric, particularly at low inlet velocities. Intermittent drying and rewetting of the upper surfaces of the tube in wavy flow and progressive drying out over long tube lengths of the upper circumference of the tube wall in annular flow have been observed during testing by many researchers. However, at higher inlet liquid velocities, the influence of gravity is less pronounced, the phase distribution becomes more symmetrical, and the flow patterns resemble those seen in vertical upward flow.

A study of the experimental data analyzed during this investigation revealed that tube orientation does not affect the heat transfer at very low quality of vapor (corresponding to high Co number), and the Eq. (16) gives satisfactory results for the case of horizontal flow also (for Co > 0.65).

For the case, when the tube walls are not completely wet and a dry-out situation occurs, the heat transfer rate drops. The greater the portion of the surface remaining dry, the lower would be the heat transfer rate. Based on this model, a correction factor is needed to represent the fraction of the tube circumference that remains dry. The Froude number has been established to be the controlling factor in wave formation, high value of Froude number will ensure wave formation and absence of stratified flows. Also, the inertia forces tend to cause film climbing, resisted by surface tension and gravitational forces. Surface tension forces are generally negligible in comparison to gravitational and inertia forces. For any particular fluid the ratio of gravitational forces to inertia forces would then determine the film climbing and tube surface wetting, which is the Froude number.

Based on the data of Mumm [10] and Gouse and Commou [11], it is found that a correction factor of $(25 \times Fr_\ell)^{0.24}$ to the Eq. (17) gives satisfactory results for horizontal flow, for Froude number less than 0.04. No correction factor is needed for Froude number greater than 0.04. However, while testing with the Chawla's data [8] for R-11 boiling in horizontal tubes, the mean error has been found to be minimum by using different individual correction factors for the convective and the boiling parts of Eq. (17). This may be explained by the fact that the film thickness along the circumference varies at low values of Fr. The temperature gradient in the film is different at different circumferential locations. The nucleation mechanism is very sensitive to the existing temperature gradient in the film, while the convective mechanism is related, rather simply, to the velocity distribution in the film. This may lead to a different influence on each of the mechanisms by varying the circumferential film thickness distribution, hopefully represented by Fr.

Finally, the correlation suggested, on the basis of the above discussion, for the flow boiling in horizontal tubes is:

For Co > 0.65,

$$h_{TP} = h_{\ell} \; [\frac{0.533}{Co^{0.79}} + 2.3 \; (Bo \times 10^4)^{0.5}] \tag{18}$$

and for Co ≤ 0.65

$$h_{TP} = h_{\ell} \; [\frac{1.876}{Co^{0.79}} + 0.7 \; (Bo \times 10^4)^{0.9}, \qquad \text{For Fr} \geq 0.04$$

$$= h_{\ell} \; [\frac{1.876(25 \times Fr_{\ell})^{0.3}}{Co^{0.79}} + 0.7(Bo \times 10^4)^{0.9} (25 \times Fr_{\ell})^{0.1} \tag{19}$$

$$\text{For Fr}_{\ell} < 0.04$$

where, Fr_{ℓ} is the Froude number assuming all the mass flowing is in liquid form, and is given by:

$$Fr_{\ell} = \frac{V^2}{gD} \tag{20}$$

4.   COMPARISON WITH EXPERIMENTAL DATA

The proposed correlation has been tested against the data for water and organic fluids (including refrigerants). The comparisons with the data of (i) Guerrieri and Talti [1], (ii) Wright [9], (iii) Chawla [8], (iv) Mumm [10], and (v) Gouse and Coumou [11] are shown in Figs. 1-5, respectively, and the results obtained are tabulated in Table 1.

Table 2 shows the results of comparison with the existing correlations for vertical flow. The mean error is 16.6%.

Comparing the results with Jallouk's data for R114, large discrepancies were observed. At this point, a variable coefficient for the nucleate boiling term in the Eq. (14), as originally suggested by Jallouk, was introduced. For R-114, the multiplier to the right hand side of Eq. (14) is 2.21. This resulted in marked improvement, with a mean error of only 16.3%.

Thus the following equation is recommended in place of Eq. (14) for the case of R-114 boiling in vertical tubes:

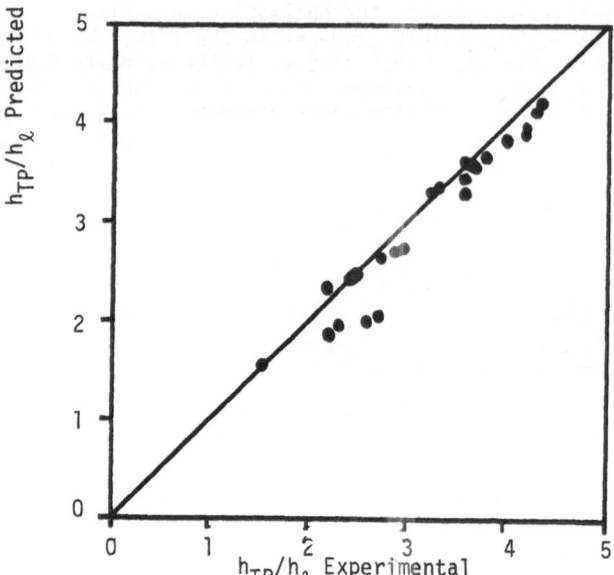

Fig. 1: Comparison of the correlation with the experimental
         data of Guerrieri and Talti [1].

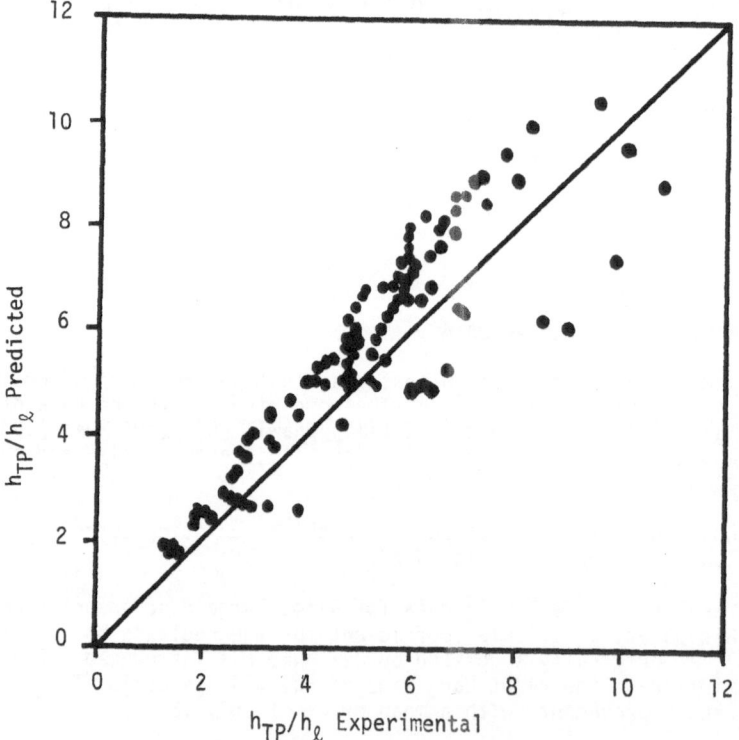

Fig. 2: Comparison of the correlation with the experimental
         data of Wright [9].

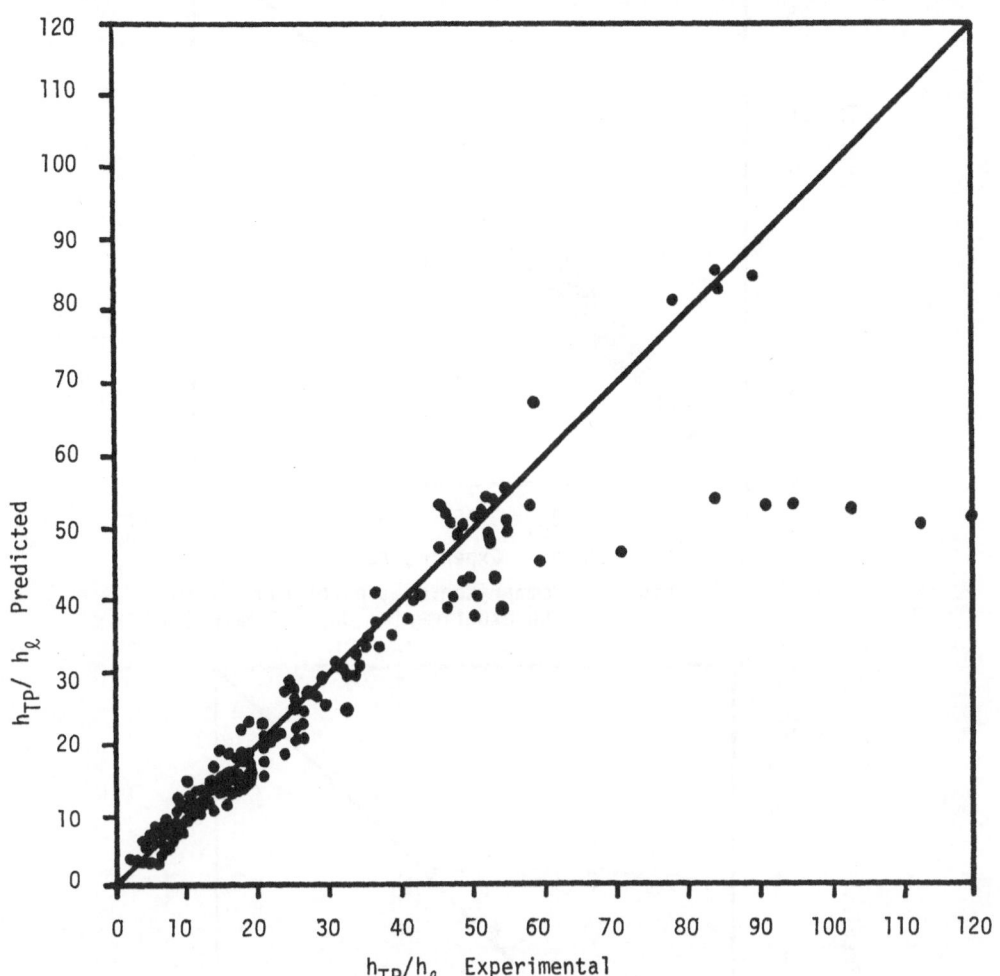

Fig. 3: Comparison of the correlation with the experimental data of Chawla [8].

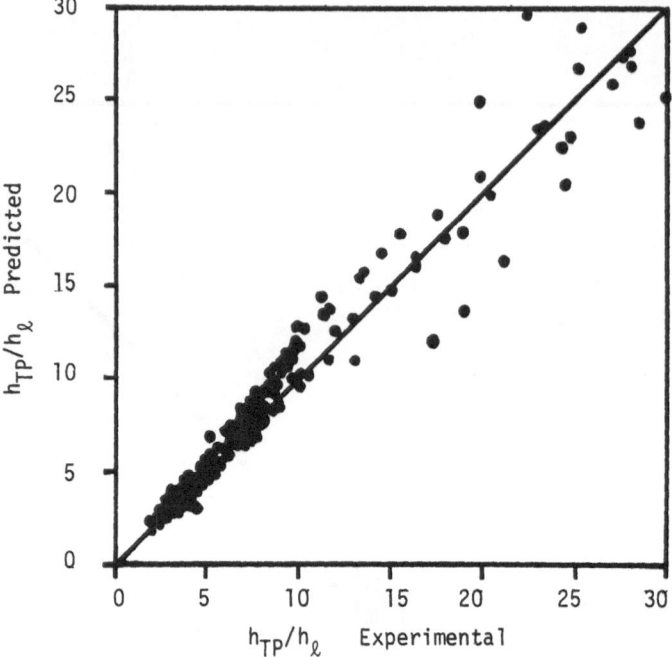

Fig. 4: Comparison of the correlation with
the experimental data of Mumm [10]

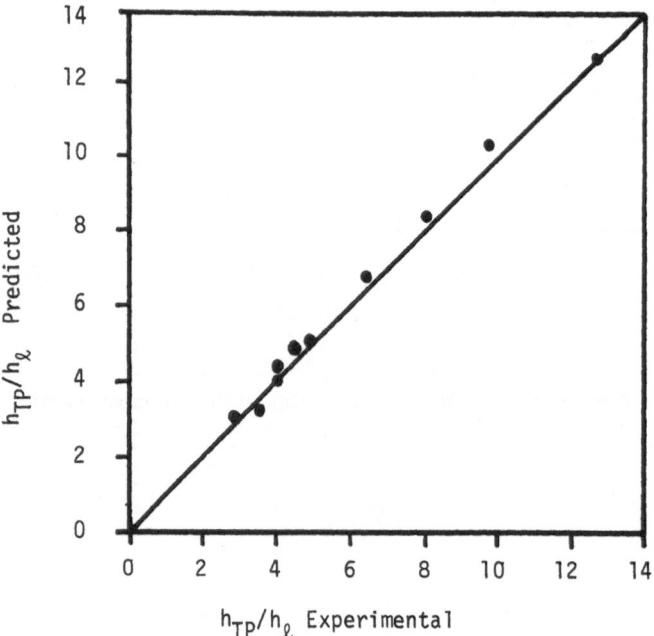

Fig. 5: Comparison of the correlation with
the experimental data of Gouse and Coumou [11].

Table 1.   Results obtained for the experimental data
with the proposed correlation

Data	No.of Data	Fluid	Flow Condition	Test Condition	% Mean Deviation
Guerrierri and Talti	23	Cyclo-hexane	Vert.up	Sat. Pressure 103 kPa	7.1
Wright	80	Water	Vert. Down	18.3mm ID, Test Sec.-1	18.4
Wright	27	Water	Vert. Down	11.98mm ID, Test Sec.-2	19.7
Chawla	51	R-11	Horizontal	25mm ID, Sat. Temp 0°C	16.1
Chawla	33	R-11	Horizontal	25mm ID, Sat. Temp. 20°C	13.0
Chawla	31	R-11	Horizontal	14mm ID, Sat. Temp. 10°C	13.6
Chawla	66	R-11	Horizontal	6mm ID, Sat. Temp 10°C	16.3
Mumm	94	Water	Horizontal	11.81mm ID, Sat. Pres. 310 kPa	9.9
Mumm	56	Water	Horizontal	11.81mm ID, Sat Pres 620 kPa	10.8
Mumm	20	Water	Horizontal	11.81mm ID, Sat. Pres. 1034 kPa	6.9
Mumm	19	Water	Horizontal	11.81mm ID, Sat. Pres. 1379 kPa	9.3
Gouse and Coumou	10	R-113	Horizontal	10.92mm ID	5.1

Total No. of Data Points = 510
Mean Deviation for all data = 13.3%

Table 2.   Comparison of Correlations, vertical flow

Data	Dengler and Addams	Guerrieri and Talti	Bennet et al	Schrock and Grossman	Chen	Shah	This Correlation
Guerrieri and Talti (Cyclohexane)	39.8	11.1	65.9	50.7	13.6	16.3	7.3
Wright (water)	24.9	75.8	30.4	51.7	15.4	13.4	18.7

Number of data points considered = 130
Mean deviation for all data this correlation: giving equal weight to
                                                each set of data = 13.0%
                                                giving equal weight to
                                                each data point = 16.6%

Table 3.   Comparison of Correlations, horizontal flow

Data	Fluid	Test Condition	Mean % Deviation Shah	This Correlation
Chawla	R-11	25mm ID	16.0	14.8
		14mm ID	10.3	13.6
		6mm ID	28.5	16.3
Mumm	Water	0.465" ID	12.8	9.8
Gouse and Coumou	R-113	0.43" ID	10.7	5.1

Number of data points considered = 300

Mean deviation for all data this correlation: giving equal weight to
                                                each set of data = 10.0%
                                                giving equal weight to
                                                each data point = 12.2%

$$h_b = 2.21 \times 2.3 \ h_\ell \ (Bo \times 10^4)^{0.5} \qquad \text{for } Co > 0.65$$

(21)

$$= 2.21 \times 0.7 \ h_\ell \ (Bo \times 10^4)^{0.9} \qquad \text{for } Co \leq 0.65$$

Table 3 shows a comparison of different correlations for horizontal flow data. It clearly shows the superiority of the present equation, with a mean error of 12.2%, over the other correlations.

5. CONCLUSIONS

The proposed correlations Eqs. (16)-(19) are based on the assumption that the nucleate boiling and the convective mechanisms are present at all times during flow boiling, and that their contribution is additive. The individual mechanism contribution has been determined by analyzing the data of five investigators. The overall mean errors are 16.6% for vertical flow, and 12.2% for horizontal flow.

Two observations made during this investigation need special mention. Firstly, in the case of horizontal tubes, the influence of Froude number on the two mechanisms, nucleate boiling and the convective, is not the same, as seen from Eq. (19). Secondly, the introduction of a variable coefficient, for individual fluids, in the nucleate boiling term significantly improves the applicability of the correlation to other fluids, as seen for Jallouk's data. It is recommended that further investigation be undertaken to substantiate the above two observations.

6. NOMENCLATURE

A   Surface area through which heat is transferred.

A_c Cross-sectional area

Bo   Boiling number q/Gh_{fg}

C_p Specific heat at constant pressure

Co   Convection number, $(1/x-1)^{0.8} \ (\rho_g/\rho_1)^{0.5}$

D   Diameter of pipe

Fr   Froude number V^2/gD

Fr_ℓ Froude number assuming all mass flowing in liquid form

g   Acceleration due to gravity

G   Total mass flux or mass velocity, W/A_c

h   Heat transfer coefficient

h_{fg} Latent heat of vaporization

h_{TP} Two-phase heat transfer coefficient local

h_1 Heat transfer coefficient for liquid phase

k Thermal conductivity

Nu Nusselt number, hD/k

p Pressure

p_c Critical pressure

Pr Prandtl number, $C_p \mu/k$

q Heat flux, Q/A

Q Total heat transferred in unit time

Re Reynolds number, GD/μ

Re_1 Superficial Reynolds number of liquid phase

T Temperature

W Total mass flow rate

x Weight fraction vapor

X_{tt} Martinelli parameter $\left(\dfrac{1-x}{x}\right)^{0.9} (\rho_g/\rho_\ell)^{0.5} (\mu_\ell/\mu_g)^{0.1}$

ψ Ratio h_{TP}/h_ℓ

μ Viscosity

ρ Density

σ Vapor-liquid surface tension

Subscripts

g For the gas phase or vapor phase

$1,\ell$ For the liquid phase

sat saturation

w wall

REFERENCES

1.   Guerrieri, S.A. and Talti, R.D., "A Study of heat transfer to organic
     liquids in single tube natural circulation boilers," Chemical Engineering
     Progress Symposium Series, Heat Transfer, Louisville, No. 18, Vol. 52,
     1956, pp. 69-77.

2. Dengler, C.E. and Addoms, J.N. "Heat transfer mechanism for vaporization of water in vertical tube," Chemical Engineering Progress Symposium Series, Vol. 52, No. 18, 1956, pp. 95-103.

3. Bennet, J.A.R., Collier, J.C., Pratt, H.T.C., and Thornton, J.D., "Heat transfer to two-phase gas-liquid systems, AERE-R 3159, 1959.

4. Schrock, V.E. and Grossman, L.M., "Forced convection boiling in tubes," Nuclear Science Engineering 12, 474-81, 1962.

5. Chen, J.C., "A correlation for boiling heat transfer to saturated fluids in convective flow," I&EC Process Design and Development, Vol. 5, No. 3, July 1966, pp. 322-329.

6. Jallouk, P.A., "Two-phase flow pressure drop and heat transfer characteristics of refrigerants in vertical tubes," Dissertation presented for the Doctor of Philosophy Degree, The University of Tennessee, Dec. 1974.

7. Shah, M.M., "A new correlation for heat transfer during boiling flow through pipes," ASHRAE Transactions, Vol. 82, part 2, pp. 66-86.

8. Chawla, J.M., "Warmeubergang und druckabfall in Waagrechten Rohren bei derstromung von verdampfenden Kaltemitteln," VDI - Forschungsheft 523, 1967.

9. Wright, R.M., "Downflow forced convection boiling of water in uniformly heated tube," University of California at Berkeley, Report UCRL 9744, 1961.

10. Mumm, J.F., "Heat transfer to boiling water forced through a uniformly heated tube," ANL-5276, 1954.

11. Gouse, S.W. and Coumou, K.G., "Heat transfer and fluid flow inside a horizontal tube evaporator phase 2", ASHRAE Transactions, Vol. 71, part 2, 1965.

12. Thakur, B.K., "A new correlation for heat transfer during flow boiling," M.S. Thesis, Rochester Institute of Technology, Rochester, NY, June 1981.

Soviet and Chinese Research Works on Vapor-Liquid Two-Phase Flows

Z.H. LIN
Xian Jiao-Tong University
Xian, The People's Republic of China

ABSTRACT

Vapor-liquid two-phase flow applications are found in a wide range of engineering systems. A lot of research work on this topic has been conducted in every industrial country. This paper plans to describe a general aspect of the research works on this topic in the power engineering region of the Soviet Union and China.

I. INTRODUCTION

Vapor-liquid two-phase flow applications are found in a wide range of engineering systems, such as boiling water reactors, breeder reactors, conventional steam boilers, evaporators of refrigeration systems, and evaporative and condensive heat exchangers in chemical and petroleum industries. Therefore, a lot of research work on this topic has been conducted in every industrial country. The Soviet Union is one of the well-known industrial countries; and in the Soviet Union, research projects on vapor-liquid two-phase flow have been conducted extensively, almost in every region.

China is a developing country and is establishing her own modern industrial systems. The first high-flux atomic reactor that has recently been put into operation, the establishment of a nuclear power plant, the ability to design and manufacture large capacity conventional power plants and other rapidly developing industries, represent her significant achievements in science and technology. Many research projects on the two-phase flow topic have also been conducted in China. This paper will describe the general aspect of research projects on this topic in the power engineering region of the Soviet Union and China.

Due to the multiplicity of published papers on this topic, it is impossible to refer to each of them. Therefore, in this paper only some of the most important and interesting problems in two-phase flow, such as two-phase frictional pressure drop, the void fraction and the two-phase flow instability problems are mentioned.

2. RESEARCH WORKS ON TWO-PHASE FLOW IN THE SOVIET UNION

2.I. Research Work on Two-Phase Frictional Pressure Drop

Several of the important correlations for two-phase frictional pressure drop in the Soviet Union are as follows:

I. Method proposed by the standard of calculating water circulation in steam boilers (1950) (I):

With this method, two-phase flow is considered as a homogeneous flow, and its frictional pressure drop can be calculated from the following correlation:

$$\Delta P_{TP} = \lambda \frac{L}{d} \frac{\gamma_L}{2} \frac{W_0^2}{g} \left[1 + (1 - \frac{\gamma_G}{\gamma_L}) \frac{W_0''}{W_0} \right] \tag{1}$$

where ΔP_{TP} is the two-phase frictional pressure drop, L is the tube length, d is the tube internal diameter, W_0'' is the superficial velocity of vapor. W_0 is the velocity considering the whole flow as liquid. γ_G, γ_L are the specific weights of the vapor and liquid respectively. λ is the frictional factor for single phase flow.

The general overview of equation (I) is as follows: For the horizontal tube, the error is about 20 percent; for the vertical tube, especially in the case of high and super-high pressure, the predicted value is much higher than the experimental data.

2. Almand correlation (2):

Almand conducted experiments in a 56 mm diameter horizontal tube. The tested pressure ranged from 10 to 90 bar, while the steam quality ranged from 0 to 90 percent. His correlation is:

$$\frac{\Delta P_{TP}}{\Delta P_0} = \frac{A(1-X)^2}{(1-\alpha)^n} \tag{2}$$

where ΔP_0 is the frictional pressure drop when the total flow is liquid. X is the vapor quality, α is the void fraction. A and n are factors determined by experimental data and can be calculated as follows:

When the vapor volumetric flow fraction β ranges from 0 to 0.25, A = I, n = 2.2; when β ranges from 0.25 to I.0, A = 0.48, n = I.9 + I.48 $\cdot$ 10^{-3}P.

For horizontal tubes, correlation (2) is more accurate than correlation (I). For vertical tubes, it has an error similar to correlation (I). Correlation (2) applies to a steam/water mixture.

3. Milopolski Method (3):

Milopolski proposed the use of the following relationship to calculate the two-phase frictional pressure drop:

$$\Delta P_{TP} = \psi \Delta P_{HO} \tag{3}$$

where ΔP_{HO} is the two-phase frictional pressure drop obtained from the homogeneous model correlation (I), ψ is a coefficient and can be obtained from figure I.

4. The Method proposed by the standard of hydraulic calculation of steam boilers (1961) (4):

The main correlation of this method is as follows:

$$\Delta P_{TP} = \psi\lambda \frac{L}{d} \frac{\gamma_L W_0^2}{2\,g} \left[1+X \left(\frac{\gamma_L}{\gamma_G}-1\right)\right] \tag{4}$$

where X is the average quality of the two-phase fluid in the tube, ψ is a corrective factor, and is a function of x, p. and W_o.

ψ can be determined from figure 2.

5. The method proposed by the standard of hydraulic calculation of steam boilers (1973) (5):

The main correlation of this method is as follows:

$$\Delta P_{TP} = \lambda \frac{L}{d} \frac{\gamma_L W_0^2}{2\,g} \left[1+X\psi\left(\frac{\gamma_L}{\gamma_G}-1\right)\right] \tag{5}$$

For the unheated tube, ψ can be found from figure 3. For the heated tube:

$$\psi = \frac{\psi_e X_e - \psi_i X_i}{X_e - X_i} \tag{6}$$

where X_e and X_i are the exit and inlet qualtiy of the heated tube, ψ_e and ψ_i can be obtained from figure 4, according to X_e and X_i, respectively.

In reality, correlation (5) is a modification of equation (I). It can be derived from the homogeneous model. Coefficient ψ is a corrective factor and considers the effect of quality, pressure and mass velocity ρW.

At lower qualities, the frictional pressure drop is increased in the presence of a heat flux over that in an unheated tube. Tarasova (6) has given an empirical equation relating the value of ΔP_{TP} in heated tubes and unheated tubes for the steam/water system as follows:

$$\left[\frac{\Delta P_{TP}}{\Delta P_0}\right]_H = \left[\frac{\Delta P_{TP}}{\Delta P_0}\right]_{UH} \left[1+4.4\cdot10^{-3}\left(\frac{q}{\rho W}\right)\right]^{0.7} \tag{7}$$

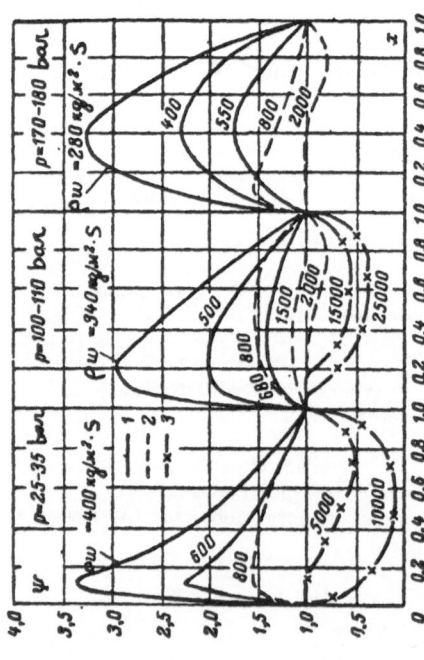

Figure 1. The relationship between ψ, X, ρW and P. 1- d=48mm, θ=10-30; 2 -d= 8.0mm, θ=0; 3- d=1-2.6mm, θ=90.

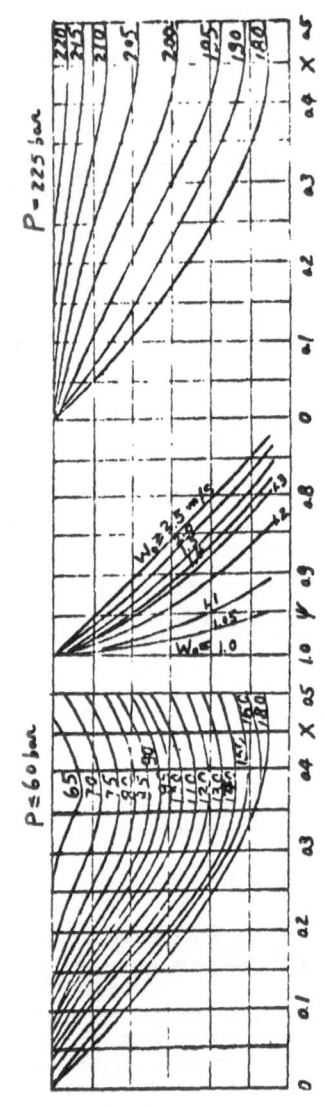

FIGURE 2. DIAGRAM FOR DETERMINING ψ

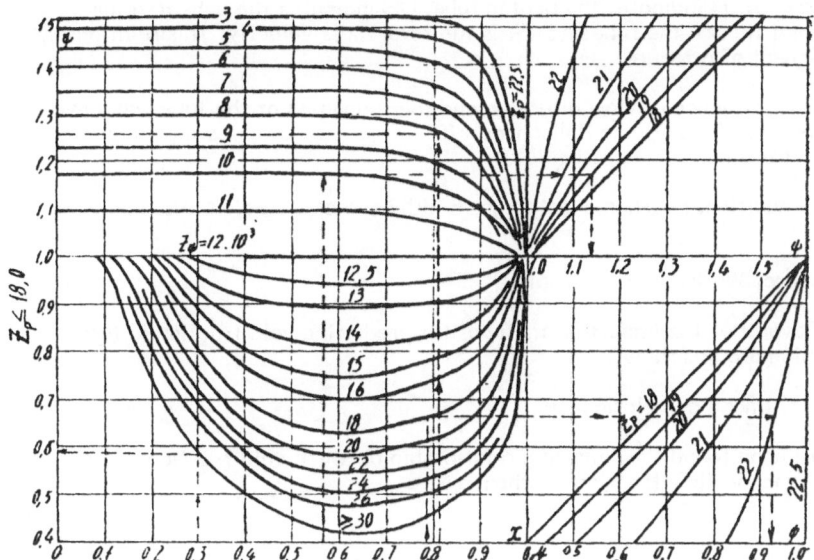

Figure 3. ψ for steam-water mixture flowing in unheated tubes. Z_ψ $(\gamma w)Z_P$; $Z_P = \dfrac{P}{10}$ (the unit of P is bar).

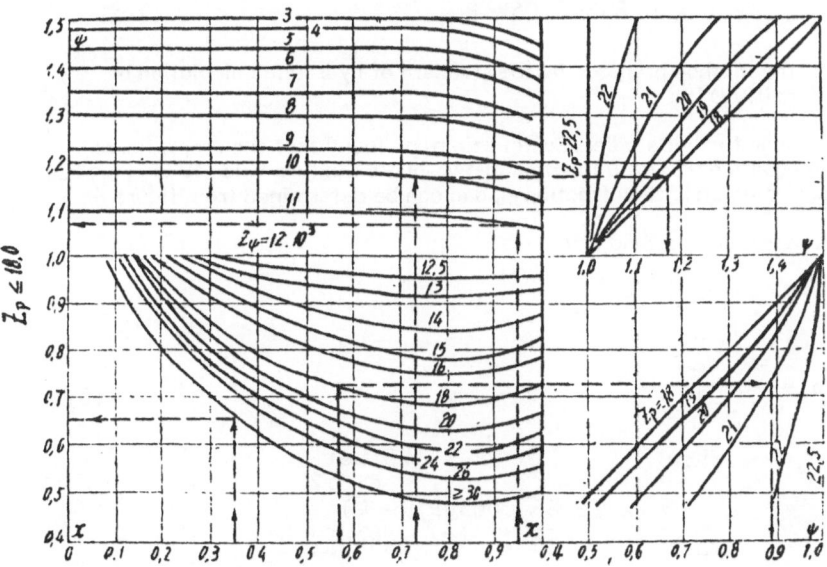

Figure 4. ψ for steam-water mixture flowing in heated tubes $Z_\psi = (\gamma w)Z_P$, $Z_P = \dfrac{P}{10}$ (the unit of P is bar).

where subscript H indicates the heated tube, UH indicates the unheated tube. q is the heat flux on the tube and ρW is the total mass velocity. This relationship is independent of pressure.

The influence of flow direction and the roughness of the tube wall have been studied by Tehaninko (7) and Bolischanski (8). Figure 5 expresses the study results of Tehaninko.

In addition, Lokshen, Kostelin, etc. (10) have also studied this topic.

2.2. Research Works on Void Fraction

Several of the important methods for predicting void fraction in the Soviet Union are as follows:

1. Almand correlation (1):

Almand found the following relationship between void fraction α and vapor volumetric flow fraction β, when $\beta < 0.9$:

$$\alpha = C\beta \tag{8}$$

where c is a proportional factor and can be calculated as follows:

$$C = 0.833 + 0.05 LnP \tag{9}$$

2. The Method proposed by the standard of hydraulic calculation of steam boilers (1973) (5):

Through further studies, Soviet scientists found that the proportional factor c in equation (8) was a function of two-phase mixture velocity W_m, pressure P and vapor volumetric flow fraction β, and can be determined from figure 6.

3. Kramelov method (11):

$$\alpha = \frac{W_0''}{W_0' + W_0'' + a} \tag{10}$$

When $\quad 11 < P < 125 \text{ bar}$
$$a = (0.65 - .0039P)(\frac{d_e}{63})^{0.2}$$
When $\quad 125 < P < 185 \text{ bar}$
$$a = (0.33 - 0.00135P)(\frac{d_e}{63})^{0.2}$$

where d_e is the equivalent diameter, W_0' and W_0'' are the superficial velocity of water and steam respectively.

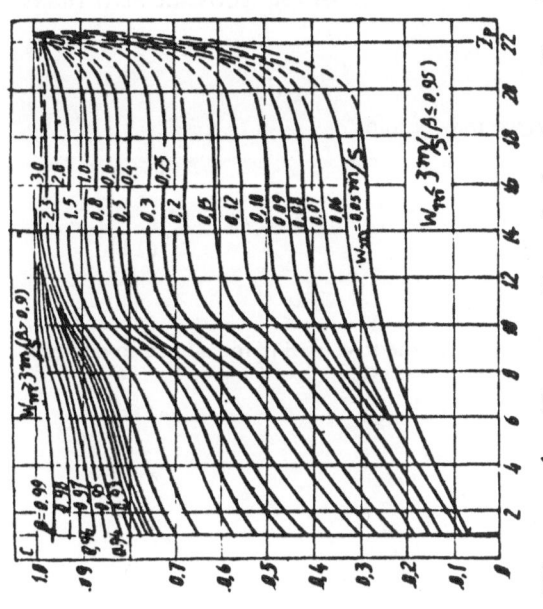

Figure 6. Proportional factor C of a vertical riser (for steam-water) $Z_P = \dfrac{P}{10}$ (the unit of P is bar).

Figure 5 The relationship between X and P_{TP} P in a vertical tube. d=33 mm. P=29.2bar. 1-3rdative roughness =10^{-4}; =8.5·10 . a- W =0.2, b-W_o=0.5, c-W_o=1.0, d- W_o=2.0 m/s

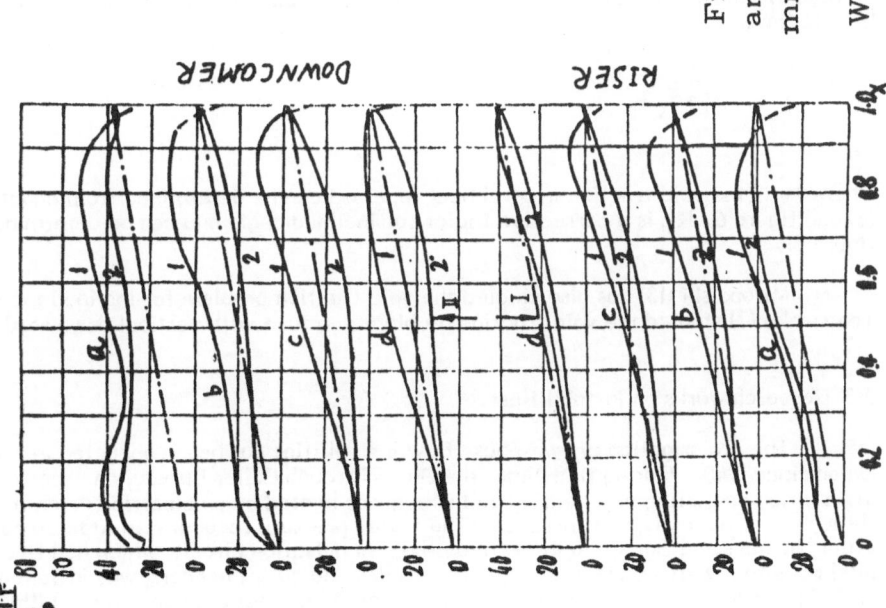

4. Osmagiken method (12):

From the two-phase flow theory, we may obtain

$$\alpha = \frac{1}{1+\frac{S\rho_G(1-X)}{\rho_L\,X}} \tag{11}$$

where ρ_L and ρ_G are densities for liquid and gas respectively, S is the slip ratio and can be predicted as follows:

$$S=1+\frac{0.6+1.5\beta^2}{4\sqrt{F_r'}}(1-\frac{P}{P_c}) \tag{12}$$

where P_c is the critical pressure, F_r' is the Froude number of water:

$$F_r'=\frac{(\rho W)^2}{9.81d_e(\rho_L)^2} \tag{13}$$

In the Soviet Union, methods for predicting the void fraction α of the inclined tube have also been studied. One of the better methods is the method proposed in the standard of hydraulic calculation of steam boilers (1973). The method is as follows:

$$\alpha_I=\alpha_V K_\theta \tag{14}$$

where α_V is the void fraction in a vertical tube and can be determined from equation (8) and figure 6. K_θ is a corrective factor for inclined angle and can be determined from figure 7.

Miropolski (13) has also studied the void fraction problem for inclined tubes. Tehonenko (7) studied the void fraction problem in a tube with vertical downward flow.

2.3 Research works on Instabilities:

The investigation of two-phase flow instabilities has been conducted in the Soviet Union since 1940. During that time, the Once-Through Boiler Bureau, an experimental installation consisting of a heating coil which was heated by an electric current passing through external coils, was installed. The tested pressure ranged from 10 to 20 bar. Later, at the same Bureau, a large steam-heated pulsation circuit, which had five parallel coils that were connected by means of inlet and outlet headers, was installed. The tested pressure ranged upward to 90 bar. On these installations, Davidov (14) (15) conducted many experiments and theoretical studies. Semnovker (16) made a series of experimental studies on small four-parallel coils, a forced-fed circulation boiler, with test pressures up to 90 bar. Petrov (17) studied theoretically both the instabilities in a parallel channel upward flow system and a horizontal flow system. Morozov (18), by using a simplified model, studied theoretically, the boiling flow in a coil of a once-through boiler. Petrov and Morozov tried to solve the problem of choosing the correct throttling rate of the evaporating tube of a forced circulating loop. Klaziakova and Gloskel (19) experimentally studied the instability of the boiling flow in vertical U tubes. The tested pressure ranged from 140 bar to 230 bar.

Treshev (20) conducted experimental studies of steam/water flow oscillation in heated tubes. The study showed that when the resistance was great enough before the boiling point was reached, the entering temperature was the saturated temperature, or greatly below it, and if the mass flow rate was high enough, then oscillation did not occur, or was very small. Celov, Smilnov and Zerkov (21) conducted an experimental study on the flow stability boundary of steam-generating tubes connected parallel with each other, with uneven heat input. Dolgov and Sudnitsyn (22) studied experimentally the hydrodynamic instability in boiling water reactors. They found that the amplitude of oscillation was reduced by increasing the inlet pressure drop, reducing the inlet subcooling temperature, or raising the pressure.

According to the accumulated experimental data, the standard of hydraulic calculations of steam boilers (1961) proposed two figures for detemining oscillations in horizontal tubes. Figure 8 expresses the relationship between the ratio of pressure drop A and mass velocity ρW. Figure 9 expresses the relationship between the ratio of pressure drop A and Pressure P.

The ratio of pressure drop is

$$A=\frac{\Delta P_T + \Delta P_H}{\Delta P_E} \tag{15}$$

where ΔP_T is the pressure drop of the throttling device at the entrance of a tube, ΔP_H the pressure drop of the heated tube section with liquid only. ΔP_E is the pressure drop of the heated tube section with evaporation. According to pressure and mass velocity, two values of A may be obtained. The large A will be chosen to calculate the required pressure drop of throttling devices at the entrance of a tube that may prevent the flow from oscillating. Thus, the required size of the throttling devices can also be determined.

Baldina and Kalinin (23) (24) experimentally studied the boiling flow instabilities in four parallel horizontal tubes and three W type tubes. Tested pressure ranged from 21 bar to 235 bar. The test medium was water. The limit curve of stability obtained from experimental data is shown in figure 10.

In figure 10, A has the same meaning as that in correlation (15). B_d is the dynamic capacity coefficient,

$$B_d = \frac{\delta B}{\delta G \tau} \tag{16}$$

where δB is the change of the mass of fluid in the tube. τ is the time that the fluid needs to pass through the tube and δG is the change of flow rate per unit time.

During the designing, B_d can be calculated, and then the required A for avoiding oscillations can be determined from figure 10.

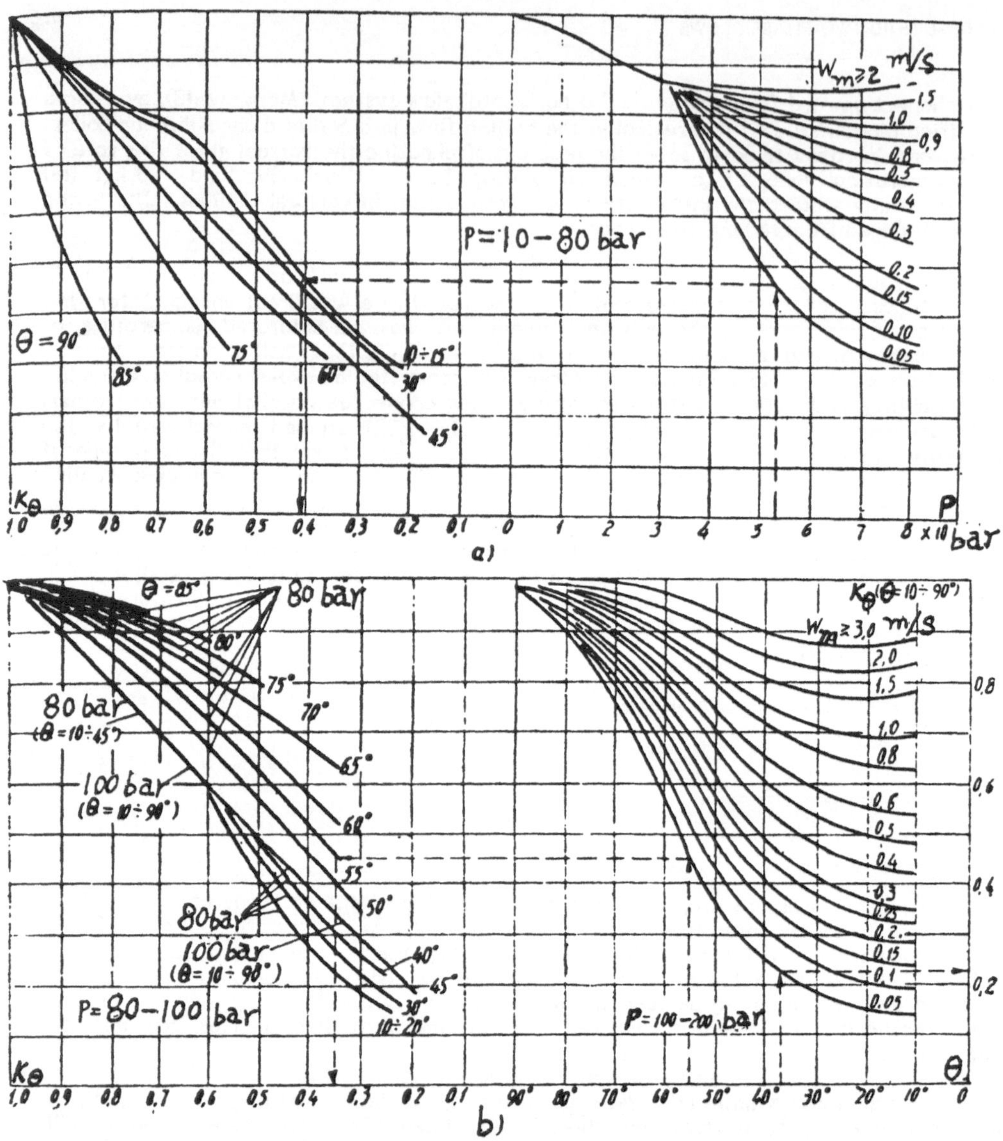

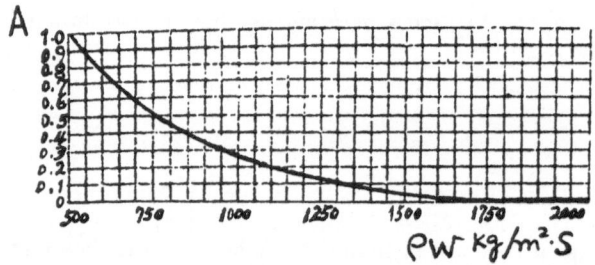

Figure 7. Corrective factor for inclined angle K_θ.

Figure 8. The relationship between
A and

Figure 10 applies for $\rho W > 500$ $Kg/m^2.S.$
when $\rho W \lessapprox 500$ $Kg/m^2.S.$ The required A should be modified by multiplying A and the term 500/ W.

Habenski, Baldina, Kalinin and Betelson (25) (26) (27) (28) made an analysis of discharge pulsation in the system of boiling tubes that were parallel. By numerical integration of a system of equations describing unsteady heat transfer and hydrodynamics in a vapor generating channel on a computer, a diagram for determining the lowest limit of mass velocity that could prevent the flow from oscillating was obtained (figure 11).

From figure 11, according to mass velocity W, the enthalpy inlet subcooling we may obtain the required resistance factor ζ at the entrance for preventing the flow from oscillating; or according to $\zeta, \Delta i_o$, we may obtain the required critical mass velocity ρ W that may prevent the flow from oscillating.

Figure 11 applies to horizontal tubes and was obtained under the following conditions: P = 100 bar, q = 200 . 10^{3} $Kcal/m^2$, internal tube diameter d = 20 MM, heated length L = 18.6 m. For other cases, the required critical mass velocity ρW may be modified as follows:

$$(\rho W)_c = 5.38 \times 10^{-6} k_p \ (\rho W)_c^{P=100} \ \frac{q}{d} \frac{L}{d} \quad (17)$$

where $(\rho W)_c$ is the modified critical mass velocity, $(\rho W)_c^{P = 100}$ is the critical mass velocity obtained under pressure P = 100 bar. K_p is the pressure factor and can be determined from figure 11, according to the given ζ and P. Figure 11 applies for steam/water, two-phase flow when the range of frequency of flow rate oscillation is smaller than 0.2 - 0.25 HZ. The error for determining ζ from figure 11 is smaller than 20 percent.

For the vertical tube, the critical mass velocity can be predicted as follows:

$$(\rho W)_c^V = C (\rho W)_c^H \quad (18)$$

where $(\rho W)_c^V$ is the critical mass velocity for a vertical tube, $(\rho W)_c^H$ the critical mass velocity for a horizontal tube, c is a proportional factor obtained from figure 12.

Koshelev, Sulnov and Nikitina (29) reported the test results of three heated tubes constructed parallel. The phase shifts of flow oscillation will occur even under supercritical pressure conditions. Leleev (30) studied theoretically the oscillation phenomena occurring when a medium passed through the tube of a steam generator under supercritical pressure. Schevolz, Kozmin and Shevaly, Kozmin and Glaglev (31) studied the flow stability experimentally in U type tube and η type tube elements at supercritical pressure. They also studied the influence of heat input and hydraulic unevenness on flow stabilities. Thermal-acoustic oscillations have also been studied.

3. RESEARCH WORKS ON TWO-PHASE FLOW IN CHINA

3.1 Research Works on Two-Phase Frictional Pressure Drop

By correlating the published experimental data, Lin (33) found that the vapor-liquid two-phase flow frictional pressure drop could be predicted as follows:

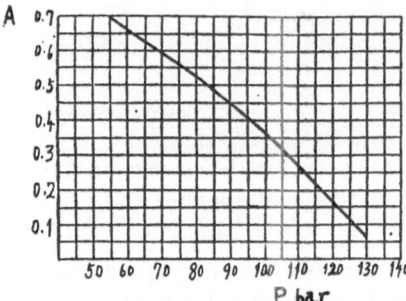

Figure 9. The relationship be
-tween A and P.

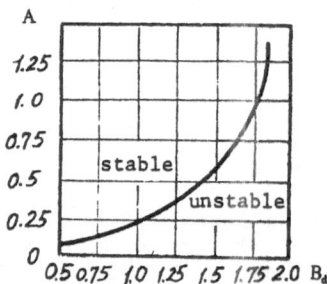

Figure 10. The limit curve
Of stability.

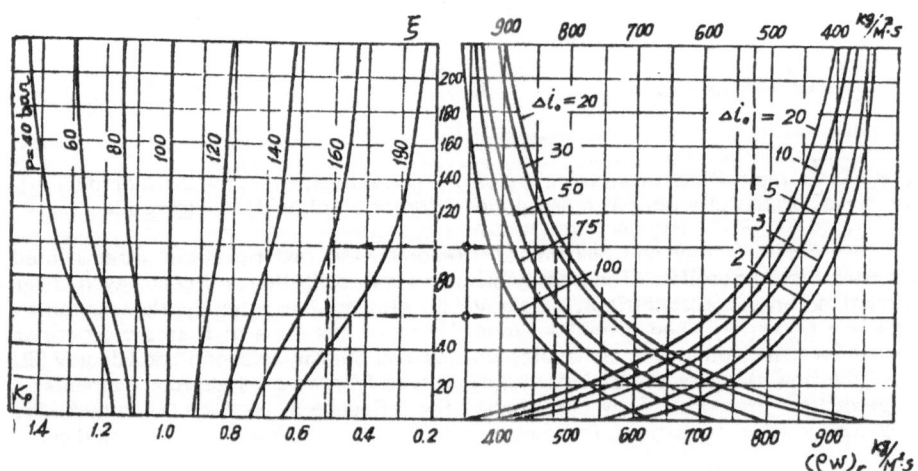

Figure 11. The critical mass velocity for parallel horizontal tubes.

$$\Delta P_{TP} = \lambda_{TP} \gamma_m \frac{L \ W_m^2}{d2g} \tag{19}$$

where λ_{TP} is the two-phase frictional factor obtained from experimental data and is a function ot pressure and mass velocity. W_m is the mean velocity of two-phase mixture. γ_m is the mean specific weight of the two-phase flow. L and d are the length and diameter ameter of the tube respectively. Tien (34) experimentally studied the frictional pressure drop of water at the state of surface boiling. Chin (35) studied the frictional pressure drop of high pressure water passing through bundles under the condition of saturated boiling and subcooled boiling. Cao (36) studied the two-phase flow pressure drop in a bundle channel. Chen et al., (37) studied the pressure drop when the two-phase fluid was passing through a Y tube. Gi (38) studied the frictional pressure drop of sub-cooled water boiling in a jacket tube. Chang (39) studied the frictional pressure drop of high pressure water under the conditions of saturate boiling. Chang (40) used a semi-homogeneous model to predict the two-phase frictional pressure drop.

The two-phase frictional pressure drop in an internal helical tube and the frictional pressure drop when a mixture of oil, gas and water were passed through a vertical tube were also studied.

3.2 Research Works on Void Fraction and the Measurements of Vapor Quality

The measurements and predictions of void fraction and vapor quality are very important problems in two-phase flow problems. In relationship to this, Cao (41) measured the steam void fraction by using the radiation attenuation method. He used C_s^{137} as the source and NaI as the scintillation crystal. Lin (42) studied the predicting method of void fraction. For inclined tubes, Lin used the relationship of equation (14). K is a function of W_o, P, and inclined angle and can be determined from figure 13.

Lin (43), Wang and Han (44) and Chen (45) studied the measurements of vapor quality on the flow rate with an orifice. Lin proposed the following correlation to predict the quality or flow rate of a two-phase flow:

$$\sqrt{\Delta P_{TP}/\Delta P_0} = \theta + X(\sqrt{\rho_L/\rho_G} - \theta) \tag{20}$$

where ΔP_{Tp} is the pressure drop across the orifice for two-phase flow, ΔP_0 is the pressure drop across the orifice, assuming that the total flow is liquid, λ_{Tp} is a corrective coefficient obtained experimentally. X is the vapor quality, and ρ_L and ρ_G are the density of the liquid and gas respectively.

Under the condition of given pressure and measured ΔP_{Tp} we may determine quality X from equation (20). In equation (20), θ can be determined from figure 14 according to the given pressure or ρ_G/ρ_L

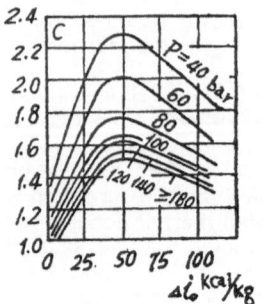

Figure 12. Proportional
factor C.

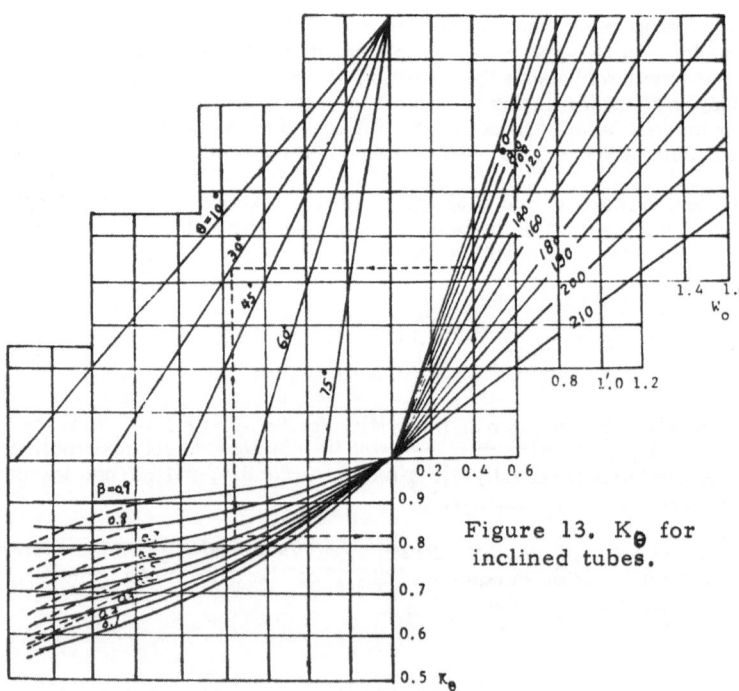

Figure 13. K_θ for
inclined tubes.

3.3. Research Works on Two-Phase Instabilities

In this respect, Chen and Chen (46) studied the flow characteristics in a boiling tube when a two-phase fluid was flowing upwards and downwards. Bi (47) experimentally investigated the hydrodynamic instability in a once-through evaporator. Fan (48) experimentally studied the two-phase instability in a low-pressure natural circulating boiling loop. Wang and Wang (49) theoretically studied the multivalue problem of the hydrodynamic characteristics of La-Mont boiler. The experiments and improvements on large once-through boilers were also reported (50) (51).

REFERENCES

1. Standard of Calculating Water Circulation in Steam Boilers, CKTI report, U.S.S.R. 1950.

2. Almand, A.A., Information of VTI, No. 4, U.S.S.R., 1947.

3. Milopolski, E.L., Teploenergitika, No. 5, U.S.S.R., 1965.

4. The Standard of Hydraulic Calculation of Steam Boilers, CKTI report, U.S.S.R., 1961.

5. The Standard of Hydraulic Calculation of Steam Boilers, CKTI report, U.S.S.R., 1973.

6. Tarasova, N.V., and Leont'ev, A.I., Teplo. Vyso Temp. No. I, 1965.

7. Tehaninko, L.K., Achievements in the Studying Region of Two-Phase Heat Transfer and Hydraulics in Elements of Power Devices, "Navk" Publishing House, U.S.S.R., 1973.

8. Bolischavski, V.M., Achievements in the Studying Region of Two-Phase Heat Transfer and Hydraulics in Elements of Power Devices, "Nauk" Publishing House, U.S.S.R., 1973.

9. Lokshen, V.A., Teploenergitika, No. 8, U.S.S.R., 1959.

10. Kostelin, S.I., Teploenergitika, No. 6, U.S.S.R., 1958.

11. Kramelov, A. Y., Engineering Calculations of Nuclear Reactor, "Atom" Publishing House, U.S.S.R., 1964.

12. Bartolma, K.K., Teploenergitika, No. 9, 1974.

13. Miropolski, E.L., Teploenergitika, No. 5, 1971.

14. Davidov, A.A., Elektricheskie Stanzu, No. 3, U.S.S.R., 1956.

15. Davidov, A.A., and Shenin, B.I. Teploenergetika, NO. II, U.S.S.R., 1959.

16. Semenovkel, I.I., Hydrodynamics and Boiling Heat Transfer in High Pressure Boilers, U.S.S.R., 1955.

17. Petrov, B.A., Hydrodynamics of Once-Through Boilers, U.S.S.R., 1960.

18. Morozov, I.I., Inzhenev, Fiz Zhur, Akad. Nauk., U.S.S.R., No. 8, 1961.

19. Klaziakova, L.V., and Gloskel, B.N., Teploenergetika, No. I.I, U.S.S.R., 1963.

20. Teshev, G.G., Energo-Machinostroenie, NO. 3, U.S.S.R., 1964.

21. Celov, I.B., Smilnov, O.K., and Zerkov, L.A., Teploenergetika, No. 10, U.S.S.R., 1964.

22. Dolgov, V.V., and Sudnitsyn, O.A., Teploenergetika, No. 10, U.S.S.R., 1965.

23. Baldina, O.M., and Kalinin, R.I., Energo-Machino-Stroenie, No. 7, U.S.S.R.,1966.

24. Baldina, O.M., Kalinin, R.I., Saburova, R.I. and Baitina, T.Z.M., Teploenergetika, No. 8, U.S.S.R., 1968.

25. Habenski, V.B., Baldina, O.M., Inzhener Zhur. Akad. Nauk., No. 5, U.S.S.R., 1969.

26. Habenski, V.B., and Betilson D.V., Teploenergetika, No. 7, U.S.S.R., 1970.

27. Habenski, V.B., Baldina,O.M., and Kalinin, R.I., Energo-Machinostroenie, No. 3, U.S.S.R., 1971.

28. Habenski, V.B., Baldina, O.M., and Kalinin, R.I., Achievements in the Studying Region of Two-Phase Heat Transfer and Hydraulics in Elements of Power Devices, "Nauk" Publishing House, U.S.S.R., 1973.

29. Koshelov, I.I., Sulnov, A.V., and Nikitina, L.V., Heat Transfer Soviet Research, No. 2, 1970.

30. Leleev, N.S., Izvestiia Vysohikh Uchebnykh Zavedenii, No. II, U.S.S.R., 1973.

31. Shevaly, A.L., Kozmin, V.V., and Glagolev, A.S., Teploenergetika, No. 3, U.S.S.R., 1971.

32. Geliga, V.A., and Prohopov, U.F., Information of Akad. Nauk., Energy and Transport, No. 6, U.S.S.R., 1974.

33. Lin, Z.H., Technology of Electric Power, No. I, China, 1965.

34. Tien, G.A., a paper presented at the 2nd National Fluid Mechanics Conference, China, 1979.

35. Chin, G.W., a paper presented at the 2nd National Fluid Mechanics Conference, China, 1979.

36. Cao, F.J., a paper presented at the 2nd National Fluid Mechanics Conference, China, 1979.

37. Chen, S.H., a scientific and technical report of the Shanghai Mechanical Institute, 1977.

38. Gi, G.H., a paper presented at the 2nd National Fluid Mechanics Conference, China, 1979.

39. Chang, B.Y., a paper presented at the 2nd National Fluid Mechanics Conference, China, 1979.

40. Chang, W.M., a paper presented at the 2nd National Fluid Mechanics Conference, China, 1979.

41. Cao, F.J., a paper presented at the 2nd National Fluid Mechanics Conference, China, 1979.

42. Lin, Z.H., a scientific and technical report of the Xian Jiao-Tong University, China, 1976.

43. Lin, Z.H., Journal of Xian Jiao-Tong University, No. 4, China, 1979.

44. Wang, M.H., and Han, B.P., report of the Shanghai Boiler Institute of China BS78-I, China, 1978.

45. Chen, S.T., Report of the Harbin Boiler Institute of China CB2-092, China, 1979.

46. Chen, X.J., and Chen, L.X., a scientific and technical report of Xian Jiao-Tong University, China, 1963.

47. Bi K.S., a paper presented at the 2nd National Fluid Mechanics Conference, China, 1979.

48. Fan, S.S., a scientific and technical report of Qing-Hua University, China, 1979.

49. Wang, Z.M., and Wang, B.H., a paper presented at the 3rd National Power Conference, China, 1979.

50. An internal report of Shanghai Central Experimental Institute of Electric Power Industry, China, 1973.

51. An internal report of Beijing Central Experimental Institute of Electric Power Industry, China, 1975.

TWO-PHASE FLOW INSTABILITIES

Effect of Heat Transfer on Density-Wave Oscillations— A Finite Difference Analysis

K.M. AKYÜZLÜ
University of New Orleans
New Orleans, Louisiana 70148, USA

T.N. VEZIROĞLU
University of Miami
Coral Gables, Florida 33124, USA

ABSTRACT

A classification of the flow oscillations in forced convection boiling is made. The physical nature of the most common type of oscillations, i.e. density-wave oscillations, is explained in detail. Experiments relating to this type of oscillations are described and the typical recordings of these oscillations are given.

Governing equations are developed using the homogeneous and phase equilibrium assumptions. The momentum equation is integrated over the system to filter out the acoustic oscillations, then, all the equations are solved simultaneously by finite difference techniques. The results show that the model is satisfactory in simulating the basic characteristics of density-wave type oscillations.

Using numerical experimentation techniques, an investigation of the effects of various parameters on density-wave oscillations is carried out. It is found out that the system exhibits the characteristics of a buckling problem when the heat input into the fluid is assumed to remain constant. In this case, there appears to be only one value of flow rate where sustained density-wave oscillations are generated. Increasing the flow rate at this same operating point results in transient oscillations which eventually die out. However, a decrease in flow rate results in oscillations with exponentially increasing amplitudes. At very low mass flow rates, however, no oscillations are generated. The effect of variations in heat transfer is also studied by numerical experimentation. The results indicate that an increase in the dependence of heat transfer on mass flow rate has a beneficial influence on the stability of the boiling system.

INTRODUCTION

The problems of oscillatory two-phase flow instabilities have been a challenge to many investigators over the past two decades. Such instabilities are found in a variety of engineering systems, such as boiling water reactors, conventional steam boilers, evaporators of refrigeration systems, and evaporative and condensive heat exchangers in chemical and petroleum industries. They are undesirable for several reasons. If sustained, flow oscillations may cause forced mechanical vibrations in the components of a two-phase flow system. Flow oscillations may also cause system control problems which might complicate the safe operation of the system. And, flow oscillations affect the local heat transfer characteristics and may induce boiling crises, such as critical heat flux and burnout.

671

Because of the reasons given above, it is important to be able to predict the conditions under which a two-phase flow system will be subject to instabilities. This can be done in two ways. The most common one is to predict the threshold of the particular instability in question. In this method of approach, usually a simplified mathematical model is used to obtain the governing set of equations. These equations then are treated by a perturbation technique and standard stability criteria. As a result, a marginal dynamic stability criterion and the associated frequency for threshold instability are obtained. Almost every type of dynamic instabilities cited in literature has been studied by this method. The acoustic oscillations, for example, were analyzed by Bergles, et al. [1]*. Maulbetsch and Griffith [2] used the same approach in their analysis of pressure drop type oscillations. Density-wave type oscillations were analyzed in a similar manner by Boure' [3]. This kind of approach is more suitable for parametric studies and understanding the physics of the phenomena.

The other method of approach is to simulate the instabilities by using appropriate mathematical models. This kind of approach makes it possible to study the transient instabilities and predict the distribution of properties, such as pressure, density and velocity across the system, and, therefore, is more suitable for design purposes. In fact, this is why the two-phase flow literature is full of computer codes which can simulate various dynamic instabilities. The SABRE code by Nahavandi and Hollen [4] for example, can predict the density-wave oscillations in a Boiling Water Reactor (BWR) core or in a recirculation steam generator. Solverg and Bakstad [5] developed the RAMONA computer code which predicts the density-wave oscillations in a natural circulation BWR. The code FLASH by Margolis and Redfield [6] was developed for digital simulation of the Loss of Coolant Accident (LOCA); it also predicts the density-wave oscillations and the acoustic oscillations.

This study was carried on primarily to meet the following objectives:

(i) To find the mathematical model and the solution method that simulates the density-wave type oscillations and predict the amplitudes and periods of the sustained oscillations at different heat input rates and.inlet subcoolings.

(ii) To use this model to predict the stability-instability boundaries for a simple geometry upflow system and study the effect of heat transfer on these boundaries.

MODES OF TWO-PHASE FLOW OSCILLATIONS

There is a great variety of two-phase flow instabilities. This requires that the instability is well identified and the physical mechanism is well described for the type under study. Therefore, a classification criteria and a thorough understanding of the physical nature of the basic modes of oscillations is required.**

Classification

The classification proposed here is based on the following general definitions. A flow is said to be "stable", when perturbated momentarily, its new

---

*The numbers in brackets refer to the references at the end of the article.

**An excellent review of two-phase flow oscillations is given in reference 7.

operating conditions tend asymptotically toward the initial ones. Practically, steady two-phase flow operating conditions undergo fluctuations due to perturbations introduced through the boundaries of the system. These fluctuations represent a very small scale disturbance, usually periodic in nature. Under certain conditions of system operation these fluctuations can describe a completely stable flow. On the other hand, in certain domains of flow, the fluctuations can trigger larger scale disturbances, periodic in nature. Instabilities caused by these fluctuations can be classified as static or dynamic. A "static instability" occurs when the fluctuations cause the flow to attain another steady state which is different from the original state. Whereas, a "dynamic instability" occurs when the fluctuations are produced by a feedback from the inertia and other effects. Therefore, the system behaves like a servo-mechanism causing the flow to oscillate periodically around the original state.

Static Instabilities

A static instability can be identified as compound or fundamental. It is said to be "compound" when several elementary mechanisms interact in the process and cannot be studied separately. It is said to be "fundamental" in the opposite sense.

There are two types of "fundamental static instabilities" - flow excursion and boiling crisis. The "flow excursion" (or Ledinegg Instability) occurs when the slope of the channel demand pressure drop - flow rate curve becomes algebraically smaller than the loop supply pressure drop - flow rate curve. The "boiling crisis" occurs when the governing heat transfer mechanism in a boiling flow changes to a different type. It is characterized by a sudden change in wall temperature and usually occurs simultaneously with flow oscillations.

The "flow pattern transition instability" is also considered a fundamental instability and has been postulated as occurring when the flow conditions are close to the point of transition between bubbly flow and annular flow.

Bumping, geysering, and chugging are "compound relaxation instabilities". These instabilities involve fundamental static instability mechanisms which are coupled so as to produce repetitive behavior. They are not necessarily periodic.

Dynamic Instabilities

The cause of dynamic instabilities may be due to a fundamental or a compound mechanism, as it is in the static instabilities.

The "fundamental dynamic instabilities" are due to two basic mechanisms, both of which involve propagation of disturbances in two-phase flow. If the disturbances are transported by pressure waves, the instability is called an "acoustic instability". It is called a "density-wave instability" when the disturbances are propagated by density waves. The density-wave instabilities are low frequency oscillations in which the period is approximately the order of magnitude of the time required for a density wave (continuity wave or kinematic wave) to travel through the channel.

The "compound dynamic instabilities" occur in a variety of forms and have different characteristics. "Boiling Water Reactor instability" is due to a strong coupling between the thermo-hydrodynamic instabilities and feedback effects characterized by the void-reactivity-power link. "Parallel channel instabilities" are observed in tubes connected by common headers at both ends. "Thermal oscillations" are associated with the thermal response of the heating wall after dryout and is due to flow oscillating between film boiling and transition boiling. "Pressure-drop oscillations" occur in systems having a compressible volume upstream of the heated section and are triggered by a static instability.

DENSITY-WAVE TYPE OSCILLATIONS

The most common oscillations in two-phase upflow systems are density-wave type oscillations. The regions where these oscillations are encountered can be shown on a plot of pressure drop versus mass flow rate. Figure 1 shows such a plot for a single channel heater with a surge tank at the upstream side and an exit restriction at the downstream side [8]. In a typical case where the heat input is 1500 Btu/hr and the inlet fluid temperature is 25°F, the fluid is all liquid for a mean mass flow rate of 2.60 lbs/min. As the flow rate is reduced, the pressure drop across the system also decreases until a flow rate of 1.93 lbs/min is reached. At this point the liquid leaving the heater is almost saturated. Further reduction in flow rate is accompanied by an increase in pressure drop as bulk boiling commences. For flow rates between 1.93 lbs/min and 0.80 lb/min, the slope of the pressure drop versus flow rate curve is negative. At a flow rate of 1.33 lb/min, point "a" of Figure 1, sinusoidal flow rate and pressure oscillations of very small amplitude are observed with a period of 36 seconds. These oscillations with low frequencies are named "pressure-drop" oscillations (Figure 2). At a flow rate of 0.85 lb/min, the test point "e" in Figure 1, the flow becomes stable and remains stable until the flow rate is reduced to 0.39 lb/min, the test point "g" in Figure 1, where sustained higher frequency oscillations commence with a period of 3.0 seconds. These higher frequency oscillations persist at lower flow rates. Their periods decrease with decrease in flow rate as shown in Figure 3. These oscillations are named "density-wave" oscillations.

Physical Description of Density-Wave Oscillations

The principles that govern the mechanism of density-wave oscillations have been explained by considering a simple model consisting of an evaporator followed by a duct and a flow restriction. It was developed by Stenning and Veziroglu [9] and is summarized below.

It is assumed that the pressure drop in the evaporator is very small, while the pressure drop across the restriction is relatively large. Furthermore, the rate of vapor generation in the evaporator is assumed to be constant. For constant pressure drop across the restriction, the volume flow rate through the restriction varies inversely as some power of the mixture density of the fluid entering the restriction. When a two-phase mixture with higher density than the steady-state valve reaches the restriction, the volume flow rate through the system becomes very small. The change of density may be initiated by turbulence or nucleation that takes place in the evaporation process. The net results of the decrease in flow rate will be the accumulation of low density mixture in the evaporator. As soon as the dense mixture passes through the restriction, the accumulated low density mixture will appear at the exit restriction. It will move through the restriction very rapidly, so that the mixture that follows will spend very little time in the evaporator. Therefore, it will end up in a more dense state. When this high density mixture arrives at the exit restriction, the whole cycle will start over again.

This model has been found to represent the density-wave oscillations quite well. In this study the model is extended to the case where the flow resistance is distributed throughout the system.

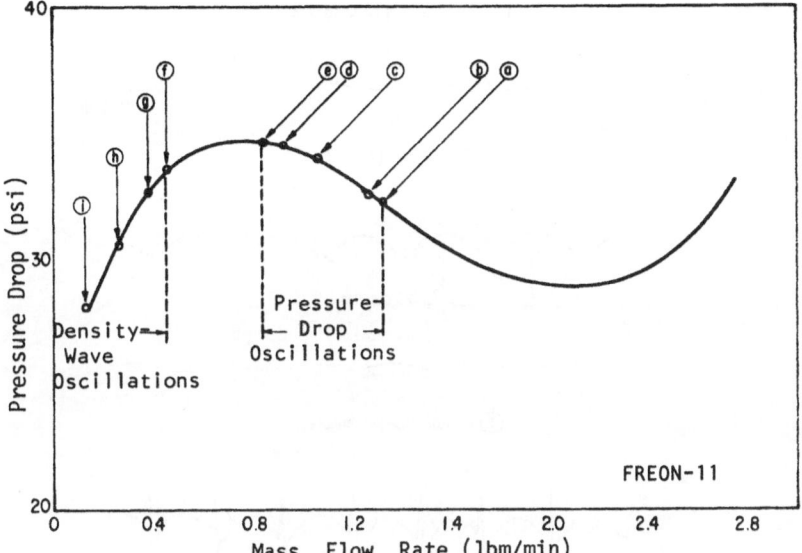

Fig.1 – System Pressure Drop vs. Mass Flow Rate for Constant Heat
Input of 1500 Btu/hr and 25°F Inlet Temperature

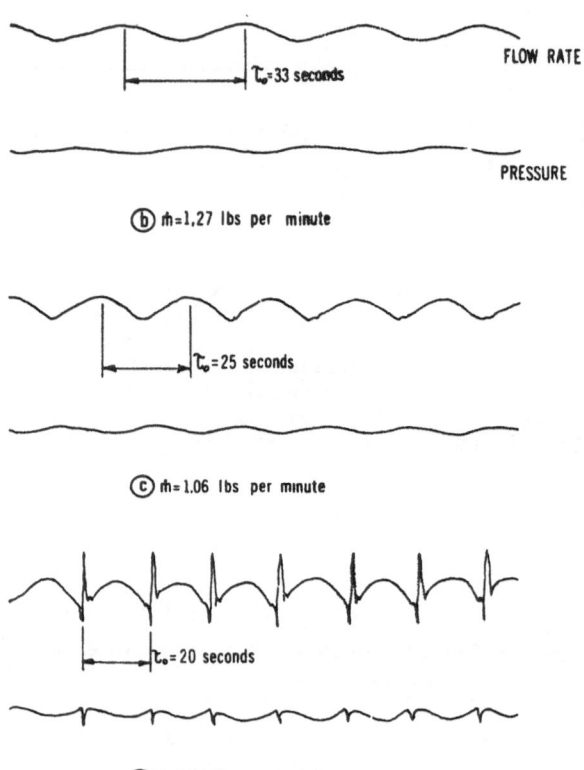

Fig.2 - Recordings of Pressure-Drop Type Oscillations

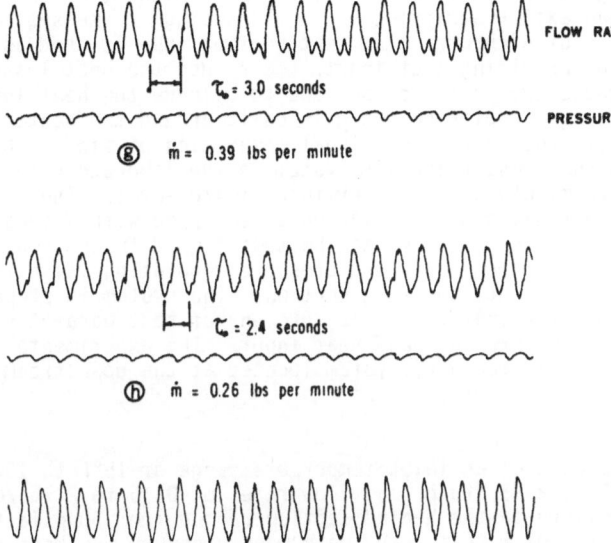

Fig.3 - Recordings of Density-Wave Type Oscillations

EXPERIMENTAL STUDY

The experimental study can appropriately be divided into three, viz., Exprimental Apparatus, Experiments and Experimental Results.

Experimental Apparatus

The experimental set-up is an open loop, forced convection, boiling upflow system operating between constant inlet and exit pressures. The loop is schematically illustrated in Figure 4. It consists of a fluid container, a control valve, a surge tank, an inlet valve, an inlet plenum, a test section, an exit plenum, and an exit restriction. All the tubing in the system including the heater is made of nichrome. The test section tube is used as an electrical resistance for providing heat input. To reduce the heat losses a vacuum chamber is used to house the test section, and to balance the heat losses a guard heater incorporating a radiation guard is built around the heater. Glass sight tubes are included in the system for visual inspection of flow. The surge tank is connected to the pressurized air system in the laboratory in order to supply air as required to maintain a perdetermined liquid level. The test fluid inlet temperatures are controlled by a cooling unit, equipped with automatic temperature controls. During the experiments the test liquid in the container is pressurized by high pressure nitrogen gas. A pressure regulating valve is used to maintain constant flow rate into the test section. The system is properly instrumented to provide adequate control and measurement of test parameters, viz., the flow rate, temperature, pressure and heat input. The experimental set up also includes a test fluid recovery system located at the downstream side of the loop.

Experiments

The tests covered an inlet temperaure range of 15°F to 150°F, heat input rates of 1023 to 2389 Btu/hr and flow rates of 0.15 to 2 lbs/min. During the tests the pressure in the heated section ranged from 30 psia to 70 psia depending on the particular test. The main tank pressure was kept constant at about 85 psia. The exit pressure was also constant at about 1.5 psig. Forty sets of experiments, corresponding to combinations of four different heat inputs and ten different inlet temperatures, were conducted. The stability boundaries were determined for each test. Short life transients were disregarded and only sustained oscillations were considered in determining the stability boundaries. The experiments were conducted with Freon-11 because it makes it possible to cover the whole range of conditions from nucleate boiling to film boiling without encountering excessive heater wall temperatures.

The tests were conducted in accordance with the following procedure:
   (i)   The main tank is filled with test liquid up to three quarters level and then pressurized by nitrogen gas.
   (ii) Inlet fluid temperature is set to a predetermined value by using the subcooling controller.
   (iii) Flow rate and heat input are gradually increased to the desired starting point, and the system is allowed to become steady.
   (iv) Measurements of temperature, pressure, flow rate and heat input are recorded.
   (v)   Then the mass flow rate is reduced by a small amount using the flow control valve and all the readings and recordings are taken. This is continued as far as the system allows.
   (vi) Above procedure is repeated for different inlet temperatures and various heat input rates.

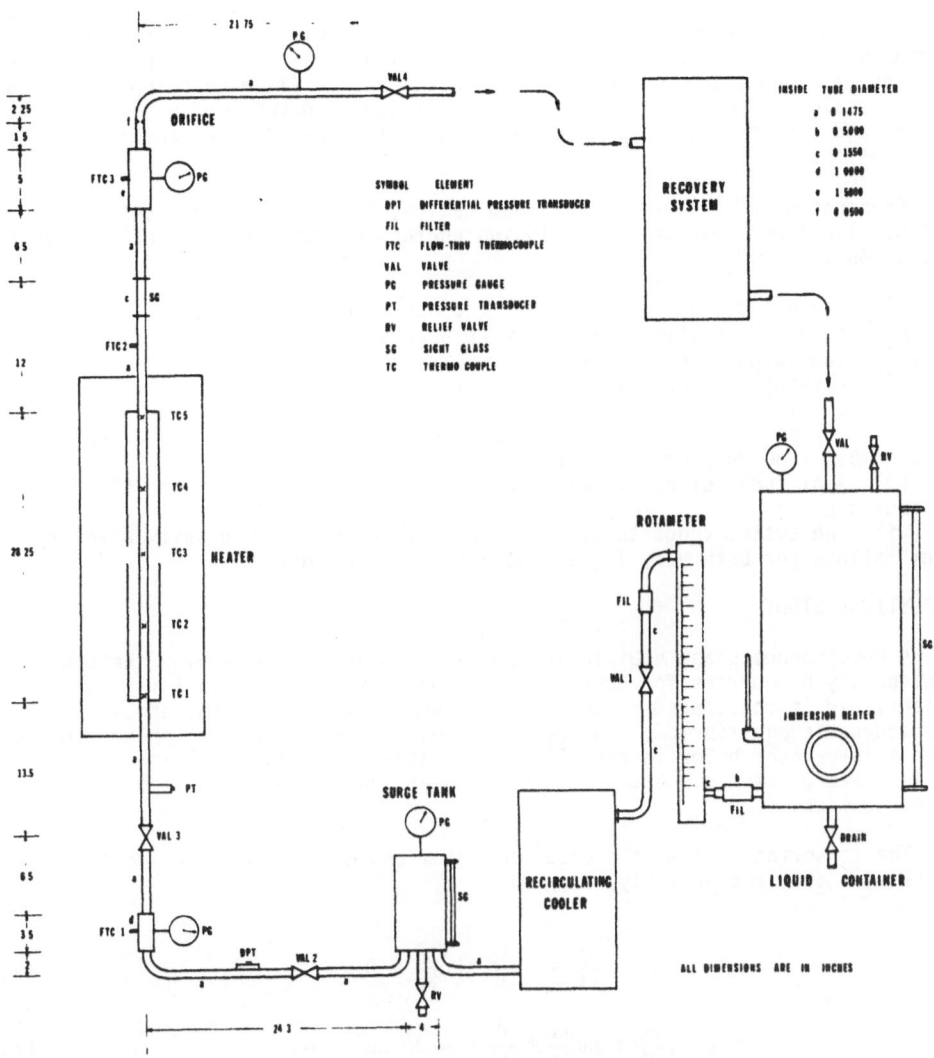

Fig.4 – SCHEMATIC DIAGRAM OF UPWARD BOILING FLOW APPARATUS

Experimental Results

Some important experimental readings and data calculated for selected test
points where density-wave oscillations were observed are presented in Table 1.
The mean mass flow rates, the inlet temperatures, system pressure drops, the
net heat inputs, the periods and the amplitudes of the oscillations are included
in this table. The expreimental results 'have been plotted on a pressure drop
versus mass flow rate plane for various heat input rates. These plots are shown
in Figures 5 and 6 for various inlet temperatures at fixed heat input rates of
1023 and 1706 Btu/hr, respectively. Experimentally determined regions of stability and instability are indicated on these figures for the density-wave type
oscillations.

Examination of the above mentioned tables, figures, and the recordings of
the oscillations indicate the following characteristics of the density-wave oscillations.

(i)   These oscillations appear on the positive slope and low flow rate branch
of the pressure drop versus mass flow rate curves.
(ii)   The surge tank pressure remains almost constant.
(iii)   Although the exit fluid temperature oscillates by a small amount, only
regional oscillations are observed in the heater wall temperature.
(iv)   Periods of oscillations are in the order of one second and they decrease
as the mass flow rate or the inlet subcooling is decreased.
(v)   Amplitudes of oscillations are higher for heat inputs and high inlet
subcoolings.
(vi)   The system tends to be more stable with regard to density-wave type
oscillations for both very low and very high subcoolings.

ANALYTICAL STUDY

A homogeneous phase equilibrium model is adopted to determine the stability-
instability boundaries for density-wave type oscillations. In this model two-
phase fluid is considered to be a homogeneous mixture where the phases are in
thermodynamic equilibrium. The governing equations for the two-phase flow region is presented below in detail. The equations for the single-phase flow
regions are omitted since they can be deduced very easily from the two-phase
flow equations [10].

The conservation laws for mass, momentum and energy for one-dimensional two-
phase flow are, respectively,

$$\frac{\partial \rho}{\partial t} + u \frac{\partial \rho}{\partial z} + \rho \frac{\partial u}{\partial z} = 0 \tag{1}$$

$$\rho \frac{\partial u}{\partial t} + \rho u \frac{\partial u}{\partial z} + \frac{\partial P}{\partial z} = -2\frac{f}{D} \rho u^2 - \rho g \tag{2}$$

$$\rho \frac{\partial h}{\partial t} + \rho u \frac{\partial h}{\partial z} = \phi \tag{3}$$

where ϕ is the heat input rate per unit volume, and f is the two-phase friction
factor. The above equations are based on the following assumptions:
(i)   the flow is assumed to be one-dimensional and homogeneous with no variation of flow parameters across the flow cross-section, (ii) thermodynamic
equilibrium is assumed to exist between phases, (iii) the effect of subcooled
boiling is ignored, (iv) superheat in the boiling region is assumed to be

TABLE 1.   SUSTAINED INSTABILITY DATA (FREON-11)
FOR HEAT INPUT RATE OF 1023, 1365, 1706,
and 2389 Btu/hr

Inlet Temp. (Deg. °F)	Net Heat Input (Btu/hr)	Mean Mass Flow (lbm/min)	System Pressure Drop(psi)	Dens.-Wave Osc. Period (Sec)	Amp. (psi)
15	999.7	0.19	18.0	2.0	4.8
25	1003.8	0.21	16.6	2.0	2.0
50	1002.5	0.31	20.8	2.0	2.4
75	999.7	0.31	23.0	2.2	2.0
100	997.7	0.35	30.0	1.7	1.8
15	1342.9	0.27	27.4	2.0	4.2
25	1339.9	0.30	28.3	2.3	3.8
50	1339.6	0.31	32.0	2.3	3.8
75	1336.5	0.37	35.0	1.9	2.4
100	1335.1	0.47	41.6	1.7	1.6
15	1680.1	0.35	37.7	2.0	5.4
25	1674.9	0.38	38.3	2.1	4.8
50	1671.9	0.41	38.5	2.0	5.4
75	1668.8	0.44	49.7	1.8	6.0
100	1666.4	0.56	49.4	1.3	4.4
15	2361.8	0.46	60.4	2.3	7.2
25	2351.9	0.44	58.0	2.0	11.2
50	2339.6	0.49	60.8	2.0	6.4
75	2341.6	0.54	64.7	1.4	4.4
100	2335.1	0.67	74.7	1.2	5.4

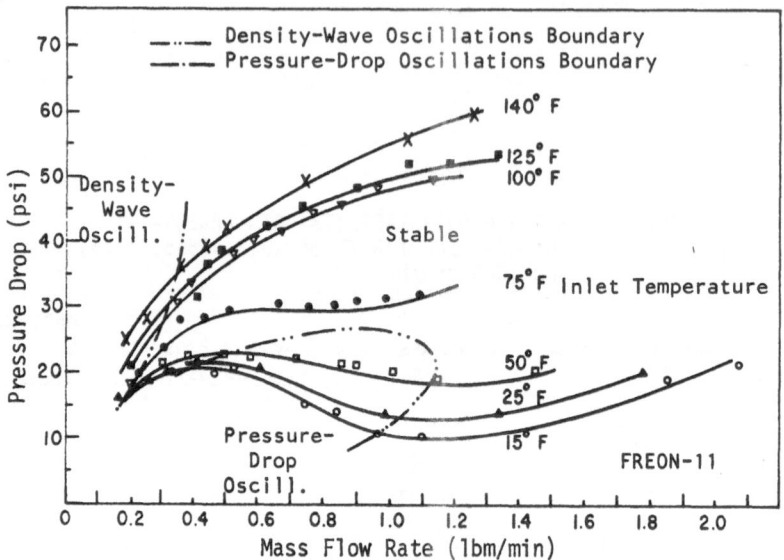

Fig.5 – Stability Map Showing Stable And Unstable Regions for Various
Inlet Temperatures at 1023 Btu/hr Heat Input

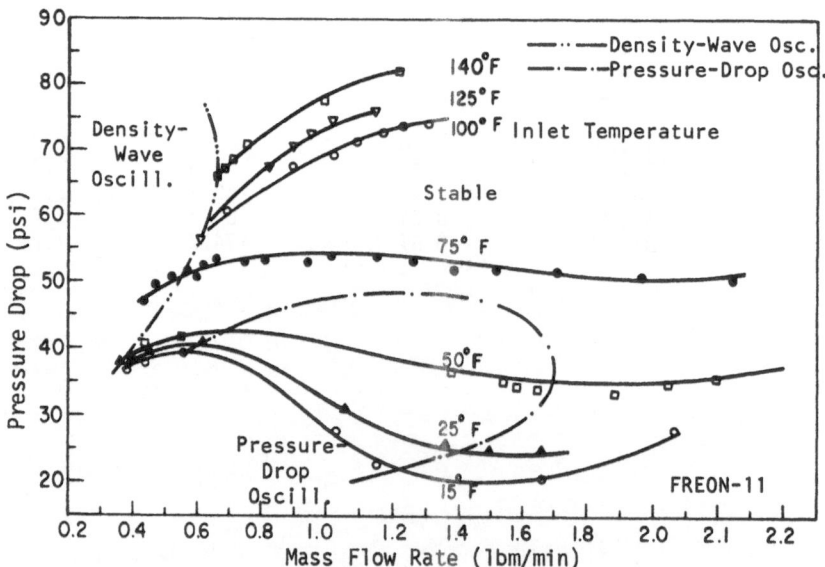

Fig.6 – Stability Map Showing Stable and Unstable Regions for Various
Inlet Temperatures at 1706 Btu/hr Heat Input

negligible, (v) all contributions to fluid enthalpy change due to dissipation, kinetic and potential energy changes are ignored, and (vi) compressiblity effects and pressure drop due to convective acceleration are ignored in the exit restriction.

In addition to the conservation equations, the equation of state is necessary to close the equation set. The homogeneous model requires that the equation of state is defined at the phase boundaries. Therefore, saturated liquid and vapor properties are defined explicitly in terms of polynomials as functions of temperature.

The conservation equations for homogeneous equilibrium flows are similar to single phase flow equations, but the constitutive equations are quite different. These relations have to account for the two-phase phenomena.

Friction Correlation

The friction between the pipe wall and the liquid-vapor mixture is modelled by using the Moody friction factor f which is redefined for two-phase flow as

$$f = C \left(\frac{\rho u D}{\mu_{tp}} \right)^{-0.25} \tag{4}$$

where μ_{tp} is the effective two-phase viscosity and C is a constant that depends on the roughness of the pipe. There are various expressions for effective viscosity in literature. Dukler's [11] definition is found to be in better agreement with the experimental data and is given by the following expression,

$$\mu_{tp} = (1 - x) \frac{\rho}{\rho_\ell} \mu_\ell + x \frac{\rho}{\rho_v} \mu_v \tag{5}$$

Heat Transfer Correlation

The heat transfer in two-phase flow is complicated by the fact that different heat transfer regimes may exist depending on the flow conditions inside the channel. Therefore, the estimation of the heat transfer coefficient requires multistep procedures which use emprical correlations. A review of the available correlations may be found in two-phase heat transfer literature [12]. No effort to find the best correlation is made here; however, a simple correlation is adopted with the following assumptions: It is assumed that the heat transfer rate into the fluid is the sum of two components, pool boiling and forced convection and that the ratio of pool boiling to heat transfer remains constant during the oscillations. Furthermore, it is assumed that the forced convection heat transfer is linearly dependent on the mass flow rate. These assumptions yield the following relation for the unsteady state heat input rate,

$$q = q_s \left(b + (1 - b) \frac{G}{G_s} \right) \tag{6}$$

Here, Q_s is the steady state heat input rate, G_s is the steady state flow rate, and b is the ratio of pool boiling to total heat transfer. The value of b may range from zero to unity and can be evaluated experimentally for a given system or predicted from available correlations of boiling heat transfer [9].

Steady State Characteristics

The initial conditions for both types of oscillations are determined by the steady-state solution of the conservation equations. Therefore, a complete set of relationships, corresponding to different heat inputs and different inlet fluid temperatures for a wide range of fluid mass flow rates, are developed as follows.

The system shown in Figure 7 is isolated from the rest of the experimental set-up. Fluid enters the system at the point designated as "I". The exit of the surge tank is "0". The part of the system from the surge tank to the inlet of the heater, including the inlet plenum is designated as region 1. Since this interval includes valves and fittings an equivalent length is defined for this part of the system. The rest of the system is divided into several segments of equal lengths. The conservation equations are written separately for each region, then they are approximated by finite difference techniques at every segment [10]. This results in as many algebraic equations as the number of segments. Tne numerical solution scheme to solve these equations has been developed and then embodied into a computer program.

Using this program, steady-state system pressure drop for a constant heat flux and an inlet temperature is calculated for various mass flow rates. A plot of these points using different correlations of effective viscosity is shown in Figure 8. The model, TFROFA-2, which uses the Dukler's definition for the effective viscosity and assumes a value of 0.16 for the constant C matches the experimental results better than any other model. Therefore, the same model is adopted in the present study. Using the computer program similar plots are generated for various inlet temperatures at constant heat input rates. One such plot is shown in Figure 9. Plots for idfferent heat inputs are given in Reference 10. All these figures reveal that the meathematical model proposed here predicts the experimentally observed steady-state pressure drop closely for a wide range of mass flow rates. Especially, a very good matching is obtained for low inlet temperatures.

Density-Wave Type Oscillations

In the theoretical analysis of density-wave oscillations, the region between the test liquid tank and the surge tank is excluded from the system since the liquid level inside the surge tank was observed to be constant during this type of oscillations. The conservation equations for the rest of the system is then written considering the necessary assumptions for each flow region. The set of partial differential equations so obtained are non-linear in nature, therefore an analytical solution is impossible. Numerical methods need to be considered only. One such method which employs an explicit finite difference technique had been introduced by Akyuzlu [13]. Oscillations generated by this method is shown in Figs.10 and 11.In this study, a modification is applied to the momentum-equation to reduce the computer time so that a parametric study can be conducted. The modification is summarized below and is based on the assumption that the density is function only of enthalpy, for both the single phase and the two-phase flow regions.

The conservation of mass and energy for the two-phase mixture are given by Eqs. 1 and 3 as they are. The momentum equation, however, is modified as follows [14]. Sonic effects, i.e., pressure disturbances are less important than the

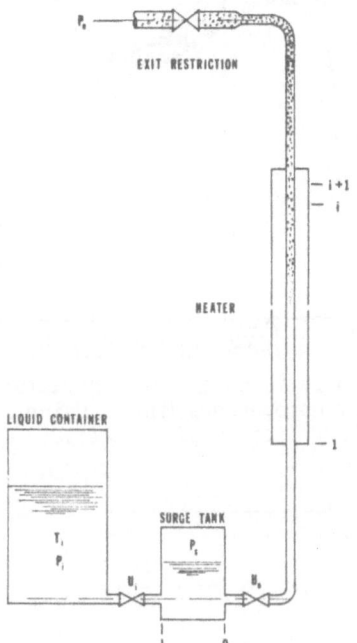

Fig.7 - Schematic of the Boiling
Upward Flow   System

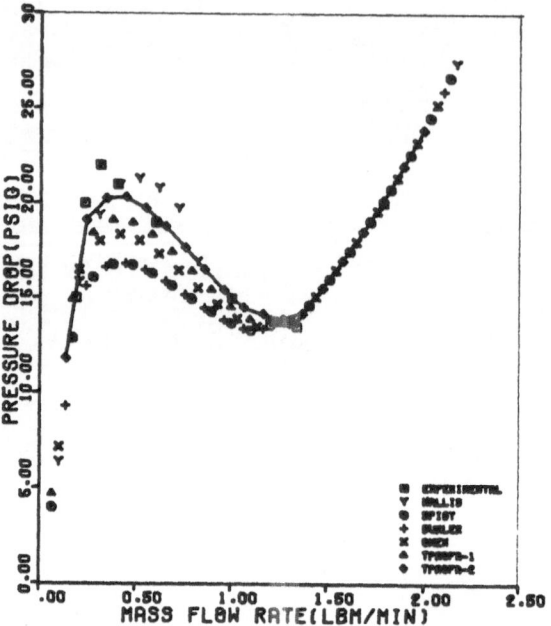

Fig.8 – Comparision of Flow Characteristics Estimated
        by Homogeneous Models Using Different
        Correlations of Effective   Viscosity

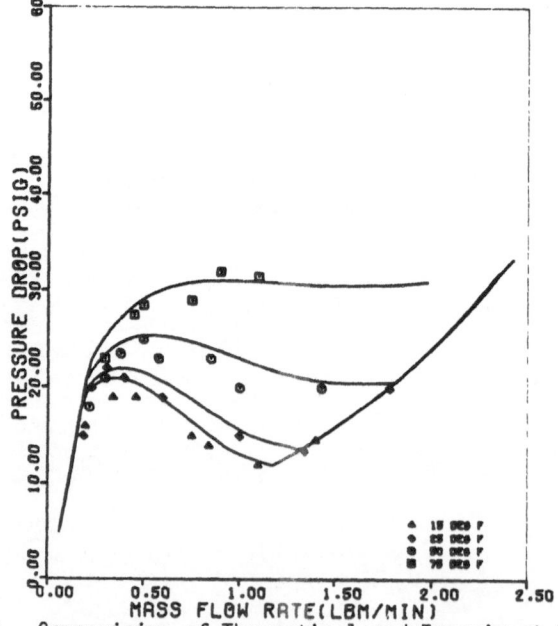

Fig.9 – Comparision of Theoretical and Experimental
        Steady State Flow Characteristics   at
        1023 Btu/hr Heat Input

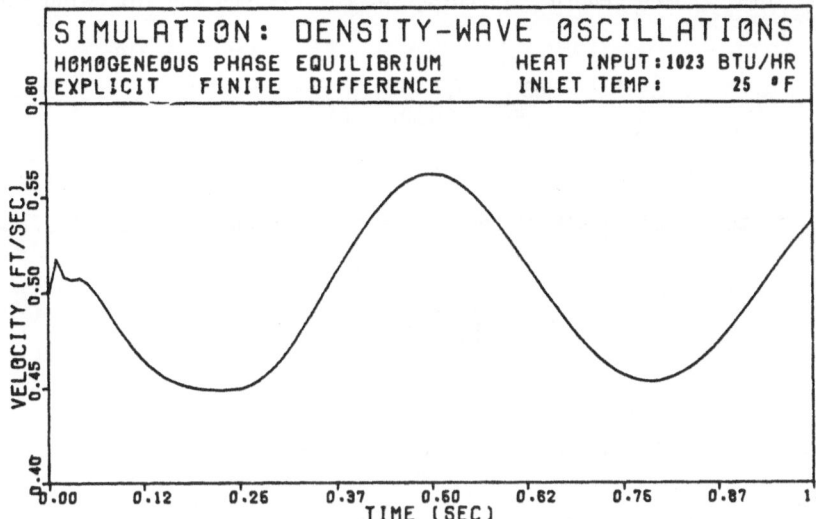

Fig.10 - Numerical Simulations of Density-Wave Oscillations by Explicit Finite Difference Method - Velocity at System Inlet

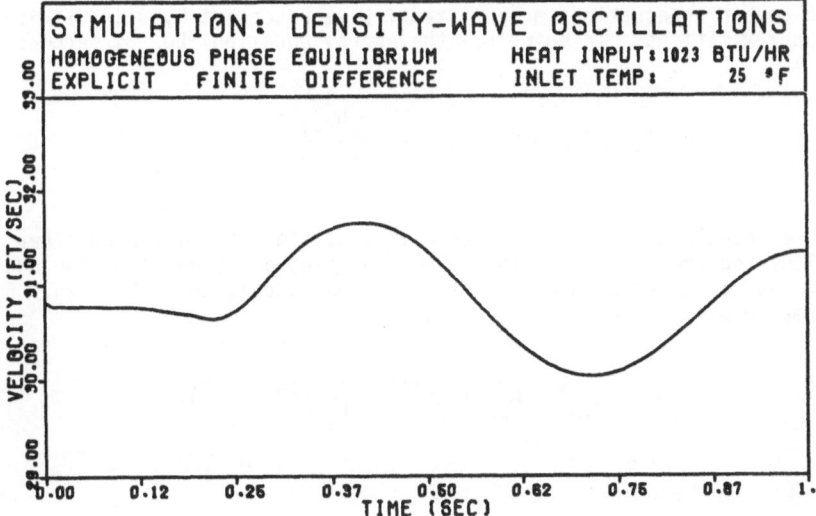

Fig. 11- Numerical Simulations of Density- Wave Oscillations by Explicit Finite Difference Method - Velocity at System Inlet

density effects in the case of density-wave type oscillations. These second
order disturbances can be filtered out when the momentum equation is integrated
over the total tubing length. In doing so, Eq. 2 takes the following form

$$\frac{dG_{av}}{dt} = \frac{1}{L_t}(\Delta P_s - R) \tag{7}$$

where L_t is the total length of the pipe and G_{av} is the average mass flux:

$$G_{av} = \frac{1}{L_t} \int_0^{L_t} \rho u \ dz \tag{8}$$

R in Eq. 7 represents the total resistance to the fluid and is given by

$$R = \rho_e u_e^2 - \rho_o u_o^2 + 2 \int_0^{L_t} \frac{f}{D} \ \rho u^2 \ dz + g \int_0^{L_v} \rho dz \tag{9}$$

and ΔP_s is the steady state pressure drop between the surge tank and the exit
restriction.

The governing equations are written for each segment by using finite difference approximations. In doing so, the energy equation, Eq. 3, becomes

$$h_i^{j+1} = h_i^j - \Delta t^{j+1}[u_i^{j-1}(\frac{h_i^j - h_{i-1}^j}{\Delta z_i})] + \Delta t^{j+1}(\frac{\phi_i^j}{\rho_i^j}) \tag{10}$$

Since the inlet enthalpy is known at all times, the enthalpies at other nodes
are calculated from this equation. The densities are found from the equation
of state using the reference pressures and the calculated enthalpies. Eq. 8 is
approximated by

$$G_{av}^{j+1} = G_{av}^j + \Delta t^{j+1} \cdot \frac{1}{L_t} (\Delta P_s - R^j) \tag{11}$$

and R^j (Eq. 10) becomes

$$R^j = \rho_e^j(u_e^j)^2 - \rho_o^j(u_o^j)^2 + I_f^j + I_g^j \tag{12}$$

The integrals I_f and I_g are approximated by

$$I_f^j = \frac{2}{D} \sum_{i=1}^{n-1} f_{i+\frac{1}{2}}^j \; \rho_{i+\frac{1}{2}}^j \; (u_{i+\frac{1}{2}}^j)^2 \Delta z_i + 2\frac{L_1}{D}f_o^j \rho_o (u_o^j)^2 \tag{13}$$

and

$$I_g^j = g \sum_{i=1}^{n-1} \rho_{i+\frac{1}{2}}^j \Delta z_i + g\rho_o L_{1v} \tag{14}$$

where

$$\rho_{i+\frac{1}{2}}^j = \frac{1}{2}(\rho_{i+1}^j + \rho_i^j) \tag{15}$$

and similarly

$$u_{i+\frac{1}{2}}^j = \frac{1}{2}(u_{i+1}^j + u_i^j) \tag{16}$$

The continuity equation, Eq. 1, is written as

$$\frac{\partial G}{\partial z} = -\frac{\partial \rho}{\partial t} \tag{17}$$

This equation is approximated by

$$\frac{(G_{i+1}^{j+1} - G_i^{j+1})}{\Delta z_{i+1}} = -\frac{(\rho_{i+\frac{1}{2}}^{j+1} - \rho_{i+\frac{1}{2}}^j)}{\Delta t^{j+1}} \tag{18}$$

Solving Eq. 18 for the mass flux, one obtains

$$G_{i+1}^{j+1} = G_i^{j+1} + \xi_i^{j+1} \tag{19}$$

where ξ is defined to be

$$\xi_i^{j+1} = -\frac{\Delta z_{i+1}}{\Delta t^{j+1}}(\rho_{i+\frac{1}{2}}^{j+1} - \rho_{i+\frac{1}{2}}^j) \tag{20}$$

Eq. 19 is written in terms of the inlet mass flux as

$$G_i^{j+1} = G_o^{j+1} + \beta_i^{j+1} \tag{21}$$

where

$$\beta_i^{j+1} = \beta_{i-1}^{j+1} + \xi_i^{j+1} \tag{22}$$

with

$$\beta_o^{j+1} = 0 \tag{23}$$

The unknown G_o in Eq. 22 is defined in terms of G_{av} by

$$G_o^{j+1} = G_{av}^{j+1} - \frac{1}{2L} \sum_{i=0}^{n-1} (\beta_{i+1}^{j+1} + \beta_i^{j+1}) \Delta z_{i+1} \tag{24}.$$

Once the mass fluxes are known at each node, then the velocities are calculated from

$$u_i^{j+1} = G_i^{j+1} / \rho_i^{j+1} \tag{25}$$

The numerical scheme is found to be stable when the time increment is calculated by

$$\Delta t^{j+1} \leq \frac{\Delta z_i}{(u_i^j)_{max}} \tag{26}$$

The above equations are modified for different regions of the flow channel. The reader should refer to Final Report NSF ENG 75-16681 [10] for the details of the finite-difference analysis.

Numerical simulations of density-wave oscillations have been generated by the above technique. In the case of a constant heat input of 1023 Btu/hr and an inlet temperature of 25°F the oscillations are sustained only for an inlet velocity of 0.5 ft/sec. This can be seen in Figure 12. At this point an increase in mass flow rate while heat input is kept constant results in transient oscillations that eventually die out. These transient are theoretically experienced up to 0.9 ft/sec inlet velocity. Further increase in inlet velocity results in no oscillations. This behavior is consistent with the experimental observations. A different behavior is observed when the inlet velocity is decreased from 0.5 ft/sec. A decrease in flow rate at this point results in transient oscillations with increasing amplitudes. Further decrease in mass flow rate is accompanied by transient oscillations with decreasing amplitudes. This behavior is in contrast with experimental observations where oscillations are always sustained for low mass flow rates. The theoretical behavior is believed to be the result of constant heat input assumption. The transient behavior continues until the velocity reaches to 0.25 ft/sec. At lower velocities that this value no oscillations are generated. It is not known at this time whether the oscillations cease out at low inlet velocities or not, since it was not possible to conduct experiments at very low mass flow rates.

The above numerical experimentation was carried out for various inlet

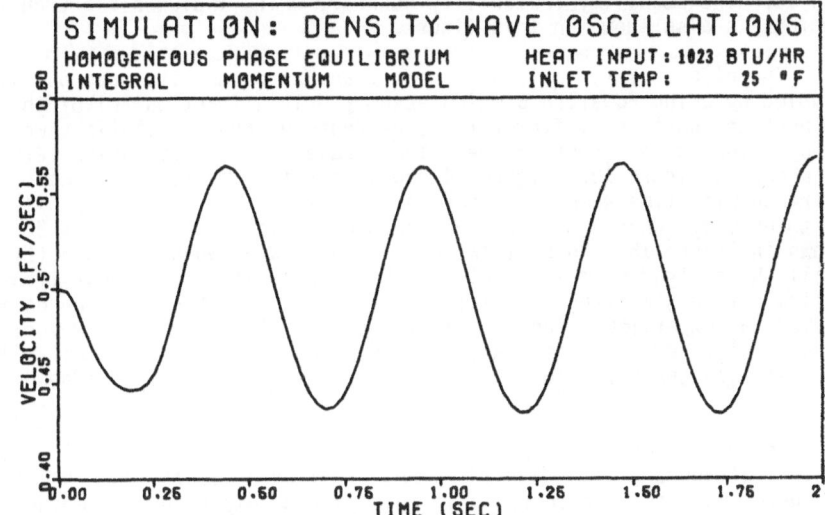

Fig.12 - Effect of Heat Transfer on Density-Wave Oscillations - Heater
Inlet Velocity for b = 1.0

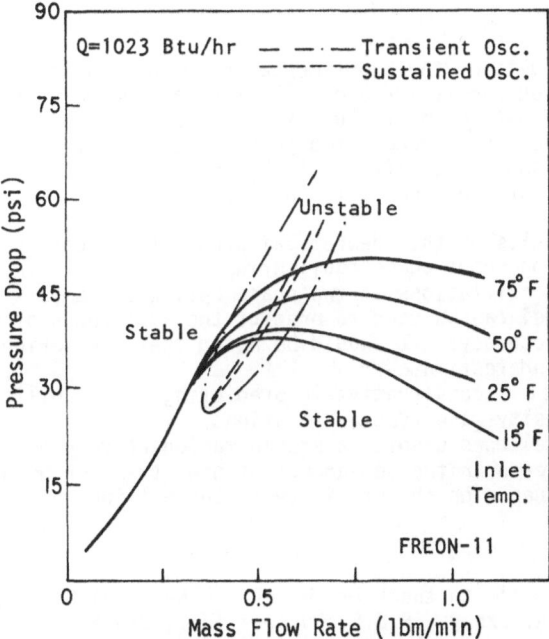

Fig.13 - Transient and Sustained Instability Boundaries
Predicted by the Homogeneous Model

temperatures. The points where the sustained oscillations were observed are
then connected to form the unstable region and is shown with a dashed line on
Figure 13. The stability-instability boundary which includes the transient oscillations is also shown on this figure.

The effect of variations in heat transfer on density-wave oscillations
is studied by using Equation 6 which assumes only a fraction (b) of the total
heat input as function of flow rate. The density-wave oscillations were generated by assuming b=1 (constant heat input case). The oscillations are sustained as shown in Figure 12. Figure 14 shows that the amplitude of the oscillations
do decrease with time when all other parameters are kept the same while b is
taken to be 0.9. When b is 0.8 then the oscillations die out after 2 or 3 cycles
as shown in Figure 15. No oscillations are generated when b is 0.6 (Figure 16).
From all these figures it is concluded that the dependence of heat transfer on
mass flow rate has a strong stabilizing effect on density-wave type oscillations.
The numerical experimentation was carried out at different inlet velocities for
different inlet temperatures. As a result, transient stability-instability
boundaries were generated for different values of b and they are shown in Figure
17.

The effect of heat input and the effect of inlet subcooling on density-
wave oscillations are also studied by numerical experimentation. As can be seen
from Figures 18 and 19 an increase in inlet fluid temperature, keeping all other
parameters constant, results in increase of amplitudes and decrease in periods
of oscillations. The same kind of behavior is obtained when the heat input rate
is increased. This can be seen in Figures 20 and 21 for heat input rates of
1706 and 2389 Btu/hr, respectively. All these results are consistent with the
experimental observations.

CONCLUSIONS

From the results of the experimental study one can conclude that:
(i) Sustained density-wave oscillations are observed for all inlet temperatures provided that the heat flux is sufficiently high and/or the flow rate
is sufficiently low to produce a two-phase mixture of relatively high quality.
(ii) Period and the amplitude of the oscillations are dependent on the inlet fluid temperature and the heat input.

From the results of the theoretical study it is concluded that:
(i) The homogeneous phase equilibrium model is satisfactory in simulating
the density-wave oscillations in two-phase upflow systems.
(ii) The model can be used to predict the amplitudes of the oscillations
with reasonable accuracy. It should be noted that the period of density-wave
oscillations is underestimated.
(iii) The model is conservative in predicting the stability-instability
boundaries of density-wave type oscillations.
(iv) The model does predict a stable region at very low mass flow rates.
(v) An increase in the dependence of heat transfer on mass flow rate has
a beneficial influence on the stability of the boiling upflow system.

ACKNOWLEDGEMENTS

The authors wish to thank Dr. Win Aung, National Science Foundation, Dr.
T. Dogan and Dr. N. Ozboya, E.D.S. Nuclear Inc., San Francisco. The secretarial
work of Ms. M. Aguiluz is also acknowledged.

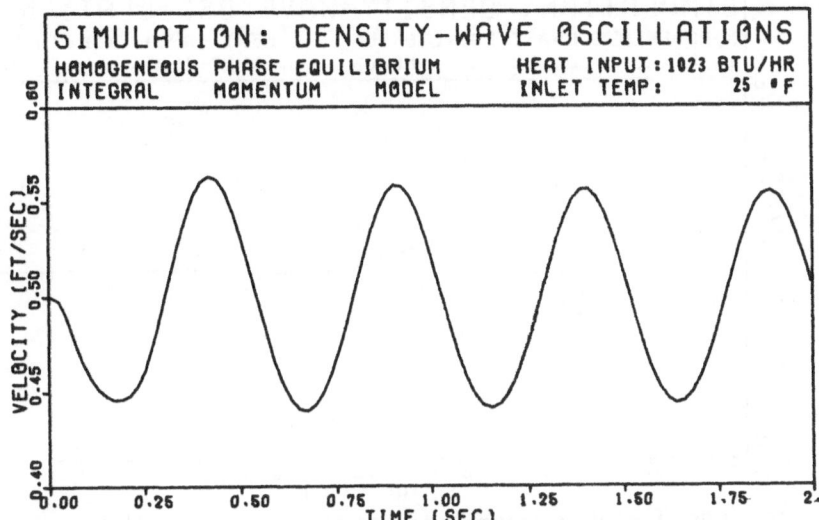

Fig.14 - Effect of Heat Transfer on Density-Wave Oscillations - Heater
Inlet Velocity for b = 0.9

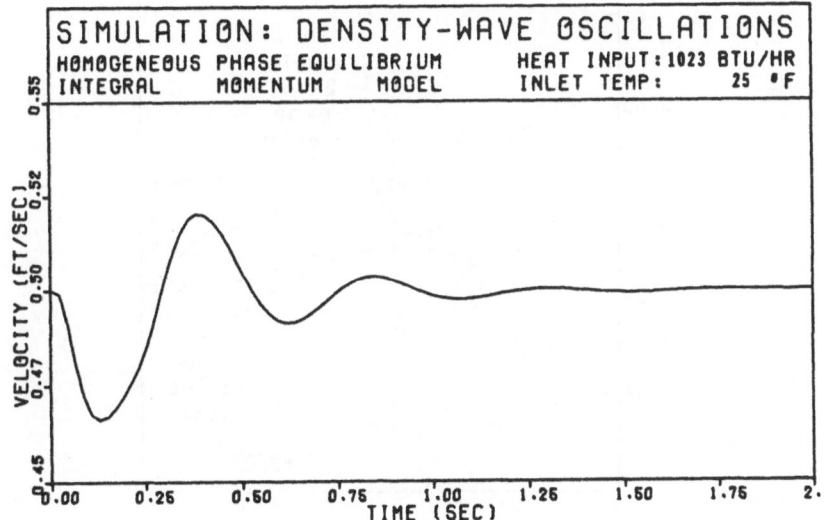

Fig.15 - Effect of Heat Transfer on Density-Wave Oscillations - Heater
Inlet Velocity for b = 0.8

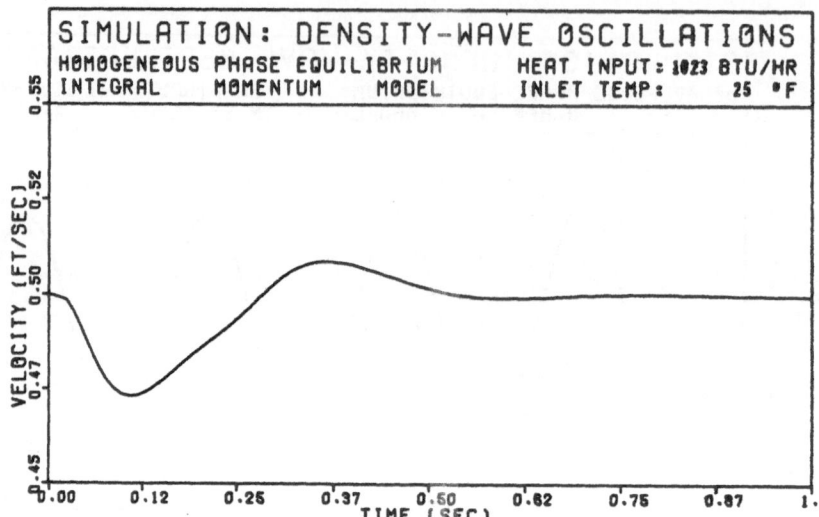

Fig.16 – Effect of Heat Transfer on Density-Wave Oscillations – Heater
Inlet Velocity for b = 0.6

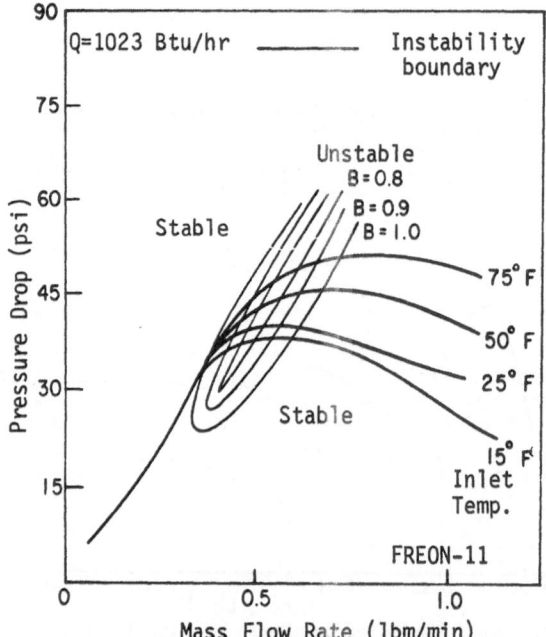

Fig.17 – Effect of Heat Transfer on the Instability
Boundaries Predicted by the Homogeneous
Model

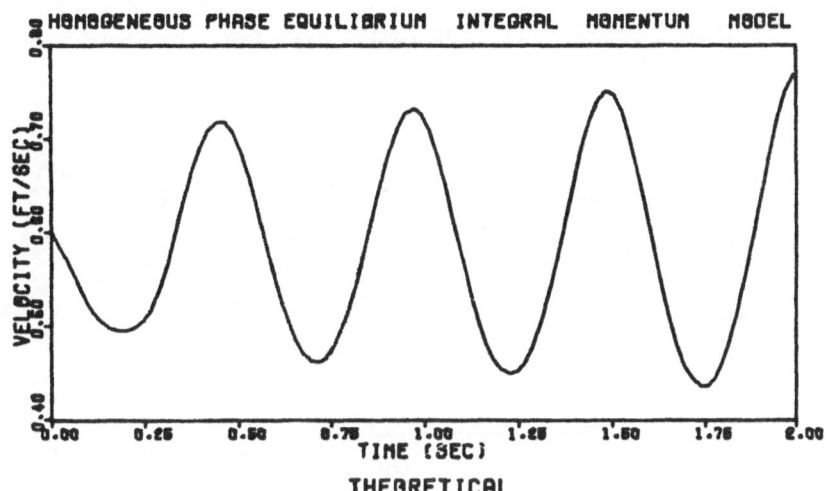

HEAT INPUT: 1023 BTU/HR INLET TEMP: 50 °F

Fig.18 - Numerical Simulation of Density-Wave Oscillations - Heater
Inlet Velocity (T_i = 50°F)

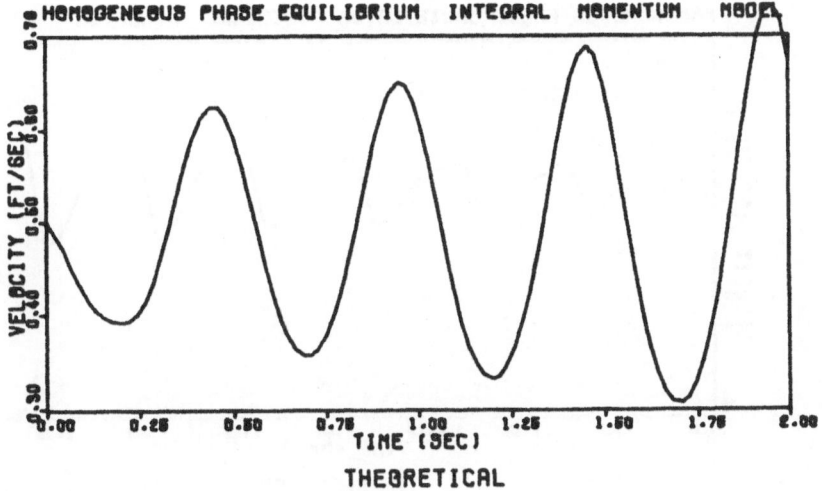

HEAT INPUT: 1023 BTU/HR INLET TEMP: 75 °F

Fig.19 - Numerical Simulation of Density-Wave Oscillations - Heater
Inlet Velocity (T_i = 75°F)

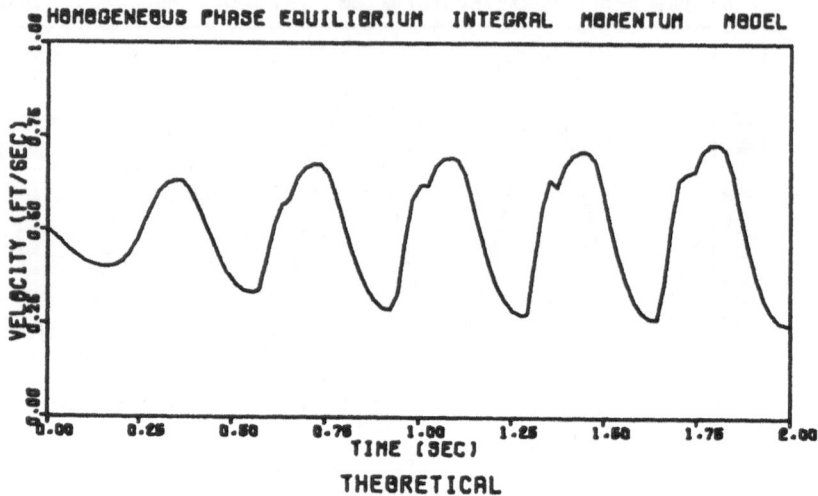

HEAT INPUT: 1706 BTU/HR   INLET TEMP: 25 °F

Fig. 20- Numerical Simulations of Density-Wave Oscillations - Heater
Inlet Velocity (Q = 1706 Btu/hr)

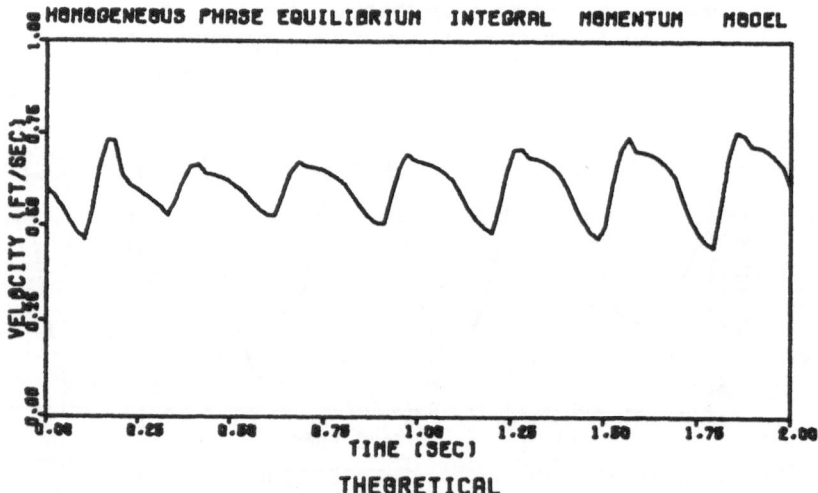

HEAT INPUT: 2389 BTU/HR   INLET TEMP: 25 °F

Fig. 21 - Numerical Simulation of Density-Wave Oscillations - Heater
Inlet Velocity (Q = 2389 Btu/hr)

NOMENCLATURE

Symbols	Definition
A	Area
b	fraction of heat input independent of flow
D	diameter of pipe
f	Fanning friction coefficient
g	gravitational acceleration
G	mass flux
h	specific enthalpy
L	tubing length
n	number of segments
p	pressure
Q	power input into the fluid
R	total resistance to the fluid
t	time
T	equilibrium temperature
u	velocity
z	space coordinate

Greek Symbols

ϕ	power density
μ	viscosity
ρ	thermodynamic density

Subscripts

av	average
e	exit
f	saturated liquid; fluid
h	heater
i	inlet of surge tank; i^{th} segment
o	exit of surge tank
s	steady state; surge tank
t	total
tp	two-phase
v	vertical
w	wall

Superscripts

j	j^{th} time

REFERENCES

1. Bergles, A.E., Goldberg, P., and Maulbetsch, J.S., "Acoustic Oscillations in a High Pressure Single Channel Boiling System," EURATOM Report, Proceedings Symposium on Two-Phase Flow Dynamics at Eindhoven, pp. 535-550 (1967).

2. Maulbetsch, J.S., and Griffith, P., "A Study of System - Induced Instabilities in Forced-Convection Flows With Subcooled Boiling", MIT Engineering Projects Lab Report 5382-35, (1965).

3. Boure', J.A., "The Oscillatory Behavior of Heated Channels", Part I and II, French Report CEA-R 3049, Grenoble (1966).

4. Nahavandi, A.N., and von Hollen, R.F., "A Space-Dependent Dynamic Analyisis of Boiling Water Reactor Systems," Nuclear Eng. and Sci. 20, pp. 392-413 (1964).

5. Solberg, K.O., and Bakstad, P., "A Model for Dynamics of Nuclear Reactors with Boiling Coolant with A New Approach to the Vapor Generation Process", ERATOM Report, Proceedings of Symposium on Two-Phase Flow Dynamics, Eindhoven, (1967).

6. Margolis, S.G. and Redfield, J.A., "FLASH: A Program for Digital Simulation of the Loss of Coolant Accident," WAPD-TM-534, Bettis Atomic Power Laboratory (1966).

7. Boure'- J.A., Bergles, A.E., and Tong, L.S., "Review of Two-Phase Flow Instability", Nuclear Engineering and Design, North Holland Publishing Company (1973).

8. Veziroglu, T.N., "Oscillations in Boiling Upward Flow", Union Carbide Corporation, Nuclear Division Subcontract No. 2785, Final Report, (1967).

9. Stenning, A.H., Veziroglu, T.N., "Flow Oscillations in Forced Convection Boiling", NASA-CR-72122, Final Report, Vol. II, (1967).

10. Veziroglu, T.N., and Kakac, S., "Two-Phase Flow Instabilities and Effect of Inlet Subcoolings", Final Report, NSF-ENG 75-1668, (1980).

11. Dukler, A.E., et. al., "Pressure Drop and Hold-up in Two-Phase Flow", A.I.Ch.E.J., Vol. 10, No. 1, pp. 38-51 (1961).

12. Bjornard, T.A., and Griffith, P., "PWR Blowdown Heat Transfer", Proceedings of the Symposium on the Thermal and Hydraulic Aspects of Nuclear Reactor Safety, Vol. 1, (1977).

13. Akyuzlu, K.M., Veziroglu, T.N., Kakac, S., and Dogan, T., "Finite Difference Analysis of Two-Phase Flow Pressure-Drop and Density-Wave Oscillations", Warme-and Stoffubertragung (Thermo- and Fluid Dynamics). Vol. 14, p. 253-267, (1980).

14. Meyer, J.E. and Rose, R.P., "Application of a Momentum Integral Model to the Study of Parallel Channel Boiling Flow Oscillations", ASME Paper No. 62-HT-41 (1962).

15. Akyuzlu, K.M., "Mathematical Modelling of Two-Phase Flow Oscillations", Ph.D. Thesis, University of Miami, Miami, Florida, U.S.A. (1979).

Effect of Heat Transfer Augmentation on Two-Phase Flow Instabilities in a Vertical Boiling Channel

A. MENTES, H. GÜRGENCI, O.T. YILDIRIM, S. KAKAÇ,
and T.N. VEZIROĞLU
University of Miami
Coral Gables, Florida 33124, USA

ABSTRACT

The effect of different heater surface configurations on two-phase flow instabilities has been investigated in a single channel, forced convection, open loop, up-flow system. Freon-11 is used as the test fluid and six different heater tubes with various inside surface configurations have been tested at five different heat inputs. In addition to temperature and pressure recordings, high speed motion pictures of the two-phase flow were taken for some of the experiments to study the two-phase flow behavior at different operating points. Experimental results are shown on system pressure drop versus mass flow rate curves, and stability boundaries are also indicated on these curves. Comparison of different heater tubes is made by the use of the stability boundary maps and the plots of inlet throttling necessary to stabilize the system versus mass flow rate. Tubes with internal springs were found to be more stable than the other tubes.

1. INTRODUCTION

Two-phase flow instabilities may cause operational and safety problems in many heat exchange equipment involving two-phase flow such as boiling water nuclear reactors, steam generators, boilers, evaporators, heat exchangers, rocket engines, cryogenic equipment and various chemical process units.

Sustained oscillations of the mass flow rate, system pressure and temperature are undesirable as they can cause mechanical vibrations, control problems and in extreme circumstances do disturb the heat transfer characteristics so that a heat transfer surface may burn out.

Basically there are three identifiable types of oscillations, namely, pressure-drop type, density-wave type and thermal oscillations, which may be encountered in two-phase flow systems. Extensive work has been carried out to find the basic mechanisms and the predictions of these oscillations. A good review of the subject may be found in [1] and [2].

Up to now the effects of various parameters, such as inlet and exit restrictions, inlet subcooling, thermal flux and parallel

A. Mentes and H. Gürgenci are on leave from Middle East Technical University, Ankara, Turkey.

channel geometry have been studied experimentally and theoretically [1 through 6].

In recent years, there has been increased emphasis on techniques to enhance two-phase flow heat transfer; an excellent survey of the subject is presented by Bergles [7]. The importance of the subject and the increase in the use of heat transfer augmentation with two-phase flow systems have led us to study the effect of different heater surface configurations on two-phase flow instabilities. During our studies, the effect of surface configurations on heat transfer coefficient has also been examined and a correlation has been obtained [8].

2. EXPERIMENTAL SYSTEM

A schematic diagram and the basic dimensions of the experimental setup are shown in Figure 1. The setup was specially designed and built to generate the main types of two-phase flow oscillations, and allows the changing of the heater tubes without disturbing other parts of the system.

All the tubing except the recovery section is made of 7.493 mm

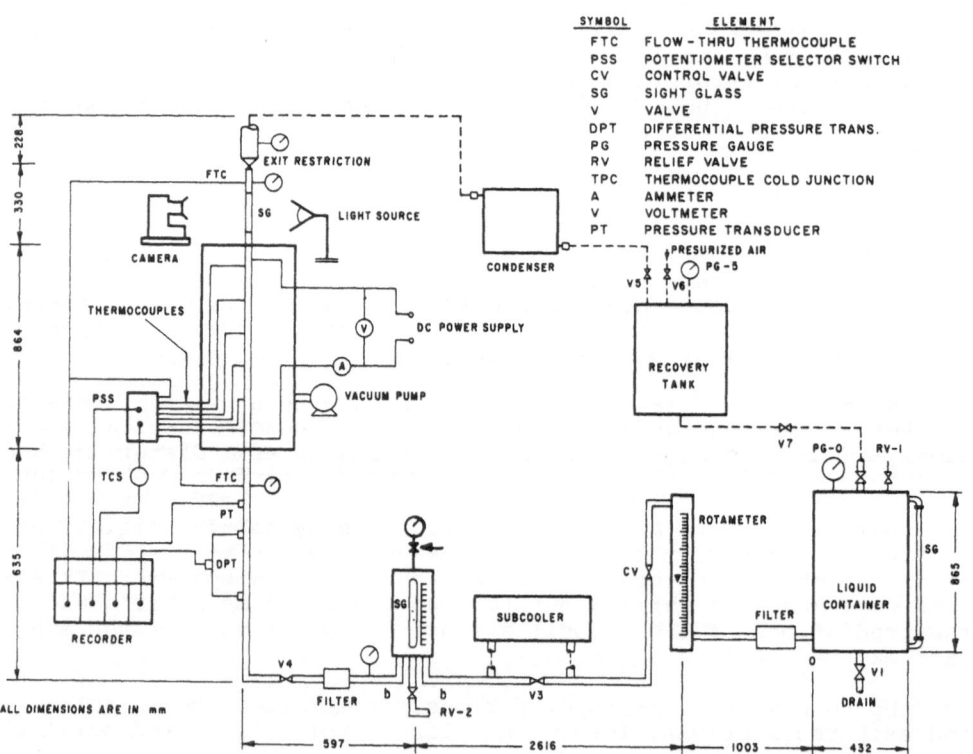

Fig. 1. Schematic Drawing of the Experimental System

ID Nichrome tube, and the connections are made by Swagelok-type joints. In the recovery section, tubing of 19.05 mm diameter is used to minimize pressure losses, thus maintaining a nearly constant level of exit pressure.

A 7.62 cm long transparent piece of tubing is incorporated into the loop right after the heater. This makes it possible to observe and to record the two-phase flow oscillations in their various modes by means of a high-speed camera.

Basically, the experimental setup is an open loop, forced convection boiling, upflow system operating between constant inlet and exit pressures. The loop can be divided into three parts: fluid supply system, test section and the recovery section. Fluid supply system includes a main tank, a filter, an inlet control valve and a subcooling controller installed in that order. A pressurized nitrogen supply connected to the main tank provides the necessary pressure to pump the liquid through the system. Two series of experiments were conducted at different main tank pressures of 5.15 bars and 6.53 bars.

Test section is the part of the loop where two-phase flow oscillations occur under controlled conditions. It includes a surge tank, an inlet throttling valve, the heater tube and an exit restriction. The main body of instrumentation is also clustered in this section.

The surge tank is basically a cylinder of 10.16 cm ID with a height of 22.86 cm. During the experiments, it is partially filled with nitrogen and acts as a capacitance.

The heater tube, 60.5 cm long, is used as the electrical resistance to provide the heat input. Power is supplied through a controlled D.C. transformer-rectifier unit, which has a capacity of 20 kW and is equipped with a continuous current setting. The heater is connected to the loop by CAJON-VCO couplings which needs no axial clearance for dismantling and reinstallation of different heater tubes. To minimize heat losses, a radiation guard heater is built around the heater tube and then the whole assembly is housed in a vacuum jacket. A vacuum pump connected to this jacket maintains a vacuum of less than 1.0 mm Hg absolute. The test section is terminated by an exit restriction, which is a sharp edge orifice with an inner diameter of 1.5875 mm.

Recovery section consists of a condenser and a recovery tank. The recovery tank is kept open to atmosphere to insure a constant level of exit pressure.

Flow rate and heat input, along with pressures and temperatures at appropriate locations, are measured during each run. Temperature measurements are made by standard Copper-Constantan thermocouples. Five of these, fixed to the outer surface of the heater at 13.75 cm intervals, are used for heater wall temperature measurements. Two flow-through thermocouples, one before the heater and one after the heater, are used to measure the fluid bulk temperatures at these locations. Pressures are measured at four different points by Bourdon type pressure gauges. In addition to those, a strain gauge type pressure transducer is used to

record pressure variations at the heater inlet. Flow rate measurements are made using a calibrated rotameter for mean values and
a differential pressure transducer for instantaneous variations in
mass flow rate at the heater inlet.

A four channel amplifier-recording unit was used to record the
mass flow rate, inlet pressure, exit fluid temperature, and any one
of the heater wall temperatures continuously. The heat input to
the system is obtained by measuring the current and the voltage
drop across the heater tube.

Six different heater tubes, which are shown in Figure 2, were
prepared and used during the experiments.

Each heater was made of 7.493 mm inside and 9.525 mm outside
diameter, 605 mm long of Nichrome tubes. Tubes are classified
according to their effective diameters, which is defined as,

$$d_e = \sqrt{\frac{4V}{\pi L}}$$ (1)

where, V is the net inside volume and L is the length of the
heater tube.

The tube A is the bare tube which is used for comparison
purposes. Descriptions of tubes are given in Table 1.

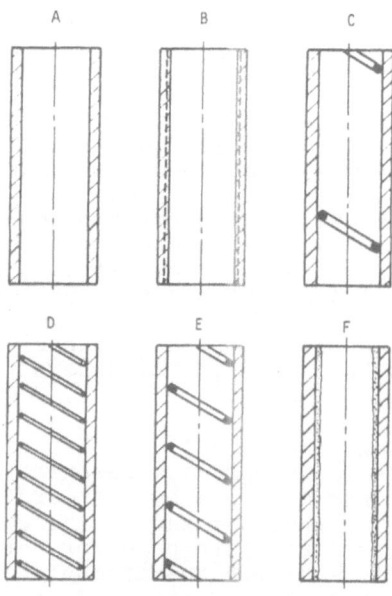

Fig. 2. Heater Surface Configurations

Table 1.   Description of the Heater Tubes

TUBE	DESCRIPTION OF THE TUBE	d_e mm
A	Bare	7.493
B	Threaded, 7.938 mm--16 threads per 25.4 mm	7.619
C	With internal spring of 0.794 mm wire diameter and 19.05 mm pitch	7.446
D	With internal spring of 0.432 mm wire diameter and 3.175 mm pitch	7.401
E	With internal spring of 1.191 mm wire diameter and 6.350 mm pitch	7.192
F	Coated with: Union Carbide Linde High Flux Coating	7.073

Transparent part of the system is used to observe and to take pictures of the two-phase flow system behavior during operation. A HYCAM 40-0007 (100 foot) high-speed camera was used for this purpose.

More detailed information on the setup and the instrumentation can be found in reference [2].

3.   EXPERIMENTS

For each heater tube, six different experiments, one without heat input, five with different heat inputs, were carried out. Each experiment was composed of a sufficient number of tests to cover a wide range of boiling regimes, from single phase liquid up to very high qualities.

Oscillations were identified by the cyclic variation of the test section pressure, the mass flow rate and the temperatures. Instability thresholds were located by disregarding short life transients and considering only sustained oscillations.

3.1.   Experimental Procedure

For each heater tube, an initial experiment was conducted without heat input to find the single phase characteristics. Experiments with heat input were started with high mass flow rate and continued to lower mass flow rates to cover the whole boiling region.

The test procedure can be outlined as follows:

1.   With enough liquid in the main tank, the tank was

pressurized using nitrogen gas.

2. Surge tank was half filled with liquid Freon-11 and
pressurized to a predetermined value by nitrogen gas.

3. Flow rate and heat input were increased gradually to the
desired starting point, and the system was allowed to become
steady, as indicated by recordings of system pressure, temperature
and flow rate.

4. Measurements of temperature, pressure, flow rate and heat
input were taken and critical observations were noted.

5. The mass flow rate was reduced by a small amount using
the inlet control valve and the system was allowed to become
steady, and the readings were taken. This procedure was repeated
starting from step 4 until sustained oscillations were observed.
After reaching to the unstable region mass flow rate was first
increased and then decreased very slowly to locate the boundaries.

6. While operating in the unstable region, first recordings
of the flow rate, the heater inlet pressure, the heater exit fluid
temperature and a selected heater wall temperature were made.
Then, the system was stabilized by closing the inlet throttling
valve slowly and readings were taken. After taking the readings,
the inlet throttling valve was brought into full open position and
mass flow rate was reduced by a small amount.

The above procedure was repeated for each tube and heat input.
Experiments were stopped after reaching the dry-out.

A continuous film was made during some of the tests, while
the recorder simultaneously kept a continuous history of the flow
variables during the filming.

3.2. Experimental Results

Two basic types of oscillations, namely the pressure-drop
type and the density-wave type, and various forms of superimposi-
tion of these oscillations were observed during the experiments.
Experimental results are presented in graphical forms. Figures 3-8
show steady-state characteristics of the various tubes, with system
pressure drop and mass flow rate as coordinates. Oscillation
boundaries are also indicated on these figures. These are the
typical sets of curves showing the total system pressure drop
versus mass flow which is parabolic with zero heat input and
S-shaped for two-phase flow for various values of the heat input.

Boundaries for the pure pressure-drop and the superimposed
oscillations for different tubes were collected in Figures 9 and
10. An examination of these figures shows that the curves for
different heater tubes exhibit a similar pattern, with internally
springed tubes having narrower unstable region. The order of
tubes is the same for both figures.

A similar order is also observed at the plots of inlet
throttling necessary to stabilize the system against the mass flow

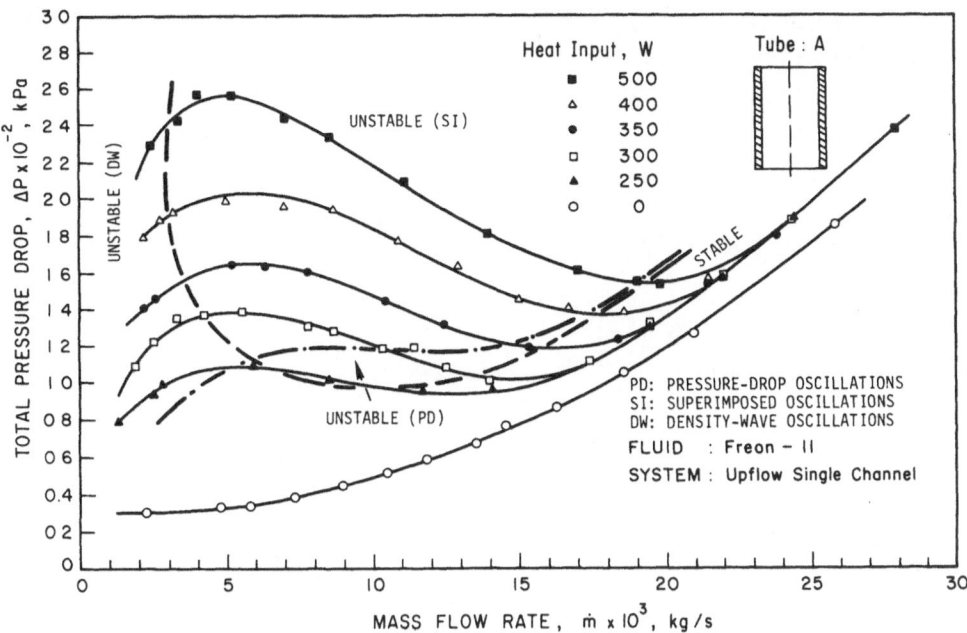

Fig. 3. Steady-State Characteristics and Stability Boundaries.
         Inlet Temperature: 23°C

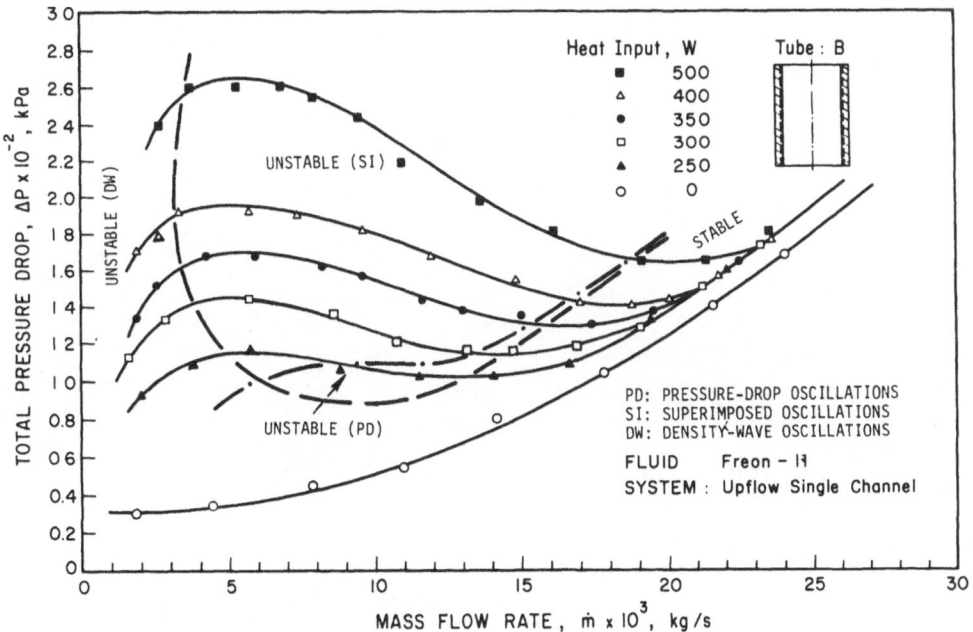

Fig. 4. Steady-State Characteristics and Stability Boundaries.
         Inlet Temperature: 23°C

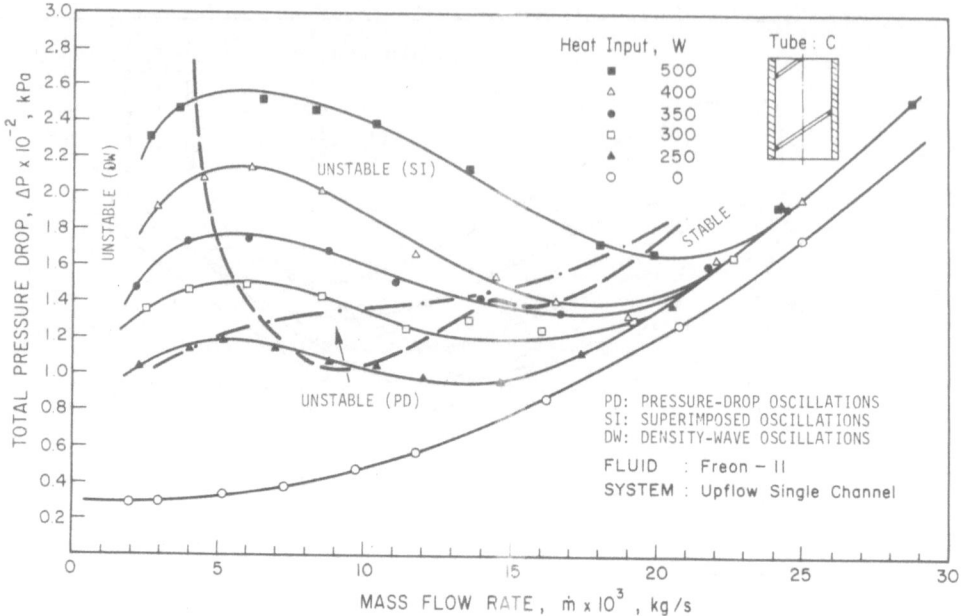

Fig. 5.   Steady-State Characteristics and Stability Boundaries.
          Inlet Temperature: 23°C

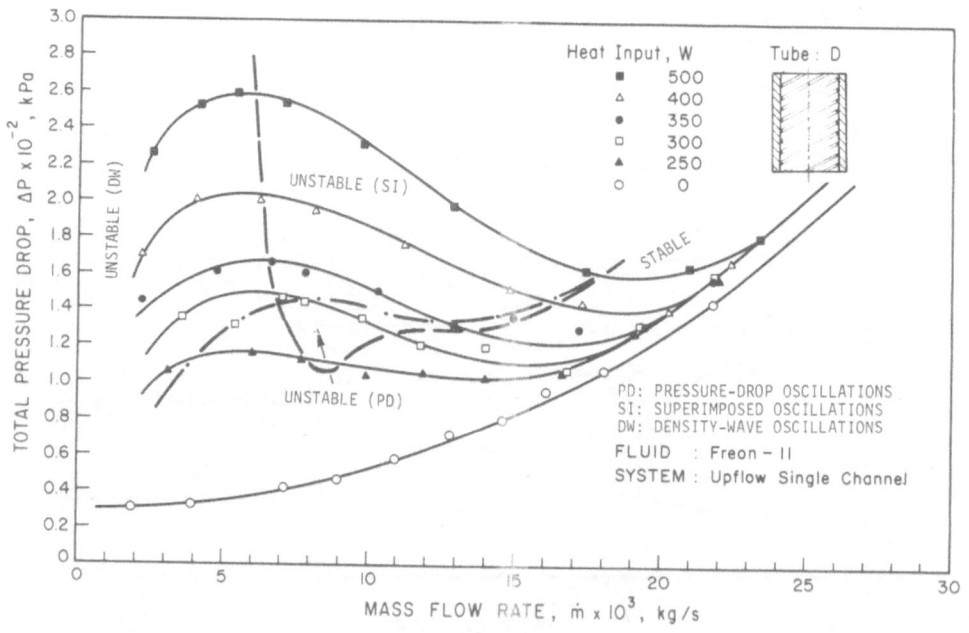

Fig. 6.   Steady-State Characteristics and Stability Boundaries.
          Inlet Temperature: 23°C

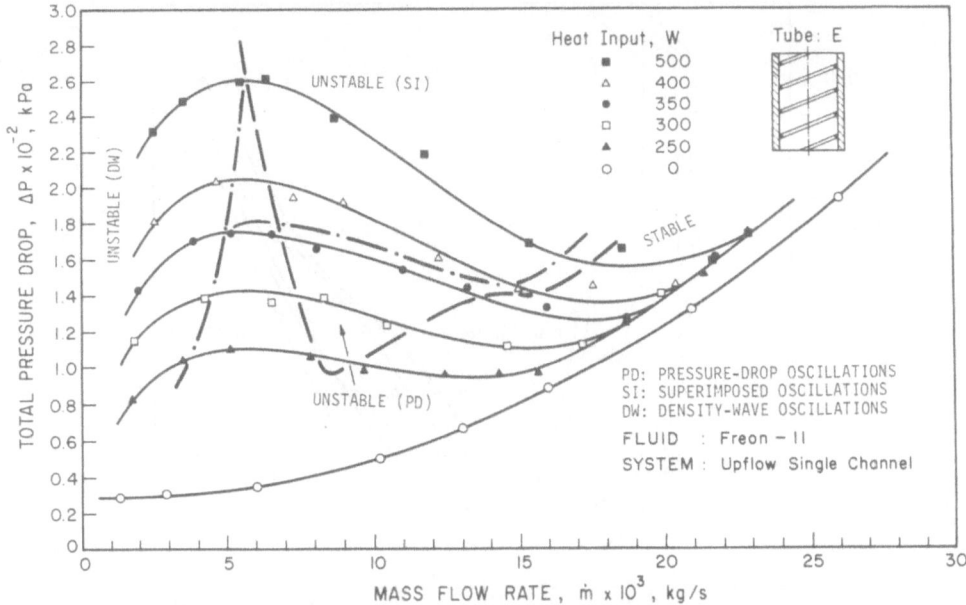

Fig. 7.   Steady-State Characteristics and Stability Boundaries.
          Inlet Temperature: 23°C

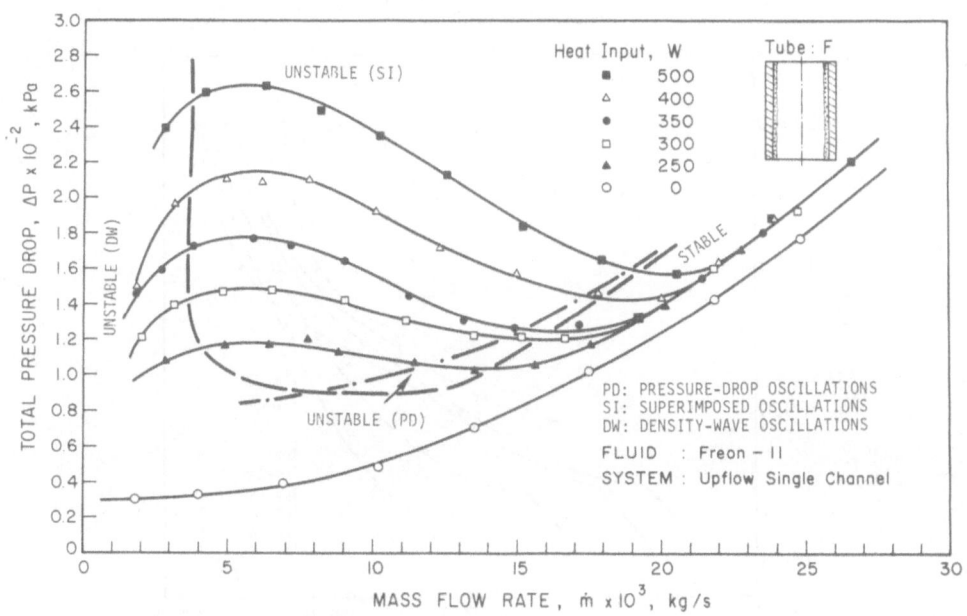

Fig. 8.   Steady-State Characteristics and Stability Boundaries.
          Inlet Temperature: 23°C

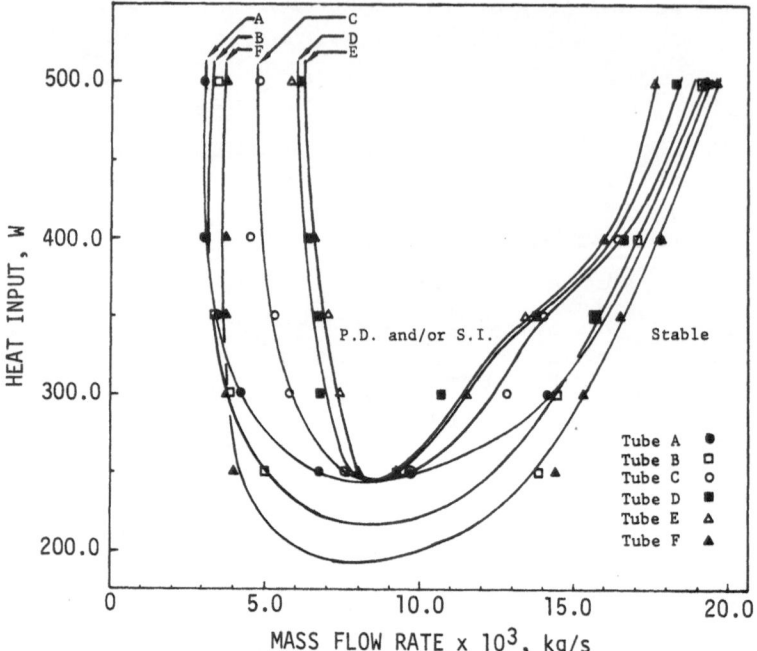

Fig. 9.   Boundaries of the Pressure-Drop Type Oscillations

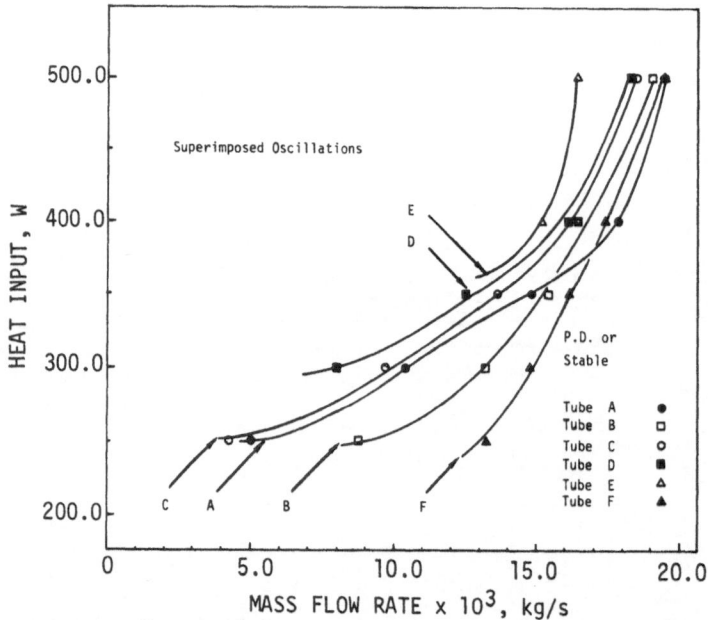

Fig. 10.   Boundaries of the Pressure-Drop Type Oscillations with
           Superimposed Density-Wave Type Oscillations

rate as shown on Figures 11, 12, 13 and 14 for various values of
heat inputs on the internally different augmented tubes.

Inlet throttling necessary to stabilize the system during the
oscillations is found by subtracting the inlet pressure drop,
which would be observed if the inlet throttling valve were full
open, from the inlet pressure drop found after stabilizing the
system by closing the inlet throttling valve. The former ficti-
tious pressure drop was calculated using equation (2), which is
obtained by the least squares polynomial fitting of the inlet
frictional drop data for zero heat input experiments.

$$\Delta P_{i,f} = 9.450 \times 10^3 + 74.031 \, \dot{m} + 14.390 \, \dot{m}^2 \tag{2}$$

Where, $\dot{m}$ is the mass flow rate in gr/sec and $\Delta P_{i,f}$ is the ficti-
tious pressure drop in N/m^2.

From the Figures 11 through 14 it is seen that the additional inlet
throttling necessary to stabilize the system rises sharply after the
onset of oscillations and attains a maximum which corresponds to
an operating point approximately at the center of the negative
slope region of total system pressure drop versus mass flow rate
curve. The curve starts to rise after dropping to a minimum value,
which is around the maximum of the total pressure drop versus mass
flow rate curve.

Some of the results obtained by analyzing the output of the
data reduction program [1] and the recordings of the flow parameters are presented in Tables 2 and 3. Table 2 shows the results
obtained during the superimposed oscillations and Table 3 shows
the results obtained during the density-wave oscillations. It was
not possible to supply a similar table for pure pressure-drop type
oscillations, since the region for pure pressure-drop type
oscillations was very narrow.

Recordings of the inlet pressure, mass flow rate and temperatures were used to find out the amplitudes and the frequencies of
the oscillations. Oscillation characteristics were changing from
tube to tube. Figures 15 through 20 show the typical recordings
of the superimposed and the density-wave oscillations for heater
tubes A, E and F.

Films taken during some of the experiments were analyzed using
a Lafayette-00200 film analyzer. Analysis showed that all of the
five two-phase flow regimes occurred during a pressure-drop oscillation cycle. Figure 21 shows photographs of different flow
regimes and a pressure-drop oscillation cycle of the heater inlet
pressure, with superimposed density-wave oscillations. As seen in
Figure 21, bubbly flow is observed around the flat region of the
cycle (points a and b). As time passes and the pressure increases
bubbles become progressively larger, forming slugs (c) and passing
through the transparent section at regular intervals. With the
passing of time, the slugs get longer and longer and the flow
becomes a chaotic mixture of liquid and vapor (d). Near the
maximum of the cycle the flow regime changes from churn to annular (e). During the annular flow there is a momentary stopping
and even reversing of the flow, which is clearly observed in the
films. After the reversal, the liquid surges and mist fills the

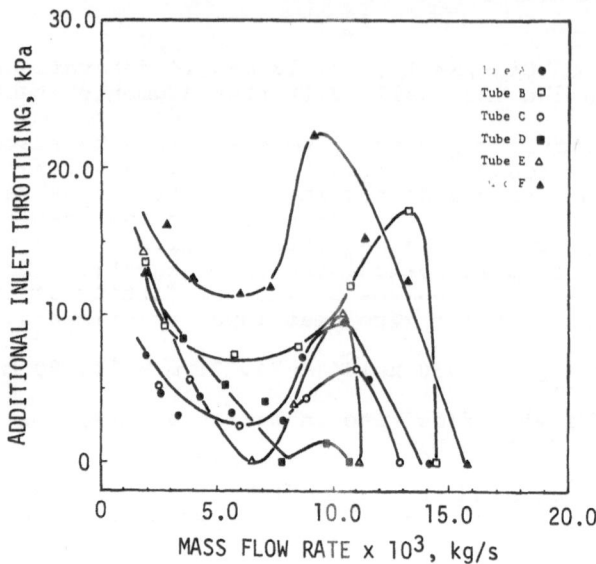

Fig. 11. Inlet Throttling Necessary to Stabilize the System
(single channel, upflow boiling; fluid: Freon-11,
heat input: 300 W)

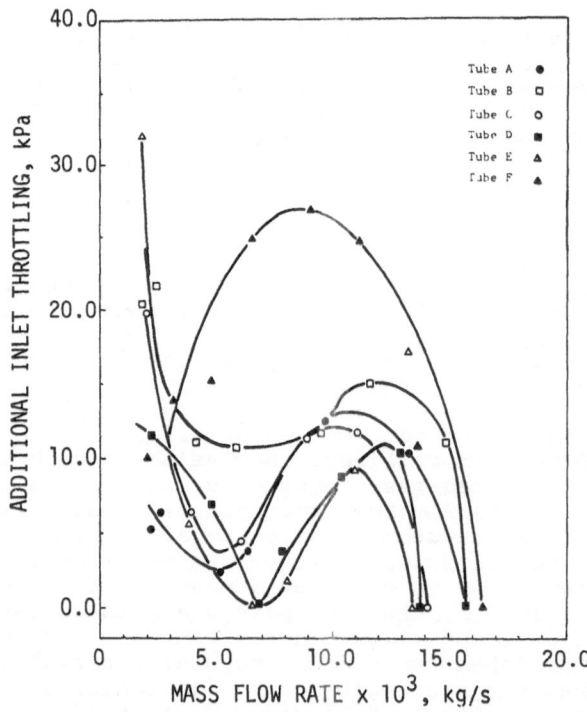

Fig. 12. Inlet Throttling Necessary to Stabilize the System
(single channel, upflow boiling; fluid: Freon-11,
heat input: 350 W)

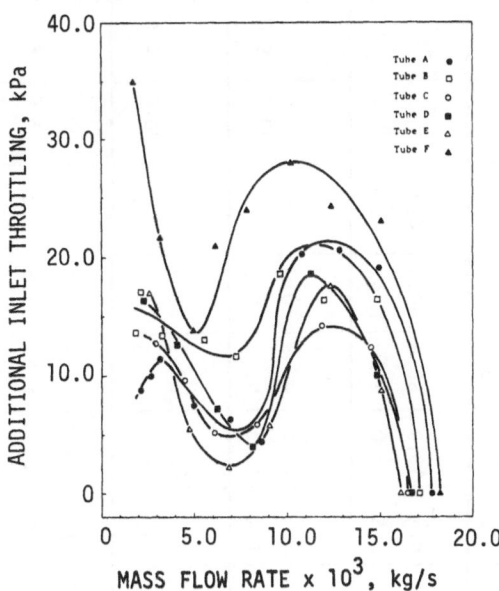

Fig. 13.   Inlet Throttling Necessary to Stabilize the System
(single channel, upflow boiling; fluid: Freon-11,
heat input: 400 W)

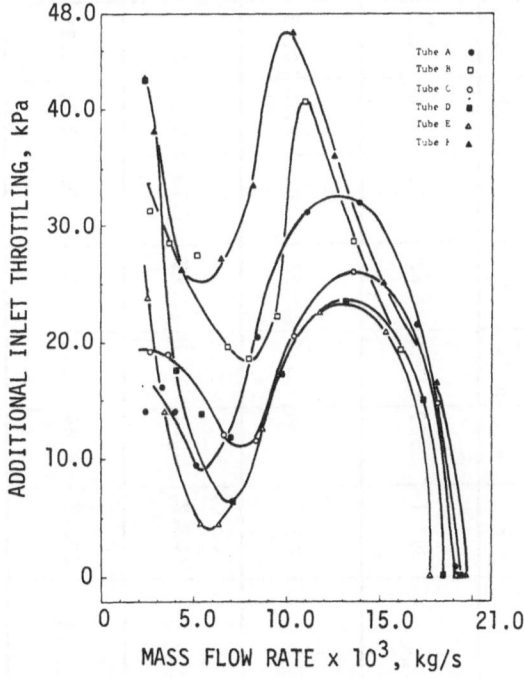

Fig. 14.   Inlet Throttling Necessary to Stabilize the System
(single channel, upflow boiling; fluid: Freon-11,
heat input: 500 W)

Table 2. Sustained Oscillation Data (Density-Wave type oscillations superimposed on Pressure-Drop type oscillations, single boiling channel, upflow, test fluid: Freon-11, heat input: 400.0 W) P.D.: Pressure-Drop; D.W.: Density-Wave; N.A.: Not Available

Tube & Experiment Number	Actual Heat Input W	Exit Quality %	Mass Flow Rate gr/s	Pressures, bars				Temperatures, Nominal and Amplitudes, °C							Periods, sec	
				Heater Inlet P_i	System Exit P_e	$\Delta P_i \times 10$		Heater Inlet	Heater Wall					Heater Exit	P.D.	D.W.
						P.D.	D.W.	T_1	T_2	T_3	T_4	T_5	T_6	T_7		
A AV-400-3/6	397.1	2.89	12.9	2.62	1.03	4.40	6.60	22.6	55.6	N.A.	58.9	60.6	60.6	48.9	30.0	30.0
								N.A.	N.A.	N.A.	5.0	N.A.	N.A.	5.0		
B CV-400-1/7	399.9	4.44	12.0	2.58	1.01	2.28	3.03	23.1	53.9	56.7	57.8	58.9	60.0	49.4	30.0	9.0
								N.A.	N.A.	N.A.	N.A.	N.A.	5.0	5.8		
C BV-400-1/6	403.2	4.21	11.8	2.73	1.19	3.10	2.07	23.8	56.7	61.1	62.2	63.9	63.9	52.8	36.0	4.0
								N.A.	13.0	10.0	7.5	5.5	5.5	5.0		
D EV-400-1/6	398.8	4.84	11.3	2.67	1.10	2.76	2.28	23.1	53.9	57.2	58.3	61.1	62.8	51.1	52.0	8.0
								N.A.	N.A.	N.A.	N.A.	N.A.	5.0	7.6		
E DV-400-1/5	398.0	1.92	12.2	2.41	1.14	4.21	2.62	23.8	47.8	55.0	58.3	62.2	61.7	51.1	64.0	8-9
								1.0	10.0	11.0	N.A.	N.A.	5.0	6.2		
F FV-400-1/6	401.9	18.9	12.3	2.65	1.03	1.79	1.86	24.3	56.7	57.2	56.7	56.7	55.6	50.6	14.0	Irreg.
								3.0	3.0	3.0	2.6	2.6	2.6	4.4		

Table 3. Sustained Oscillation Data (Density-Wave type oscillations, single boiling channel upflow, test fluid: Freon-11, heat input: 400.0 W) P.D.: Pressure-Drop; D.W.: Density-Wave; N.A.: Not Available

Tube & Experiment Number	Actual Heat Input W	Exit Quality %	Mass Flow Rate gr/s	Pressures, bars Heater Inlet P_i	System Exit P_e	$\Delta P_i \times 10$ P.D.	$\Delta P_i \times 10$ D.W.	Temperatures, Nominal and Amplitudes, °C Heater Inlet T_1	Heater Wall T_2	T_3	T_4	T_5	T_6	Heater Exit T_7	Periods, sec P.D.	Periods, sec D.W.
A AV-400-3/13	405.6	55.1	3.20	2.86	1.03	2.0	8.8	23.8 / 2.5	61.7 / N.A.	N.A. / N.A.	66.7 / N.A.	68.9 / N.A.	67.8 / N.A.	56.1 / 2.5	Irregular	Irregular
B CV-400-1/11	393.4	51.8	3.29	2.87	1.03	–	3.79	24.6 / N.A.	60.0 / N.A.	63.3 / N.A.	63.3 / N.A.	62.8 / N.A.	65.0 / 5.0	55.0 / 5.0	–	4.0
C BV-400-1/9	400.3	62.1	2.92	2.94	1.05	–	3.28	25.0 / 2.5	65.6 / 1.0	68.3 / 1.0	67.2 / 1.0	67.2 / 1.0	62.8 / 1.0	56.1 / 2.5	–	2.5
D EV-400-1/9	401.6	84.6	2.28	2.76	1.14	–	4.41	25.0 / N.A.	60.0 / N.A.	62.8 / N.A.	63.3 / N.A.	63.9 / N.A.	66.1 / 1.5	53.9 / 3.5	–	6.1
E DV-400-1/9	403.1	73.9	2.55	2.80	1.06	–	2.62	24.6 / N.A.	62.2 / N.A.	65.0 / N.A.	65.0 / N.A.	65.6 / N.A.	64.4 / N.A.	55.0 / 2.5	–	3.3
F FV-400-1/11	401.9	56.7	3.13	2.91	1.01	1.83	2.55	24.8 / 3.8	61.1 / N.A.	60.6 / N.A.	61.1 / N.A.	60.0 / N.A.	60.6 / N.A.	56.1 / 3.8	14.5	4.5

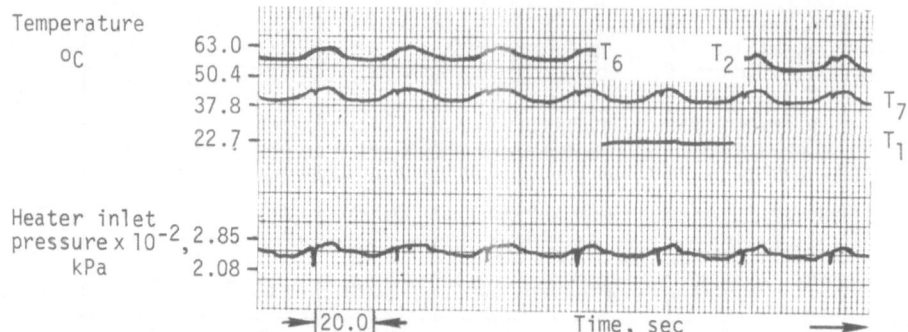

Fig. 15. Superimposed Oscillations (Tube A; heat input: 400 W; inlet temperature: 22.6°C; mass flow rate: 12.9 gr/sec)

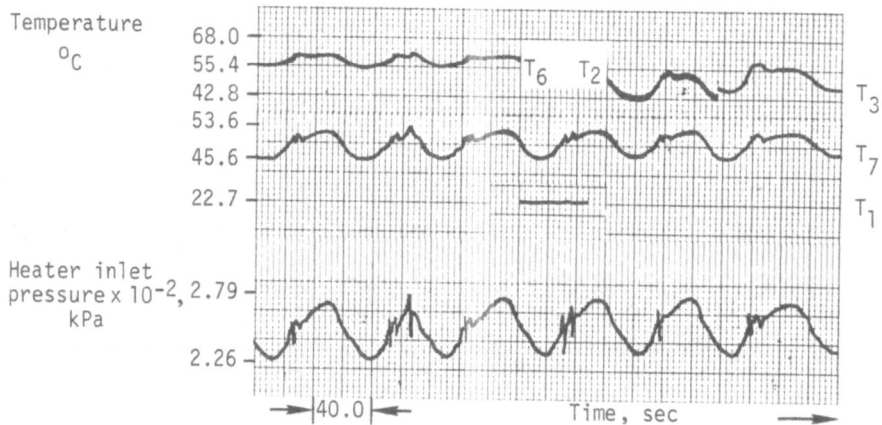

Fig. 16. Superimposed Oscillations (Tube E; heat input: 400 W; inlet temperature: 23.8°C; mass flow rate: 12.2 gr/sec)

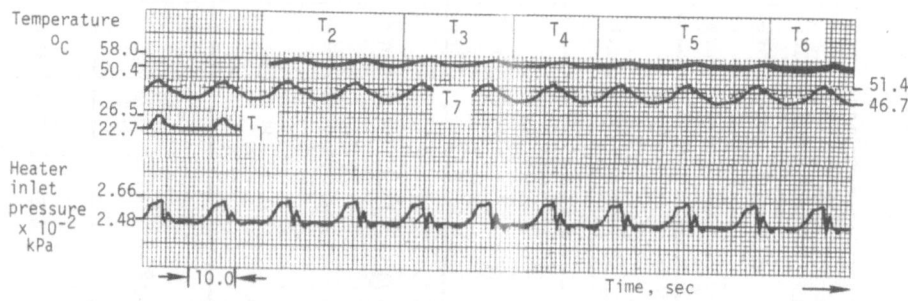

Fig. 17. Superimposed Oscillations (Tube F; heat input: 400 W; inlet temperature: 24.3°C; mass flow rate: 12.3 gr/sec)

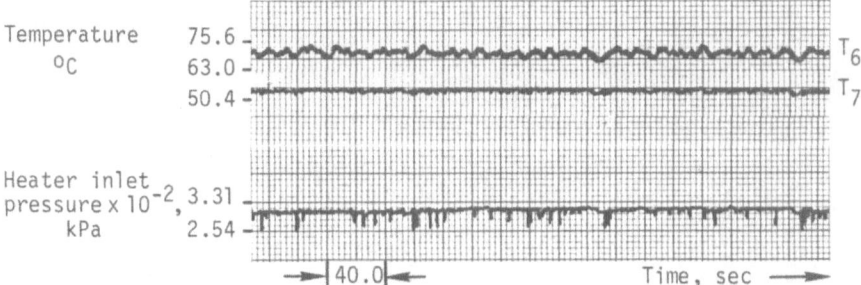

Fig. 18. Transition from Pressure-Drop Oscillations to Density-Wave Oscillations (Tube A; heat input: 400 W; inlet temperature: 23.8°C; mass flow rate: 3.20 gr/sec)

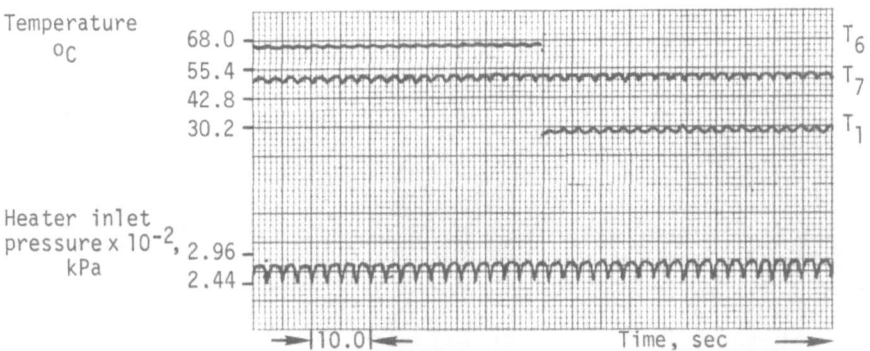

Fig. 19. Density-Wave Oscillations (Tube C; heat input: 400 W; inlet temperature: 25.0°C; mass flow rate: 2.92 gr/sec)

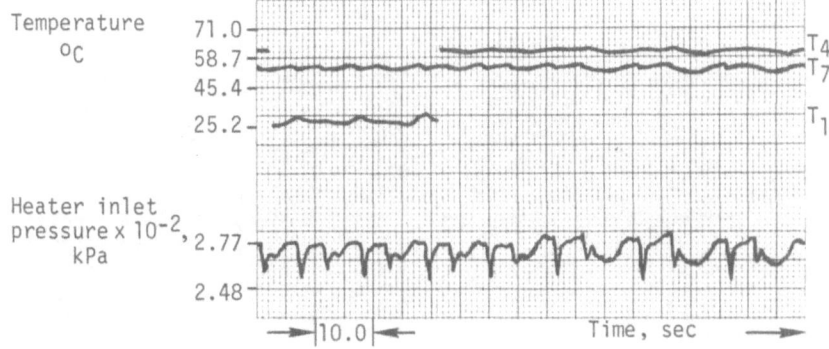

Fig. 20. Transition from Pressure-Drop Oscillations to Density-Wave Oscillations (Tube F; heat input: 400 W; inlet temperature: 24.8°C; mass flow rate: 3.13 gr/sec)

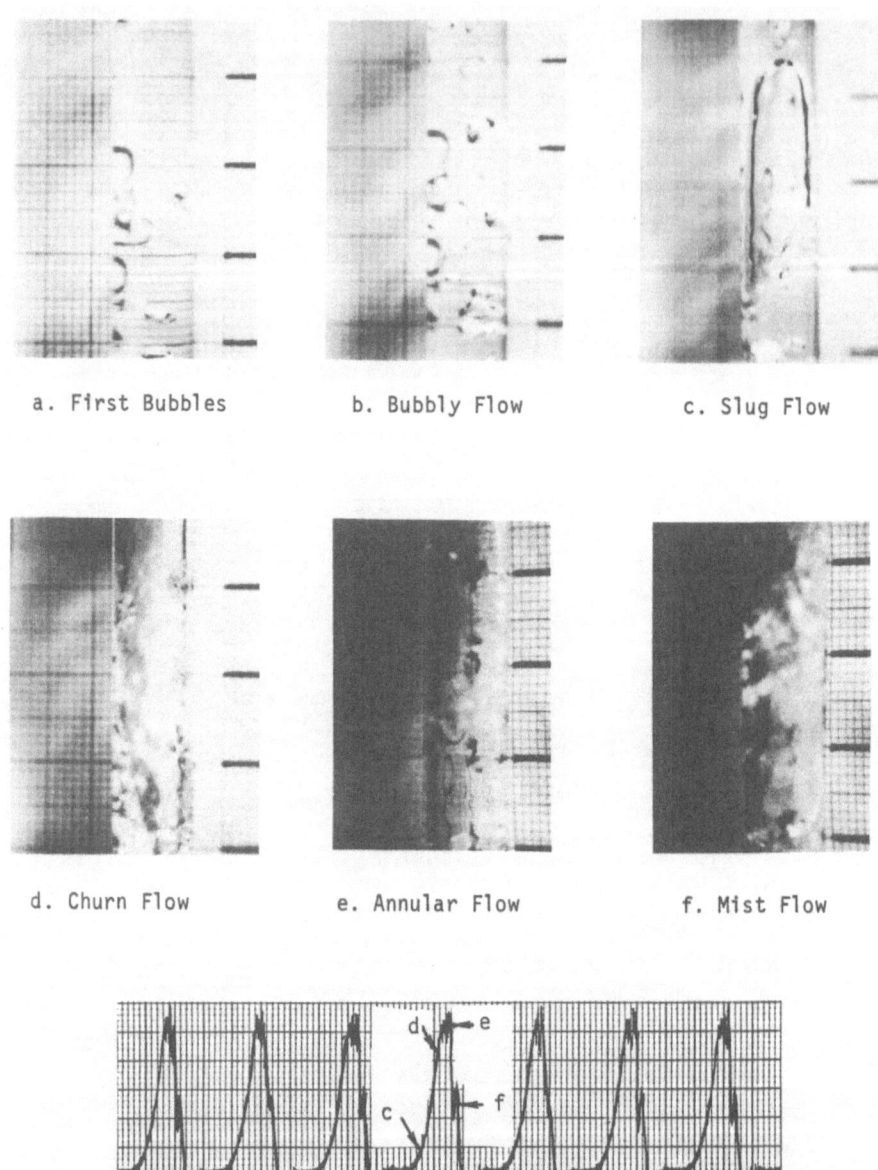

Fig. 21. Observed Flow Regimes and Their Locations on a Heater Inlet Pressure Oscillation Cycle

transparent tube (f). This passes very quickly. After a few slugs the flow becomes almost liquid for a brief period of time, around the minimum of the cycle, and then the sequence repeats itself.

4. DISCUSSION AND CONCLUSIONS

Different heater tubes with various augmented heat transfer surfaces were tested and the results of experiments have been presented in graphical forms and in two tables.

From the Figures 3 through 8, it can be seen that for all of the tubes with different internal augmentation, pressure-drop oscillations occur first as the mass flow rate is decreased. These oscillations are sinusoidal at the onset, but change shape with the decrease of mass flow rate, and a gradual transform to density-wave oscillations takes place.

An examination of Figures 9 through 14 and the Table 1 shows that the heater tubes with internal springs, namely tubes C, D, and E, have narrower unstable regions in this order and thus, more stable than the others, which are bare, threaded and the coated ones. Order of the curves in Figures 9 through 11 suggests that for the tubes with internal springs, the stability increases with decreasing effective diameter, but there is not a consistent pattern for the other tubes.

From the investigation of Figures 9, 10 and 11 through 14, one can deduce that among the tested tubes, tube with Linde High Flux Coating is the most unstable one. For this tube, for every heat input oscillation boundaries were located at higher mass flow rates than the other tubes, but on the other hand the amplitudes of the pressure-drop and the density-wave oscillations were smaller than the other tubes tested.

Figures 15 through 20, which are recordings taken during the experiments, show that the nominal values and the amplitudes of the heater wall temperatures change from tube to tube. There is not a regular pattern, but the nominal values and the amplitudes are smaller for the coated tube, and the wall temperature distribution is uniform for this tube.

Heat transfer coefficients change during the oscillations and among the tubes. Comparing the bare tube with the other tubes it has been found out [8] that the time average local heat transfer coefficients during oscillations increased 1 to 8% for different spring configurations, about 20% for the internally threaded tube, and up to 94% for the coated tube.

Another interesting observation is the change in oscillation periods: For the density-wave oscillations every tube had similar periods, but for the superimposed oscillations there was a great difference among the tubes, although the amount of air in the surge tank kept constant for each tube. Period of the oscillations in Tube E was more than four times the period of the oscillations in Tube F.

These interesting behaviors may be due to the wave propagation lags and feedback effects in the system as a result of the combined interaction between mass flow, the vapor generation pressure drop and the variation of compressible volume within and upstream of the two-phase flow system. Further research is being conducted in these lines and this is one of the series of papers planned on the subject of two-phase flow instabilities with internally augmented surfaces.

ACKNOWLEDGEMENTS

The authors gratefully acknowledge the financial support of the National Science Foundation and the NATO Scientific Affairs Division.

REFERENCES

1. Veziroğlu, T. N. and Kakaç, S. 1980. Two-phase flow instabilities and effect of inlet subcooling. University of Miami, Florida, Clean Energy Research Institute, final report, NSF project ENG 75-16618.

2. Veziroğlu, T. N. and Kakaç, S. 1981. Two-phase flow instabilities. University of Miami, Florida, Clean Energy Research Institute, annual report, NSF project CME 79-20018.

3. Akyüzlü, K., Veziroğlu, T. N., Kakaç, S. and Doğan, T. 1980. Finite difference analysis of two-phase flow pressure-drop and density-wave oscillations. Wärme-und Stoffübertragung. Vol. 14, pp. 253-267.

4. Doğan, T., Kakaç, S. and Veziroğlu, T. N. 1982. Lumped parameter analysis of two-phase flow instabilities. 7th International Heat Transfer Conference, September 6-10, 1982, München, F.R. Germany.

5. Cumo, M., Palazzi, G. and Rinaldi, L. 1981. An experimental study on two-phase flow instability in parallel channels with different heat flux profile. Comitato Nazionale Energia Nucleare, report ref. CNEN-RT/ING(81)1.

6. Aritomi, M., Aoki, S. and Inoue, A. 1979. Instabilities in parallel channel of forced-convection boiling upflow system, (III)--system with different flow conditions between channels. Japan. Journal of Nuclear Science and Technology, Vol. 16, No. 5, pp. 343-355.

7. Bergles, A. E. 1981. Principles of heat transfer augmenta-tion--II, two-phase heat transfer. Heat Exchangers-Thermal Hydraulics Fundamentals and Design. Eds. Kakac, S., Bergles, A. E. and Mayinger, F. McGraw-Hill, pp. 857-914.

8. Lin, Z. H., Veziroğlu, T. N., Kakaç, S., Gürgenci, H. and Menteş, A. 1982. Heat transfer in oscillating two-phase flows and effect of heater surface conditions. 7th International Heat Transfer Conference, September 6-10, 1982, München, F.R. Germany.

Pressure-Drop and Density-Wave Instability Thresholds in Boiling Channels

H. GÜRGENCI, T.N. VEZIROĞLU, and S. KAKAÇ
Department of Mechanical Engineering
University of Miami
Coral Gables, Florida 33124, USA

ABSTRACT

In this study, a criterion for linearized stability with respect to both the pressure-drop and the density-wave oscillations is developed for a single-channel upflow boiling system operating between constant pressures with upstream compressibility introduced through a surge tank. Two different two-phase flow models, namely a constant-property homogeneous flow model and a variable-property drift-flux model, have been employed. The conservation equations for both models and the equations of surge tank dynamics are first linearized for small perturbations and the stability of the resulting set of equations for each model are examined by use of Nyquist plots. As a measure of the relative instability of the system, the amounts of the inlet throttling necessary to stabilize the system at particular operating points have been calculated. The results are compared with experimental findings. Comparisons show that the drift-flux formulation offers a simple and reliable way of determining the instability thresholds. The homogeneous flow model is good to acquire a physical understanding of the problem.

INTRODUCTION

The analysis of the instabilities, the prediction of the threshold conditions for instabilities and their limit cycles, has been one of the main concerns of the two-phase flow studies. In general, two-phase flow instabilities may occur whenever heat is added to a flowing fluid and produces a change in volume by boiling. The existence of such instabilities in two-phase mixtures has been known for some time and the appearance of these instabilities at either subcritical or supercritical pressures is undesirable. Such instabilities may not only degrade the performance of boiling flow systems, but also can result in premature burnout and control problems that can be destructive. Therefore, it becomes important to have simple stability criteria that can be used for design purposes. Flow stability becomes of particular importance in water-cooled and water-moderated nuclear reactors and steam generators. The safe operating power of a boiling water reactor can be determined by the instability threshold values of such system parameters as flow rate, pressure, wall temperatures, exit mixture quality, etc. The designer's job is to predict the threshold of flow instability so that he can design around it or compensate for it.

Two of the most common two-phase flow instabilities have been identified by Stenning and Veziroglu [1] as the density-wave and the pressure-drop oscillations. Density-wave oscillations are associated with kinematic wave propagation phenomena. Fluid waves of alternatively higher and lower density mixtures travel across the system during the oscillations. It takes one high and one low wave to make one cycle so it can be concluded that the periods of the density-wave oscillations are roughly equal to twice the transit time of a fluid particle through the system.

Stenning and Veziroglu [1] found that the pressure-drop oscillations occurred only when the pressure drop across the test section decreased with increasing flow rate and their periods were governed by the volume and the compressibility of the vapor in the system, including the compressibility introduced at the upstream of the heater through a surge tank partially filled with air. These oscillations are compound dynamic instabilities generated by the operation of static excursive effects in cooperation with the effects of the compressible volume inertia and system delayed feedbacks. The periods are usually much larger than the residence time of a fluid particle in the heater. However, as the amount of compressibility is reduced, the oscillation periods decrease proportionally.

A number of studies have been conducted in the past dealing with the pressure-drop and the density-wave oscillations. Zuber [2], Boure [3], Ishii [4], Nakanishi [5], Unal [6] investigated the density-wave type oscillations in boiling channels. Stenning and Veziroglu [1], Fukuda and Kobori [7], Ozawa et al [8], Veziroglu and Kakac [9] studied various aspects of pressure-drop and density-wave oscillations.

In all of the past studies on two-phase flow instabilities, the pressure-drop and the density-wave oscillations have been treated separately with different sets of assumptions for each oscillation type. However, it is desirable to have a unified treatment of these two main types of the dynamical two-phase flow instabilities which, under certain conditions, may appear together as superimposed oscillations. In this study, a criterion for linearized stability with respect to both the pressure-drop and the density-wave oscillations will be developed for a single-channel upflow boiling system operating between constant pressures with upstream compressibility introduced through a surge tank.

NONLINEAR REPRESENTATION OF THE SYSTEM

A schematic view of the upward-flow boiling system is represented in Fig. 1. The system may be divided into three parts: inlet tubing between the supply and the surge tanks, the surge tank, and the test section from the surge tank up to the system exit. A detailed description of the loop and the experimental procedure is given in [10].

In unsteady flows, the fluid velocity through the inlet tubing, u_i, may differ from the heater inlet velocity, u_0, due to the accumulation of the working fluid in the surge tank. The pressure drop from the supply tank up to the surge tank may be separated into one frictional and one inertial component (no gravitational pressure drop since there is no elevation difference between the supply tank outlet and the surge tank inlet). The frictional component is assumed to be concentrated at the inlet flow control valve. The pressure drop across this section can be written as

$$p_i - p_t = K_i \rho_0 u_i^n + \rho_0 L_i \frac{du_i}{dt} \qquad (1)$$

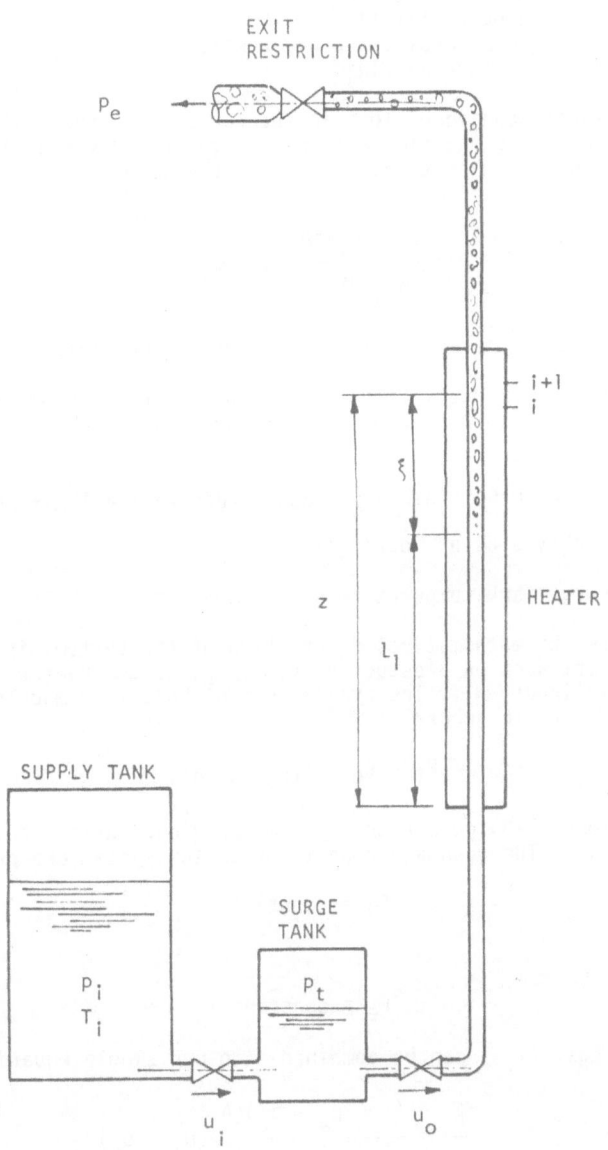

Fig.1. Schematic view of the upward-flow boiling system

where n is taken as 2 and K_i depends on the valve opening.

The surge tank provides for the upstream compressibility. Acting as a capacitance, it is an important dynamic component of the loop, especially during the pressure-drop oscillations. The surge tank is partially filled with air so that the compressible volume consists of a mixture of air and the saturated vapor of the working fluid.

From continuity considerations in the surge tank, the following equation can be written relating the time rate of change of the surge tank pressure to the surge tank inlet and exit velocities and the amount of the surge tank compressibility.

$$\frac{dp_t}{dt} = \frac{(p_t - p_v)^2 A}{p_{tas} V_{cs}} (u_i - u_o) \tag{2}$$

The assumptions inherent in this equation may be stated as follows:

(i)   Thermodynamic equilibrium conditions are assumed between the vapor and the liquid phases of the working fluid (Freon-11 in the present case).

(ii) Air and working fluid vapor constitute an ideal mixture.

(iii) Air behaves as an ideal gas.

(iv) The surge tank temperature stays constant.

To complete the mathematical description of the system, it is necessary to relate the test section pressure drop, P_t-P_e, to the heater inlet flow rate, u_o, and the heat input, q'. The available two-phase flow models may be used for this purpose and in general it may be written as

$$P_t - P_e = \psi(u_o, u_o, q') \tag{3}$$

The equations (1-3) govern the system behaviour during steady-state and oscillatory flows. The boundary conditions on the system are prescribed as

$$P_i = \text{constant} \tag{4}$$

and

$$P_e = \text{constant} \tag{5}$$

Note that Eqs. (1-3) can be combined into one single equation as

$$\frac{d\psi}{dt} = \frac{(\psi + P_e - p_v)^2 A}{p_{tas} V_{cs}} (u_i - u_o) \tag{6}$$

The inertial pressure drop in the inlet tubing can be ignored when compared with the frictional drop so that

$$u_i = \left(\frac{P_i - P_e - \psi}{K_i} \right)^{1/2} \tag{7}$$

and

$$\frac{d\psi}{dt} = \frac{(\psi + P_e - P_v)^2 A}{P_{tas} V_{cs}} \left[\sqrt{\frac{P_i - P_e - \psi}{K_i}} - u_o \right] \tag{8}$$

As it can be seen, Eq.(8) is a highly nonlinear ordinary differential equation with u_o being the dependent variable.

LINEARIZED EQUATIONS

The equations of the preceding section can be linearized by assuming the time-varying flow variables to be of the form $u_i = u_{is} + \tilde{u}_i$, $u_o = u_{os} + \tilde{u}_o$ $P_t = P_{ts} + \tilde{P}_t$, etc., where $\tilde{u}_i$, $\tilde{u}_o$, $\tilde{P}_t$ are the small perturbations on the steady-state values of these quantities. By substituting the perturbed variables, cancelling the steady-state terms and ignoring the products of the perturbations, one obtains the following set of the linearized state equations:

$$\tilde{P}_t = g(t) \tilde{u}_o \tag{9}$$

$$\frac{d\tilde{P}_t}{dt} = K_t (\tilde{u}_i - \tilde{u}_o) \tag{10}$$

$$- \tilde{P}_t = 2K_i \rho_o u_{is} \tilde{u}_i + \rho_o L_i \frac{d\tilde{u}_i}{dt} \tag{11}$$

where

$$K_t = P_{tas} A / V_{cs} \tag{12}$$

Taking the Laplace transforms, substituting $u_{is} = u_{os}$, and rearranging the terms, one obtains the following equations in the complex domain which describe the linearized relations between the main system variables, u_i, u_o and P_t.

$$G(s)U_o(s) - P_t(s) = 0 \tag{13}$$

$$K_t U_i(s) - K_t U_o(s) - sP_t(s) = 0 \tag{14}$$

$$(2K_i \rho_o u_{os} + \rho_o L_i s) U_i(s) + P_t(s) = 0 \tag{15}$$

where $U_i(s)$, $U_o(s)$ and $P_t(s)$ are the Laplace transforms of $\tilde{u}_i$, $\tilde{u}_o$ and $\tilde{P}_t$, respectively. $G(s)$ is a transfer function for the test section defined as

$$G(s) = \frac{P_t(s)}{U_o(s)} \tag{16}$$

At the oscillation thresholds where the amplitudes are very small, the linearized model gives a good description of the physical reality. Thus it is possible to obtain the instability thresholds by investigating the instability of the linearized state equations. The stability of these linearized equations is necessary and sufficient for the stability of the system in this context. On the other hand, it is not possible to obtain the limit cycles of the oscillations by linearized models because when the operating point is deep in the unstable region, the assumption of small perturbations is no longer valid and nonlinear restraints unaccounted for in the linearized model govern the system behavior.

A linearized stability criterion can be obtained from the coefficient matrix of the Eqs. (13) through (15). The characteristic equation for the system is found by making the determinant of this matrix vanish, i.e.,

$$\begin{vmatrix} 0 & G(s) & -1 \\ K_t & -K_t & -s \\ (2K_i\rho_o u_{os}+\rho_o L_{is}) & 0 & 1 \end{vmatrix} = 0 \tag{17}$$

or

$$K_t[G(s) + 2K_i\rho_o u_{os} + \rho_o L_{is}s] + sG(s)(2K_i\rho_o u_{os}+\rho_o L_{i}s) = 0 \tag{18}$$

The system is stable when all the roots of this equation have negative real parts. When at least one of the roots is purely imaginary, then the system is at the instability threshold. When one or more of the roots of the characteristic equation lie in the right-hand side of the complex plane, then the old system is in the unstable region; the amplitudes of the oscillations grow in time until they are checked by nonlinear restraining effects.

This characteristic equation applies to both the pressure-drop type oscillations and the higher-frequency oscillations such as of the density-wave type. However, it is too complicated to use in parametrical studies as a stability test, even for the most simple two-phase flow model. Fortunately, it is possible to simplify the problem further. The pressure drop-flow rate characteristics of the section between the supply and the surge tanks are very stiff in comparison with the other parts of the system. This makes it possible to assume the mass flow rate at the surge tank entrance to stay constant during the oscillations. This was indeed the case observed during the experiments when the inlet mass flow rate as measured by the rotameter would not oscillate beyond ±5% of its nominal value even during most violent oscillations.

With this assumption, the number of governing equations are reduced by one, so that one has

$$P_t - P_e = \psi(u_o,\dot{u}_o,q') \tag{19}$$

for the pressure drop from the surge tank up to the exit, and

$$\frac{dp_t}{dt} = \frac{(p_t - p_v)^2 A}{p_{tas} V_{cs}} (u_i - u_o)$$ (20)

for the surge tank, where u_i is constant and equal to the steady-state value of the inlet velocity, u_{os}. Then as before the linearized state equations are written as

$$\tilde{p}_t = g(t) \tilde{u}_o$$ (21)

and

$$\frac{d\tilde{p}_t}{dt} = -K_t \tilde{u}_o$$ (22)

Taking the Laplace transforms, one obtains

$$P_t(s) = G(s) U_o(s)$$ (23)

and

$$sP_t(s) = -K_t U_o(s)$$ (24)

The above equations define an ordinary feedback-control system, the characteristic equation of which can be written as

$$1 + \frac{sG(s)}{K_t} = 0$$ (25)

where K_t is as given by Eq. (12) and $G_{(s)}$ by Eq. (16). The capital letters stand for the Laplace transforms; thus $P_{t(s)}$ and $U_{o(s)}$ are the Laplace transforms of the surge tank pressure perturbation, $\tilde{p}_t$, and the inlet velocity perturbation $\tilde{u}_o$, respectively.

As explained before, the overall system stability can be checked by examining the roots of the characteristic equation, Eq. (25). Depending on the two-phase flow model used the transfer function, $G_{(s)}$, relating the system pressure drop to the inlet velocity may or may not be expressed in an explicit analytical form. In either case, the topology of the roots of Eq. (25) can be obtained easily by using well-known frequency-domain techniques such as Nyquist plots. In this study, first, an explicit expression for $G_{(s)}$ will be used as obtained by a homogenous-flow model with concentrated pressure-drops; and later, a numerical method will be introduced to compute the transfer function, $G_{(s)}$ for more general cases using a drift-flux formulation.

HOMOGENEOUS FLOW MODEL

The homogeneous-flow model dates back as far as the beginnings of two-phase flow research. Its applicability mostly depends on the flow regines experienced; for those flow regimes such as bubbly, churn or mist flow where the two phases are expected to be well-mixed, the model is most applicable. For others, such as annular or slug flow where there is sufficient reason to believe that the phases flow as if in two separate streams, the model is not as applicable.

The primary underlying assumptions of the model are stated as follows:

(i)   The flow is one-dimensional.

(ii)   The phases are in thermodynamic equilibrium

(iii)   The pressure and the other properties are uniform across each section taken along the channel.

(iv)   The heat input is constant and uniform along the heater length.

(v)   The energy equation is limited to a thermal balance.

(vi)   The flow is not allowed to reverse.

(vii)   The wall heat capacity is neglected.

(viii) The fluid properties are evaluated at a representative system pressure and they are assumed to be independent of pressure for further considerations.

The three conservation equations for the homogeneous flow model can be written as follows [12]

$$\frac{\partial \rho}{\partial t} + \frac{\partial (\rho u)}{\partial z} = 0 \tag{26}$$

$$\rho \frac{\partial u}{\partial t} = -\rho u \frac{\partial u}{\partial z} - \frac{\partial p}{\partial z} - 2\frac{f}{d}\rho u^2 - \rho g \tag{27}$$

$$\frac{\partial h}{\partial t} = -u \frac{\partial h}{\partial z} + \frac{q'}{\rho A} \tag{28}$$

These add up to three equations with four dependent variables, ρ, u, p and h. The additional equation is the equation of the state. In the actual state of affairs, the state equation is of the form

$$\rho = \rho(p,h) \tag{29}$$

However, by neglecting the pressure dependence in this equation, the homogeneous flow model can be simplified to a large extent. In this case, the state equation is written as

$$\rho = \rho(h) \tag{30}$$

Under conditions of thermodynamic equilibrium and constant-property flow, the specific form of Eq. (30) becomes

$$\rho = \rho_f \qquad\qquad\qquad , \text{ when } h \leq h_f$$

(31)

$$1/\rho = 1/\rho_f + v_{fg}(h - h_f)/h_{fg} \qquad , \text{ when } h_f < h \leq h_g$$

The continuity equation can be written as

$$\frac{\partial \rho}{\partial h} \left[\frac{\partial h}{\partial t} + u \frac{\partial h}{\partial z} \right] + \rho \frac{\partial u}{\partial z} = 0$$

(32)

Substituting Eq. (28) into Eq. (32), one gets the following equation

$$\frac{\partial u}{\partial z} = \frac{q'}{A} \frac{d(1/\rho)}{dh}$$

(33)

A time constant θ is defined such that

$$\frac{\partial u}{\partial z} = \frac{1}{\theta}$$

(34)

where

$$\theta \rightarrow \infty \qquad\qquad \text{when } h \leq h_f$$

(35)

$$\theta = Ah_{fg}/v_{fg}q' \qquad \text{when } h_f < h \leq h_g$$

By combining Eqs. (26) and (34), one obtains

$$\frac{1}{\rho} \left[\frac{\partial \rho}{\partial t} + u \frac{\partial \rho}{\partial z} \right] = -\frac{1}{\theta}$$

(36)

Then the field equations can be written as

$$\frac{\partial u}{\partial z} = \frac{1}{\theta}$$

(37)

$$\frac{1}{\rho} \left[\frac{\partial \rho}{\partial t} + u \frac{\partial \rho}{\partial z} \right] = -\frac{1}{\theta}$$

(38)

These two equations define a system with $\rho(z,t)$ and $u(z,t)$ as the dependent variables. The boundary condition on this system is provided by the momentum equation, Eq. (27).

The equations (37),(38) and (27) are made dimensionless by defining new variables as follows:

Length : $\bar{z} = z/L_h$ (39)

Time : $\bar{t} = t/\theta_s$ (40)

Density: $\bar{\rho} = \rho/\rho_0$ (41)

The other terms are non-dimensionalized accordingly, to give

$$\frac{\partial \bar{u}}{\partial \bar{z}} = \frac{1}{\bar{\theta}} \tag{42}$$

$$\frac{1}{\bar{\rho}} \left[\frac{\partial \bar{\rho}}{\partial \bar{t}} + \bar{u} \frac{\partial \bar{\rho}}{\partial \bar{z}} \right] = - \frac{1}{\bar{\theta}} \tag{43}$$

and as the boundary condition

$$\frac{\partial \bar{p}}{\partial \bar{z}} = -\bar{\rho} \frac{\partial \bar{u}}{\partial \bar{t}} - \bar{\rho}\bar{u} \frac{\partial \bar{u}}{\partial \bar{z}} - 2 \frac{f}{\bar{d}} \bar{\rho}\bar{u}^2 - \bar{\rho}\bar{g} \tag{44}$$

Since the heat capacity of the wall is neglected, the heat input does not vary with time so that $\bar{\theta} = \theta/\theta_s = 1$ in the above equations. The two-phase friction factor, f, is assumed to be constant along the channel and equal to the liquid-phase friction factor in the subcooled region. In order to have workable equations, the distributed pressure losses are ignored. In that case, the pressure drop across the system from the surge tank up to the exit can be written as

$$\bar{P}_t - \bar{P}_e = \bar{\psi}(\bar{t}) = \Delta\bar{p}_o(\bar{t}) + \Delta\bar{p}_e(\bar{t}) \tag{45}$$

where $\Delta\bar{p}_o(\bar{t})$ is the dimensionless pressure drop across the fluidic resistance elements in the inlet tubing and is expressed as

$$\Delta\bar{p}_o(\bar{t}) = K_o \bar{u}_o^2(\bar{t}) \tag{46}$$

The second term at the right-hand side of Eq. (45) is the pressure drop across the exit restriction which is expressed as follows

$$\Delta\bar{p}_e(t) = F_M(x_e) K_e \bar{G}_e^2 \tag{47}$$

where F_M is a two-phase friction multiplier which may be expressed as

$$F_M = c_1 x_e^2 + c_2 x_e + 1 \tag{48}$$

where the coefficients have been computed from the experimental data as $C_1 = 121$ and $C_2 = 30$.

The time-dependent velocity and density distributions are obtained from Eqs. (42) and (43). After linearizing Eq. (45), substituting the velocity and the density terms, and taking the Laplace transforms, the open-loop transfer function for the test section is obtained as

$$G(s) = 2 K_o \bar{u}_{os}$$

$$+ K_e \bar{u}_{os}^2 e^{-s\bar{t}_3} \left\{ \left[\frac{2F_{Ms}}{\bar{u}_{os}} - \frac{2F_{Ms}}{\bar{u}_{hes}} - \frac{2c_1 x_{hes} + c_2}{\bar{\rho}_{hes}\bar{u}_{os}\bar{v}_{fg}} \right] \frac{1 - e^{-s\bar{t}_1}}{s} \right.$$

$$+ \left[\frac{2F_{Ms}}{\bar{u}_{os}} - \frac{2c_1 x_{hes} + c_2}{\bar{\rho}_{hes} u_{os} \bar{v}_{fg}} \right] \frac{e^{-s\bar{t}_1}}{s-1} [1 - e^{-(s-1)\bar{t}_2}] + \frac{2F_{Ms}}{\bar{u}_{hes}} \Bigg\} \tag{49}$$

The parameter, $G_{(s)}$, so defined is the transfer function for the test section when the distributed pressure losses are ignored. Substitution of Eq. (49) into Eq. (25) leads to the general system characteristic equation which allows one to investigate the overall system atability with respect to both the pressure-drop and the density-wave oscillations.

It is observed from Eq. (49) that the test section transfer function, $G_{(s)}$, involves three time lags, t_1, t_2 and t_3. The first one is associated with the subcooled region, the second one with the two-phase region in the heater and the third one with the exit tubing. There is no time lag associated with the superheated region since the fluid is assumed to leave the heater as saturated mixture. The time lags are represented by exponential terms after Laplace transformation. Exponential functions have infinite number of roots which preclude the use of the conventional algebraic stability criteria, such as Routh-Hurwitz. In this study, a Nyquist-plot technique will be used to examine the topology of the roots of the characteristic equation.

The equation (49) can be rewritten as

$$G_{(s)} = 2K_o \bar{u}_{os} + G_{h(s)} \tag{50}$$

where $G_{h(s)}$ is the transfer function for the heater and the post-heater sections and the first term is associated with the inlet tubing between the surge tank and the heater.

Substitution of Eq. (50) into Eq. (25) gives

$$1 + \frac{s}{K_t} [2K_o \bar{u}_{os} + G_h(s)] = 0 \tag{51}$$

This equation can be rewritten as

$$K_o + \left[\frac{G_h(s)}{2\bar{u}_{os}} + \frac{\bar{K}_t}{2s\bar{u}_{os}} \right] = 0 \tag{52}$$

This form is more useful to express the relative stability of the system in terms of the inlet throttling coefficient, K_o. The first term between the parentheses represents the heater and the exit sections; the second term represents the effects of the surge tank compressibility, $\bar{K}_t$ being a dimensionless factor related to the surge tank compressibility as defined by

$$\bar{K}_t = (P_{tas} A/V_{cs}) \rho_o L_h / \theta^2 \tag{53}$$

The Nyquist plots of the term in parentheses for various mass flow rates and heat fluxes are sufficient to determine the relative stability of the system. Since $G_{h(s)}$ involves exponential functions, the graphs will intersect the real axis for an infinite number of times. The outmost intersection gives the amount of inlet throttling necessary to stabilize the sytem. If this is lower than the actual throttling, the system is stable under these conditions; if not, it is unstable. The Nyquist plot of the characteristic equation at an unsteady operating point is presented in Fig. 2.

After testing the stability of a sufficient number of steady-state operating points by Nyquist plots, it is concluded that the system is unstable whenever there is boiling. Inlet throttling constants required to stabilize the system in these cases are plotted in Fig. 3. The magnitudes are unreasonably high although the qualitative behavior agrees with experimental findings. On the basis of the present results, it is concluded that the homogeneous flow model is not much useful for quantitative predictions of the oscillation thresholds. However, it can still be used for a qualitative analysis of the instability problem.

DRIFT-FLUX MODEL

For a number of two-phase flow regimes such as annular flow or slug flow, the assumption of equal phase velocities is not very reasonable. One may take into account the relative velocity between the phases by using a drift-flux formulation, such as the one proposed by Zuber and Findlay [11]. The primary assumptions of the drift-flux model are stated as follows:

(i)   The flow is one-dimensional.

(ii)   The phases are in thermodynamic equilibrium.

(iii)   The pressure and other fluid properties are uniform across each section taken along the channel. The effect of the radial distribution of the void fraction is represented by a distribution parameter, C_0.

(iv)   The heat input may be non-uniformly distributed along the channel and the effect of the heat capacity of the wall is taken into account.

(v)   The energy equation is limited to a thermal balance.

(vi)   In the boiling region, the fluid is treated as a mixture of saturated liquid and vapor phases traveling at different velocities.

(vii)   The axial conduction in the wall is neglected.

(viii) The flow is not allowed to reverse.

In addition to these basic assumptions, the liquid phase is assumed to be incompressible so as to simplify the ensuing algebra. The dimensionless conservation equations are written as follows [12].

Dimensionless continuity equation:

$$\frac{\partial}{\partial \bar{t}} \left[\bar{\rho}_f(1-\beta) + \bar{\rho}_g \beta \right] + \frac{\partial}{\partial \bar{z}} \left[\bar{\rho}_f u_f(1-\beta) + \bar{\rho}_g \bar{u}_g \beta \right] = 0 \tag{54}$$

Dimensionless energy equation:

$$\bar{q}' = \frac{\partial}{\partial \bar{t}} \left[\bar{\rho}_f \bar{h}_f(1-\beta) + \bar{\rho}_g \bar{h}_g \beta \right] + \frac{\partial}{\partial \bar{z}} \left[\bar{\rho}_f \bar{u}_f \bar{h}_f(1-\beta) + \bar{\rho}_g \bar{u}_g \bar{h}_g \beta \right] \tag{55}$$

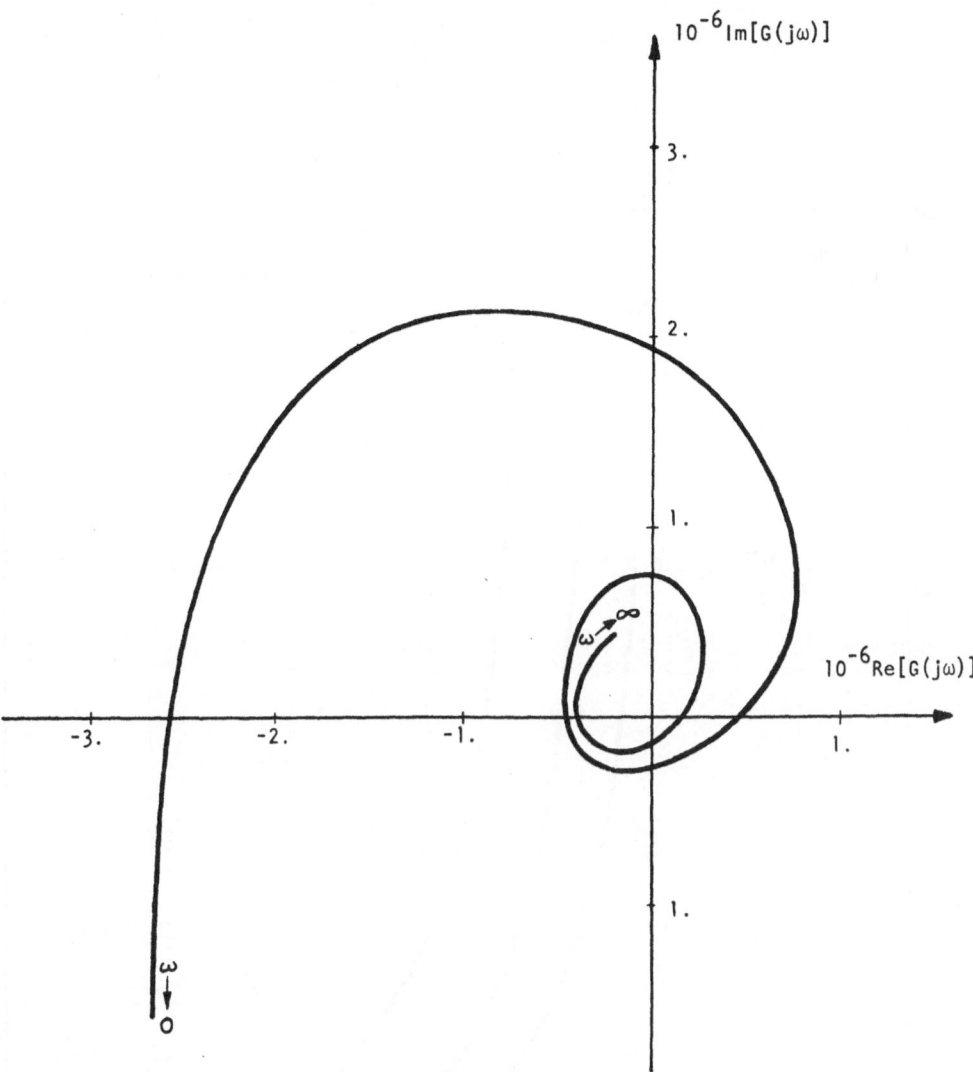

Fig. 2. Nyquist plot of characteristic equation obtained by Homogeneous-Flow Model at an unsteady operating point ($Gd/\mu_f = 4500$, $q'/h_{fg}\mu_f = 13.0$, $T_i = 23^\circ C$)

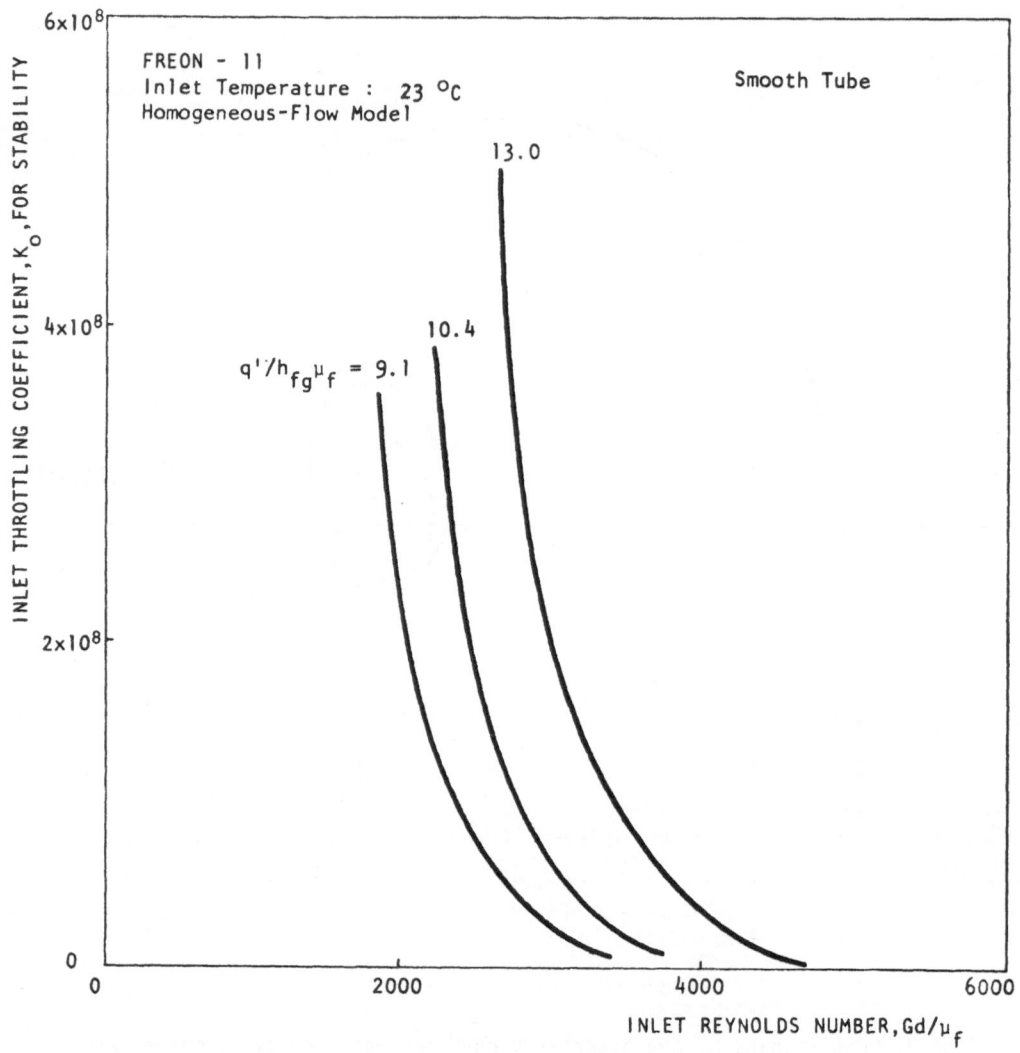

Fig. 3. Inlet throttling required to stabilize system as a function of inlet
Reynolds number and dimensionless heat input as predicted by a linearized homogeneous-flow model with concentrated pressure-drops

Dimensionless momentum equation:

$$-\frac{\partial \bar{p}}{\partial \bar{z}} = \frac{\partial}{\partial \bar{t}} [\bar{\rho}_f \bar{u}_f (1-\beta) + \bar{\rho}_g \bar{u}_g \beta] + [\partial \bar{p} / \partial \bar{z}]_{tp,fric}$$

$$+ \frac{\partial}{\partial \bar{z}} \quad \bar{G}^2 [\frac{(1-x)^2}{\bar{\rho}_f(1-\beta)} + \frac{x^2}{\bar{\rho}_g \beta}] \quad + g [\bar{\rho}_f(1-\beta) + \bar{\rho}_g \beta] \qquad (56)$$

where

$$\bar{G} = \bar{\rho}_f \bar{u}_f (1-\beta) + \bar{\rho}_g \bar{u}_g \beta \qquad (57)$$

The dimensionless variables in these equations are related to the actual variables through the following relations:

Time	: $\bar{t} = t/\tau$ where $\tau = \rho_0 h_{fgo} A/q'_s$	(58)
Length	: $\bar{z} = z/L_h$	(59)
Density	: $\bar{\rho} = \rho/\rho_0$	(60)
Enthalpy	: $\bar{h} = h/h_{fgo}$	(61)
Velocity	: $\bar{u} = u\tau/L_h$	(62)
Pressure	: $\bar{p} = p\tau^2/\rho_0 L_h^2$	(63)
Gravitational acceleration	: $\bar{g} = g\tau^2/L_h$	(64)
Heat input	: $\bar{q}' = q'/q'_s$	(65)
Temperature	: $\bar{T} = T c_{po}/h_{fgo}$	(66)

Since the effect of the wall heat capacity is included in the model, a separate energy equation should be written for the wall. Neglecting the axial heat conduction term, an energy balance for the tube wall leads to the following dimensionless equation:

$$\lambda_w \frac{d\bar{T}_w}{d\bar{t}} = 1 - \bar{q}' \qquad (67)$$

where the dimensionless constant, λ_w, is defined as

$$\lambda_w = \frac{\rho_w c_w A_w}{\rho_0 c_{po} A} \qquad (68)$$

Thus, Eqs. (54), (55), (56) and (67) constitute the governing equations for the flow. The quality, x, and the void fraction, β, in these equations are defined in the following way:

$$x = \frac{\rho_g u_g A_g}{\rho_f u_f A_f + \rho_g u_g A_g} \qquad (69)$$

and

$$\beta = A_g / A \qquad (70)$$

where A_f and A_g are the flow areas for the liquid and vapor, respectively, and A is the total cross-sectional area.

The vapor velocity, u_g, is related to the volumetric flux, j, by the following expression of Zuber and Findlay:

$$U_g = C_0 + j\, u_{gj} \qquad (71)$$

where C_0 is a distribution parameter and u_{gj} is the drift velocity.

The two-phase frictional pressure gradient in the momentum equation is related to the single-phase pressure gradient through the use of a two-phase friction multiplier which is defined by Eq. (48).

It is assumed that each flow variable can be considered as consisting of one steady-state and one oscillating part so that, for example,

$$u_0(t) = u_{os} + \tilde{u}_0(t) \qquad (72)$$

where $u_0(t)$ is the time-varying perturbation on the steady-state inlet velocity, u_{os}, and it is assumed to be comparatively small in magnitude. Other variables such as u_f, u_g, β and q' are also represented in the same form. These perturbed variables are then substituted into the conservation equations during which process the products of the perturbations are ignored so that the perturbation equations should turn out to be linear. This analysis is carried out for the subcooled flow region, heater two-phase flow and exit regions. The exit quality range is from 0.0 to 1.0, i.e. there is no superheated region in the test section. This arrangement is sufficient in investigating the thresholds of the pressure-drop and the density-wave oscillations.

After some straightforward but cumbersome algebra, the following expression is obtained for the open-loop transfer function of the system:

$$G(s) \equiv \Delta P_{tot}(s)/U_0(s) = 2(K_0'-1)\bar{u}_{os} + s\bar{L}_0 + E_1 G_{Je}(s) + E_2 G_{Be}(s)$$

$$- \int \left\{ (1-r)[s\bar{u}_{gs} + \frac{2\bar{u}_{gs}}{\bar{u}_{os}}(\,d\bar{p}/d\bar{z}\,)_{fric,s} + \bar{g}] - \frac{(2c_1 x_s + c_2)x_s^2 j_s}{r\bar{u}_{gs}\beta_s^2} \right\} G_B(s,z)dz$$

$$+ \int \left\{ [1-\beta_s C_0(1-r)][s+\frac{2}{2\bar{u}_{os}}(d\bar{p}/d\bar{z})_{fric,s}] - \frac{(2c_1 x_s + c_2)x_s^2 \bar{u}_{gj}}{r\bar{u}_{gs}^2 \beta_s} \right\} G_J(s,z)dz \quad (73)$$

where the integrals are to be taken over the whole two-phase region from the boiling point up to the exit restriction. Also,

$$E_1 = 2\bar{u}_{fes}(1-\beta_{es}C_0) + 2r_e\bar{u}_{ges}\beta_{es}C_0$$

$$+ \Delta\bar{p}_{es}\left\{ \frac{2}{\bar{u}_{os}}[1-\beta_{es}C_0(1-r_e)] - \frac{(2c_1 x_{es} + c_2)\bar{u}_{gj}x_{es}^2}{r\bar{u}_{ges}\beta_{es}^2 F_{Mes}} \right\} \quad (74)$$

and

$$E_2 = \bar{u}_{fes}^2 - 2\bar{u}_{fes}\bar{u}_{ges} + r_e\bar{u}_{ges}^2$$

$$+ \Delta\bar{p}_{es}\left[\frac{(2c_1 x_{es} + c_2)x_{es}^2 \bar{J}_{es}}{r_e\bar{u}_{ges}\beta_{es}^2 F_{Mes}} - \frac{2\bar{u}_{ges}(1-r_e)}{\bar{u}_{os}} \right] \quad (75)$$

The void-to-flow rate transfer function, $G_B(s,z)$, and the volumetric flux-to-flow rate transfer function, $G_J(s,z)$, are defined as follows:

$$G_B(s,z) \equiv B(s,z)/U_0(s) \quad (76)$$

and

$$G_J(s,z) \equiv J(s,z)/U_0(s) \quad (77)$$

where $B(s,z)$, $J(s,z)$ and $U_0(s)$ are the Laplace transforms of the perturbations on the void fraction, $\beta(z,t)$, volumetric flux, $j(z,t)$, and the heater inlet velocity, $u_0(t)$, respectively.

Given the steady-state operating conditions, it is not difficult to solve numerically for $G_B(j\omega,z)$ and $G_J(j\omega,z)$ from the linearized continuity and momentum equations. Then, by substituting them into Eq. (73) and performing the numerical integrations, it is possible to compute the value of the open-loop transfer function, $G(j\omega)$, at any frequency, ω. Thus, the Nyquist plots for the characteristic equation, Eq. (25), are generated numerically as the frequency, ω, is increased from zero to infinity.

A sufficient number of steady-state operating points at different heat inputs and mass flow rates have been tested before concluding that the system is unstable whenever there is boiling. This conclusion is in agreement with both the experimental findings and the predictions of the homogeneous model. However, the drift-flux model is quantitatively much better than the homogeneous model. This is clearly evident from Fig. 4 where the experimental and theoretical coefficients of inlet throttling at instability thresholds are plotted against the mass flow rates at different heat inputs.

The inlet throttling coefficient, K_0, can be considered as a measure of closing the inlet control valve; the case when it goes to infinity corresponds

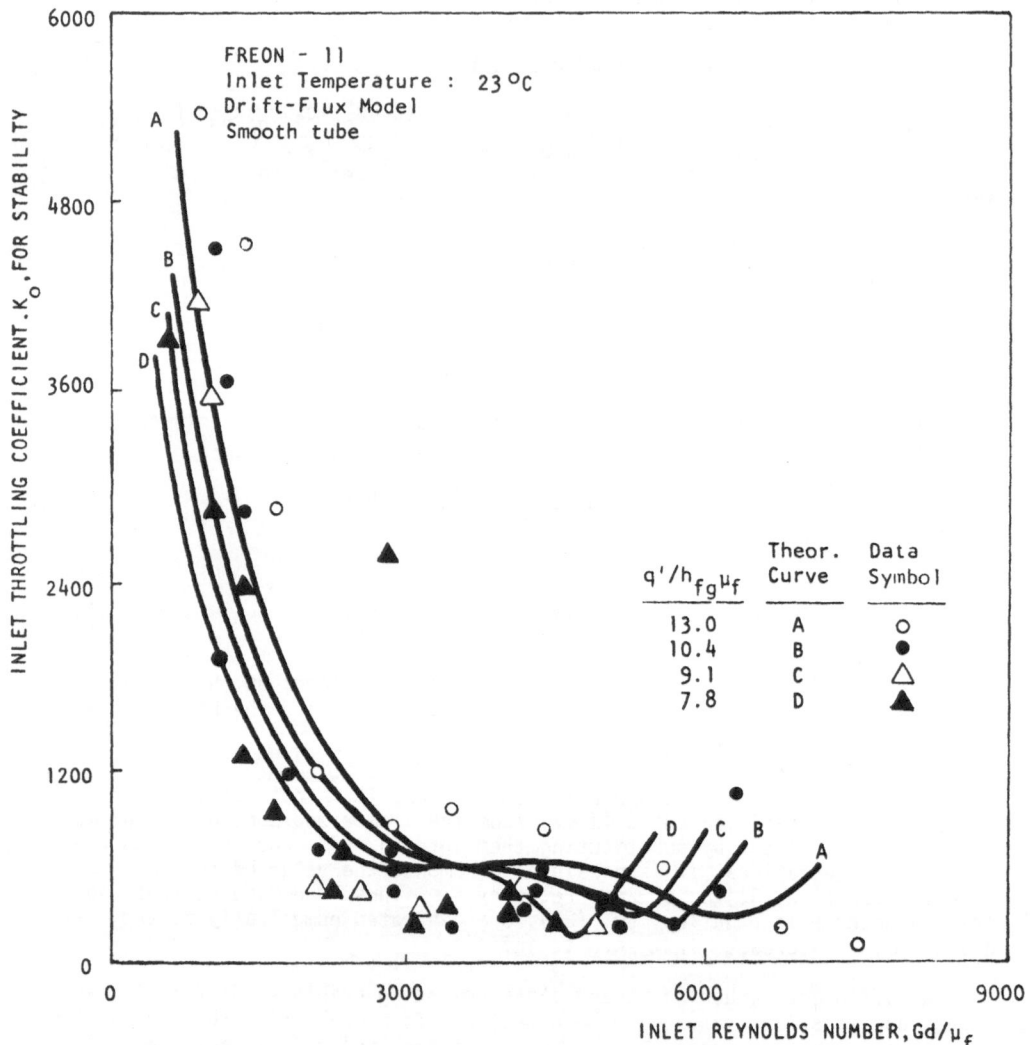

Fig. 4. Inlet throttling required to stabilize system as a function of inlet Reynolds number and dimensionless heat input ; comparison between predictions of drift-flux model and experimental results

to shutting off the valve. This coefficient is zero when the valve is totally open and when there is no inlet restriction coming from any other source. An examination of Fig. 4 shows that the stability of the system slightly decreases at very low exit qualities. For the intermediate flow rates, the stabilizing valve position stays nearly constant. However, as the flow rate is reduced below a certain value, the system stability starts to decrease steeply so that more and more inlet throttling is required to stabilize the system.

The effect of exit quality on the system stability is more clearly demonstrated in Fig. 5, where the stabilizing inlet throttling coefficients are plotted against the exit qualities at various heat input and flow rate values. As can be seen from that figure, the exit restriction dominates the system stability characteristics, so that it becomes sufficient to correlate the stabilizing inlet throttling coefficients with exit quality, for this particular system.

CONCLUSIONS

The characteristic equation which is given on Eq. (25) makes it possible to investigate the system stability with respect to both the pressure-drop and the density-wave oscillations at the same time. The open-loop transfer function, $G_{(s)}$, in this equation may be found by using the available two-phase flow models.

Two different two-phase flow models, namely a constant-property homogeneous flow model and a variable-property drift-flux model, have been developed and the results of the theoretical analysis are compared with experimental findings. Conditions of thermodynamic equilibrium are assumed for each model. In the homogeneous model, the effects of the wall heat storage and the variation of fluid properties with pressure are neglected and the system pressure drop after the surge tank is assumed to be concentrated at the two restrictions placed before and after the heater. In the drift-flux model, the heat flux may change in time and along the channel, the distributed pressure drop terms due to two-phase gravitational, frictional, accelerational and inertial pressure drops are included in momentum equation and all the fluid properties evaluated at local pressures.

The conservation equations for both models and the equations of surge tank dynamics are first linearized for small perturbations and the stability of the resulting set of equations for each model are examined by use of Nyquist plots.

As a measure of the relative instability of the system, the amounts of inlet throttling necessary to stabilize the system at particular operating points have been calculated. Obviously, the bigger the magnitude of the stabilizing inlet throttling coefficient at a particular operating point, the less stable is the system at that point. The theoretical results have been compared with experimental findings. On the basis of this comparison, it is concluded that the drift-flux formulation offers a simple and reliable way of determining the instability thresholds.

NOMENCLATURE

A	Area
$B_{(s,z)}$	Laplace transform of void fraction perturbation
C_o	Void fraction distribution parameter in drift-flux model
Cpo	Specific heat of entering liquid

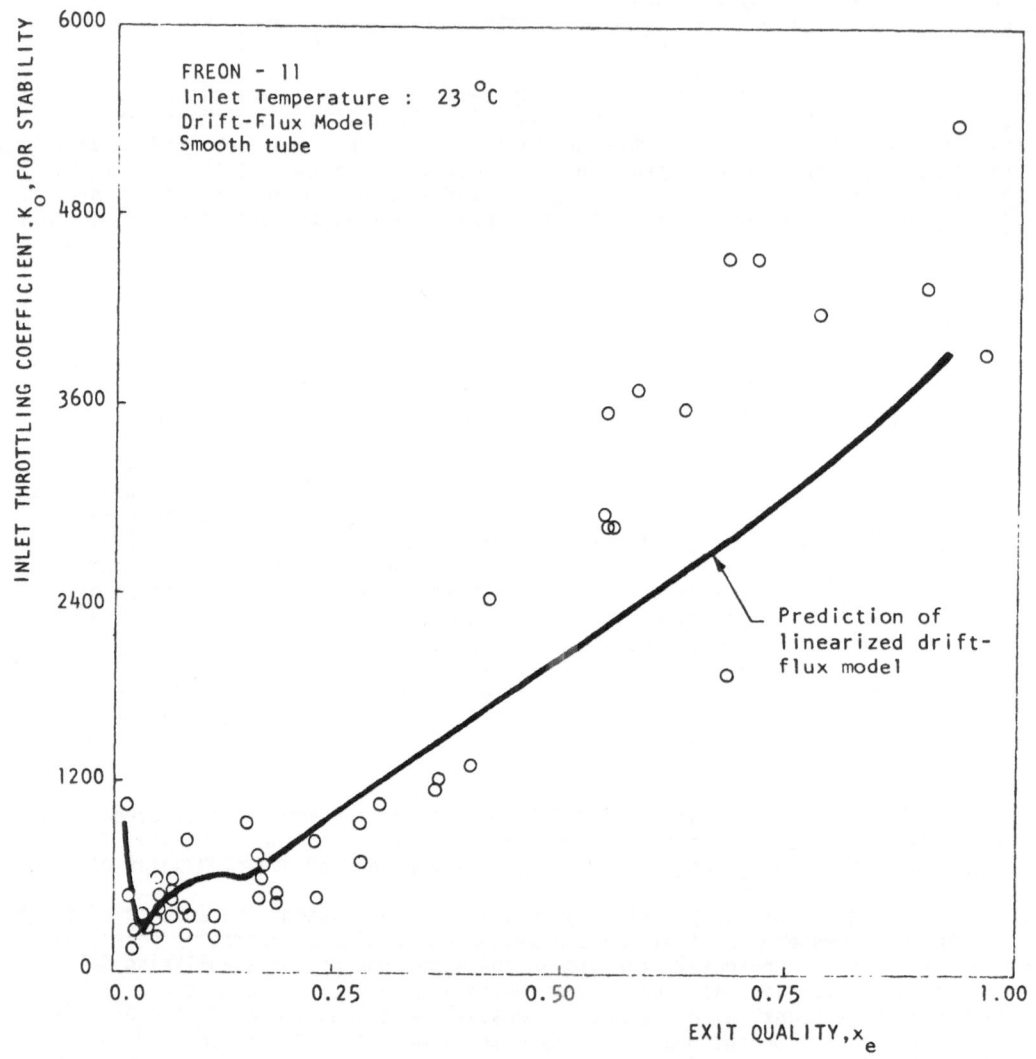

Fig. 5. Inlet throttling required to stabilize system as a function of exit
quality ; comparison between predictions of drift-flux model and experimental data at five different heat inputs

C_w	Specific heat of wall material
d	Diameter
F_m	Two-phase friction multiplier
f	Friction factor
G	Mass flux
$G_{(s)}$	Inlet flow-to-pressure drop transfer function
$G_B(s,z)$	Inlet flow-to-void fraction transfer function
$G_J(s,z)$	Inlet flow-to-volumetric flux transfer function
g	Gravitational acceleration
h	Enthalpy
$J(s,z)$	Laplace transform of volumetric flux
j	Volumetric flux
K_i	Inlet restriction pressure drop coefficient
K_e	Exit restriction pressure drop coefficient
K_t	Surge tank constant, $P_{tas}A/V_{cs}$
L_i	Length of tubing between supply and surge tanks
L_o	Length of tubing between surge tank and heater inlet
L_1	Length of subcooled region
P	Pressure
P_i	Supply tank pressure
P_e	System exit pressure
P_t	Surge tank pressure
P_v	Saturation pressure
q'	Heat input per unit length
r	Density ratio, ρ_g/ρ_f
S	Complex variable
T	Temperature
t	Time
t_1	Residence time of a fluid particle in subcooled region
t_2	Residence time of a fluid particle in boiling heater region
t_3	Residence time of a fluid particle in exit tubing
$U_o(s)$	Laplace transform of inlet velocity perturbation
u	Velocity
u_{gj}	Vapor drift velocity in drift-flux model
V	Volume
V_c	Surge tank compressible volume
v	Specific volume
x	Quality
z	Distance from heater inlet

Greek letters

β	Void fraction
ψ	System pressure drop, P_t-P_e
λ_w	Dimensionless wall constant, $\rho_w C_w A_w/\rho_o C_{po}A$
ρ	Density
τ	Reference time for drift-flux model, $\rho_o h_{fgo}A/q'_s$
θ	Reference time for homogeneous model, $A h_{fg}/V_{fg}q'_s$

Subscripts

a	Air
e	System Exit
f	Saturated liquid
fric	Frictional
g	Saturated vapor

i   Surge tank inlet
o   Surge tank exit and heater inlet
s   Steady-state
t   Surge tank
tp   Two-phase
w   Wall

ACKNOWLEDGEMENTS

The authors gratefully acknowledge the financial support of the National
Science Foundation. The authors wish to extend their thanks to Dr. Win Aung,
National Science Foundation, A. Mentes and O. T. Yildirim, University of Miami.
The secretarial work of Ms. A. Raffle is also acknowledged.

REFERENCES

1.   Stenning, A. H. and Veziroglu, T.N., "Flow Oscillation Modes in Forced
      Convection Boiling", Proc. Heat Transfer and Fluid Mechanics Institute,
      Stanford University Press, 301, (1965).

2.   Zuber, N., "Analysis of Thermally Induced Flow Oscillations in the Near
      Critical and the Supercritical Thermodynamic Region", NASA 8-11422 (1966).

3.   Boure, J., "The Oscillatory Behavior of Heated Channels – An Analysis of
      Density Effects", CEAR 3049, Centre d'Etudes Nucleaires de Grenoble, France
      (1966).

4.   Ishii, M., "Study on Flow Instabilities in Two-Phase Mixtures" ANL-76-23
      (1976).

5.   Nakanishi, S., Ishigai, S., Ozawa, M., and Mizuta, Y., "Analytical Investi-
      gation of Density-Wave Oscillation", Technology Reports of the Osaka
      University, V.28, No. 1421 (1978).

6.   Unal, H.C., "Density-Wave Oscillations in Sodium-Heated Once-Through Steam
      Generator Tubes", Trans. ASME, J. Heat Transfer, 103, 485-491 (1981).

7.   Fukuda, K. and Kobori, T., "Two-Phase Flow Instability in Parallel Channels",
      6th International Heat Transfer Conference, Toronto, Ontario, Canada (1978).

8.   Ozawa, M., Akagawa, K., Sakaguchi, T., Tsukahara, T. and Fujii, T.,
      "Oscillatory Flow Instabilities in Air-Water Two-Phase Flow System - 1.
      Pressure Drop Oscillation", Bull. JSME, 22 (174), 1763-1770 (1979).

9.   Veziroglu, T.N. and Kahai, S., "Two-Phase Flow Instabilities and Effect
      of Inlet Subcooling", NSF Project-Eng 75-16618, Final Report (1980).

10.   Mentes, A., Gurgenci, H., Yildirim, O.T., Kakac, S. and Veziroglu, T.N.,
      "Effect of Heater Surface Configurations on Two-Phase Flow Instabilities
      in a Vertical Boiling Channel". Proceedings of the 16th Southeastern Seminar
      on Thermal Sciences, 19-21 April 1982, Miami, Florida U.S.A.

11.   Zuber, N. and Findlay, J., "Average Volumetric Concentration in Two-Phase
      Flow Systems", Trans. ASME, Series C, J. Heat Transfer, 87, 453 (1965).

12.   Veziroglu, T.N. and Kakac, S., "Two-Phase Flow Instabilities", NSF Project
      CME 79-20018, First Annual Report (1981).

Simplified Nonlinear Descriptions of Two-Phase Flow Instabilities in a Vertical Boiling Channel

H. GÜRGENCI, T.N. VEZIROĞLU, and S. KAKAÇ
Department of Mechanical Engineering
University of Miami
Coral Gables, Florida 33124, USA

ABSTRACT

A constant-property homogeneous flow model is developed to generate the limit cycles of pressure-drop and density-wave oscillations in a single-channel upflow boiling system operating between constant pressures, with upstream compressibility introduced through a surge tank. In the model, thermodynamic equilibrium conditions are assumed and the effects of the wall heat storage and the variation of the fluid properties are neglected. For the purposes of analyzing the density-wave oscillations, the system pressure drop is assumed to be concentrated at the two restrictions placed before and after the heater. The distributed pressure drop terms are included in the treatment of the pressure-drop oscillations, while, in this case, quasi steady-state conditions are assumed to persist throughout the system. Satisfactory agreement with the experimental cycles is noted for the pressure-drop oscillations. As for the density-wave oscillations, the agreement with the experiments is reasonably good regarding the periods of the oscillations, but not so good for the amplitudes.

1. INTRODUCTION

The two-phase flow instability problems can be approached in two ways: One may either work with the nonlinear conservation equations, solve them for certain initial conditions and see whether the solutions are asymptotically stable or not, or one may first linearize the governing equations and then analyze resulting set of the linearized equations by well-tried linearized stability tests. Most of the two-phase flow instability studies published in recent years use linearized models to predict the instability thresholds and to generate stability maps for a given system. The linearized stability criteria are simpler to use--though the derivation of the criteria might be complicated. The effects of various physical parameters on stability are readily evident. However, for certain two-phase flow applications, the unstable operation may be the normal mode of operation, the frequencies and amplitudes being required to lie within specified limits. Then, linearized methods are of little use since they cannot provide information regarding the oscillation limit cycles. In such cases, solutions to the nonlinear equations are needed which can be obtained only by

---

H. Gürgenci is on leave from Middle East Technical University, Ankara, Turkey.

numerical means.

The development of new numerical codes for various two-phase
flow applications has progressed in parallel to the advances in
digital computer technology in the last two decades. One of the
first successful attempts in this direction is by Meyer and
Rose [1]. Since then, a multitude of computer codes have been
developed and put into use in a diversity of industrial two-phase
flow applications such as the operation characteristics of an array
of parallel channels (e.g. BWR bundles or steam generator tubes),
steam condensation instabilities in BWR suppression pools, loss-
of-coolant-accident (LOCA) processes in PWRs, countercurrent two-
phase flow phenomena encountered in certain operation phases of
PWRs and BWRs. A review of various computer codes developed by
the national laboratories and private companies is given by
Bouré [2]. Probably the best of them are proprietary.

An inherent problem of these nonlinear finite-difference
solution schemes is that they are usually more useful for fore-
casting the system behavior under prescribed operating conditions
than for parametric studies. The effects of changing various
parameters on the subsequent behavior of the system are not
readily evident. In order to achieve this purpose, a sufficient
number of computer runs, or so-called "numerical experiments,"
have to be performed, which may be time-consuming and prohibitively
expensive in most cases.

In this study, two very simple numerical methods will be
developed to simulate pressure-drop and density-wave oscillations
in a single-channel upflow boiling system operating between
constant pressures, with upstream compressibility introduced
through a surge tank. The pressure-drop and the density-wave
oscillations constitute two major types of two-phase flow instabilities. This terminology has been first proposed by Stenning
and Veziroglu [3] and it has been generally accepted.

A homogeneous equilibrium model will be developed in which
the effects of the wall heat storage and the variation of the
fluid properties are neglected. For the purposes of analyzing the
density-wave oscillations, the system pressure drop is assumed to
be concentrated at the two restrictions placed before and after
the heater. The distributed pressure drop terms are included in
the treatment of the pressure-drop oscillations, while, in this
case, quasi steady-state conditions are assumed to persist throughout the system. The theoretical cycles will be compared with
experimental cycles recorded on a single-channel vertical boiling
flow system.

2. GOVERNING EQUATIONS

A schematic view of the upward-flow boiling system is
represented in Fig. 1. A detailed description of the loop is
given in [4]. The governing equations for this system can be
written as follows [5]:

Pressure-drop characteristics of the section between the
supply and the surge tanks are given as

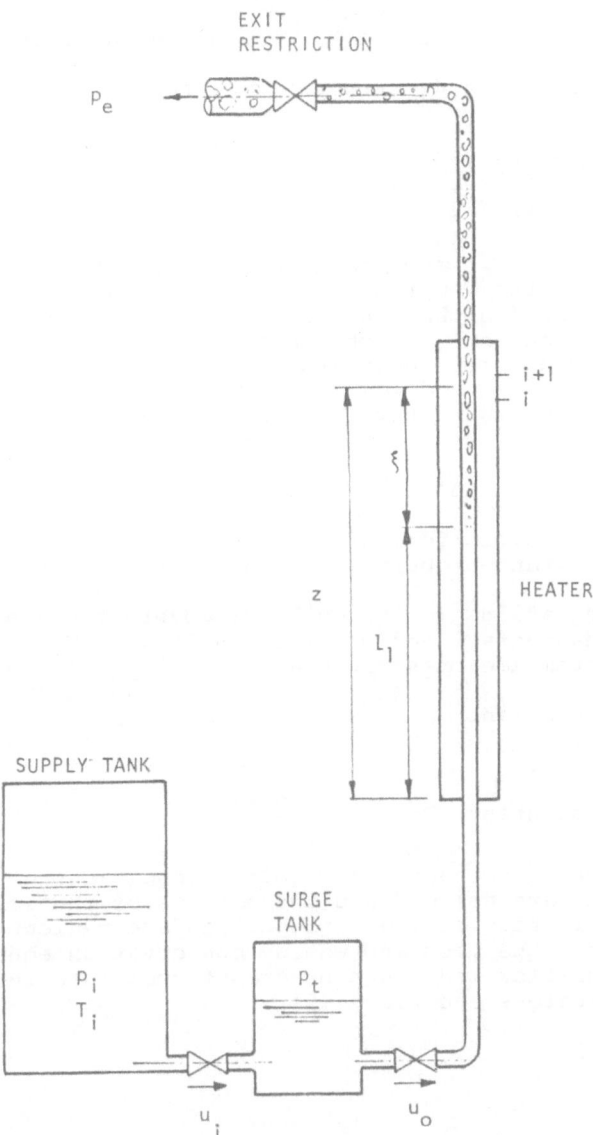

Fig. 1. Schematic diagram of the boiling upward-flow system

$$P_i - P_t = K_i \rho_o u_i^n + \rho_o L_i \frac{du_i}{dt} \tag{1}$$

where n is taken as 2 and K_i depends on the valve opening.

The time rate of change of the surge tank pressure is written as

$$\frac{dp_t}{dt} = \frac{(p_t - p_v)^2 A}{p_{tas} V_{cs}} (u_i - u_o) \tag{2}$$

where p_{tas} and V_{cs} are the steady-state values of the air partial pressure and the compressible volume in the surge tank, respectively. Assuming that the air-vapor mixture behaves ideally, $p_{tas} V_{cs}$ depends only on the mass of air trapped in the surge tank and the fluid inlet temperature.

Pressure drop after the surge tank:

$$P_t - P_e = \psi (u_o, \dot{u}_o, q') \tag{3}$$

where the specific form of the function, ψ, will be found by using a constant-property homogeneous-flow model.

The equations (1) through (3) govern the system behavior during steady-state and oscillatory flows. The boundary conditions on the system are prescribed as

$$P_i = \text{constant} \tag{4}$$

and

$$P_e = \text{constant} \tag{5}$$

In the homogeneous flow model, conditions of thermodynamic equilibrium are assumed and the effects of the wall heat storage and the variation of fluid properties are neglected. With these assumptions, the mass and energy conservation equations for the homogeneous-flow model can be transformed into the following pair of dimensionless equations [5]:

$$\frac{\partial \bar{u}}{\partial \bar{z}} = \frac{1}{\theta} \tag{6}$$

$$\frac{1}{\bar{\rho}} \left[\frac{\partial \bar{\rho}}{\partial \bar{t}} + \bar{u} \frac{\partial \bar{\rho}}{\partial \bar{z}} \right] = - \frac{1}{\theta} \tag{7}$$

and the momentum equation is written as

$$\frac{\partial \bar{p}}{\partial \bar{z}} = - \bar{\rho} \frac{\partial \bar{u}}{\partial \bar{t}} - \bar{\rho}\bar{u} \frac{\partial \bar{u}}{\partial \bar{z}} - 2 \frac{f}{\bar{d}} \bar{\rho}\bar{u}^2 - \bar{\rho}\bar{g} \tag{8}$$

where the dimensionless variables are defined as follows:

$$\bar{z} = z/L_h \; ; \quad \bar{t} = t/\theta_s \; ; \quad \bar{\rho} = \rho/\rho_o \; ; \quad \bar{u} = u\theta_s/L_h \; ;$$

$$\bar{p} = p\theta_s^2/\rho_o L_h^2 \; ; \quad \bar{g} = g\theta_s^2/L_h \; ; \quad \bar{\theta} = \theta/\theta_s \tag{9}$$

The reference time, θ, is defined as

$$\theta \to \infty \qquad \text{when } h \le h_f$$

$$\theta = Ah_{fg}/v_{fg}q' \qquad \text{when } h_f < h \le h_g \tag{10}$$

Since the heat capacity of the wall is neglected, the heat input does not vary with time so that $\bar{\theta} = \theta/\theta_s = 1$ in the above equations. The two-phase friction factor, f, is assumed to be constant along the channel and equal to the liquid-phase friction factor in the subcooled region.

The Eqs. (6) and (7) give the density and the velocity distribution along the system while Eq. (8) can be integrated to give the boundary condition on Eqs. (6) and (7). The singular pressure drop across the exit restriction is given by

$$\Delta \bar{p}_{e(t)} = F_{M(x_e)} K_e \bar{G}_e^2 \tag{11}$$

where F_M is a two-phase friction multiplier which may be expressed as

$$F_M = c_1 x_e^2 + c_2 x_e + 1 \tag{12}$$

where the coefficients have been computed from the experimental data as $c_1 = 121$ and $c_2 = 30$.

3. PRESSURE-DROP OSCILLATIONS

The periods of the pressure-drop oscillations are very large compared with the residence time of a fluid particle through the system. Therefore, quasi steady-state conditions can be assumed along the test section. Then, one can write

$$\bar{\rho}_{(z,t)} \bar{u}_{(z,t)} = \bar{\rho}_o \bar{u}_{o(t)} = \bar{u}_{o(t)} \tag{13}$$

Substituting this into Eq. (8) and performing the integration, the dimensionless time-dependent pressure drop across the system from the surge tank up to the exit is found as

$$P_t - P_e = \psi_{[u_o(t)]}$$

$$= \frac{2f_o u_o^2}{d} \left[L_o + \frac{t_1 u_o d}{d_e} + \frac{u_o d}{2d_e} \left(\frac{1}{u_o} - t_1\right) \left(\frac{1}{u_o} - t_1 + 2\right) + L_e \left(\frac{1}{u_o} - t_1 + 1\right) \right]$$

$$+ g \left[L_{oz} + t_1 u_o + u_o \ell n \left(\frac{1}{u_o} - t_1 + 1\right) + Le \left(\frac{1}{u_o} - t_1 + 1\right)^{-1} \right]$$

$$+ u_o^2 \left(\frac{1}{u_o} - t_1\right) + [K_o + K_e F_M(u_o)] \, u_o^2 \tag{14}$$

where the bars indicating non-dimensionality have been dropped for convenience. The quasi steady-state conditions have been assumed to prevail throughout the system and the inertial pressure drop terms have been neglected. This equation is applicable for $0 \le x_e \le 1$ or $1/t_1 \ge u_o \ge 1/(t_1 + v_{fg})$ since

$$x_e = (1/\rho_e - 1)/v_{fg} \tag{15}$$

or, assuming quasi steady-state conditions,

$$x_e = (1/u_o - t_1)/v_{fg} \tag{16}$$

When u_o is greater or equal to $1/t_1$, the flow is single-phase flow throughout and the system pressure drop in this case is given by

$$\psi[u_{o(t)}] = 2\frac{f_o}{d} u_o^2 \left(L_o + \frac{d}{d_e} + L_e\right) + g(L_{oz} + 1 + L_{ez}) + (K_o + K_e) u_o^2 \tag{17}$$

The surge tank pressure, p_t, is related to the surge tank inlet and outlet flow rates as given by the dimensionless form of Eq. (2),

$$\frac{dp_t}{dt} = C_t (p_t - p_v)^2 (u_i - u_o) \tag{18}$$

where C_t is a dimensionless constant defined as

$$C_t = \frac{\rho_o L_h^3 A}{p_{tas} V_{cs} \theta_s^2} \tag{19}$$

The mass flow rate at the surge tank entrance can be assumed to stay constant during the oscillations. This assumption is based on the fact that the pressure drop - flow rate characteristics of the section between the supply and the surge tanks are very stiff in comparison to the other parts of the system. During the experiments, it was observed that the surge tank inlet velocity, u_i, would not oscillate beyond $\pm 5\%$ of its nominal value even during the most violent oscillations. Therefore, Eq. (18) can be rewritten, by substituting

$$u_i = u_{os} \tag{20}$$

as

$$\frac{dp_t}{dt} = C_t (p_t - p_v)^2 (u_{os} - u_o) \tag{21}$$

This is the governing equation for the pressure-drop oscillations. To examine the behavior of a particular steady-state operating point, the inlet velocity, u_{os}, is given a small perturbation supplying the initial condition for Eq. (21) as

$$u_o = u_{os} + \varepsilon \quad \text{at } t = 0 \tag{22}$$

The solution will revert to the steady-state operating point if the system is stable at that point. Otherwise, sustained oscillations will be attained after a short transient. A simple finite-difference scheme to obtain the time-dependent solution is constructed as follows.

3.1. Finite-Difference Solution Scheme

The surge tank pressure function, $p_{t(t)}$, at the time step $j+1$ is related to the value at the previous time step by

$$p_t^{j+1} = p_t^j + \Delta t C_t (p_t^j - p_v)^2 (u_{os} - u_o^j) \tag{23}$$

and the heater inlet velocity is given in implicit form by

$$p_t^{j+1} - p_e = \psi(u_o^{j+1}) \tag{24}$$

where either Eq. (14) or Eq. (17) should be used for ψ, corresponding to the value of the exit quality. There is a slight problem here. The system pressure drop, $\psi(u_o)$, as given by Eqs. (14) and (17), is a multi-valued function of the inlet velocity, u_o, for certain points when $\psi_{min} \leq \psi \leq \psi_{max}$, where the limits are the lower and the upper peaks of the pressure drop-flow rate curve. Thus, for these points, there are three possible values for u_o^{j+1} at each time step, all of which satisfy Eq. (24). However, due to the inertia of the fluid volume inside the system, it is reasonable to expect the flow to follow a path of least resistance and to proceed to that value of u_o^{j+1} which is nearest to the one at the previous time step. In other words, at each time step $j+1$, out of all possible solutions for u_o^{j+1} given implicitly by Eq. (24), that value is chosen for which $|u_o^{j+1} - u_o^j|$ is minimum.

It should be noted that, in those parts of the solution domain, where the pressure function, ψ, increases or decreases monotonously, a simpler way of doing things is introduced as follows. Both Eqs. (14) and (17) describe differentiable functions and it is trivial to obtain

$$\psi'(u_o) \equiv d\psi/du_o \tag{25}$$

so that

$$dp_t/dt = \psi' \, du_o/dt \tag{26}$$

and Eq. (21) can be rewritten as

$$\frac{du_o}{dt} = C_t \frac{(p_t-p_v)^2}{\psi'} (u_{os}-u_o) \tag{27}$$

Then a simple finite-difference scheme is constructed as follows,

$$u_o^{j+1} = u_o^j + C_t(\Delta t) \frac{(p_t^j-p_v)^2}{\psi'(u_o^j)} (u_{os}-u_o^j) \tag{28}$$

and

$$p_t^{j+1} = \psi(u_o^{j+1}) \tag{29}$$

At each time step, compute

$$A \equiv (p_t^{j+1} - p_t^j) / \Delta t \tag{30}$$

and

$$B \equiv (dp_t/dt)^j = C_t(p_t^j-p_v)^2(u_{os}-u_o^j) \tag{31}$$

When A and B are of the same sign, the solution proceeds to the next time step through Eq. (28). When A and B are of opposite signs, the flow undergoes an excursion to another flow rate at the same pressure as governed by Eq. (24). It should be noted that this scheme is essentially the same as the first one, only more convenient because it is not necessary to solve u_o^{j+1} at each time step from the implicit relation of Eq. (24). An implicit iterative scheme for the velocity is sought for only immediately after the flow excursions.

3.2. Results

When the preceding finite-difference scheme is applied at different steady-state operating points, the subsequent behavior of the solution turns out to be in agreement with the predictions of the linearized analysis [5]. Those points for which the steady-state pressure drop increases with increasing flow rate have been found to be stable, the solution in such points reverts to steady-state conditions after a short transient. But when the starting point is in that part of the pressure drop-flow rate curve when the slope is negative, the initial perturbation slowly grows in size and leads to sustained pressure-drop oscillations.

A comparison with the experimental results is shown in Fig. 2. As seen from that figure, there is a fairly good agreement between the experimentally obtained and theoretically predicted amplitudes of the oscillations. However, the predicted periods are twice greater than the experimental periods. The main reason for this discrepancy between the experimental and theoretical periods is considered to be the way the flow excursions are treated. In the

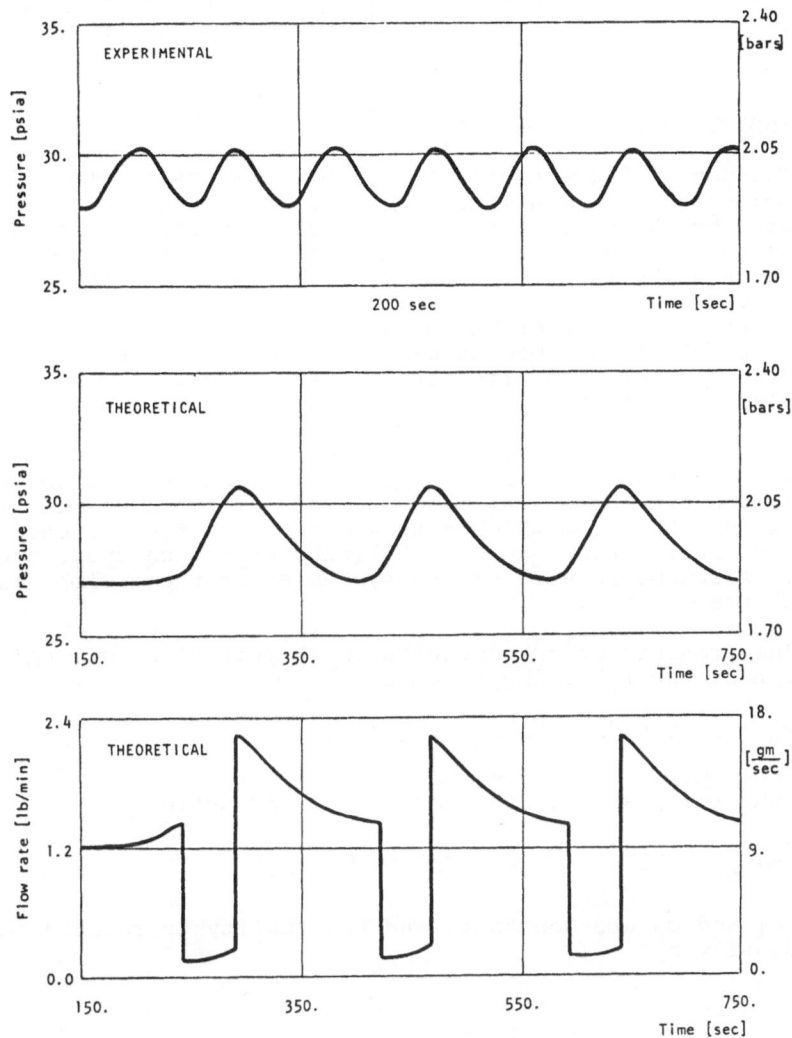

Fig. 2. Pressure-drop oscillations at $\dot{m} = 9.0$ gm/sec, q = 300 W, $T_i = 23^\circ C$, $m_a = 1.0$ gm. Comparison between experimental cycles and results of homogeneous-flow model.

present model, it is assumed that the flow excursions occur instantaneously and at constant pressure. With a better explanation for the mechanism of flow excursions, it should be possible to get a fairly good agreement between the experimental and theoretical frequencies even with as simple a model as the present one.

4. DENSITY-WAVE OSCILLATIONS

The density-wave oscillations have periods in the order of the residence time of a fluid particle in the heater. They are generated by the operation of complicated feedback mechanisms between the test section pressure drop and the fluid velocity. What is meant by the test section pressure drop is the pressure drop before and across the heater, across the exit tubing and the exit restriction. In order to simplify the analysis, this pressure drop will be assumed to be concentrated at two restrictions placed just before and after the heater. The system then consists of an inlet restriction, a heater tube where the fluid density changes through boiling and an exit restriction after the heater. The inlet and exit restrictions are represented by the throttling coefficients K_o and K_e, respectively. The heater and the exit restriction are the essential components of the nonlinear time-delay feedback system which generate the density-wave oscillations. The inlet restriction has a stabilizing effect on these oscillations. A schematic view of the system after these simplifications is presented in Fig. 3.

The pressure at the incipient bulk boiling point, $p_b(t)$, is given by the following relation

$$p_b = p_t - K_o u_o^2 = K_e F_{M(x_e)} G_e^2 + p_e \qquad (32)$$

where the two-phase friction multiplier is defined as

$$F_{M(x_e)} = c_1 x_e^2 + c_2 x_e + 1 \qquad (33)$$

where c_1 and c_2 are constants and the quality is related to the exit density by

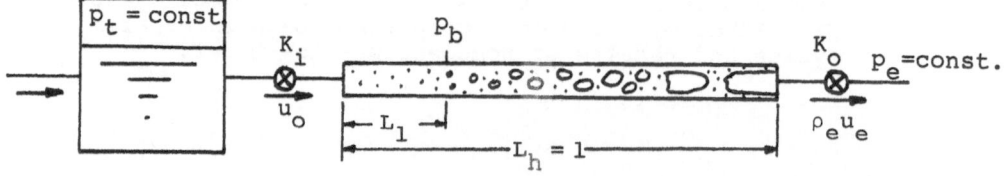

Fig. 3. Simplified system diagram for density-wave oscillations

$$x_e = (1/\rho_e - 1)/v_{fg} \tag{34}$$

Integrating Eq. (7), the following implicit relation for the exit density, ρ_e, is obtained

$$1 = e^t \int_{t+\ln\rho_e}^{t} u_{o(t-t_1)} e^{-t} dt + L_{1(t)} \tag{35}$$

and the subcooled length, $L_{1(t)}$, is given by

$$L_{1(t)} = \int_{t-t_1}^{t} u_{o(t)} dt \tag{36}$$

where the time lag, t_1, is defined as

$$t_1 = \rho_o v_{fg}(h_f-h_i)/h_{fg} \tag{37}$$

which represents the dimensionless residence time of a fluid particle in the subcooled region of the heater.

The mass flux through the exit restriction, $G_{e(t)}$, is defined as

$$G_e = \rho_e u_e = \rho_e(u_o - L_1 + 1) \tag{38}$$

The surge tank pressure, p_t, will be assumed to be constant in the analysis. This assumption is in close agreement with the experimental observations where the high-frequency density-wave oscillations are easily absorbed by the inertia of the surge tank volume.

The two governing equations (one forward and one feedback relation) for the density-wave oscillations are written from Eq. (32) as

$$p_b = K_e F_{M(x_e)} G_e^2 + p_e \tag{39}$$

and

$$u_o = \sqrt{\frac{p_t - p_b}{K_o}} \tag{40}$$

As in the previous section on the pressure-drop oscillations, to examine the behavior of a particular steady-state operating point, the inlet velocity at that point, u_{os}, will be given a small perturbation at time, t=0.

$$u_o = u_{os} + \varepsilon \quad \text{at } t = 0 \tag{41}$$

This will be the initial condition for Eqs. (39) and (40). As time progresses, the solution will either revert to the steady-state operating point, if the system is stable at that point, or sustained oscillations will be achieved after a short transient. A simple finite-difference scheme to obtain the time-dependent solution is constructed as follows.

4.1. Finite-Difference Solution Scheme

The pressure, p_b, at the (j+1)th time step is given by

$$p_b^{j+1} = K_e F_M(x_e^{j+1}) \ (G_e^{j+1})^2 + p_e \tag{42}$$

where the quality is to be calculated from

$$x_e^{j+1} = (1/\rho_e^{j+1} - 1)/v_{fg} \tag{43}$$

The time rate of change of the exit density may be obtained from Eq. (35) as

$$\frac{d\rho_{e(t)}}{dt} = \rho_{e(t)} \left[\frac{\rho_{e(t)} u_{e(t)}}{u_{o(t-t_1+\ln\rho_e)}} - 1 \right] \tag{44}$$

so that

$$\rho_e^{j+1} = \rho_e^j + \Delta t \rho_e^j \left[\frac{\rho_e^j u_e^j}{u_o^{j-n}} - 1 \right] \tag{45}$$

where

$$n = \left[\!\left[\frac{t-t_1+\ln\rho_e}{\Delta t} \right]\!\right] \tag{46}$$

The double brackets, $[\![\cdot]\!]$, indicate the integer operator.

Similarly, Eq. (36) yields

$$\frac{dL_{1(t)}}{dt} = u_{o(t)} - u_{o(t-t_1)} \tag{47}$$

so that

$$L_1^{j+1} = L_1^j + \Delta t \left[u_o^j - u_o^{j-m} \right] \tag{48}$$

where

$$m = [\![(t-t_1)/\Delta t]\!] \tag{49}$$

and G_e^{j+1} is to be found from

$$G_e^{j+1} = \rho_e^{j+1} \, u_e^{j+1} \tag{50}$$

A simple algorithm to generate the solution may be stated as follows:

1. Compute L_1^{j+1} from Eq. (48).

2. Compute ρ_e^{j+1} from Eq. (45).

3. Compute x_e^{j+1} from Eq. (43).

4. Use the following predictor formula for u_e^{j+1}

$$\left(u_e^{j+1}\right)^P = 2u_e^j - u_e^{j-1} \tag{51}$$

5. Compute G_e^{j+1} from Eq. (50).

6. Compute p_b^{j+1} from Eq. (42).

7. Compute u_o^{j+1} from the following

$$u_o^{j+1} = \left[\frac{p_t - p_b^{j+1}}{K_o} \right]^{\frac{1}{2}} \tag{52}$$

Reverse flow is not allowed so that $u_o^{j+1}=0$ for $p_t < p_b^{j+1}$.

8. Compute $\left(u_e^{j+1}\right)^C$ from the following

$$\left(u_e^{j+1}\right)^C = u_o^{j+1} - L_1^{j+1} + 1 \tag{53}$$

9. If $\left(u_e^{j+1}\right)^C \cong \left(u_e^{j+1}\right)^P$, go to the next step. Otherwise, correct $\left(u_e^{j+1}\right)^P$ and return to the 5th step. The corrector formula is

$$\left(u_e^P\right)^{new} = \left(u_e^P\right)^{old} + 0.1 \left[u_e^C - \left(u_e^P\right)^{old} \right] \tag{54}$$

10. Go to the next time increment and start from the first step.

4.2. Results

The algorithm explained in the preceding section has been applied at different steady-state operating points. The limit cycles for the density-wave oscillations have been obtained at

various heat input and mass flow rate values. A typical example
of experimental density-wave oscillation cycles is presented in
Fig. 4. The predicted cycles at the same operating point are
shown in Fig. 5. As seen from this figure, the frequency of the
theoretical cycles is about 5 cps which is in good agreement with
the experimental frequencies (4-6 cps). However, the amplitudes
of the theoretical cycles are bigger than the actual experimental
amplitudes. From Fig. 4, it is observed that the heater inlet
pressure apparently drops as low as the system exit pressure during
the theoretical cycles. This is due to the fact that the system
pressure drop is assumed to be concentrated at two restrictions.
Thus, when the flow is blocked at the exit orifice as happens
during a cycle, the pressure drop after the heater inlet simply
vanishes according to the model. Actually, a part of the system
pressure drop is distributed along the heater and the exit tubing.
This pressure drop, which is composed of frictional, gravitational,
accelerational and inertial components, is never zero; therefore,
the lower limit for the actual heater inlet pressure cycles is
higher than the one predicted by this analysis. However, when
compared with the more elaborate models of the previous studies
[7, 8], the results of the present simplified model seem to be
equally reasonable. The model can be improved by considering a
distributed pressure drop across the system, instead of the present
singularities. This would require a few additional manipulations
during the numerical computations but the final results would be
more realistic.

5.   CONCLUSIONS

A constant-property homogeneous equilibrium model was developed to generate the limit cycles of pressure-drop and density-
wave oscillations in a single-channel upflow boiling system
operating between constant pressures, with upstream compressibility
introduced through a surge tank. Thermodynamic equilibrium conditions were assumed and the effects of the wall heat storage were
neglected. For the density-wave oscillations, the system pressure
drop was assumed to be concentrated at two restrictions placed
before and after the heater. The distributed pressure drop terms
were included in the treatment of the pressure-drop oscillations,
while, in this case, quasi steady-state conditions were assumed to
persist throughout the system.

Regarding the pressure-drop oscillations, there is a fairly
good agreement between the experimental and the theoretical cycles.
The discrepancy in the periods is presumably due to the incomplete
description of the flow excursions which play an important role in
the pressure-drop oscillations.

As for the density-wave oscillations, the theoretical
frequencies are in good agreement with the experimental frequencies. However, the amplitudes differ, which fact is basically due
to neglecting the distributed pressure drops along the system.

The proposed solution scheme is very easy to apply on a
digital computer and little computer time is required. The effects
of various factors on the amplitudes and the frequencies of the
oscillations are readily evident due to simplicity of the model.

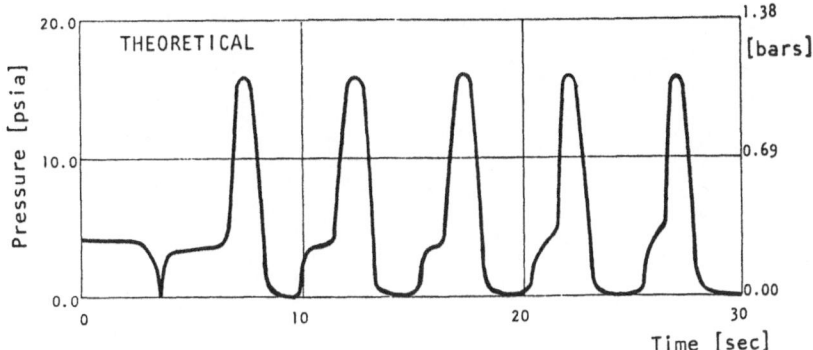

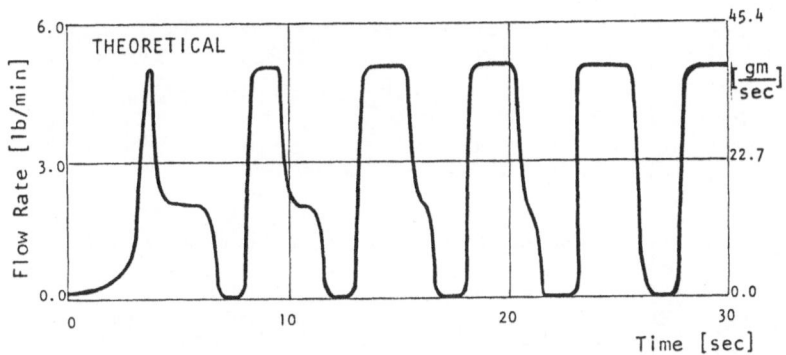

Fig. 4. Development of density-wave oscillation cycles as predicted by homogeneous-flow model; ṁ = 1.5 gm/sec,
q = 350 W, T_i = 23°C.

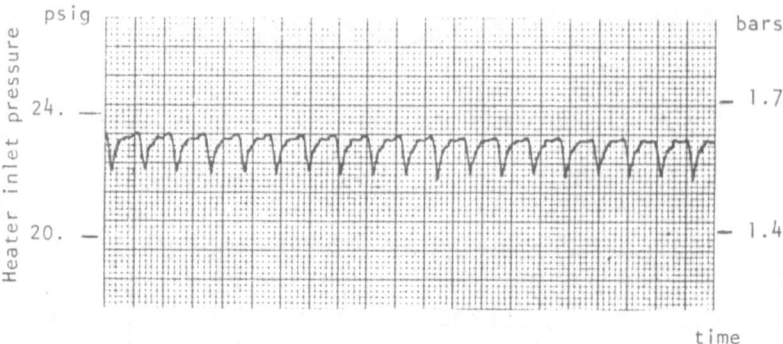

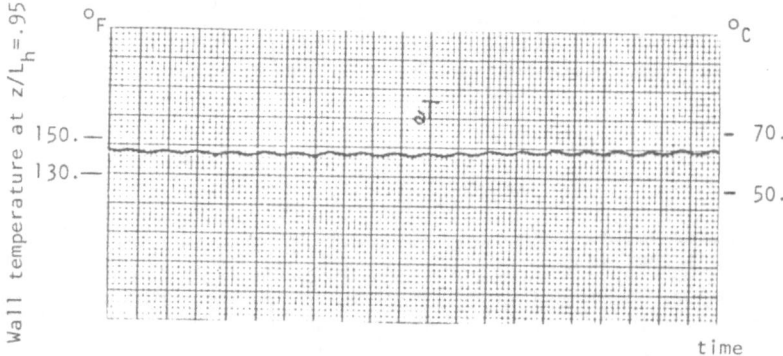

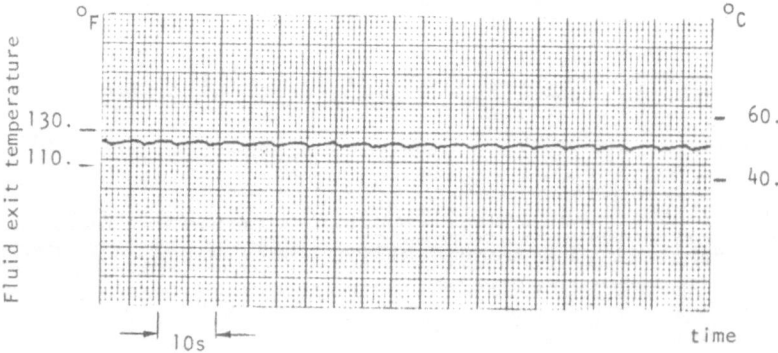

Fig. 5.   Density-wave oscillations ($\dot{m}$ = 1.5 gm/sec, q = 350 W,
T_i = 23°C).

The model is especially useful for parametrical studies to clarify
the basic nonlinear mechanisms of the oscillations, without using
too much computer time.

ACKNOWLEDGEMENTS

The authors gratefully acknowledge the financial support of
the National Science Foundation.

The authors extend their thanks to Messrs. A. Menteş and
O. T. Yıldırım for their very valuable suggestions. The secretarial work of Ms. Norma Sage is also acknowledged.

NOMENCLATURE

A   Area

C_t Dimensionless surge tank constant, $\rho_o L_h^3 A / p_{tas} V_{cs} \theta_s^2$

d   Diameter

d_e Equivalent diameter

F_M Two-phase friction multiplier

f   Friction factor

G   Mass flux

G(s)   Inlet flow-to-pressure drop transfer function

g   Gravitational acceleration

h   Enthalpy

K_i Throttling constant between supply and surge tanks

K_e Exit restriction pressure drop coefficient

K_o Throttling constant between surge tank and heater

L_i Length of tubing between supply and surge tank

L_o Length of tubing between surge tank and heater inlet

L_1 Length of subcooled region

p   Pressure

p_i Supply tank pressure

p_e System exit pressure

p_t Surge tank pressure

p_v Saturation pressure

q' Heat input per unit length

s Complex variable

T Temperature

t time

t_1 Residence time of a fluid particle in subcooled region

t_2 Residence time of a fluid particle in boiling heater region

t_3 Residence time of a fluid particle in exit tubing

$U_o(s)$ Laplace transform of inlet velocity perturbation

u Velocity

V Volume

V_c Surge tank compressible volume

v Specific volume

x Quality

z Distance from heater inlet

Greek letters

ψ System pressure drop

ρ Density

θ Reference time, $Ah_{fg}/v_{fg}q'_s$

Subscripts

a Air

b Incipient boiling point

e System exit

f Saturated liquid

g Saturated vapor

i Surge tank inlet

o Surge tank exit

s Steady-state

t Surge tank

REFERENCES

1. Meyer, J., and Rose, R., "Application of a Momentum Integral
    Model to the Study of Parallel Channel Boiling Flow
    Oscillations," J. Heat Transfer, 85, 1 (1963).

2. Bouré, J. A., Bergles, A. E., and Tong, L. S., "Review of Two-
    Phase Flow Instability," Nuclear Engineering and Design, 25,
    165 (1973).

3. Stenning, A. H., and Veziroğlu, T. N., "Flow Oscillation Modes
    in Forced Convection Boiling," Proc. 1965 Heat Transfer and
    Fluid Mechanics Institute, 301-316, Stanford University Press,
    Stanford, California, U.S.A. (1965).

4. Menteş, A., Gürgenci, H., Yıldırım, O. T., Kakaç, S., and
    Veziroğlu, T. N., "Effect of Heater Surface Configurations on
    Two-Phase Flow Instabilities in a Vertical Boiling Channel,"
    Proc. 16th Southeastern Seminar on Thermal Sciences,
    19-21 April 1982, Miami, Florida, U.S.A.

5. Gürgenci, H., "Two-Phase Flow Instabilities," Ph.D. Thesis,
    University of Miami, Coral Gables, Florida, U.S.A. (1982).

6. Veziroğlu, T. N., and Kakaç, S., Two-Phase Flow Instabilities,
    NSF Project ME 79-20018, Final Report (1982).

7. Akyüzlü, K., Veziroğlu, T. N., Kakaç, S., and Doğan, T.,
    "Finite Difference Analysis of Two-Phase Pressure-Drop and
    Density-Wave Oscillations," Wärme-und Stoffübertragung, 14,
    253-267 (1980).

8. Veziroğlu, T. N., and Kakaç, S., Two-Phase Flow Instabilities
    and Effect of Inlet Subcooling, NSF Project Eng 75-16618,
    Final Report (February 1980).

The Analysis of Two-Phase Flow Instabilities in a Horizontal Single Channel

O.T. YILDIRIM
University of Miami
Coral Gables, Florida 33124, USA

H. YÜNCÜ
Middle East Technical University
Ankara, Turkey

S. KAKAÇ
University of Miami
Coral Gables, Florida 33124, USA

ABSTRACT

A study of the stability of an electrically heated single channel, forced convection horizontal system was conducted by using Freon-11 as the test fluid. Two major modes of oscillations, namely, density-wave type (high frequency) and pressure-drop type (low frequency) oscillations have been observed. The steady-state operating characteristics and stable and unstable regions are determined as a function of heat flux, exit orifice diameter and mass flow rate. Different modes of oscillations and their characteristics have been investigated. The effect of the exit restriction on the system stability has also been studied.

By using a simple steady-state solution, a mathematical model has been developed to predict the transient behavior of boiling two-phase systems. The model is based on homogeneous flow assumption and thermodynamic equilibrium between the liquid and vapor phases. The transient characteristics of boiling two-phase flow horizontal system are obtained for various heat inputs, flow rates and exit orifice diameters by solving the governing equations. Theoretical and experimental results have been compared.

1. INTRODUCTION

Within the last two decades, the expected heat fluxes in heat exchangers have been increased considerably. Because of this fact, boiling two-phase flow systems have been a subject of interest. This is mainly due to the very large heat transfer coefficients obtained in boiling heat transfer.

Utilizing a boiling fluid in heat exchangers solves the problem of obtaining high heat transfer rates for reasonable temperature differences, but a new problem of instabilities arises, which are encountered in two-phase flow systems. These instabilities should be avoided since they can cause mechanical vibrations, control problems and, in some cases, they disturb the heat transfer characteristics so that heat transfer surface may burn out.

There is a great variety of mechanisms which can lead to unstable behavior of two-phase flows. Depending on these mechanisms, various types of static and dynamic instabilities may arise. Nucleation instabilities, flow pattern instabilities, Ledinegg instabilities, thermal oscillations, density-wave type oscillations, pressure-drop type oscillations are some of them. Among them, the most common dynamic instabilities are the density-wave and pressure-drop type oscillations which are identified by Stenning and Veziroglu [1].

Pressure-drop type oscillations are induced due to S-shaped pressure-drop versus mass flow rate characteristics of two-phase systems. They can occur in systems that have a compressible volume which may be distributed within or upstream of the heated section. These oscillations are observed when the slope of pressure drop versus mass flow rate curve is negative.

Density-wave type oscillations, which are the most common type instabilities in two-phase flow systems, are due to the multiple feedbacks between the flow rate, the vapor generation rate, and the pressure drop in a boiling channel. Density-wave type oscillations have small periods in the order of the time required for a fluid particle to travel through the heated channel. It was found experimentally by Yadigaroglu and Bergles [2] and analytically by Ishii and Zuber [3] that several modes of density-wave instabilities are possible.

One of the first studies on two-phase flow instabilities was that of Quandt [4]. In his study, Quandt assumed some spatial forms for the flow and enthalpy perturbations. Wallis and Heasley [5] formulated the same problem in Lagrangian terms. Meyer and Rose [6] used a computerized difference-equation method of solution for a so-called "momentum integral" model. Maulbetsch and Griffith [7] showed the effect of dP/dV associated with the compressible volume, on the pressure-drop type oscillations. Shotkin [8] suggested a spatial-averaged model of boiling flow to investigate the instabilities in vertical channels. Stenning and Veziroglu [9] developed a linearized model to predict the pressure-drop type oscillations. Kakaç and Veziroglu [10]-[13] studied the sustained and transient oscillations both in single and multichannel systems. Excellent reviews of technical literature on two-phase flow instabilities are given by Bouré, Bergles and Tong [14], Yadigaroglu [15], Bergles [16], and Veziroglu and Kakaç [17], and Kakaç and Veziroglu [18]. This study deals with two-phase flow instabilities in a horizontal channel and the effect of exit restriction on the system stability. An analytical model based on the homogeneous flow assumption is also presented.

2. EXPERIMENTAL APPARATUS

The experimental apparatus is schematically illustrated in Figure 1. It has been specially designed to generate and investigate the pressure-drop and density-wave type oscillations in a horizontal single channel flow system. It is an open loop system with constant inlet and exit pressures.

The test section and the necessary instruments are assembled on a steel frame. The piping system is constructed from 5 mm inner diameter nichrome tube except between the orifice and recovery tank

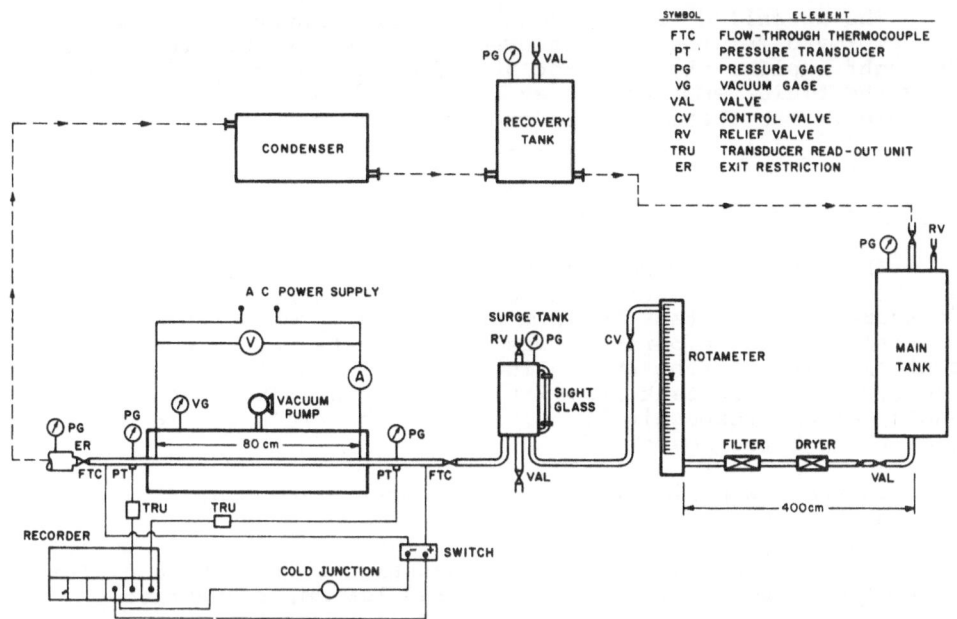

Fig. 1.   Schematic Drawing of the Experimental System

where copper tube with 10 mm inner diameter is used to minimize the pressure drop in the recovery system.   This is necessary to maintain a constant exit pressure.   80-cm portion of the tube serves as the resistance heater.   It is insulated electrically from the rest of the system by using teflon unions at both ends and thermally, from the surroundings, by a vacuum jacket.

The surge tank, placed before the heater, serves as an energy storage element in the system.   It is made of a stainless steel sheet with a thickness of 3 mm.   It has a cylindrical shape with a height of 275 mm and diameter of 197 mm.   A sight glass and a Bourdon pressure gauge, attached to the surge tank, allow us to observe the pressure and the level of the liquid inside the surge tank.   A valve is also assembled between the heater and the surge tank.   The inlet throttling through this valve has a stabilizing effect on the system stability.

The rotameter that is installed before the surge tank is manufactured by Rotameter Mfg. Co. Ltd. (Series 1000, range 0.3-2.4 lbs/m) and calibrated for Freon-11.   The needle valve is placed between the surge tank and the rotameter.   The flow rate is controlled by means of this valve.

To measure the pressure variations at the inlet of the heater, a pressure transducer manufactured by Southern Instruments Ltd. is used.   The transducer is type T500 with a range of 0-100 psig.   It is designed for use with the Southern Instruments type M1861 read-out unit.   The read-out unit is connected to a Hewlett-Packard 7706B recorder to record the variations of pressure.

The so-called "Flow-Through" copper-constantan thermocouples are employed for the temperature measurements. The term "Flow-Through" implies that the hot junction is placed inside the tube. Only the fluid temperatures at the inlet and exit of the heater are measured. The thermocouples are also connected to the same recorder. The detailed information is given in Ref. [19].

3. EXPERIMENTAL PROCEDURE

The experiments were carried out in three series with Freon-11 as the test fluid. During these series of experiments, inlet temperature and main tank to exit pressure drop were kept constant. The inlet fluid temperature was found to stay at 20°C without any control unit. So the subcooling controller unit was excluded from the system. Main tank and exit pressures were kept at 7 kgf/cm^2 and 2 kgf/cm^2 respectively. This way, the fluid was forced to flow through the heater.

In these experiments, exit restriction and heat fluxes were chosen as test parameters while mass flow rate was varied in small decrements. A different orifice was installed for each series of experiments. Instabilities were investigated for three different orifice diameters, 1.4 mm, 1.6 mm, and 1.8 mm, respectively.

The heat fluxes were changed from 300 W to 600 W. The flow rates were reduced incrementally between 2.4 and 0.3 lb/min to determine the steady state pressure drop, Δp_s, versus mass flow rate curve and the boundaries of pressure-drop and density-wave type oscillations. The system geometry was not altered for good repeatability.

The experimental procedure can be summarized as follows:

i)   The main tank and the recovery tank were pressurized after the electronic instruments were warmed up and calibrated.

ii) The control valve was opened to let the fluid flow at the highest rate the rotameter could read.

iii) The heat input was increased to the required level.

iv) The system was allowed to become steady.

v)   The measurements of surge tank pressure and mass flow rate were taken. The inlet pressure and the exit fluid temperature were recorded by the recorder. The system stability was observed and noted.

vi) The mass flow rate was reduced by a small amount using the control valve placed before the surge tank.

vii) Starting from item "iv" the procedure was repeated.

As stated earlier, the pressure drop occurring between the surge tank and the heater plays an important role in the stability of the system. Thus, it has to be kept constant for a better comparison of the experimental results. This was accomplished by

checking this pressure drop at a selected reference flow rate
before each test run.

4. EXPERIMENTAL RESULTS AND DISCUSSION

Pressure-drop and density-wave type oscillations were observed
during the experiments. Figures 2 and 3 show the typical pressure-
drop type oscillations of inlet pressure and exit fluid temperature
for two different mass flow rates at the same heat input rate. All
the pressure-drop type oscillations observed had density-wave type
oscillations superimposed on them. As the flow rate was reduced,
these density-wave type oscillations lasted for a longer time
making the period of pressure-drop type oscillations larger. Two
typical recordings of density-wave type oscillations are also given
by Figures 4 and 5.

Some of the important experimental readings and data calculated for selected test points are presented in Tables 1 through 4.
The mean mass flow rate, inlet to exit density ratio, exit quality,
system pressure drop, the periods and amplitudes of oscillations
are also included in these tables. The pressure drop from surge
tank exit to orifice exit was taken as the "system pressure drop."

The experimental results have been plotted on system pressure
drop versus mass flow rate plane, for three different orifices and
five different heat inputs. Experimentally determined regions of
stability are indicated on these curves. Figures 6, 7 and 8 are
the plots for various heat inputs with 1.4 mm, 1.6 mm, and 1.8 mm
orifice diameters, respectively. For a better understanding of the
effect of exit restriction on system stability, the same results
are plotted for various exit restrictions at the fixed heat input
rates as shown in Figures 9 through 12. The orifices are indicated
by their corresponding loss coefficients, K, in the case of a
single phase flow.

The general behavior of the curves is similar. The pressure-
drop type oscillations occur in the region where $dP_s/d\dot{m}$ is negative.
The density-wave type oscillations are induced at the left of the
maximum point of the curves except for 1.4 mm orifice diameter at
317 W heat input rate. In this case, the dry-out occurs before
the density-wave type oscillations.

5. CONCLUSIONS OF THE EXPERIMENTAL ANALYSIS

According to the experiments, the following conclusions can
be derived.

(1) The period of pressure-drop type oscillations increases
    towards the stability boundaries (Tables 1 through 4).

(2) The period of pressure-drop type oscillations increases with
    decreasing orifice diameter or increasing heat flux (Tables
    1 through 4). This fact is believed to be related to the
    compressibility in the surge tank. As the orifice diameter
    is reduced or the heat flux is increased, the pressure inside
    the surge tank becomes higher. In other words, compression

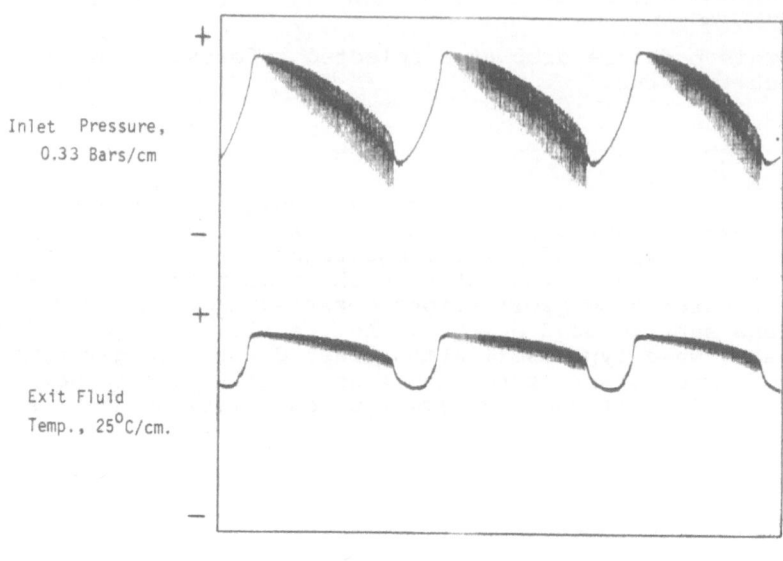

Inlet Pressure,
0.33 Bars/cm

Exit Fluid
Temp., 25°C/cm.

Time, 40 sec/cm

Fig. 2. Typical Pressure-Drop Type Oscillations (Orifice
diameter = 1.4 mm; ṁ = 0.006 kg/s; heat input = 504.3 W)

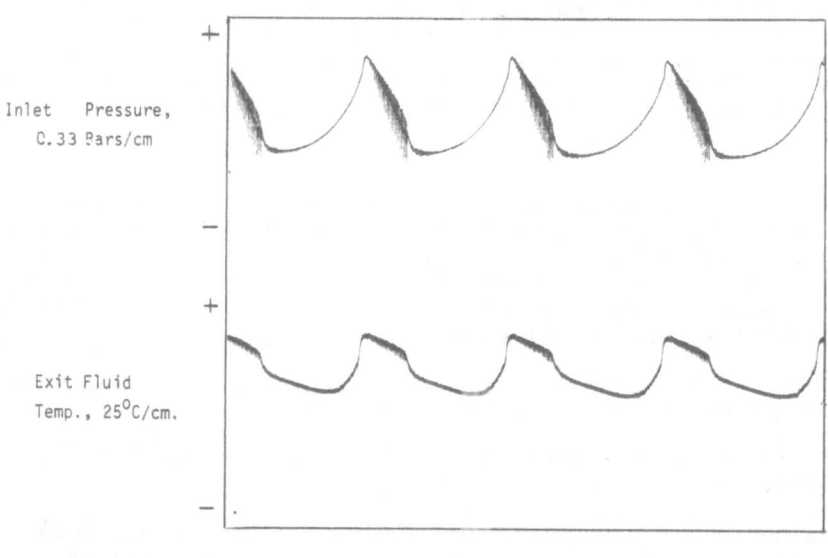

Inlet Pressure,
0.33 Bars/cm

Exit Fluid
Temp., 25°C/cm.

Time, 40 sec/cm

Fig. 3. Typical Pressure-Drop Type Oscillations (Orifice
diameter = 1.4 mm; ṁ = 0.011 kg/s; heat input = 504.3 W)

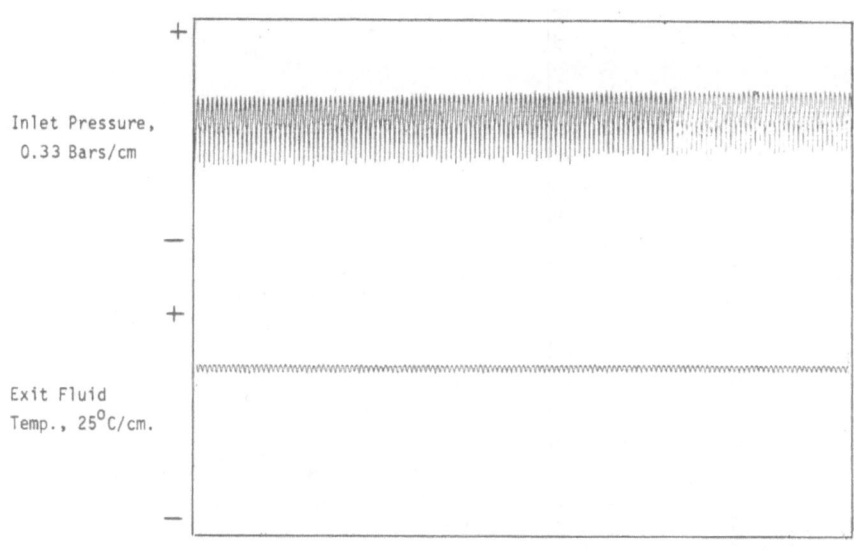

Inlet Pressure,
0.33 Bars/cm

Exit Fluid
Temp., 25°C/cm.

Time, 10 sec/cm.

Fig. 4.   Typical Density-Wave Type Oscillations (Orifice
diameter = 1.8 mm; ṁ = 0.004 kg/s; heat input = 607.5 W)

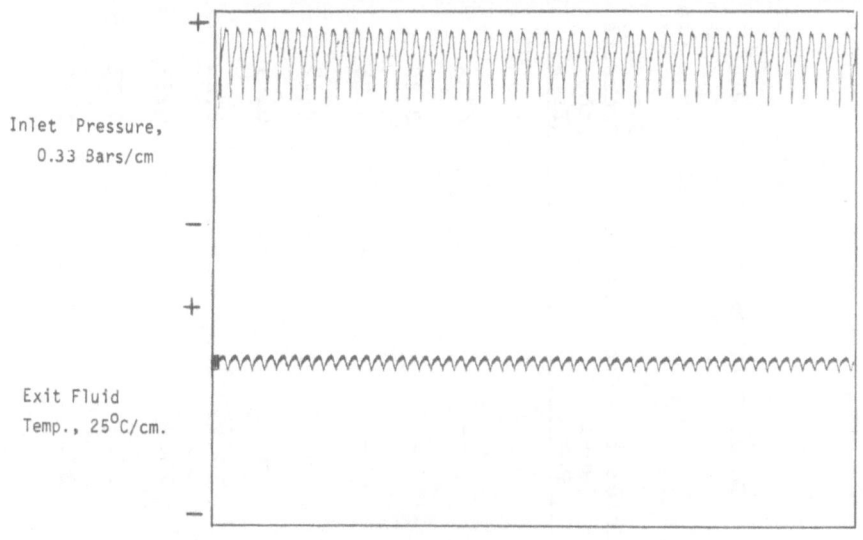

Inlet Pressure,
0.33 Bars/cm

Exit Fluid
Temp., 25°C/cm.

Time, 4 sec/cm

Fig. 5.   Typical Density-Wave Type Oscillations (Orifice
diameter = 1.4 mm; ṁ = 0.004 kg/s; heat input = 600.1 W)

Table 1 Experimental Results

Total Heat Input: 321.75 Watts   Orifice Diameter: 1.6 mm

System Exit Pressure: 3.0 Bars

Test Run No.	Mass Flow Rate (kg/s)	Inlet to Exit Density Ratio	Exit Quality (%)	System Pressure Drop (Bars)	Pressure-Drop Type Osc. Period (sec)	Amp. (Bars)	Density-Wave Type Osc. Period (sec)	Amp. (Bars)
1	0.0174	--	--	0.45	--	--	--	--
2	0.0154	--	--	0.36	--	--	--	--
3	0.0134	--	--	0.30	--	--	--	--
4	0.094	--	--	0.25	--	--	--	--
5	0.0082	2.87	1.9	0.19	56	0.16	--	--
6	0.0072	5.87	5.2	0.19	66	0.16	--	--
7	0.005	16.1	16.8	0.25	76	0.16	--	--
8	0.004	23.9	26.1	0.33	100	0.16	--	--
9	0.003	32.4	36	0.37	--	--	--	--
10	0.003	37.6	42.2	0.36	--	--	1.5	0.1

Table 2 Experimental Results

Total Heat Input: 504.3 Watts   Orifice Diameter: 1.4 mm

System Exit Pressure: 3.0 Bars

Test Run No.	Mass Flow Rate (kg/s)	Inlet to Exit Density Ratio	Exit Quality (%)	System Pressure Drop (Bars)	Pressure-Drop Type Osc. Period (sec)	Pressure-Drop Type Osc. Amp. (Bars)	Density-Wave Type Osc. Period (sec)	Density-Wave Type Osc. Amp. (Bars)
1	0.0178	--	--	1.02	--	--	--	--
2	0.014	--	--	0.78	--	--	--	--
3	0.012	--	--	0.76	--	--	--	--
4	0.011	3.1	2.6	0.78	132	0.7	--	--
5	0.009	7.4	8.3	0.88	108	0.7	--	--
6	0.0075	11.9	14.6	1.0	140	0.7	--	--
7	0.006	18.4	24.4	1.14	164	0.7	--	--
8	0.0045	29.4	41.2	1.28	272	0.7	--	--
9	0.004	34.7	50	1.35	--	--	--	--
10	0.0025	67	96.7	1.33	--	--	1.2	0.32

Table 3 Experimental Results

Total Heat Input: 498.15 Watts   Orifice Diameter: 1.6 mm

System Exit Pressure: 3.0 Bars

Test Run No.	Mass Flow Rate (kg/s)	Inlet to Exit Density Ratio	Exit Quality (%)	System Pressure Drop (Bars)	Pressure-Drop Type Osc.		Density-Wave Type Osc.	
					Period (sec)	Amp. (Bars)	Period (sec)	Amp. (Bars)
1	0.0177	--	--	0.5	--	--	--	--
2	0.0165	--	--	0.46	--	--	--	--
3	0.0144	--	--	0.43	--	--	--	--
4	0.0132	--	--	0.53	80	0.4	--	--
5	0.01	5.4	5.4	0.7	68	0.4	--	--
6	0.009	7.8	8.5	0.76	72	0.4	--	--
7	0.0077	11.6	13.8	0.85	88	0.4	--	--
8	0.0065	16.8	20.8	0.91	100	0.4	--	--
9	0.0044	33.0	43.1	0.97	--	--	--	--
10	0.004	38.1	50.1	0.97	--	--	--	--
11	0.0032	53.0	69.5	0.93	--	--	1.0	0.30

Table 4 Experimental Results

Total Heat Input: 504.3 Watts Orifice Diameter: 1.8 mm

System Exit Pressure: 3.0 Bars

Test Run No.	Mass Flow Rate (kg/s)	Inlet to Exit Density Ratio	Exit Quality (%)	System Pressure Drop (Bars)	Pressure-Drop Type Osc.		Density-Wave Type Osc.	
					Period (sec)	Amp. (Bars)	Period (sec)	Amp. (Bars)
1	0.0174	--	--	0.446	--	--	--	--
2	0.0156	--	--	0.43	--	--	--	--
3	0.0144	--	--	0.39	--	--	--	--
4	0.0137	--	--	0.39	--	--	--	--
5	0.01	6.7	6.7	0.53	56	0.26	--	--
6	0.0073	15.0	17.4	0.67	80	0.26	--	--
7	0.006	21.8	26.3	0.72	180	0.26	--	--
8	0.0055	25.4	31.0	0.73	--	--	--	--
9	0.0045	35.0	43.4	0.75	--	--	1.0	0.23

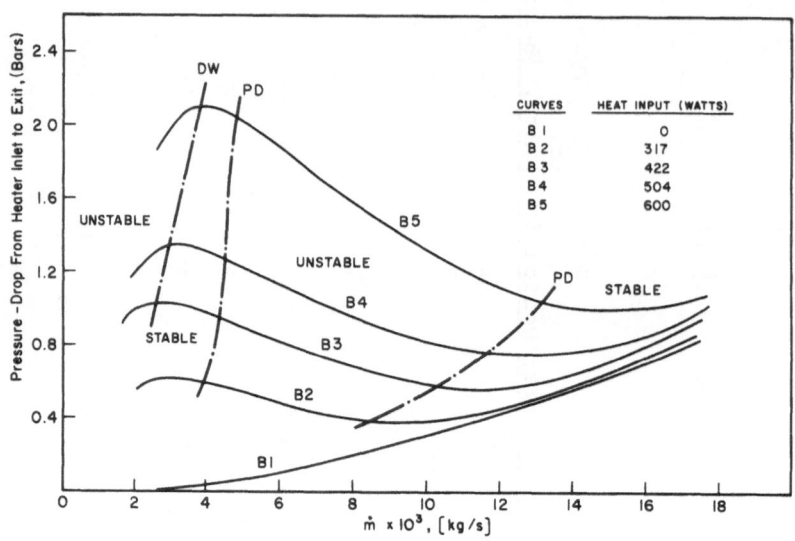

Fig. 6. Steady-State Characteristics of System at Various Heat Inputs with Stability Boundaries for 1.4 mm Orifice Diameter

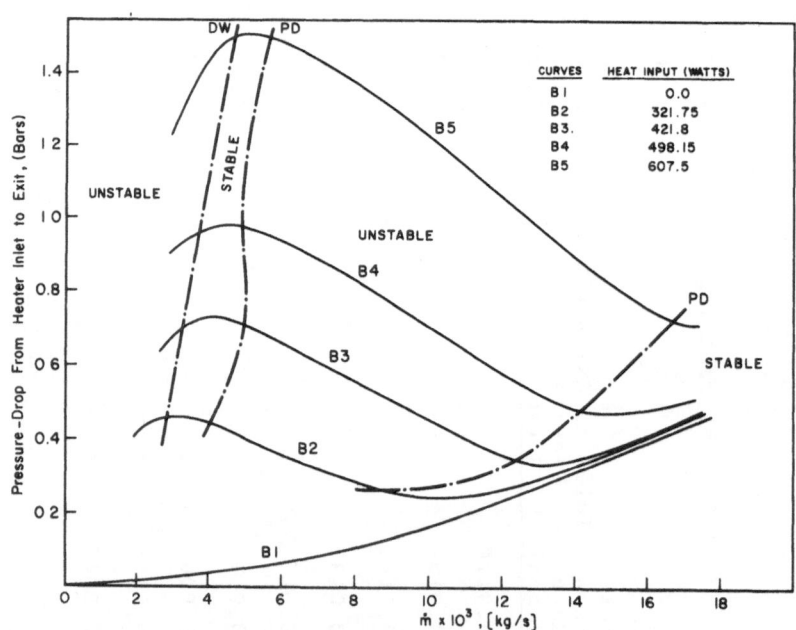

Fig. 7. Steady-State Characteristics of System at Various Heat Inputs with Stability Boundaries for 1.6 mm Orifice Diameter

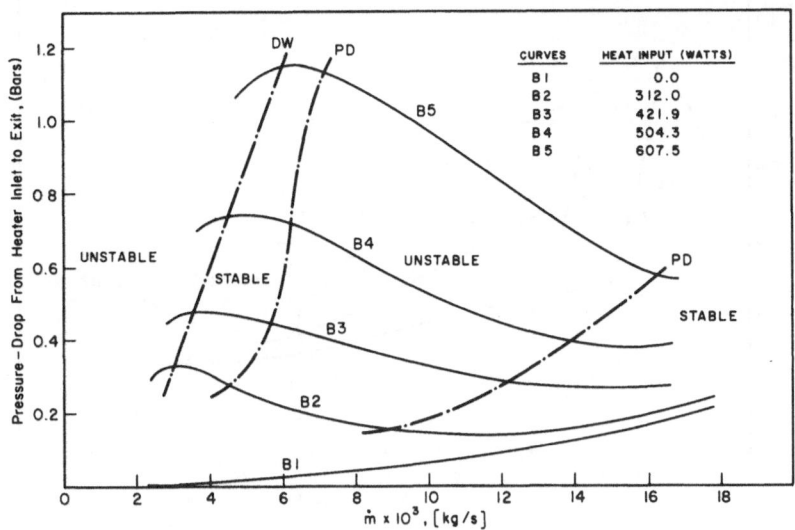

Fig. 8. Steady-State Characteristics of System at Various Heat Inputs with Stability Boundaries for 1.8 mm Orifice Diameter

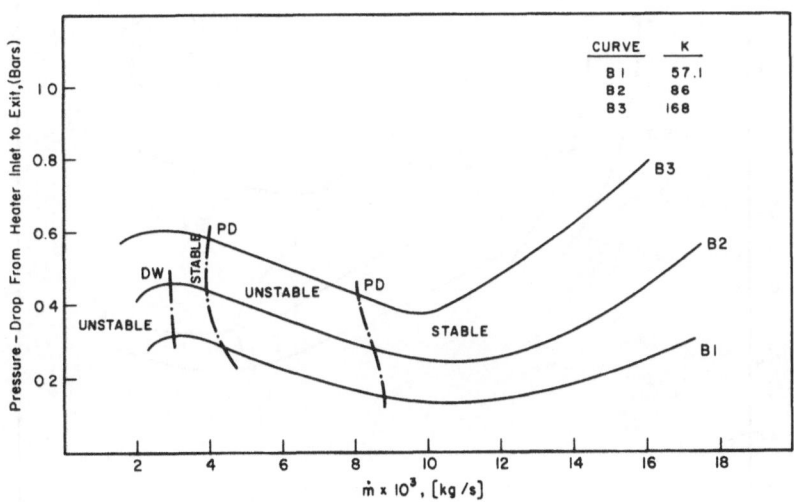

Fig. 9. Steady-State Characteristics of System at Various Exit Restrictions with Stability Boundaries for 320 W Heat Input

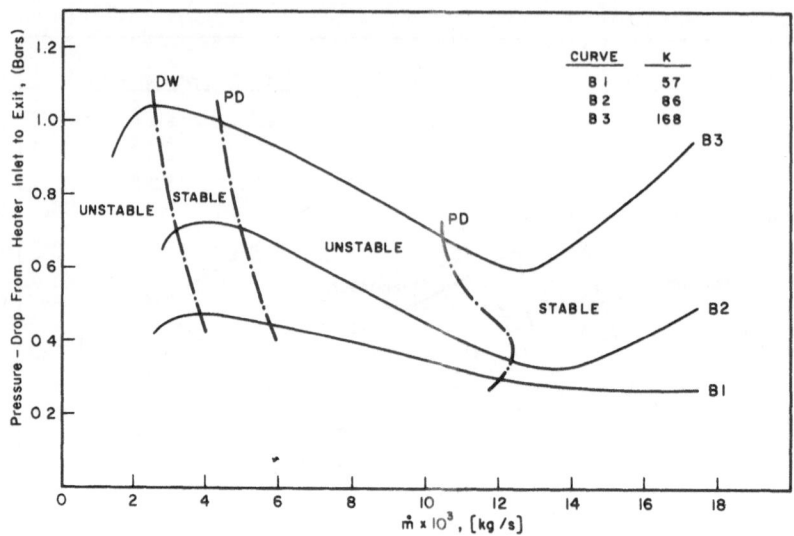

Fig. 10.   Steady-State Characteristics of System at Various Exit Restrictions with Stability Boundaries for 420 W Heat Input

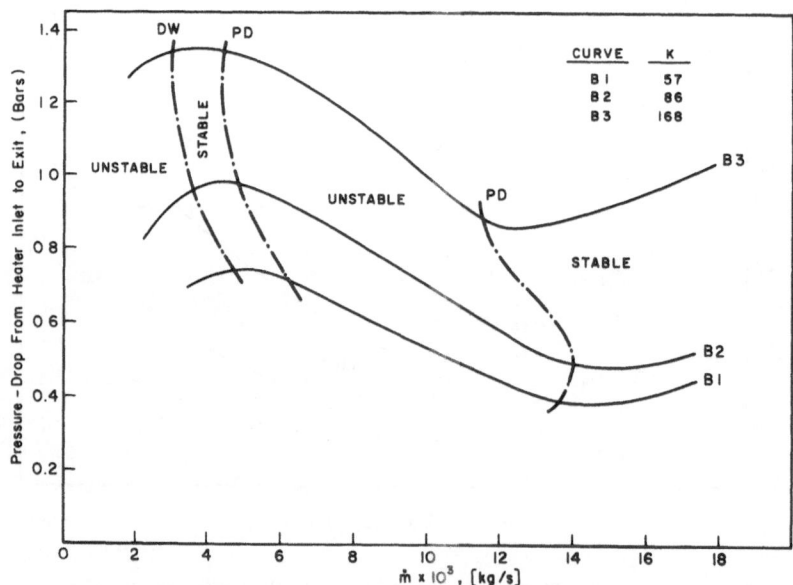

Fig. 11.   Steady-State Characteristics of System at Various Exit Restrictions with Stability Boundaries for 500 W Heat Input

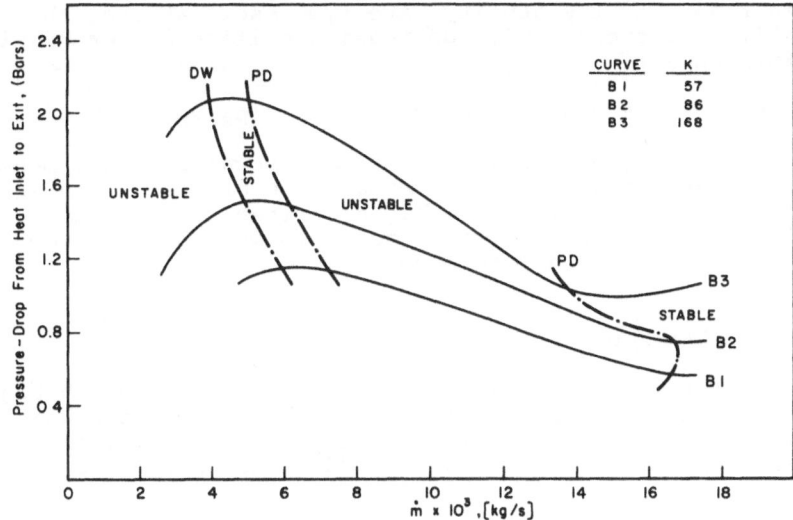

Fig. 12.   Steady-State Characteristics of System at Various Exit Restrictions with Stability Boundaries for 600 W Heat Input

of compressible volume results in an increase in the periods
of pressure-drop type oscillations.

(3) The amplitude of pressure-drop type oscillations is in the
order of the pressure drop difference between the maximum and
minimum points of Δp_S versus $\dot{m}$ curves.   This means higher
amplitudes at smaller orifice diameters or higher heat fluxes
(Tables 1 through 4).

(4) The period of density-wave type oscillations increases with
decreasing orifice diameter or decreasing heat flux (Tables 1
through 4).

(5) The amplitude of density-wave type oscillations increases with
a decrease in orifice diameter or an increase in heat flux
(Tables 1 through 4).

(6) Pressure-drop type oscillations are observed when the slope of
Δp_S versus $\dot{m}$ curve is sufficiently negative (Figures 6
through 8).

(7) Density-wave type oscillations occur at the left of maximum
point of Δp_S versus $\dot{m}$ curve (Figures 6 through 8).

(8) At a fixed orifice diameter, the system stability decreases
with respect to both instabilities with an increase in heat
input rate (Figures 6 through 8).

(9) With regard to the density-wave type oscillations, the system stability increases with decreasing orifice diameter (Figures 9 through 12).

(10) As the orifice diameter gets smaller, Δp_s versus $\dot{m}$ curve becomes steeper. This makes the rightmost stability boundary of pressure-drop type oscillations move towards higher mass flow rates. But as we further decrease the orifice diameter, the boiling is retarded due to higher pressure in the system. This causes the stability boundary to move back to lower mass flow rates (Figures 9 through 12).

6. MATHEMATICAL MODEL

The geometry considered in this analysis is shown in Fig. 13. It is assumed that the pressure drop in the heater is negligible when compared with that of exit restriction. To simplify the mathematical approach the following additional assumptions are introduced:

1. A homogeneous flow is assumed.

2. The flow is assumed to be one dimensional.

3. Thermodynamic equilibrium is assumed to exist between the phases.

4. The subcooled boiling is ignored.

5. The heat loss from the heater to the surrounding is considered to be negligible.

6. The heat capacity of the wall of heater is neglected. Thus constant and uniform heat flux into the fluid is assumed.

7. Density is assumed to be a function of enthalpy alone.

8. All contributions to fluid enthalpy change by pressure changes with space and time, dissipation, kinetic energy changes are neglected.

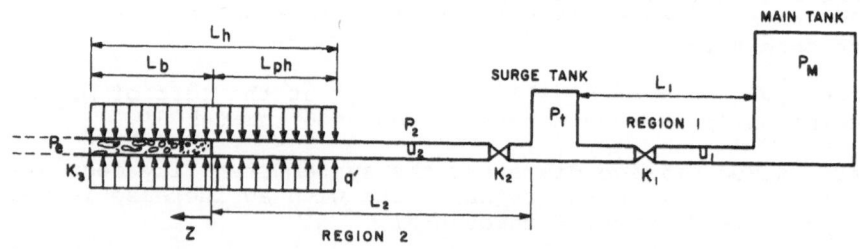

Fig. 13.   Schematic Diagram of the Boiling Flow System, Used for Mathematical Analysis

9.   Pressure drops are assumed to occur across the restrictions.

Using the above assumptions, the conservation equations for mass, energy, and momentum can be written as:

$$\frac{\partial \rho}{\partial t} + \frac{\partial G}{\partial z} = 0 \tag{1}$$

$$\rho \frac{\partial h}{\partial t} + G \frac{\partial h}{\partial z} = \frac{q}{A} \tag{2}$$

$$\frac{\partial G}{\partial t} + \frac{\partial}{\partial z}\left(\frac{G^2}{\rho}\right) = -\frac{\partial p}{\partial z} - \tau_w \frac{C}{A} \tag{3}$$

where

$$G = \rho u \tag{4}$$

After neglecting the frictional and the momentum flux terms and integrating the equation along z, Eq. (3) becomes

$$\Delta P = \int_{z_1}^{z_2} \frac{\partial G}{\partial t} \, dz + \Delta P_r \tag{5}$$

where ΔP_r is the pressure drop across the restrictions.

The equation of state is

$$\rho = \rho(h) \tag{6}$$

7.   METHOD OF SOLUTION

An analytic approach is used to study the system stability. The conservation equations are linearized and Laplace-transformed. The resulting characteristic equation of system is then analyzed for stability by means of Nyquist plots.

Region 1. For this region the density is constant (Fig. 15). Then $\overline{G_1}$ is only a function of time. The momentum equation for this region is then

$$P_M - P_t = L_1 \frac{dG_1}{dt} + \Delta P_1 \tag{7}$$

Linearization and Laplace-transformation of Eq. (7) gives

$$-\delta P_t = \left[L_1 s + \left(\frac{d\Delta P_1}{dG_1}\right)_{t=0}\right] \delta G_1 \tag{8}$$

Surge tank dynamics. Continuity equation for the surge tank can be written as

$$G_1 A - G_2 A = \rho_\ell \frac{dv_\ell}{dt} = -\rho_\ell \frac{dv_g}{dt} \tag{9}$$

Using ideal gas and constant temperature process assumption, one can obtain

$$G_1 - G_2 = \frac{P_{to} V_{go} \rho_\ell}{A \, p_t^2} \frac{dp_t}{dt} \tag{10}$$

Linearization and Laplace-transformation of the above equation gives

$$\delta G_1 - \delta G_2 = Rs\delta P_t \tag{11}$$

where

$$R = \frac{V_{go}\rho_\ell}{AP_{to}} \tag{12}$$

Boiling region. Using the equation of state (6), the following relations can be obtained:

$$\frac{\partial \rho_b}{\partial t} = \frac{\partial \rho_b}{\partial h_b} \frac{\partial h_b}{\partial t} \tag{13}$$

$$\frac{\partial \rho_b}{\partial z} = \frac{\partial \rho_b}{\partial h_b} \frac{\partial h_b}{\partial z} \tag{14}$$

where ρ and h are the homogeneous mixture properties given by

$$\frac{1}{\rho} = v_f + xv_{fg} \tag{15}$$

$$h = h_f + xh_{fg} \tag{16}$$

By the use of these mixture properties, it can be shown that

$$\frac{\partial \rho_b}{\partial h_b} = -\rho_b^2 \frac{v_{fg}}{h_{fg}} \tag{17}$$

First of all the steady state solutions of velocity and density have to be found as the initial conditions of the problem. In order to find them, the continuity and energy equations together with Eqs. (13), (14), and (17) are written in the following form:

$$\rho_{bo} \frac{du_{bo}}{dz} + u_{bo} \frac{d\rho_{bo}}{dz} = 0 \tag{18}$$

$$\frac{dh_{bo}}{dz} = \frac{q}{A\rho_{bo}u_{bo}} \tag{19}$$

$$\frac{d\rho_{bo}}{dz} = \rho_{bo}^2 \frac{v_{fg}}{h_{fg}} \frac{dh_{bo}}{dz} \tag{20}$$

After substitution of Eq. (19) into Eq. (20) and eliminating $d\rho_{bo}/dz$ in Eq. (18) by Eq. (20), one obtains

$$\frac{du_{bo}}{dz} = \frac{v_{fg}q}{h_{fg}A} = \Omega \tag{21}$$

It should be noted that the term Ω is vapor volume generation rate per unit volume of boiling region.

Eq. (21) can be solved by the following boundary condition:

$$z = 0 \qquad u_{bo} = u_{20} = u_{10} \tag{22}$$

Then Eq. (21) yields

$$u_{bo} = \Omega z + u_{10} \tag{23}$$

The density distribution can now be found from the continuity equation (18) as

$$\rho_{bo} = \frac{\rho_\ell u_{10}}{\Omega z + u_{10}} \tag{24}$$

Similarly, for unsteady state analysis, the continuity equation becomes:

$$\frac{\partial \rho_b}{\partial t} + \rho_b \frac{\partial u_b}{\partial z} + u_b \frac{\partial \rho_b}{\partial z} = 0 \tag{25}$$

Substituting Eqs. (13), (14), and (17) into Eq. (25) and using energy equation (2) to eliminate the enthalpy terms, one obtains

$$\frac{\partial u_b}{\partial z} = \Omega \tag{26}$$

Linearization and Laplace-transformation of Eqs. (25) and (26) yields the following:

$$\frac{d\delta u_b}{dz} = 0 \tag{27}$$

$$s\delta\rho_b + \frac{du_{bo}}{dz}\delta\rho_b + \rho_{bo}\frac{d\delta u_b}{dz} + u_{bo}\frac{d\delta\rho_b}{dz} + \frac{d\rho_{bo}}{dz}\delta u_b = 0 \tag{28}$$

Eq. (27) implies that δu_b is not a function of z. Then,

$$\delta u_b = C_1 \tag{29}$$

Since z is measured from the point where boiling starts at

steady state, the boundary conditions can be specified at
$z = \Delta L_{ph}(t)$. The first boundary condition is

$$u_b(\Delta L_{ph},t) = u_2(t) \tag{30}$$

After linearization, Eq. (30) becomes

$$\Delta u_b + u_{bo} + \frac{du_{bo}}{dz} \Delta L_{ph} = u_{20} + \Delta u_2 \qquad \text{at } z = 0 \tag{31}$$

Since $(u_{bo})_{z=0} = u_{20}$, Laplace-transformation gives (at $z = 0$)

$$\delta u_b + \frac{du_{bo}}{dz} \delta L_{ph} = \delta u_2 \tag{32}$$

It can be written that

$$L_{ph} = \int_{t-\tau}^{t} u_2 dt \tag{33}$$

or

$$\frac{dL_{ph}}{dt} = u_2(t) - u_2(t-\tau) \tag{34}$$

After linearization and Laplace-transformation of Eq. (34), one
obtains

$$\delta L_{ph} = \frac{1-e^{-s\tau}}{s} \delta u_2 \tag{35}$$

Substitution of Eq. (35) for δL_{ph} into Eq. (32) and the use of
Eq. (21) yields

$$\delta u_b = \left[\frac{s-\Omega+\Omega e^{-s\tau}}{s} \right] \delta u_2 \tag{36}$$

The term δu_b can be eliminated from Eq. (28) by using Eq. (36), then
Eq. (28) becomes

$$\frac{d\delta \rho_b}{dz} + \frac{s+\Omega}{\Omega z+u_{10}} \delta \rho_b = \frac{\Omega \rho_\ell u_{10}}{(\Omega z+u_{10})^3} \left[\frac{s-\Omega+\Omega e^{-s\tau}}{s} \right] \delta u_2 \tag{37}$$

The boundary condition is

$$\rho_b(\Delta L_{ph},t) = \rho_\ell \tag{38}$$

After linearization and Laplace-transformation of Eq. (38), one
gets

$$\delta\rho_b + \frac{d\rho_{bo}}{dz} \delta L_{ph} = 0 \qquad \text{at } z = 0 \tag{39}$$

Using Eqs. (24) and (36), $\delta\rho_b$ can be found as

$$\delta\rho_b = \frac{\Omega\rho_\ell[1-e^{-s\tau}]}{u_{10}s} \delta u_2 \qquad \text{at } z = 0 \tag{40}$$

Eq. (37) can now be solved by the use of Eq. (40) to get

$$\delta\rho_b = \frac{\Omega\rho_\Omega u_{10}[s-\Omega+\Omega e^{-s\tau}]}{(s-\Omega)(\Omega z+u_{10})} \delta u_2 - \frac{\rho_\ell\Omega e^{-s\tau}\delta u_2}{(\Omega z+u_{10})(s-\Omega)} e^{-\frac{s}{\Omega}\ell n\frac{\Omega z+u_{10}}{u_{10}}} \tag{41}$$

The pressure drop from surge tank to exit can now be determined by using Eq. (5). Neglecting the inertia of the boiling length, Eq. (5) becomes

$$P_t - P_e = L_2 \frac{dG_2}{dt} + K_2 \frac{G_2^2}{\rho_\ell} + \Delta p_{or} \tag{42}$$

where Δp_{or} is the pressure drop across the exit restriction. Linearization and Laplace-transformation of Eq. (42) gives

$$\delta p_t = [L_2 s + 2K_2 u_{10}]\delta G_2 + \left(\frac{d\Delta p_{or}}{dG_e}\right)_{t=0} \delta G_e \tag{43}$$

where

$$\delta G_e = \rho_{eo}\delta u_e + u_{eo}\delta\rho_e \tag{44}$$

Since $z = L_b$ at exit, substitution of Eqs. (21), (24), (36), and (41) into Eq. (44) gives

$$\delta G_e = F(s)\,\delta G_2 \tag{45}$$

where

$$F(s) = \frac{u_{10}}{u_{eo}} + \frac{\Omega}{s-\Omega} e^{-s\tau}\left[\frac{u_{10}}{u_{eo}} - e^{-\frac{s}{\Omega}\ell n\frac{u_{10}}{u_{eo}}}\right]\delta G_2 \tag{46}$$

Then Eq. (43) becomes

$$\delta p_t - [L_2 s + 2K_2 u_{10} + \left(\frac{d\Delta p_{or}}{dG_e}\right)_{t=0} F(s)]\,\delta G_2 = 0 \tag{47}$$

The resulting system of equations are the Eqs. (8), (11) and (47):

$$\delta p_t + \left[L_1 s + \left(\frac{d\Delta p_1}{dG_1} \right)_{t=0} \right] \delta G_1 = 0$$

$$Rs\delta p_t - \delta G_1 + \delta G_2 = 0$$

$$\delta p_t - \left[L_2 s + 2K_2 u_{10} + \left(\frac{d\Delta p_{or}}{dG_e} \right)_{t=0} F(s) \right] \delta G_2 = 0$$

The characteristic equation of this system can be found by equating the determinant of coefficient matrix of the above system of equations to zero, which is

$$\left[L_2 s + 2K_2 u_{10} + \left(\frac{d\Delta p_{or}}{dG_e} \right)_{t=0} F(s) \right] + \left[L_1 s + \left(\frac{d\Delta p_1}{dG_1} \right)_{t=0} \right] \left[1 + \right.$$

$$\left. Rs \left[L_2 s + 2K_2 u_{10} + \left(\frac{d\Delta p_{or}}{dG_e} \right)_{t=0} F(s) \right] \right] = 0 \qquad (48)$$

8.   COMPARISON OF ANALYTICAL RESULTS

The characteristic equation (48) has two unknowns. These are $\left(\frac{d\Delta p_{or}}{dG_e} \right)_{t=0}$ and $\left(\frac{d\Delta p_1}{dG_1} \right)_{t=0}$. The second term is relatively easy to determine. But, the first term requires a good estimation of the steady state characteristics of exit orifice. Unfortunately, there are not any generally applicable formulae to be used in determining the pressure drop across a restriction in two-phase flow. For simplicity, it can be assumed that the pressure drop at the orifice is in the form of

$$\Delta p_{or} = p_2 - p_e = K_3 \rho_e u_e^2 \qquad (49)$$

where K_3 is the corresponding loss coefficient of orifice in single phase flow. This equation can be written for steady state by using Eqs. (23) and (24)

$$(\Delta p_{or})_{t=0} = K_3 \rho_\ell u_{10} (\Omega L_{bo} + u_{10}) \qquad (50)$$

Differentiation of this equation with respect to G_e gives

$$\left(\frac{d\Delta p_{or}}{dG_e} \right)_{t=0} = \left(\frac{d\Delta p_{or}}{dG_2} \right)_{t=0} = 2K_3 u_{10} + K_3 \Omega L_{bo} - K_3 \Omega L_{ph} \qquad (51)$$

Similarly, the pressure drop in Region 1 can be lumped into

$$(\Delta p_1)_{t=0} = K_1 \rho_\ell u_{10}^2 \qquad (52)$$

so

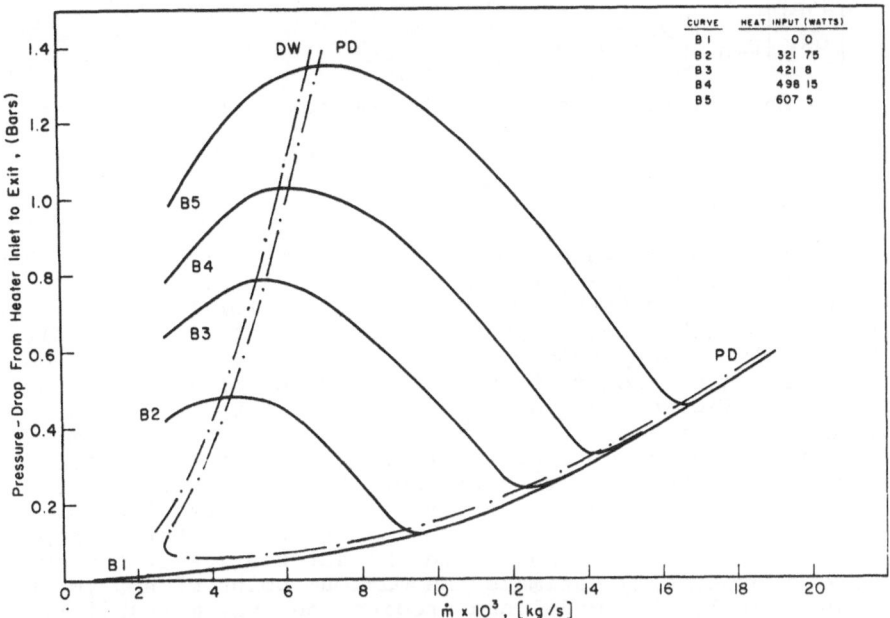

Fig. 14. Predicted Characteristics of System at Various Heat
Inputs for 1.6 mm Orifice Diameter

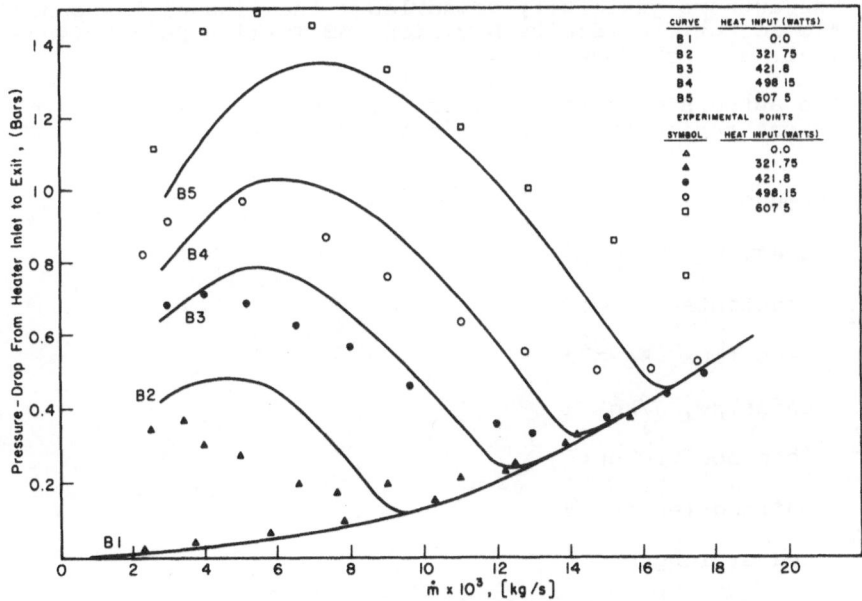

Fig. 15. Predicted Steady-State Characteristics of System at
Various Heat Inputs for 1.6 mm Orifice Diameter

$$\left(\frac{d\Delta p_1}{dG_1}\right)_{t=0} = 2K_1 \rho_\ell u_{10} \tag{53}$$

The characteristic equation is now ready to be analyzed. If the characteristic equation is divided by the highest order term, a function in the following form can be obtained.

$$1 + G(s) = 0$$

where $G(s)$ is an equivalent open loop transfer function which can be analyzed for stability by using Nyquist plots. Figure 14 gives the thresholds of unstable regions of system for 1.6 mm orifice diameter at various heat inputs. It is clear that the results are qualitatively in very good agreement with the experimental data. The stable region around the maximum point of Δp_s versus $\dot{m}$ curve is very well predicted. A quantitative comparison is very hard to make because of the approximation made in Eq. (49). A comparison of the predicted steady state characteristics of system by Eq. (49) is given in Figure 15. It shows that there is considerable error involved in determining the steady state characteristics of the system by Eq. (49). But, it is obvious that if a better approximation is available for steady state characteristics of the orifice, which will especially estimate the maximum point of Δp_s versus $\dot{m}$ curve accurately, the model can predict the system stability very well.

ACKNOWLEDGEMENTS

The authors gratefully acknowledge the financial support of the NATO Scientific Affairs Division and Turkish Scientific and Technical Research Council.

The authors wish to thank Dr. O. Yesin for his valuable suggestions during the completion of this work.

NOMENCLATURE

A   area, m^2

C_1, C_2 constants

G   mass flux, $kg/m^2 s$

h   enthalpy, J/kg

K   loss coefficient

L_b boiling length, m

L_h heater length, m

L_{ph} preheater length, m

L_1 length of region 1, m

L_2 length of region 2, m

m   mass flow rate, kg/s

p   pressure, N/m^2

Δp_s system pressure drop, from inlet of heater to exit, N/m^2

q   heat input rate per unit length, w/m

R $\dfrac{V_{go}\rho_\ell}{A\, p_{to}}$, s^2/m

s   complex variable, 1/s

τ residence time of liquid particle in preheater, s

t   time, s

u   velocity, m/s

V   volume, m^3

v   specific volume, m^3/kg

x   quality

z   space variable, m

Δ perturbated quantity

δ Laplace-transformed perturbated quantity

Ω $\dfrac{v_{fg}q}{h_{fg}A}$, 1/s

ρ density, kg/m^3

Subscripts

b   boiling region

e   exit

f   saturated liquid

fg   difference between the saturated liquid and gas

g   saturated gas

ℓ subcooled liquid

M   main tank

or   orifice

r   restriction

t   surge tank

1   region 1

2   region 2

REFERENCES

1. Stenning, A. H. and Veziroglu, T. N., Flow Oscillation Modes
    in Forced Convection Boiling, Proceedings of the 1965 Heat
    Transfer and Fluid Mechanics Institute (1965).

2. Yadigaroglu, G. and Bergles, A. E., An Experimental and
    Theoretical Study of Density-Wave Oscillations in Two-Phase
    Flow, MIT Report DSR74629-3 (1969).

3. Ishii, M., Zuber, N., Thermally Induced Flow Instabilities in
    Two-Phase Mixtures, 4th International Heat Transfer Confer-
    ence, Paris, Paper B5-11 (1970).

4. Quandt, E. R., Analysis and Measurement of Flow Oscillations,
    Chemical Engineering Progress Symposium Series, No. 32,
    Vol. 57 (1960).

5. Wallis, G. B., and Heasley, J. H., Oscillations in Two-Phase
    Flow Systems, Journal of Heat Transfer, 83, 363 (1961).

6. Meyer, J. E., and Rose, R. P., Application of a Momentum
    Integral Model to the Study of Parallel Channel Boiling Flow
    Oscillations, J. Heat Transfer, ASME, Series C, 85, 1 (1963).

7. Maulbetsch, J. S., and Griffith, P., A Study of System-Induced
    Instabilities in Forced-Convection Flows with Subcooled
    Boiling, MIT, Mechanical Engineering Department (1965).

8. Shotkin, L. M., Stability Considerations in Two-Phase Flow,
    Nuclear Science and Engineering, 28, 317-324 (1967).

9. Stenning, A. H., and Veziroglu, T. N., Oscillations in Two-
    Component Two-Phase Flow, Final Report, NASA CR-72121 (1967).

10. Kakaç, S., Veziroglu, T. N., Akyüzlü, K., and Berkol, O.,
    Sustained and Transient Boiling Flow Instabilities in a Cross-
    Connected Four Parallel Channel Upflow System, 5th Interna-
    tional Heat Transfer Conference, Paper No. B5.11, Tokyo (1974).

11. Kakaç, S., Veziroglu, T. N., Özboya, N., Lee, S. S., Transient
    Boiling Flow Instabilities in a Multi-Channel Upflow System,
    Wärme und Stoffubertragung, 10, 175-188 (1977).

12. Kakaç, S., Veziroglu, T. N., Ergür, H. S., and Uçar, I., The
    Effect of Inlet Subcooling on Sustained and Transient Boiling
    Flow Instabilities in a Single Channel Upflow System, 6th In-
    ternational Heat Transfer Conference, Toronto (1978).

13. Veziroglu, T. N., Kakaç, S., Two-Phase Flow Instabilities and Effect of Inlet Subcooling, Final Report, NSF Project Eng 75-16618.

14. Bouré, J. A., Bergles, A. E., and Tong, L. S., Review of Two-Phase Flow Instability, Nuclear Engineering and Design 25, North-Holland Publishing Company, pp. 165-192 (1973).

15. Yadigaroglu, G., Two-Phase Flow Instabilities and Propagation Phenomena, in Thermodynamics of Two-Phase Systems for Industrial Design and Nuclear Engineering, Eds. J. M. Delhaye, M. Giot and M. L. Riethmuller, Hemisphere Publishing Corporation, New York (1981).

16. Bergles, A. E. and Ishigai, S. (Eds.), Two-Phase Flow Dynamics, Hemisphere Publishing Corporation, New York (1981).

17. Veziroglu, T. N. and Kakaç, S., Two-Phase Flow Instabilities, Final Report, NSF CME 79-20018 (1982).

18. Kakaç, S. and Veziroglu, T. N., Two-Phase Flow Instabilities--A Review of the Current Position in Advances in Two-Phase Flows and Heat Transfer, Eds. Kakaç, S. and Ishii, M., Martinus Nijhoff Publishers BV, The Netherlands (1982).

19. Yıldırım, O. T., Two-Phase Flow Instabilities in a Horizontal Channel, Master Thesis, Middle East Technical University, Ankara (1981).

MASS TRANSPORT

Evaporation of Liquid Droplets: Low and High Mass Flux Data

K.F. LOUGHLIN, L. HADLEY-COATES, and K. HALHOULI
University of Petroleum and Minerals
P.O. Box 144
Dhahran International Airport
Dhahran, Saudi Arabia

ABSTRACT

Experimental data for the evaporation of water, ethyl ether, dichloro-/methane, carbon tetrachloride, acetone and n-octane into a stagnant atmosphere at varying temperatures are reported. The calculation of the wet bulb temperatures is strongly dependent on the Lewis number and on the high mass flux conditions present. The theoretical and experimental high mass flux transfer coefficients show a mean difference of $\pm$ 8.54% provided that the dimensionless groups Sh^0, Nu^0, Pr, Sc, Le and Gr are evaluated at the surface temperature and composition and that Hanna's factor for density variations is incorporated.

1. INTRODUCTION

The simultaneous heat and mass transfer by evaporation has been studied extensively both theoretically and experimentally. Theoretical approaches for spherical droplet evaporation include the use of empirical mass transfer coefficient approximations (Ranz and Marshall (1), Newbold and Amundson (2), and Pei and Gauvin (3) among many), boundary layer techniques such as Prakash and Sirignano (4), and radiative techniques such as Harpole (5) for high temperature. Surprisingly the fundamental Navier-Stokes momentum equation, energy equation and species equation do not yet appear to have been solved for this case. The mechanisms controlling evaporation appear to be, for practical sized droplets, molecular diffusion and natural convection (Ranz and Marshall (1), Pei and Gauvin (3)). Experimental measurement techniques have involved both gravimetric and photographic techniques.

For high mass flux evaporation Spalding's driving force B should be employed for both the heat and mass transfer equations to analyse the data. The mass and heat transfer coefficients should also be corrected for the high mass flux using either the stagnant film theory, laminar boundary layer theory, or penetration theory as summarized by Nienow (6). The wet bulb temperature has seldom been measured and theoretically derived simultaneously except for the low mass transfer studies of Ranz and Marshall (1), and this is a severe limitation in most of the reported studies. Finally doubt as to what mean properties to use in the evaluation of the dimensionless groups in the wet bulb temperature has been expressed by Ranz and Marshall (1), Pei and Gauvin (2) and Downing (7). Skelland (8) summarized the three possible considerations as 1) selection of the average of the bulk and surface properties 2) selection of a mean film temperature and composition to evaluate the physical properties and 3) unusual combinations of properties.

In this study we report on the high mass flux evaporation of 5 organics under natural convective atmospheric conditions at different temperatures.

Also included in some data for the evaporation of water.

2. THEORY

The liquid droplets, suspended from a thin silica fibre, evaporated into a stagnant atmosphere. For the derivation of the wet bulb temperature, we assumed (9); 1) evaporation was controlled by molecular diffusion and natural convection; 2) the droplets are spherical; 3) the temperature of the droplet rapidly attains the wet bulb temperature on being exposed to the atmospheric environment; 4) the recession of the surface promotes the rapid buildup of the steady state temperature and concentration profiles; 5) the heat and mass transfer driving forces are B_h and B_m respectively; 6) to correct for high mass flux, film theory corrections were employed; and 7) significant density variations were incorporated using Hanna's relationship (10). The resulting mass and energy balance expression for the wet bulb temperature is (9).

$$\text{Le Sh*} \left[\frac{\ln(M_o/M_\infty)}{(M_o/M_\infty -1)} \right] \ln(1 + B_m) = \text{Nu*} \ln(1 + B_h) \tag{1}$$

where Le, Sh* and Nu* are the dimensionless Lewis, limiting Sherwood and limiting Nusselt numbers respectively, the logarithmic expressions are the film theory corrections, and the molecular ratio expression is Hanna's density correction factor. In the case of low mass flux equation (1) reduces to

$$\text{Le Sh*} \, B_m = \text{Nu*} \, B_h \tag{2}$$

which is the traditional wet bulb expression given by Bird, Stewart and Lightfoot (11) and Treybal (12) among others, although here it is recast in dimensionless groups.

A macroscopic mass balance gives the rate of change of mass with time. The resulting equation is

$$\frac{d\eta^2}{d\tau} = -C \, \text{Sh*} \tag{3}$$

with $\eta(o) = 1$,

$$C = \frac{\rho}{\rho_\ell} \frac{\ln(M_o/M_\infty)}{(M_o/M_\infty -1)} \ln(1+B_m), \tag{4}$$

and $$\text{Sh*} = 2 + 0.6 \, \text{Gr}^{\frac{1}{4}} \, \text{Sc}^{1/3} \tag{5}$$

assuming we are employing Ranz and Marshall's correlation. On combining equations 3, 4 and 5, the result is

$$\frac{d\eta^2}{d\tau} = -A - B^{3/4} \tag{6}$$

with $\eta(o) = 1$, $A = 2C$, and $B = 0.6C \, \text{Gr}_o^{\frac{1}{4}} \, \text{Sc}^{1/3}$ where Gr_o is the Grashof number at time zero. Equation (6) was solved numerically.

3. EXPERIMENTAL

Data was obtained gravimetrically using a Cahn microbalance and photographically using a Hyspeed camera (J. Holland, Model 10/16). Details are provided in thesis of Halhouli (13).

4. RESULTS AND DISCUSSSION

Experimental data for the evaporation of droplets of water, acetone, carbon tetrachloride, dichloromethane, ethyl ether and n-octane are presented in figures 1 to 9 as plots of dimensionless radius squared versus dimensionless time. Approximately 4 to 6 runs were performed at each temperature indicated by the different symbols employed. All data runs are not shown due to congestion but for the data presented the reproducibility can be observed to be very satisfactory.

To calculate the theoretical curves the wet bulb temperature had first to be calculated using equation (1). For evaluation of Le, Sh* and Nu* in this expression the selection of the mean film temperature is critical. We found that these properties should be evaluated at the surface conditions based on a statistical analysis of the data (9). The procedure to calculate the wet bulb temperature was to assume a surface temperature, evaluate B_h, B_m, and Le, employ equation 5 to estimate the Sh* number and its analogue to estimate Nu*, and iterate equation (1) to convergence. In the case of water equation(2)was used instead of equation (1). The results of the estimation are presented in Table 1.

Table 1

Parameters for Evaluation of Wet Bulb Temperature

Compound	Le	Density Factor	T_∞ °C	T_{WB} °C	B_m	B_h
Water	0.99	0.995	26	17.8*	0.00387	0.00383
Ethyl Ether	0.614	0.936	30	-20.4	.237	.143
Dichloro-methane	0.589	0.931	27	-18.5	.238	.131
N-Octane	0.284	0.970	26	20	.057	.019
	0.296	0.953	40	29.7	.099	.033
Acetone	0.582	0.946	27	-9.7	.120	.072
	0.591	0.935	40	-6.1	.151	.090
Carbon Tetrach-loride	0.393	0.858	28	3.7	.334	.123
	0.406	0.825	41	9.4	.461	.165

*Relative Humidity is 43%

For the low mass flux evaporation of water, the Lewis number and density factor are approximately unity, and since the Sh* and Nu* are similar, the heat and mass transfer driving forces B_h and B_m are approximately equal. However, this is not true for the high mass flux data which are strongly dependent on the Lewis number giving rise to significant differences between the heat and mass transfer driving forces. The Lewis number for the organics varied from 0.284 to 0.614 giving up to a threefold difference in the heat and mass transfer driving forces. Accordingly the mass diffusivity is significantly less than the thermal diffusivity and is the parameter most important in calculating the wet bulb temperature for the high mass flux data. As comparison with literature data is limited, the wet bulb temperatures for the high mass flux data were compared to the adiabatic temperatures. The T_{wb} values were 4 to 13°C greater than the adiabatic temperatures being greatest for the

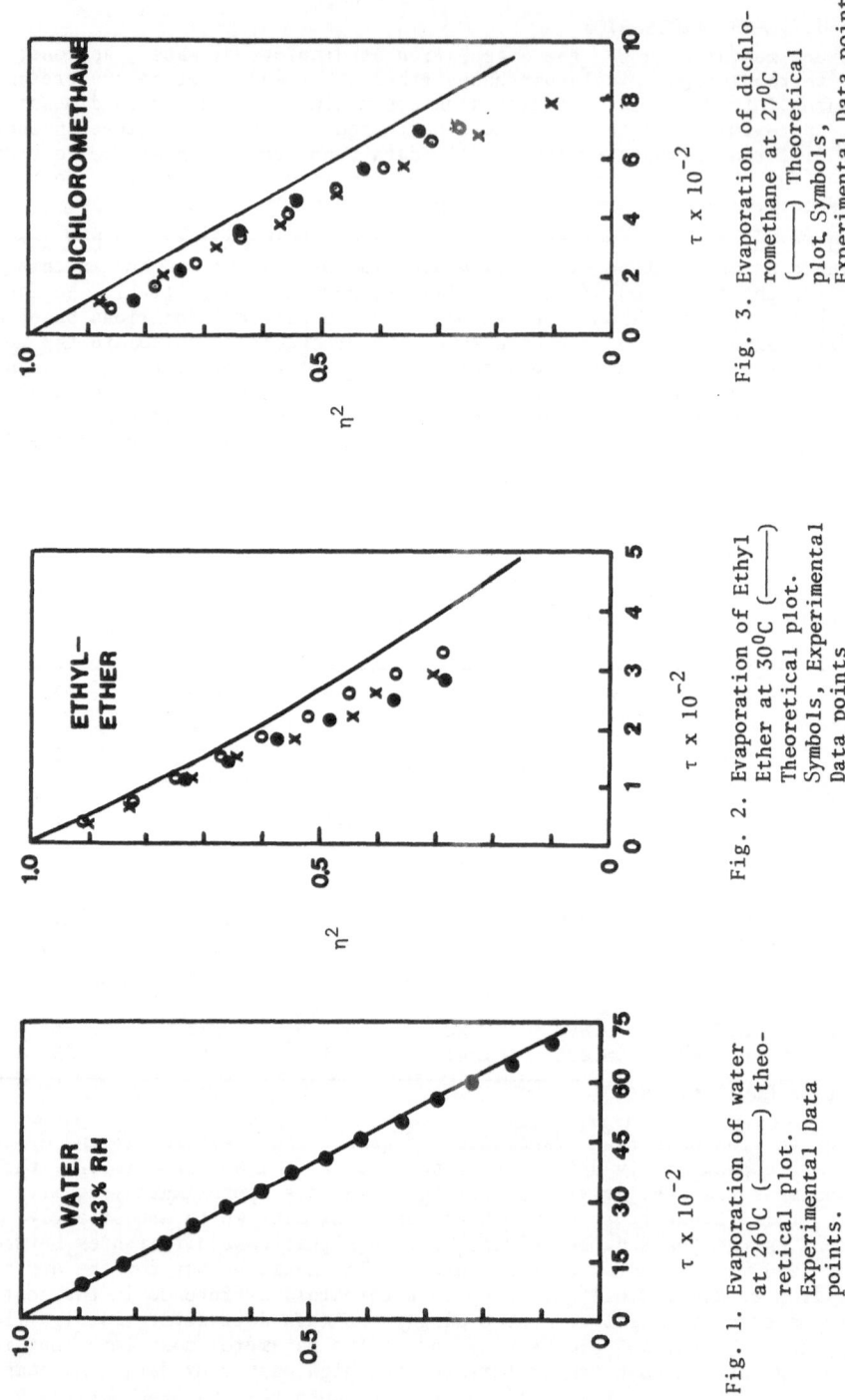

Fig. 1. Evaporation of water at 26°C (———) theoretical plot. Experimental Data points.

Fig. 2. Evaporation of Ethyl Ether at 30°C (———) Theoretical plot. Symbols, Experimental Data points

Fig. 3. Evaporation of dichloromethane at 27°C (———) Theoretical plot, Symbols, Experimental Data points.

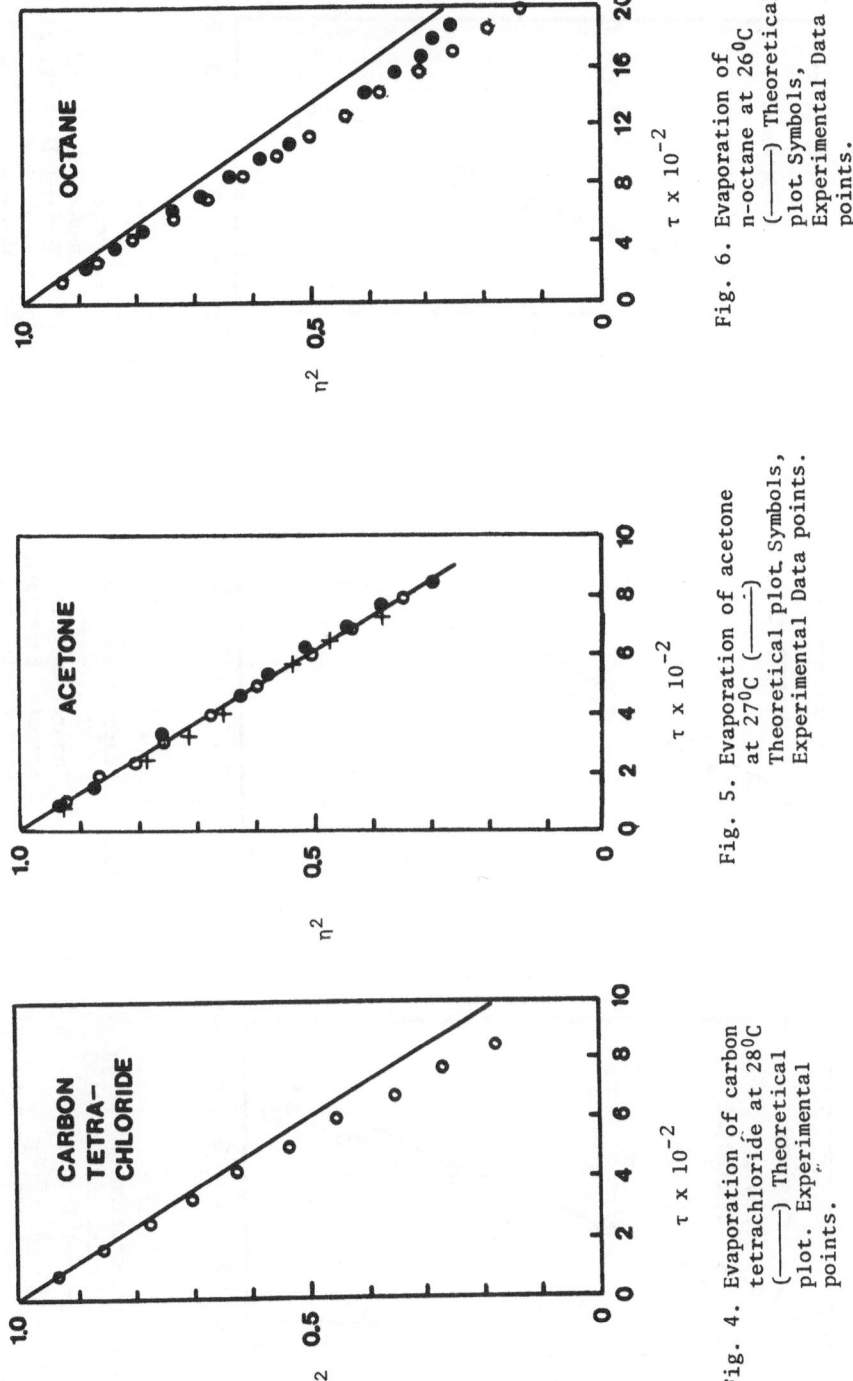

Fig. 4. Evaporation of carbon
tetrachloride at 28⁰C
(———) Theoretical
plot. Experimental
points.

Fig. 5. Evaporation of acetone
at 27⁰C (———)
Theoretical plot. Symbols,
Experimental Data points.

Fig. 6. Evaporation of
n-octane at 26⁰C
(———) Theoretical
plot Symbols,
Experimental Data
points.

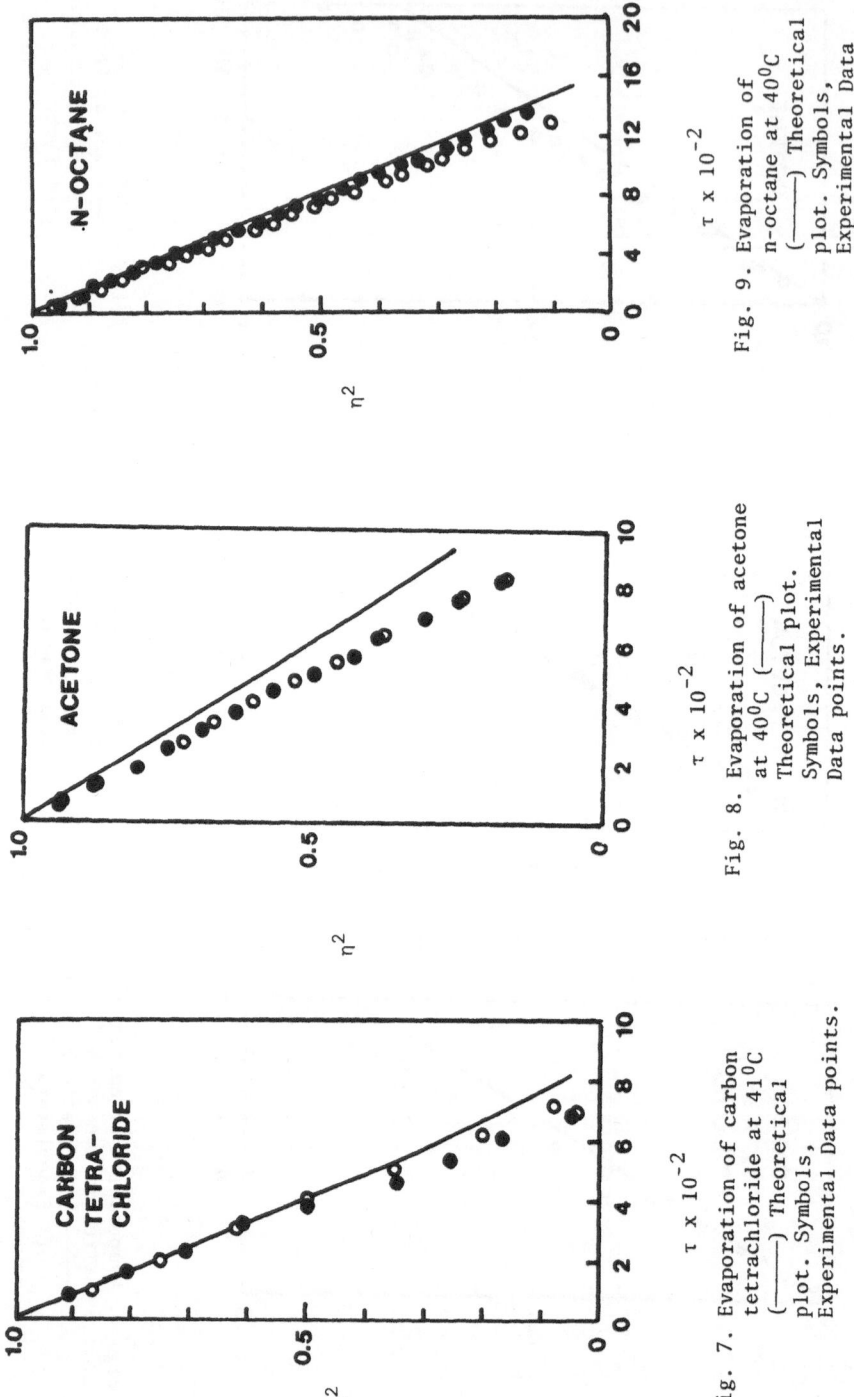

Fig. 7. Evaporation of carbon
tetrachloride at 41^0C
(———) Theoretical
plot. Symbols,
Experimental Data points.

Fig. 8. Evaporation of acetone
at 40^0C (———)
Theoretical plot.
Symbols, Experimental
Data points.

Fig. 9. Evaporation of
n-octane at 40^0C
(———) Theoretical
plot. Symbols,
Experimental Data
points.

higher molecular weight species. The Sh* and Nu* numbers were approximately 4 indicating equal contributions of diffusion and natural convection.

The density factor correction of Hanna (10) is significant for all the high mass flux data and cannot be neglected. The factor varies from 0.825 for carbon tetrachloride at $41^{\circ}C$ to 0.97 for n-octane at $26^{\circ}C$ indicating very significant variations with density. Evaporation of n-octane at high temperature has a density factor of 0.879 indicating that although low at room temperature, for higher temperatures it cannot be omitted (9).

The theoretical plots shown as solid lines in figures 1 to 9 were calculated using equations (4) and (6) after the wet bulb driving force B_m had been established. The agreement between the experimental data and the predicted plots for the low mass flux evaporation of water and for the high mass flux evaporation of the organics is satisfactory.

Mass transfer coefficients were extracted from the data using the equation.

$$k^{o}_{wexp} = \frac{n_{Ao}}{B_m} \tag{7}$$

where n_{Ao} is the measured initial mass flux. Theoretical mass transfer coefficients were derived from the expression

$$k^{o}_{w} = k^{*}_{w} \frac{\ln(1 + B_m)}{B_m} \frac{\ln(M_o/M_\infty)}{(M_o/M_\infty - 1)} \tag{8}$$

The values of experimental and theoretical mass transfer coefficients are tabulated in Table 2.

The agreement between both coefficients is of the order of the accuracy of the physical properties, and is thus very satisfactory.

Table 2

Experimental and Theoretical Mass Transfer Coefficients

	T_∞ $^{\circ}C$	k_{wexp} $\times 10^2$	k^o_w $\times 10^2$
Ethyl Ether	30	0.224	0.193
Dichloromethane	27	0.240	0.226
N-Octane	26	0.133	0.117
	40	0.131	0.133
Acetone	27	0.234	0.217
	40	0.277	0.248
Carbon Tetrachloride	28	0.224	0.212
	41	0.184	0.199

The theory for the calculation of the wet bulb temperature for high mass flux transfer in natural convection has been refined and successfully applied to explain the evaporation data for five organics. The Sherwood number for the evaporation process indicates that both the diffusive and natural convective mechanisms are of equal importance in the transfer process. The radiant energy only amounts to 1.5% of the total energy transferred for high mass flux transfer but may amount to 10 to 15% for low mass transfer natural convective evaporation of droplets.

NOMENCLATURE

A   Constant in equation 6

B   Mass or Heat Transfer drawing force

B   Constant in equation 6

C_p Heat capacity, cals/gm^{o}C

C   Constant defined by equation 4

D   Diffusivity, cm^2/sec

d   droplet diameter, cm

Gr   Grashof number ($= \dfrac{d^3 g \, \rho_\infty \, (\rho_o - \rho_\infty)}{\mu^2}$)

h   heat transfer coefficient, cals/sec.cm^2. ^{o}C

k   Thermal conductivity, gas phase, cals/sec. cm.^{o}C

k   Mass transfer coefficient, gm/cm^2. sec.

Le   Lewis number (= $\rho C_{pf} \, D_{AB}/k$)

M   molecular weight, or average molecular weight

n   flux, gm/cm^2. sec.

Nu   Nusselt number (= hd/k)

R   droplet radius, cm

Sc   Schmidt number (= $\mu/\rho \, D_{AB}$)

Sh*   Sherwood number (= $k_w^* d \, / \, \rho D_{AB}$)

T   Temperature, ^{o}C or ^{o}K

Greek Letters

η Dimensionless radius, R/R_o

μ viscosity, poise

ρ density, gm/cm^3

τ dimensionless time (= $D_{AB} t/R_o^2$)

Subscripts/Superscripts

h   heat transfer

m   mass transfer

o   surface conditions

f   film conditions

WB   wet bulb

∞ bulk conditions

*   Limiting mass flux conditions

o   High mass flux conditions

ACKNOWLEDGEMENT

The financial assistance of the University of Petroleum & Minerals is gratefully acknowledged. We are grateful to Mr. Tasudduq Husain K. Rahman for typing the manuscript.

REFERENCES

1. Ranz, W.E. and Marshall, W.R., Chem. Engg. Progr., 1952, $\underline{48}$, 141 (I), 173 (II).

2. Newbold, F.R. and Amundson, N.R., A.I.Ch.E. Journal, 1973, $\underline{19}$, 22.

3. Pei, D.C.T. and Gauvin, W.H., A.I.Ch.E. Journal, 1963 $\underline{9}$, 375.

4. Prakash S., and Sirigano W.A., Int. J. Heat Mass Transfer, 1980, $\underline{23}$, 253.

5. Harpole, G.M., Int. J. Heat Mass Transfer, 1980, $\underline{23}$, 17.

6. Nienow, A.W., Brit. Chem. Engg., 1967, $\underline{12}$, 1737.

7. Downing, C.G., A.I.Ch.E. Journal, 1966, $\underline{12}$, 760.

8. Skelland, A.H.P., 'Diffusional Mass Transfer' John Wiley and Sons, New York (1974).

9. Loughlin, K.F., Hadley-Coates, L., and Halhouli, K., submitted to Chem. Eng. Sci.

10. Hanna, O.T., A.I.Ch.E.J., 1962, $\underline{8}$, 278.

11. Bird, R.B., Stewart, W.E., Lightfoot, E. N., 'Transport Phenomena', John Wiley and Sons, New York (1960).

12. Treybal, R.E., 'Mass Transfer Operations', McGraw-Hill Co., New York (1955).

13. Halhouli, K., 'Evaporation of Liquid Droplets', M.S. Thesis, University of Petroleum & Minerals, Dhahran, Saudi Arabia (1981).

Potassium Sulfate Deposition Rate and Heat Transfer on a Single Tube in a Cross Flow

HSING CHUANG
Department of Mechanical Engineering
University of Louisville
Louisville, Kentucky 40208, USA

ABSTRACT

The deposition rate of potassium sulfate on a single tube in a cross flow of gas mixtures containing fume and submicron particles of potassium sulfate was studied experimentally. An average seed deposition rate of 1.21×10^{-5} g/cm^2-s and an average initial thermophoretic velocity of 0.05 m/s were found for an average gas temperature at 1250 K and an average tube surface temperature at 650 K. As the seed deposition time and thickness increased, the external heat transfer coefficients appeared to decrease from the initial value, which was in agreement with the McAdams' formula, and approach an asymptote.

1. INTRODUCTION

The national effort to develop the highly efficient, coal-fired, open-cycle MHD and steam power plant was undertaken by several research laboratories to establish the technology and engineering base. In particular, works on the development of the heat and seed recovery system, which include the steam bottoming plant, air heaters, and the pollution control system were commenced at the Argonne National Laboratory (ANL). The primary goal of the Heat and Seed Recovery (HSR) Technology program was to assure that the HSR equipment for the Engineering Test Facility would be available on schedule and operate effectively and reliably. In order to attain this goal, an MHD Engineering Test Unit (METU) was constructed and operated at ANL to experimentally verify the analytical models of transport processes as well as identify and test the candidate materials for the HSR components. Furthermore, a small-scale 2-MW facility was built at ANL to support the program goals.

The versatile METU was used to experimentally verify the seed deposition rate and examine the heat transfer rate from a single tube in a cross flow. Seed deposit characteristics were also observed. Technical information and knowledge obtained from this experiment were useful to the experimental program of the Argonne MHD Process Engineering Laboratory (AMPEL) as well as the modeling development and design of all HSR components.

2. APPARATUS AND PROCEDURE

A versatile METU was constructed in the Bldg. 212 at ANL for materials testing and transport process studies at high temperatures [1]. It consisted

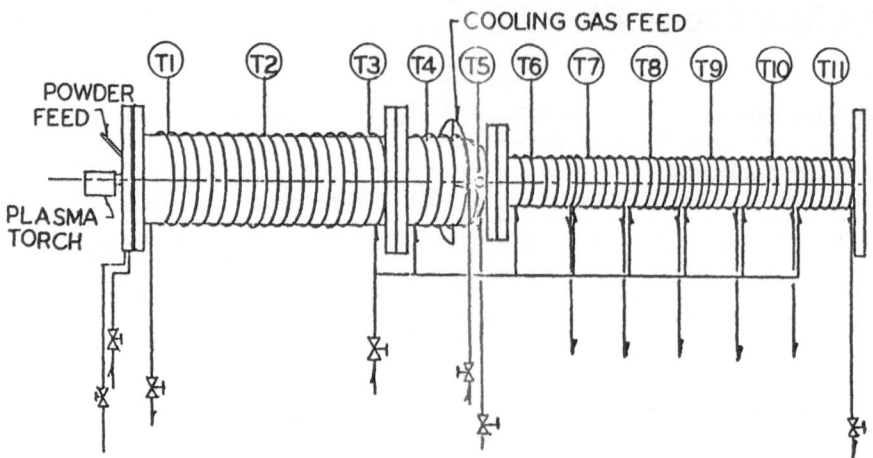

Fig. 1.   Engineering Test Unit

of a commercial plasma torch (METCO type 7MB Plasma Flame Spray System), an engineering test station (ETS), a materials test station, and support equipment in a sound-proof room. The ETS as shown in Fig. 1 is basically made of water-cooled cylindrical sections with the plasma torch at the inlet of the system.

The first section of the ETS system includes a platform attached to a water-cooled flange for the plasma torch support and several concentric cylinders of ceramic and castable insulation inside a thin metallic cylinder for the inner lining and a 20.32-cm (8-in) I.D. water-cooled steel cylinder for the outer shell. The second section has a reducing nozzle, which converges from 20.32-cm (8-in) to 10.16-cm (4-in) I.D., and has in portion of it the same inner lining as the first section. The 13-cm (5.125-in) I.D. inner lining forms a gas-filled annulus with the outer cylinder and provides for recovering some of the heat loss to the wall by means of a diluent gas flow through the annulus and discharging it into the main flow at the inlet of the system. The last section is a 10.16-cm (4-in) I.D., water-cooled, steel pipe which vents the gas mixture into a scrubber.

Several 1.9-cm (3/4-in) O.D. stainless steel tubes with four thermocouples brazed to the wall as shown in Fig. 2 were made. There are two additional thermocouples, one on each end coupling of the test tube, for measuring the inlet and outlet temperatures of the cooling nitrogen gas. For each test, a single test tube was inserted horizontally through the center line of the ETS at the reducing section. The gas inlet of the tube was connected to the cooling nitrogen tank through a flow meter while the outlet of it was joined to the cooling gas feed of the ETS system. The cooling gas can be diverted from the inlet directly to the cooling gas feed without flowing through the test tube.

An amount of K_2SO_4 equivalent to about 1.0 wt % of the gas flow was fed through the feed port on the flange by means of a powder feeder unit. The potassium sulfate seed (-50 + 150 mesh) was dried in a vacuum oven prior to

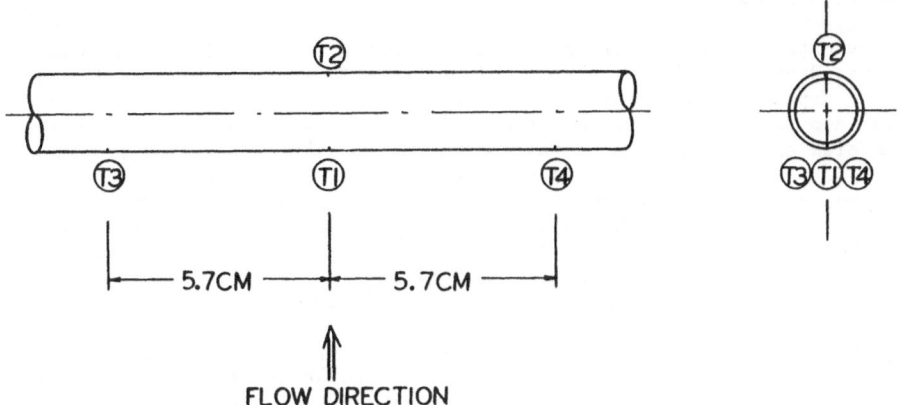

Fig. 2.   The Test Tube

the experiments because the feeding line had a tendency to plug up, particularly when the humidity was high in the feeder unit. Several runs were made to calibrate the powder feeder unit. The seed was vaporized by the nitrogen plasma torch and then mixed with additional nitrogen gas which was preheated by passing through the test tube and/or the annulus between the inner lining and the steel pipe of the ETS. Temperature of the gas in the mixing zone was in excess of 1700 K. However, the seed was condensed and became fume and/or submicron size particles before it reached the test tube.

A thermocouple placed in front of the test tube was calibrated against an optical pyrometer. The results showed that the thermocouple readings were in good agreement with the pyrometer readings. The temperature reading of the inner lining wall was also in good agreement with that of a 3.175-mm (1/8-in) alumina rod placed in the midstream of the gas mixture and measured by the pyrometer.

The test tube was cleaned and weighed before it was inserted into its position. After each test run for the duration of 10, 15, 20, 25, 30, or 45 minutes, it was cooled down to the room temperature and then the test tube was removed from the ETS to be weighed again. Because the seed deposit was loosely piled on the tube, a slight impact of it with others would lose some deposited seed. Consequently, the amount of deposited seed might be a little bit more than what was shown in the results.

The rate of heat flux were obtained by measuring the coolant temperatures at the inlet and the outlet of the test tube and the amount of coolant flow rate. Two coolant mass flow rates of $\dot{m}_c$ = 198 and 99 g/min were used in this experiment. The overall heat transfer coefficient (OHTC) is then the ratio of the heat flux per unit area and the temperature difference between the gas mixture and the tube wall.

3.   RESULTS AND DISCUSSION

The experimental results showed that about 6 to 16 x 10^{-5} kg/m^2-s of
K$_2$SO$_4$ was deposited on a single tube in a cross flow at 2.6 m/s and 1250 K.
The seed deposition was unsymmetrical, thick fluffy layer on the upstream
portion and only a thin, densely packed layer on a small portion of the downstream side of the tube, which was indicative of a turbulent flow around the
tube.   The X-ray diffraction and spectrochemical analyses showed that the
deposited material mainly consisted of K$_2$SO$_4$ with K and S as the major phases
and Cu as the possible minor phase.   Since the copper electrodes for the
plasma torch were slightly corroded, it was natural to find copper in the
deposited seed by the X-ray spectrochemical scan.   The stainless steel tube
was also barely corroded.   However, no iron was found in the scan.

For the nitrogen coolant flow rate at 198 g/min, the tube wall temperature
dropped from 680 to 530 K while for the flow rate at 99 g/min, the wall temperature dropped only from 750 to 630 K as the deposition thickness increased.
The OHTC was found to decrease with increasing deposition time and thickness.
It approached a minimum value as shown in Fig. 3 apparently because the
loosely adhering deposit fell off and/or was blown away when the deposition
time was longer than 30 minutes.   It was also very difficult to remove the tube
with all deposited seed intact when the deposition time was longer than 30
minutes.   The external heat transfer coefficient for a cylinder in a cross flow
of flue gases at 1250 K was estimated from McAdams' formula [2] to be around
140 kJ/hr-m^2-C which agreed with that measured initially and was about 50
percent higher than that measured after 30 minutes of seed deposition in this

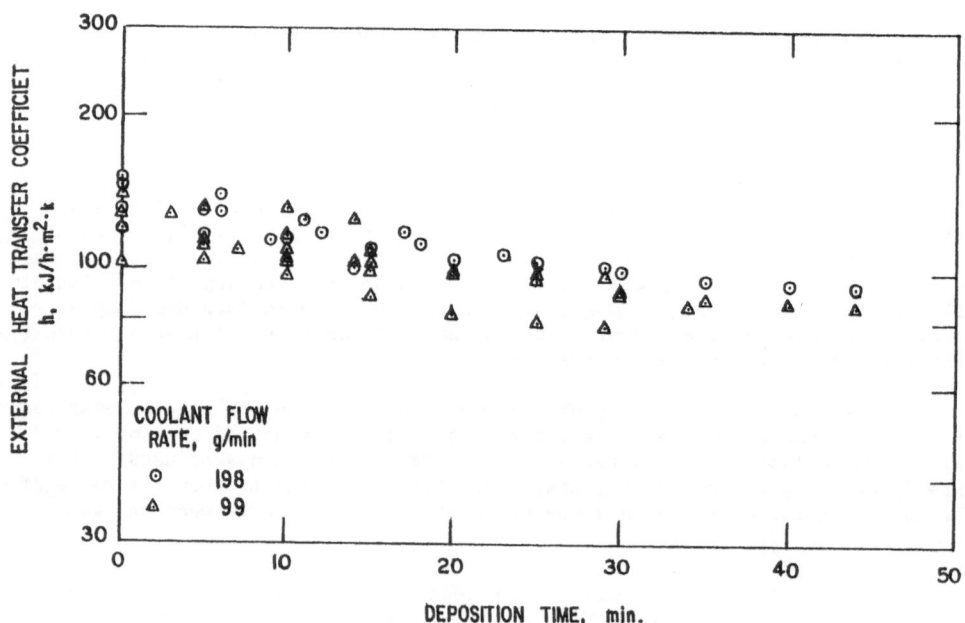

Fig. 3.   External Heat Transfer Coefficients for a Tube in
a Cross Flow with Depositing K$_2$SO$_4$ Seed

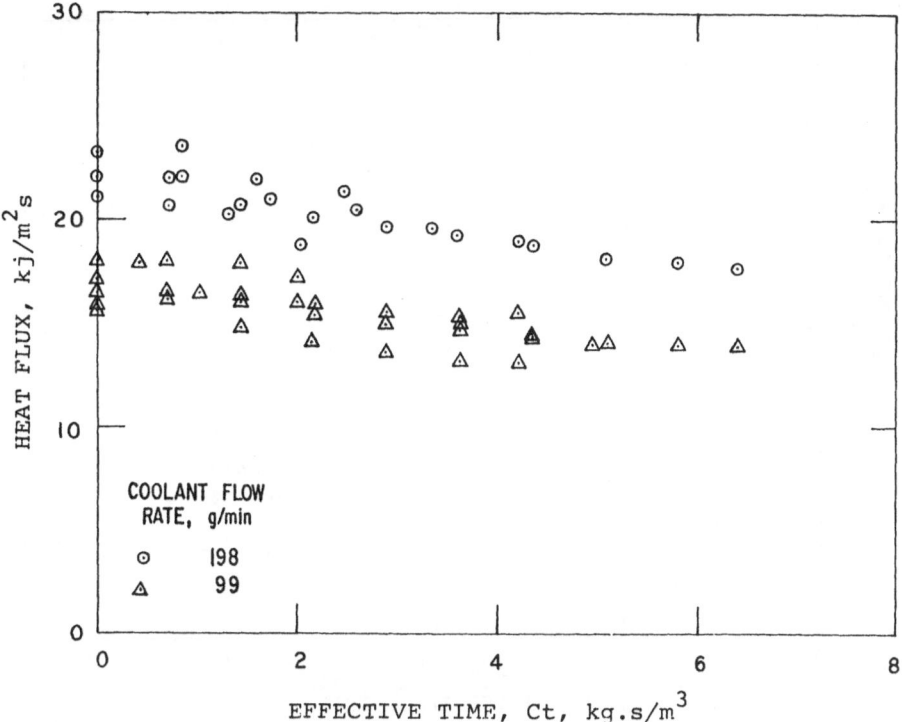

Fig 4. Heat Flux to a Transverse Tube with Depositing K_2SO_4 Seed

experiment. The OHTC for the coolant flow rate at 99 g/min was found to be slightly lower than that for the coolant flow rate at 198 g/min. Nevertheless, the heat flux through the deposited seed and the tube wall only increased by about 30 percent as the coolant flow rate was doubled as shown in Fig. 4. The heat flux obtained in this experiment was lower than that presented by Heywood and Womack [3]. This experiment used a horizontal, open-end stainless steel tube (19-mm O.D.) while Heywood and Womack had a vertical, closed-end steel tube (18-mm O.D.). The differences were apparently due to the coolant mass flow rates and flow configurations, a horizontal, open-end tube versus a vertical, closed-end tube.

The thermal conductivity of K_2SO_4 deposited on the tube wall was estimated to be 0.83 - 1.22 kJ/hr-m-C. The average surface temperature of the deposited seed was also estimated to be 630 - 740 K, the higher values for the coolant flow rate of 99 g/min and the lower values for the coolant flow rate of 198 g/min. The deposition thickness was roughly measured to be 0.4 mm for 10 minutes duration or 2.0 mm for 45 minutes duration.

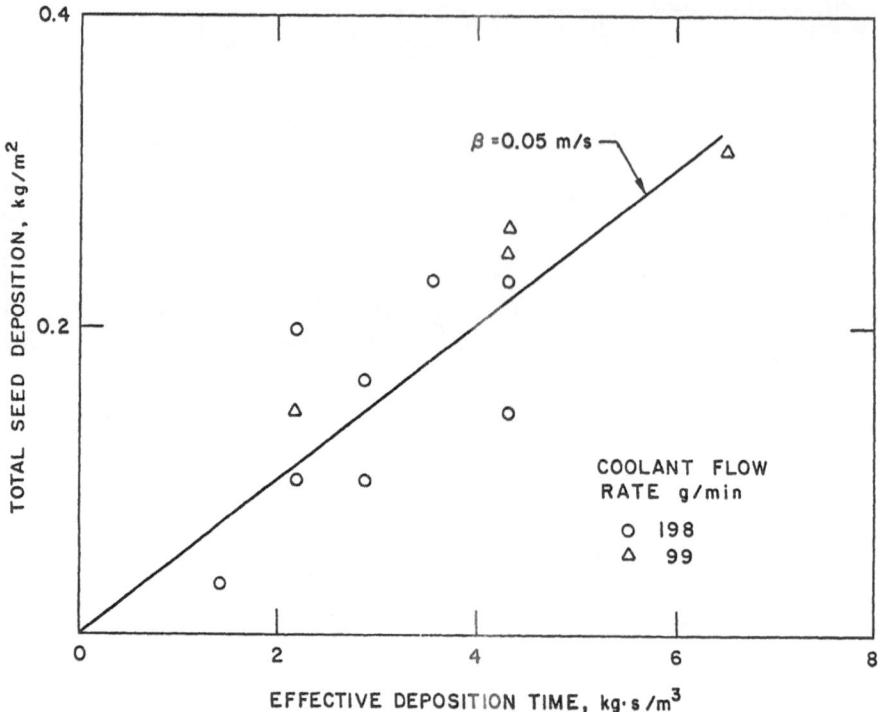

Fig. 5.   Deposition of K2SO4 Particles on a Cooled Tube

The total seed deposition on the tube can be expressed in terms of the seed concentration, C, in the gas mixture and the deposition time, t. Since the seed concentration in the gas mixture is known to be about 2.42 g/m³, the total seed deposition rate can be plotted against the effective time, Ct as shown in Fig. 5. The slope of the representative line is the thermophoretic velocity, β, which is about 0.05 m/s.

The fume of potassium sulfate may be transferred to the surface of the test tube by impingement, thermal diffusion through the boundary layer, and thermal precipitation due to the temperature gradient across the boundary layer, namely, thermophoretic forces. It has been shown [3,4] that the submicron fume has deposited on the tube surface predominantly by the thermophoretic forces.

The thermophoresis depends on the fume particle size, the mean free path of the gas molecule, and the temperature gradient in the boundary layer. The submicron particles in the gas mixture tend to migrate in the direction of decreasing temperature. The thermophoretic velocity is then used to correlate the above-mentioned variables as follows [4]:

$$\beta = \frac{1}{8} \frac{\ell U_g}{T_g} \nabla T \, \exp(-0.38 \, \frac{r}{\ell}),$$ (1)

where ℓ, U_g, T_g, and r are the man-free-path of the gas molecule, the average velocity of the gas molecule, the gas-mixture temperature, and the radius of the fume particle, respectively. The temperature gradient, ∇T, is given by

$$\nabla T = \frac{(T_g - T_s)Nu}{d_{eff}},$$ (2)

where T_s, d_{eff} and Nu are the tube wall temperature, the effective diameter of the tube with seed deposits, and the Nusselt number, respectively. The rate of seed deposition is then provided by

$$\frac{dG}{dt} = C\beta$$ (3)

where dG/dt has the dimension of mass per unit area per unit time.

A computer model using Eqs. (1)-(3) to predict the seed deposition rate on the tube bank for the AMPEL was made [5]. With the average thermophoretic velocity of 0.05 m/s obtained from Fig. 5 and the seed concentration of 2.42 g/m³, the average seed deposition rate is 1.21×10^{-5} g/cm²-s which agrees with the measured seed deposition rates of this experiment as well as a model with T_g = 1256 K, T_s = 590 K, ℓ = 1.0 μm, r = 0.1 μm, Nu = 11, and d_{eff} = 2 cm. This implies that the seed deposition rate for a single tube in a cross flow is in good agreement with that given by Eqs. (1)-(3). Heywood and Womack [3] showed that for T_g = 1345 K the initial thermophoretic velocities were about 0.03 and 0.008 m/s for T_s = 775 K and 1175 K, respectively.

Condensation rate is given by [3]:

$$\frac{dW}{dt} = 1.39 \, \Delta p \, G_m^{1/2},$$ (4)

where Δp is the partial pressure difference (in atm) of the K_2SO_4 between the main gas stream and the tube surface and G_m is the molal velocity of the gas stream. The molal velocity is the ratio of the average gas velocity to the mean molecular weight of the gas. Condensation rate at 1250 K was estimated from Eq. (4) to be about 5.6×10^{-7} g/cm²-s which was only 5 percent of the seed deposition rate. Therefore, thermophoresis was the main force for the seed deposition at this gas temperature.

4. CONCLUSION

This experimental study showed that the rate of heat transfer through the deposited K_2SO_4 on a single tube in the cross flow decreased as the thickness of the deposits increased. The rate tended to approach a minimum value for a prolonged deposition experiment. The rate was, however, found to be lower than that obtained by Heywood and Womack. The average seed deposition rate and the average initial thermophoretic velocity were found to be 1.21×10^{-5} g/cm²-s and 0.05 m/s, respectively.

ACKNOWLEDGEMENT

The author wishes to thank Drs. Terry R. Johnson and Kenneth E. Templemeyer for their encouragements and supports. Technical assistances were provided by Mr. Robert C. Peto and Mr. William B. Leniek. The financial supports of the Division of Educational Programs and the Engineering Division, Argonne National Laboratory, are gratefully acknowledged.

REFERENCES

1. Taboas, A. L., Singh, R. N., and Johnson, T. R. 1978. Plasma duct for instrumentation, materials, and process analysis. Paper presented at the ANL/DOE/ISA Symposium on Instrumentation and Control for Fossil Demonstration Plants, Newport Beach, CA. U.S.A.

2. McAdams, W. H. 1954. Heat Transmission, 3rd ed., McGraw-Hill, New York. U.S.A.

3. Heywood, J. B. and Womack, G. J. 1969. Open-Cycle MHD Power Generation, Pergamon Press, Oxford, Great Britain.

4. Styrikovich, M. A. and Mostinskii, J. L. 1976. A study of the processes resulting from the use of alkaline seed in natural gas-fired MHD facilities. Academy of Sciences, Moscow, U.S.S.R.

5. Petrick, M., Tempelmeyer, K. E., and Johnson, T. R. 1979. MHD balance of plant technology. Third Quarterly Report. ANL/MHD-78-10. Argonne National Laboratory, Argonne, IL. U.S.A.

COMBUSTION

COMBUSTION

Combustion Characteristics Associated with the Novel Fuel Mixture Comprised of Coal and Natural Gas

DUPREE MAPLES and VIC A. CUNDY
Department of Mechanical Engineering
Louisiana State University
Baton Rouge, Louisiana 70803, USA

ABSTRACT

A fuel mixture of lignite and natural gas was successfully fired exhibiting stable, compact combustion characteristics of an oil flame. The principle result of this research was the firing of 62 percent of the total energy in the flame generated by a lignite-natural gas fuel mixture was contributed by lignite. An analysis of stack gases indicated that combustion was complete, even at the 62 percent mixture ratio, at excess O_2 levels to 2.0 percent (dry volume basis). The principal conclusion of this paper is that fuel mixtures of gas and coal are feasible with coal potentially replacing large amounts of natural gas in industrial/commericial/utility units originally designed to fire only natural gas.

1. INTRODUCTION

Petroleum resources, both liquid and gaseous, have become the major sources of energy in many countries because of the availability and convenience of these fuels for both transportation and stationary power requirements. Although petroleum resources occur in almost every part of the world to some degree, the major commercially valuable resources occur in relatively few locations where geological conditions were appropriate for the formation and storage of these fuels underground.

Coal represents a politically stable energy source for the United States. Due primarily to geometric limitations, it remains economically unfeasible for industrial or utility combustion facilities originally designed to fire oil or natural gas to convert to coal burning. A potential large scale use for coal lies in the utilization of coal-oil mixtures [1-3]. In such mixtures, pulverized coal is mixed with fuel oil thereby replacing a percentage of the required conventional fossil fuel in the combustion process. Coal-oil mixtures provide feasible short term relief to the current problems required to utilize large amounts of coal.

The most significant problems of coal-oil burning is ash generation. Chambers originally designed to fire fuel oil are not equipped with either fly ash or bottom ash handling equipment. Pumping problems associated with coal-oil slurries have not completely been solved.

2. OBJECTIVE OF RESEARCH

Although coal-oil mixture technology is progressing at an accelerated rate, virtually no research has occurred concerning mixtures of coal and natural gas. The main objective of this paper is to report the results of an experimental study conducted to demonstrate the viability of mixtures involving lignite and natural gas. Maximizing the amount of lignite in lignite-natural gas fuel mixture while retaining a compact and stable flame similar to pure natural gas represent the overall goal of the research.

3. EXPERIMENTAL FACILITY

The full scale combustion facility consists mainly of a chamber, burner, and measurement system. A description of each component will be given in this section.

3.1 Combustion Chamber

Full scale experiments were performed in the combustion chamber. The chamber is a large refractory lined container with water cooled panels surrounding it. The inner dimensions of the chamber are 5.0 x 5.0 x 10.75 ft. which gives an inner volume of 269 ft^2. The water panels enclosing the chamber are on all sides of the refractory container with the exception of the front wall. The water panels were installed to remove energy generated during the combustion process. A wide-angle view port located on the rear wall allows continuous visual observation of the flame during testing and also, providing a means for photographing the flame. Stack gases exit the rear of the chamber opposite the burner housing wall. A gas sampling probe is inserted at the entrance to the stack.

3.2 Burner

An Eclipse burner (Model number 168-HGF-CGO) served as the foundation for the mixture burner which was eventually developed. The burner was originally designed to fire either fuel oil or natural gas. This is an air atomizing burner which has a turn-down ratio of 5 to 1. As shown in Figure 1, the burner consists of an inner oil flow channel fitted with an interchangeable nozzle. The oil gun is surrounded by the atomizing air flow channel which is surrounded by the natural gas passage. The primary air is supplied through an outer annulus. Swirl is achieved with sets of vanes located in both the primary and atomizing air channels. Only oil or natural gas can be fired at one time.

Burner modification adopted involved pneumatically conveying the pulverized lignite with air to the burner front where this combination was then mixed with the natural gas, additional combustion air, and then fired. As opposed to coal-oil mixture techniques, the primary fuels are not premixed and, in fact, do not combine until the burner front.

A schematic of the burner used is shown in Figure 2. As shown, the eventual design required that the lignite be transported to the burner via the atomizing air. The new burner developed for the study of lignite-natural gas studies consisted of an inner atomizing air-lignite flow channel fitted with a tapering nozzle. This pipe is surrounded by the natural gas passage and the entire configuration is enclosed by the primary air channel.

3.3 Measurements

A Fluke Model 2200B Data Logger was used to record the temperature at all locations on a regular interval. The duration of a typical test period was approximately one hour after steady conditions were obtained. The preheat period

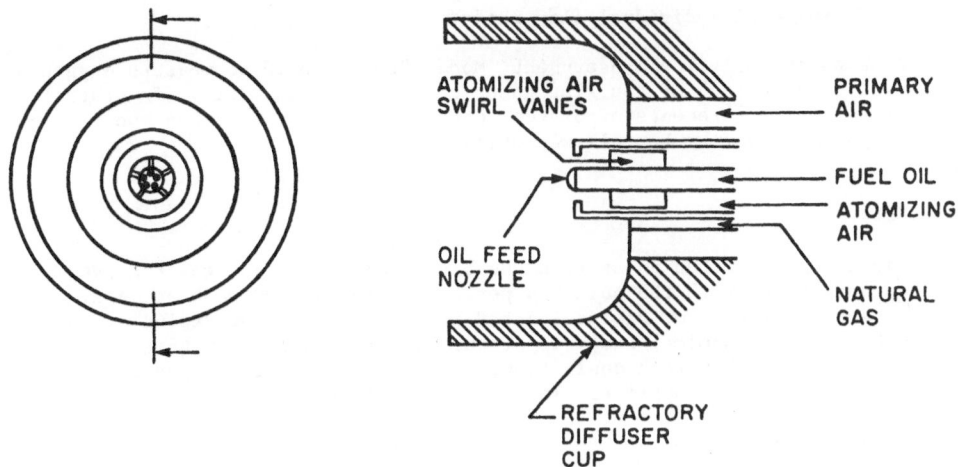

Figure 1.   Standard multifuel burner.

resulted in an average furnace temperature of 1500°F.

A gas sampling probe was situated at the inlet to the stack.   The probe was a 0.25 inch stainless steel tube through which samples of exhaust gases were drawn and further analyzed to provide information concerning combustion parameters including excess oxygen, carbon dioxide and carbon monoxide.

Several ITT Barton Series 7400 flow meters are installed into the air and natural gas lines.   An ITT Barton Series 500 Flow Totalizer Electronic System receives the signal from the flow transducer and translates the signal to provide

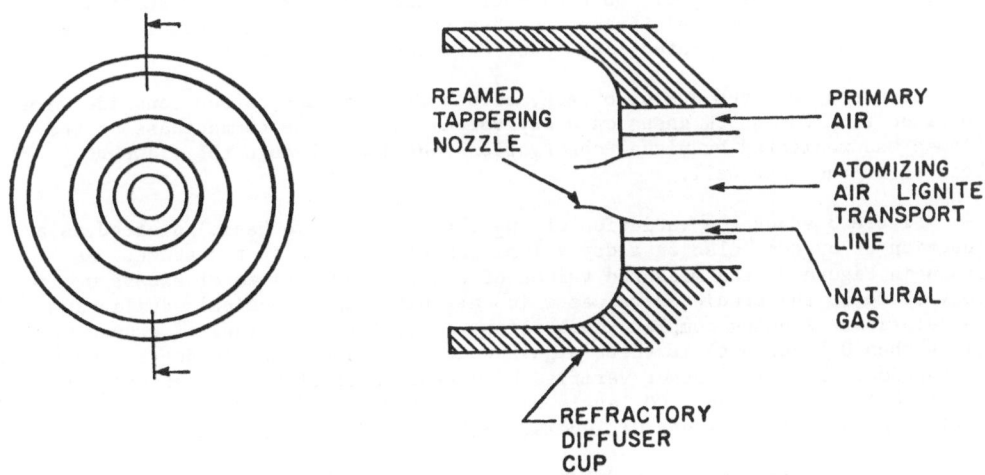

Figure 2.   Modified multifuel burner utilized in fuel mixture studies.

a visual display of individual flow rates.

A Teledyne Analytical Instrument, Model 980, is used to measure oxygen and combustible percentage of the flue gases on a dry volume basis. In order to further verify measurements of oxygen, carbon dioxide, and carbon monoxide, an "Orsat" gas analyzer by Burrell was utilized.

4. FUEL

Lignite was used on an as received basis. This lignite was received crushed and bagged with the bulk (54 percent) sized at 200 mesh or less. The lignite was obtained from fields located in the North Center region of the United States. A chemical analysis of the lignite utilized in this study was typical of average low rank coals found anywhere. The natural gas used was obtained from a local supplier. This gas was primarily methane.

4.1 Lignite Feed System

A schematic of the pneumatic lignite feed system is presented in Figure 3. The system primarily consists of a vibrating storage bin with screw feed conveyor, the mixing funnel where the atomizing air and lignite are mixed, and the feed lines connect this system to the burner. A dual jet mixer arrangement was used to promote sufficient flow along the entire length of the conveying transport line.

5. EXPERIMENTAL RESULTS

The experimental program concentrated on determining the maximum amount of lignite which could be utilized in a lignite-natural gas fuel mixture while maintaining stable, compact combustion conditions. The requirement of a compact flame was imposed since practical application of natural gas-lignite mixtures will almost certainly occur in existing combustion chambers originally designed to fire oil or natural gas. Such chambers usually do not contain sufficient combustion volume for coal firing and therefore to avoid substantial chamber modifications, the generated flame must necessarily be compact.

Standard measurements of oxygen, carbon dioxide, and carbon monoxide were obtained from the stack gases on a dry volume basis. The compactness of the flames was monitored by visual observations obtained through the viewport located in the rear wall.

Figure 3 shows the variation of CO_2 (on a dry volume basis) plotted as a function of oxygen (also on a dry volume basis) measured in the stack. As shown in Figure 3, the measured values of CO_2 for most values of excess O_2 lie very close to the predicted values which are based upon chemical equilibrium calculations assuming complete combustion. Only for low values of excess O_2 (less than 0.5 percent) is there significant variation which is due to incomplete combustion. This is further verified by measurements of CO. At excess air levels below 0.5 percent, CO levels were measured at 40,000 PPM on a dry volume basis. At all other excess O_2 levels, negligible CO was detected.

The solid lines shown in Figure 3 represent CO_2 levels predicted from a chemical equilibrium analysis based on complete combustion. Note that even though only one data point was taken for each mixture, the experimental data from near 7.0 percent excess O_2 to 2.0 percent excess O_2 lies directly on the predicted lines based upon complete combustion. This data is further

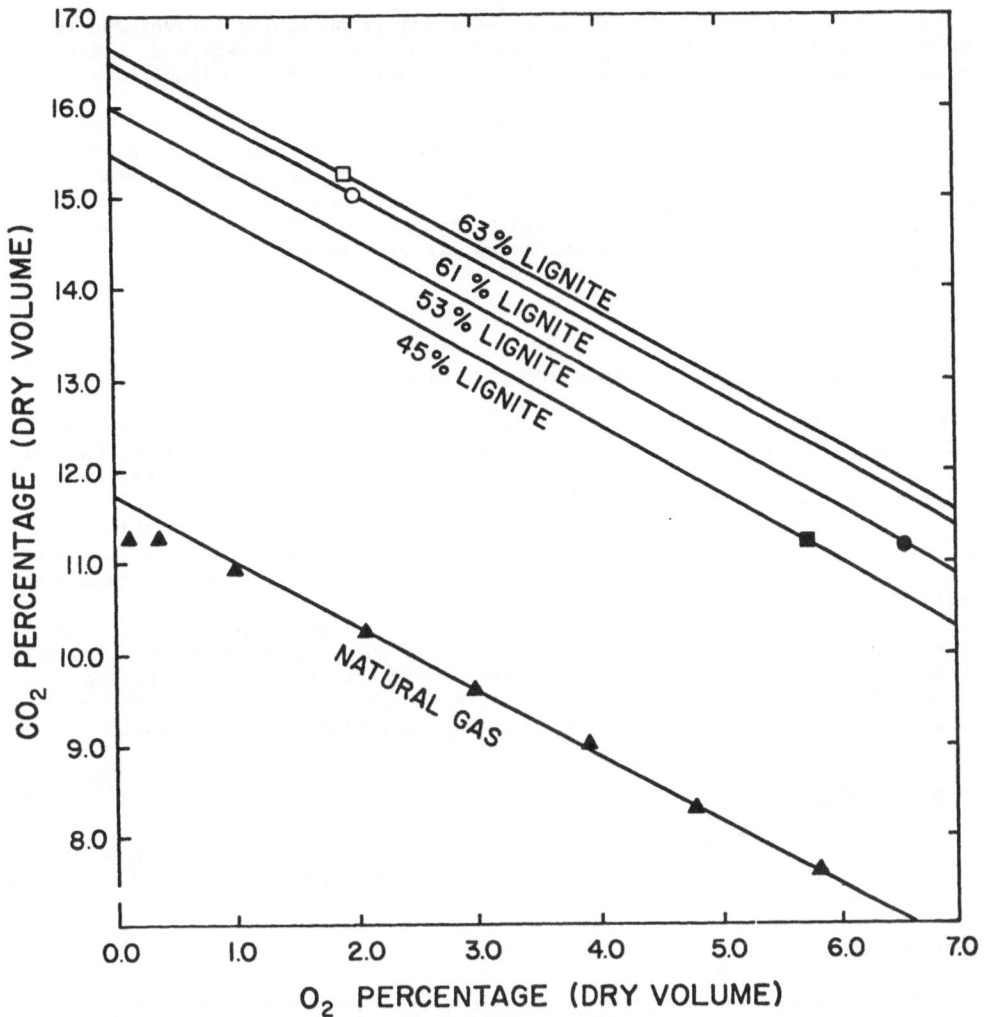

Figure 3. Dependence of CO_2 (dry volume basis) with excess O_2 (dry volume basis). Experimental data: natural gas (▲), 45% energy from lignite (■), 53 % energy from lignite (●), 61% energy from lignite (O), 63% energy from lignite (□). Solid lines represent thermochemical equilibrium calculation data based upon complete combustion.

substantiated since at each experimental data point, no detectable CO was measured.

The data of Figure 3 indicates that fuel mixtures of lignite and natural gas can be completely combusted with the lignite contributing up to 62 percent of the total energy to the flame. The flame at all levels of lignite contribution was stable and compact similar to an oil flame.

Lignite contributions beyond 62 percent may be possible; however due to equipment limitations, the maximum level has not yet been determined. Tests are being conducted to determine the maximum contribution from lignite.

The novel burner which was developed incorporated a unique design which involved the flow of pulverized lignite through the high reacting core of a small natural gas flame. The transport of the lignite utilizing the atomizing air provides for intense mixing, which further promotes the complete, compact combustion process. Clearly, high lignite loadings are feasible with such a burner system when firing fuel mixtures of lignite and natural gas.

6. CONCLUSIONS

The conclusions of this study are the following:

1. Large quantities of lignite may be fired as lignite-natural gas fuel mixtures. Simple burner modifications will produce these results, however, care should be exercised in that the exact quantities of coal which may be utilized are strongly burner dependent.
2. The results obtained in this study are directly applicable to fuel mixtures involving natural gas and the higher rank bituminous coals. Studies should be carried out to determine the maximum amounts of high rank bituminous coals that could be used in such mixtures.
3. The obvious potential of fuel mixtures involving coal and natural gas which has been demonstrated in this research should promote active research in this area. Such mixtures could clearly provide relief to the spiraling demands of industry and utility on natural gas while, providing a means to utilize the more politically stable, abundant coal reserves of the world.

ACKNOWLEDGEMENTS

The research reported in this paper was sponsored by the Board of Regents of the State of Louisiana under contract number 80-LBR/026-B12.

REFERENCES

1. Beinstock, D., and Jamgochian, E.M., Fuel, 1981, Vol. 60, pp. 851.

2. Cundy, V.A., and Maples, D., "Fuel Mixtures of Low Rank Lignite Coal with Fuel Oils", Submitted to Fuel, November, 1981.

3. Cundy, V.A., and Maples, D., ASME Paper No. 82-HT37, AIAA/ASME 3rd Joint Thermophysics, Fluid Plasma and Heat Transfer Conference, June, 1982.

Problems Encountered in the Development of a Double Cyclone Furnace

T. KREPEC and C.K. KWOK
Department of Mechanical Engineering
Concordia University
Montreal, Quebec H3G 1M8, Canada

ABSTRACT

A double cyclone furnace for the suspension burning of solid fuels based on a concept initially invented by Derek Angus, was developed. This unique vortex configuration with tangential inlets at top and bottom of the chamber was able to create a steady confined fuel residence zone and to burn the solid fuel particles requiring a long residence time for complete combustion. The position of this residence zone could be shifted to any level along the combustion chamber by changing the top and bottom inlet air-flow ratio. Several versions of the prototype chamber were investigated, and its geometrical configuration was optimized to obtain the highest exit gas temperature. It was found impractical to operate the furnace in slagging mode because of difficulties with slag evacuation. However, the experiments at non-slagging operation mode with open bottom of the chamber confirmed that the furnace could be run continuously with the ash being evacuated periodically, in a manner similar to the operation of a dust separator.

1. INTRODUCTION

Suspension burning of solid fuels in specially developed vortex chambers is now considered to be an alternative source of energy due to the forecasted shortage of hydrocarbon fuels. However, there are still several problems to be overcome before this process can be fully implemented. One of them is the relatively long time required to burn the solid fuel particles, as compared with liquid or gaseous fuels. In the existing vortex combustion chambers, there is a lack of a confined fuel residence zone where the solid fuel particles can stay until they would burn completely. The double vortex combustion chamber invented by Derek Angus (Ref. 1) was specially designed to create such a fuel residence zone by introducing two converging vortices with their tangential inlets at the top and bottom of a cylindrical chamber (Fig. 1).

With appropriate adjustment of the magnitude of flow for each individual inlet, combustion can occur at any position along the combustion chamber. Furthermore, in extreme cases it can operate in a manner similar to that of the cyclone dust separator permitting ease of ash evacuation. This concept provides the possibility of bringing the advantage of cyclone furnaces to the waste incineration field.

Initial objectives of the preliminary development were formulated as follows. The first would be to design and fabricate an experimental prototype furnace with maximum flexibility for changes of geometric configuration.

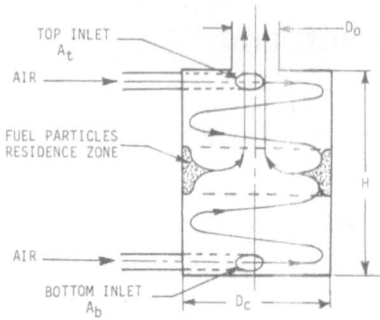

Fig. 1   Double Vortex Combustion Chamber

The hot prototype model would be tested next, initially using sawdust as the
primary fuel. The geometric configuration of the combustion chamber would
then be optimized to achieve the maximum possible temperature. The practica-
bility of operating the prototype furnace in either the slagging or non-
slagging mode would then be critically assessed. Finally, other waste-derived
fuels would be investigated.

2. PROTOTYPE AND TEST SET-UP

     In order to establish the optimum dimensions for the first prototype of
the double vortex combustion chamber, a simplified model of the gas-flow
pattern inside the chamber was defined. After determining the critical size
of the fuel particle which was assumed to be retained in the chamber by the
vortex flow, the principal dimensions of the chamber could be established
based on the balance of force calculation (Ref. 2). With the prototype combustion chamber volume assumed to be 0.028 m³ (1 ft³), the optimum chamber
diameter was found to be 0.3 m (1 ft). Since the theoretical optimization
procedure is rather crude, special effort was made to incorporate some kind of
flexibility in the prototype design. A modular approach was adopted to allow
a quick and inexpensive change of the vortex chamber configuration, as illustrated in Fig. 2. With only a few basic modular units, it was possible to

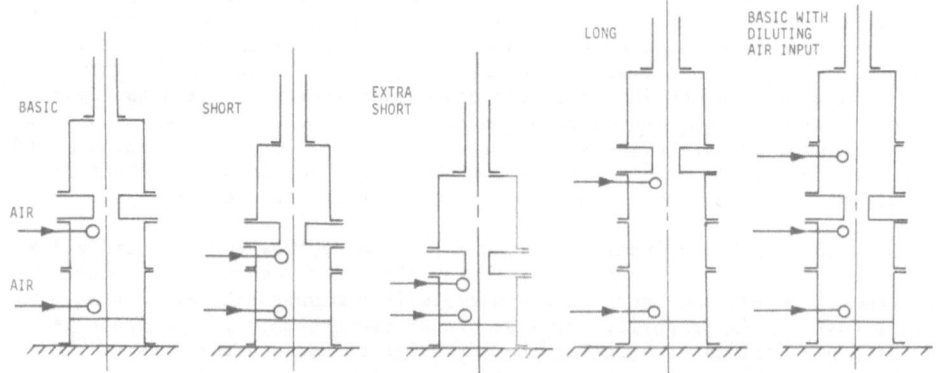

Fig. 2   Cyclone Furnace Prototype Configurations Using Modular Components

assemble several furnace versions with different heights, exit nozzle diameters, and also with secondary chamber for cooling exhaust gases as well as preheating the supply air. The combustion chamber modules were all fabricated in steel lined with 0.1 m (4 in) of casted alumina refractory.

The final configuration of the cyclone combustion chamber is shown in Fig. 3. It should be noted, that a conically shaped refractory boss was formed around the central exit nozzle to prevent the top vortex air from escaping directly through the hole. There were three viewing ports located at different levels of the vortex combustion chamber providing the facility for visual observation of fuel-particle circulation. Two additional viewing ports were placed at the bottom level of the secondary chamber to permit observation of the combustion product gases immediately after they leave the primary chamber. A refractory casted shield was placed over the nozzle outlet to direct the exit gas flow towards the circumference of the chamber. This resulted in a better mixing of the exhaust gases with the diluting air, which was introduced into the secondary chamber to protect both, the air preheating coil and the stack from overheating. This supplementary air was also considered to be helpful in completing the burning of solid fuel particles which could eventually escape from the main combustion chamber. The open bottom of the vortex combustion chamber shown in Fig. 3 was developed for the non-slagging configuration of the chamber, with ash being evacuated through a water-sealed air lock.

The schematic diagram of the cyclone furnace test set-up is shown in Fig. 4. The main air supply to the double vortex combustion chamber was

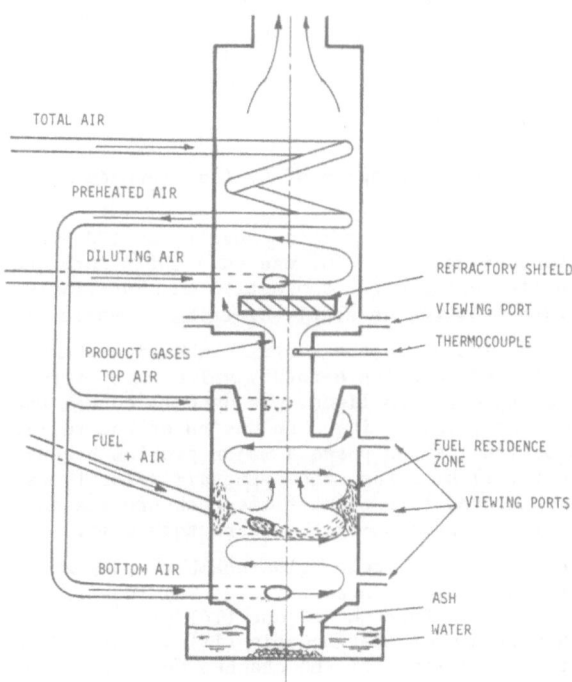

Fig. 3 Final Configuration of Cyclone Furnace Combustion Chamber

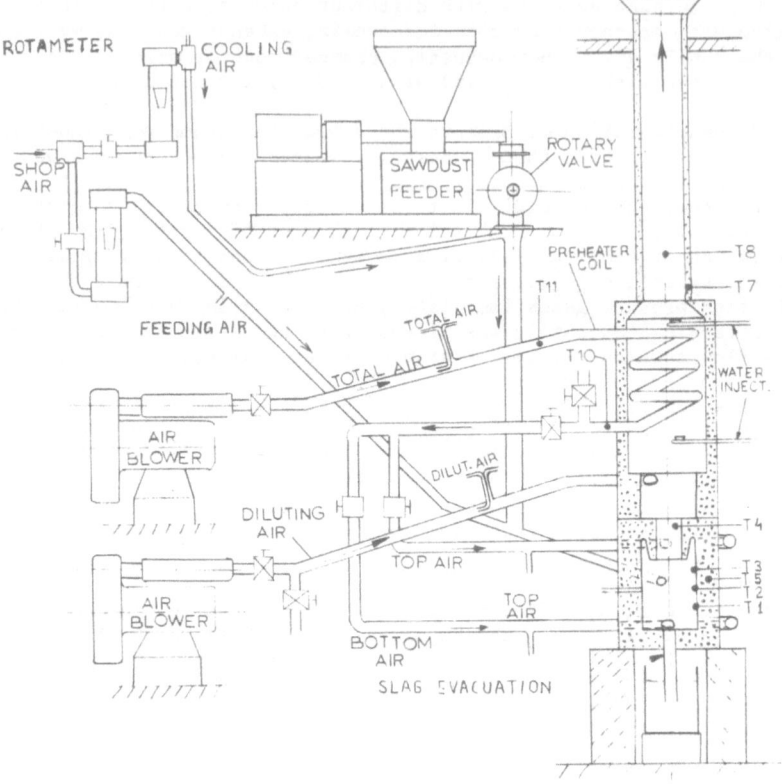

Fig. 4   Schematic Diagram of Cyclone Furnace Test Set-Up

delivered by a blower having a flow rate capacity of 170 SCFM. The air, first preheated to the temperature of 350°C, was then split into the top and bottom supply lines. Each line entered the combustion chamber via three equally spaced tangential inlets creating an evenly distributed vortex flow pattern.

Solid fuel was delivered to the vortex combustion chamber by a conventional feeder system incorporating a rotary valve which served as an air lock. Shop air, introduced through two lines, was used to convey the solid fuel particles delivered by the rotary valve. The design of the fuel-inlet configuration in the combustion chamber posed a major problem because of two conflicting requirements: 1) high fuel conveying air velocity required to avoid the inlet-hole clogging and to prevent flame back-propataion into the fuel delivery line, 2) minimum erosion of the refractory wall.

The gas temperature in the combustion chamber was measured using four platinum-rhodium thermocouples. Three (T1, T2, T3) located at three different levels at the chamber circumference and one (T4) protruding inside the exit nozzle. Another chromel-alumel thermocouple (T5) was placed inside the refractory wall at the mid-level of the chamber for monitoring of the wall

Fig. 5   Cyclone Furnace Experimental Set-Up

temperature. Several other thermo-couples were used to measure the temperature of the inlet air and of the exit gas in the stack. A view of the actual
cyclone furnace experimental set-up is shown in Fig. 5.

3. SOLID FUELS

    The major problem of energy recovery from solid wastes centers upon the
design of efficient combustion chambers which must allow retention of fuel
particles until they are completely burnt. The double cyclone combustion
chamber, as claimed by the inventor (Ref. 1), offers the best potential in the
waste incineration field. As a result, initial effort was concentrated on the
assessment of furnace performance in burning solid waste fuels, in particular
sawdust and other refuse derived fuels (RDF).

    Controlled batches of sawdust were used with the particle size held
below 6 mm (0.25 in). To obtain a homogeneous fuel, every batch of sawdust
(300 kg approximately) was thoroughly mixed and dried. The moisture content
was kept below 30%, and the high heat value as well as the chemical content
were analysed. Average values obtained from 5 consecutive batches of sawdust
are shown in Table 1. These properties of sawdust were compared with the
available data for municipal waste fuel (light fraction) as received from
AENCO Inc. (a subsidiary of Cargill Inc., New Castle, Delaware) (Ref. 3). The
properties were quite comparable, except in ash content. Therefore, in order
to experience slagging conditions when burning sawdust, granulated glass was
added (size 0.25 to 0.64 mm). The glass was not mixed with the sawdust
delivered through the feeder, but it was added directly to the rotary valve
inlet to insure uniform distribution of both components.

    In addition to the sawdust, a substitute for refuse derived fuel (RDF)
was investigated. It contained: paper (53%), wood (18%), rubber (2.5%), plas-
tic-polyethylene (12%), bread (4.5%), fresh vegetable (10%). Water represented
another percentage over all the basic components.

4. RESULTS AND DISCUSSION

    The development of the cyclone furnace combustion chamber included the
building of several versions of the prototype (as shown in Fig. 2) and the

TABLE 1:   Properties of Solid Fuels (Dry Basis Results)

Property / Fuel	Moisture Content (%)	High Heat Value (kJ/kg)	Chemical Content %				
			Carbon	Hydrogen	Nitrogen	Oxygen	Ash
Sawdust (average)	8 - 30	17,540	45.62	6.27	0.93	44.88	1.28
AENCO Refuse-Derived Fuel	23.7	16,770	35.24	6.20	0.64	40.62	17.10

testing of them at different operating conditions. As the result of that optimization work, the final constructional parameters were selected as follows:
(the initially selected values are shown in parentheses)

-combustion chamber diameter D_c = 0.305 m (not changed),
-combustion chamber height   H = 0.457 m (not changed),
-bottom air inlet area A_b = 0.00116 m² (0.002 m²),
-top air inlet area A_t = 0.00116 m² (0.002 m²),
-exit nozzle diameter D_o = 0.112 m (0.127 m);

these dimensions are indicated in Fig. 1.

As the result of that development, an exit gas temperature of over 1700°C
was reached using sawdust as fuel, with moisture content of 8%, and with an
air-to-fuel ratio holding close to stoichiometric. At that temperature the
steady-state conditions were maintained without much difficulty; some temperature fluctuations were mainly due to the lack of homogeneity of sawdust fuel.
The furnace was performing at its best when fuel was supplied at the mid-level
of the combustion chamber (as proposed in Ref. 2) and when the top and bottom
air-flow rates were maintained at the same magnitude. No particulates emmis-
sion was visually detected when operating under these conditions. With a
major variation of the top and bottom air-flow ratio, small particulates in
the form of sparks were visible in the exit gases through the viewing ports,
indicating incomplete combustion in the vortex chamber and loss of fuel. At
the same time, the exhaust gas temperature dropped significantly. Similar
fuel losses were observed when the fuel input was introduced either at the top
or at the bottom of the combustion chamber.

The swirl number "S" for the developed double vortex combustion chamber
was calculated using the formula given by Syred and Beer (Ref. 4). For isothermal conditions the value of "S" was 11.6, significantly higher than the
7.6 calculated initially for the prototype design (Ref. 2). For the combustion conditions, the swirl number was 3.4, also much higher than the 1.4 found
from preliminary calculations. The increase in the swirl number was mainly to the decrease in the air inlet flow area and to the preheating of the inlet air, both having been the result of the optimization work which aimed at
better retention of the solid fuel particles in the vortex combustion chamber.
These swirl number values, though high, still remain within the range of those
for conventional vortex combustion chambers (Ref. 4).

From flow visualization experiments in the cold double vortex chamber
(Ref. 2), it was demonstrated that by appropriate adjustment of the top and
bottom air supply, the solid fuel particles could be made to orbit at any
desirable level within the chamber. In order to confirm the above phenomenon
in the hot combustion chamber, granulated glass was added to the sawdust.

After a short run of 11 minutes at optimum operating conditions, the combustion chamber was disassembled. A well-established layer of solidified slag was found at approximately mid-level of the chamber, which was considered to be the residence zone of the fuel (Fig. 6).

The thermal efficiency of the double vortex combustion chamber was calculated without taking into account the heat-transfer losses through the wall of the chamber, which was not insulated. The numerical results indicated that the thermal efficiency was in the order of 70% to 75%, which was encouraging for the initial prototype. With different batches of sawdust having similar moisture content, the results were quite consistent. For sawdust with higher moisture content (up to 30%) continuous steady-state operation of the furnace could still be maintained but at the expense of a lower exhaust-gas temperature. The temperatures obtained with the refuse-derived fuels were at levels comparable to that of sawdust with a similar moisture content. No significant differences were observed between the burning of sawdust and the burning of refuse-derived fuel.

As mentioned in the patent description (Ref. 1), the original intention of the inventor was to operate this double vortex combustion chamber in the slagging mode for easy evacuation of the non-combustible residuals. Moreover, with the same combustion chamber volume, higher burning capacity can be achieved when operating at high temperature level which is also advantageous from the point of view of energy extraction. Initial results indicated that, with the prototype double cyclone furnace having a volume of 1 ft^3, an approximate heat-release rate of 220 kW (800,000 BTU/hour) was attained. This performance compares well with other types of vortex combustion chambers (Refs. 5 and 6).

When a furnace operates under slagging conditions, it offers all the advantages mentioned above. However, there are some disadvantages, such as high nitrogen oxide emission, increased tendency of solid fuel particles to escape from the combustion chamber, and also high cost of refractory material and maintenance associated with high temperature operation. Generally, the

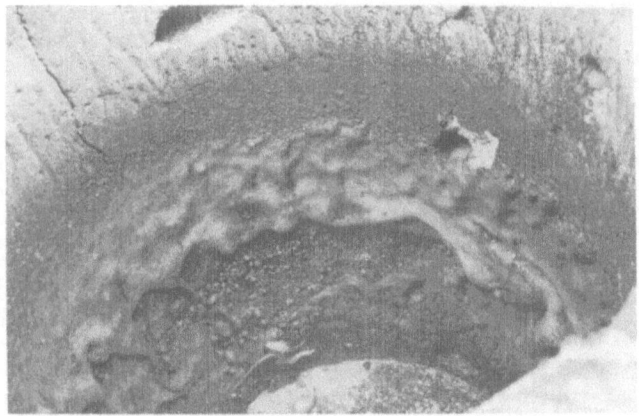

Fig. 6 Deposit of Slag Indicating Fuel Residence Zone found on the Wall of the Combustion Chamber after Short Run

advantages outweigh the disadvantages; however, in the case of the double cyclone furnace, it was found impractical to operate the furnace in the slagging
mode. The reasons were as follows:

1) As seen from Fig. 6, when a high enough temperature in the furnace was
   reached, the non-combustible additive (glass) began to melt and formed a
   layer of slag on the refractory wall in the fuel residence zone. After a
   certain period of operation, the slag accumulated and gradually found its
   way towards the bottom of the combustion chamber because of gravity. When
   the slag flow approached the bottom air inlets, the relatively cool air
   caused it to solidify around the air inlets, creating distortion of the
   bottom vortex flow pattern.

2) During the furnace operation, the solid fuel particles were conveyed by
   cold shop air and introduced to the combustion chamber at its mid-level
   which was the fuel residence zone. When fuel with glass additive was used,
   it was found that the slag deposited around the fuel inlet hole became
   solidified owing to the cold air used for conveying the fuel. This caused
   a gradual blockage of the fuel supply inlet. This problem could be solved
   by shifting the fuel inlet to the top of the chamber. However, this would
   result in losses of fuel particles escaping through the exit nozzle, as
   observed through the viewing ports, and also in lower thermal efficiency
   of the combustion chamber.

3) When burning the refuse-derived fuel with a high paper content (53%) in
   slagging mode, it was found afterwards that the combustion chamber became
   contaminated with white deposits attached to the layer of slag. These
   deposits of various irregular sizes with dimensions up to 3 cm, were found
   to contain alumina (Al_2O_3), a major component of the clay which is used as
   filler in paper production. Since the melting point of alumina is around
   1800° C, it was not transformed into slag. It was thought that the alumina
   would eventually contaminate the furnace after a longer run.

Experiments done in the non-slagging operation mode (i.e. at the temperature of 1000° C approximately), did confirm again the ability of the double
vortex combustion chamber to retain the solid fuel particles until they burnt
completely. To keep the gas temperature at a low level in order to avoid slag
formation, much excess air had to be supplied. The rate of heat release was
thus limited. The thermal efficiency of the combustion chamber was found to
be quite similar to that of the chamber at slagging temperature operation
(i.e. in the range of 70 to 75%).

For evacuation of non-combustible residuals (ashes) in the non-slagging
operation mode, the combustion chamber was modified with an open bottom using
a gas lock created by the water level (Fig. 3). The ash-evacuation procedure
followed a unique cyclic flow pattern during which the fuel flow was stopped,
the bottom air supply was temporarily terminated, and all ash remaining in the
fuel residence zone was brought by the top air flow into the sink in a manner
similar to that of a dust separator (Ref. 2). It was experimentally confirmed
that the "air grid" created in the combustion chamber by the bottom vortex is
capable of retaining the fuel in the residence zone. The amount of lost fuel
through the "air grid" was below 0.5% of the total fuel supply.

During the experiments involving the development of the cyclone furnace
the solid fuel was supplied to the vortex combustion chamber in three ways:
1) at the top of the chamber, together with the top-inlet air; 2) at the
bottom of the chamber, together with bottom-inlet air; and 3) at the mid-level
of the chamber through a specially made inlet system and conveyed by

the shop air as described in Ref. 2. **Significant erosion** of **the refractory**
wall was found after a total of 20 hours of furnace operation with
fuel introduced at the top of the combustion chamber. However, no traces of
erosion were found on the combustion chamber refractory wall after more than
20 hours of furnace operation when the improved fuel delivery system with the
inlet at the mid-level of the combustion chamber was used, as shown in Figs.
3 and 4.

5. CONCLUSIONS

The preliminary development of the double cyclone furnace was accomplished with a maximum temperature of 1700° C which was attained using sawdust
as the primary fuel. It was confirmed that the double vortex combustion
chamber is able to retain the solid fuel particles in a confined fuel residence zone until they are completely burnt. The slagging operation mode was
found impractical for this design configuration, mainly because of difficulties in slag evacuation. However, the prototype furnace could be operated
successfully in the non-slagging mode with a special procedure devised to
evacuate the non-combustible residue periodically by running the furnace as a
dust separator for a short duration of time. With further development that could
overcome the difficulties in slag evacuation, the full potential offered by
this unique design may be brought to the waste incineration field.

ACKNOWLEDGEMENTS

This research was supported by the Natural Sciences and Engineering Research Council of Canada, Grant No. G 0117, and by Boyd, Stott and McDonald
Technologies Limited, Montreal.

REFERENCES

1. Angus, D.; United States Patent No. 4.002.127, January 11, 1977, Cyclone
   Structure.

2. Krepec, T. and Kwok, C., Preliminary Investigations of a Double Vortex
   Combustion Chamber, 1981 Fall Meeting of the Combustion Institute, Eastern
   Section.

3. Sherwin, E.T. and Nollet, A.R., Solid Waste Resource, Mechanical Engineering, May 1980.

4. Syred, N. and Beer, J.M., Combustion in Swirling Flows. A Review.
   Combustion and Flame, No. 23, 1974.

5. Howell, W.J., Cyclone Furnace Firing. Babcock-Wilcox and Goldie-McCulloch
   Limited, Bulletin 59-2.

6. Mills, R.G. and Desmon, L.G., Operating Experience in the Suspension Burning
   of Waste Materials in Cyclone Incinerators. Energex Ltd., Proceedings of
   1972 ASME National Incineration Conference.

Computer Simulation of Homogeneous Combustion Parameters and Emissions Control in Gasoline Engine

NABIL Y. ELIAS and A.S. MIRZA
University of Petroleum and Minerals
Dhahran, Saudi Arabia

ABSTRACT

A generalized computer model of the combustion process in gasoline engine is presented, which can be used to determine the combustion temperature, pressure, composition of the emission species and the effect of varying different combustion parameters on constant volume combustion process. Eighteen chemical species have been considered to be present in the engine exhaust emission and their composition computed and plotted at different air-fuel ratios, pressures, temperatures and atomic ratios of the elements in reaction.

Gasoline engine used in the passenger cars is the major source of air pollution due to emitting carbon monoxide and nitrogen oxides in large quantities. This paper gives an investigation of different methods proposed to reduce the quantities of air polluting species in the exhaust. The methods are:

1. Water injection into the combustible mixture at inlet manifold.

2. Recirculation of a part of cooled exhaust gas.

3. Using methanol in place of gasoline.

The computer program developed is a versatile one and can be used for other combustion systems with minor modification. Also it can be extended to take account of other combustion species.

1. INTRODUCTION

Gasoline engines are almost exclusively the power plants for passenger cars and light trucks. The world total at present is approximately 300 million passenger cars. It has been evaluated that these automotive vehicles are the major contributor to air pollution on a total mass basis. Table I shows a comparison of emission quantities from different mobile sources. The air pollutants from the motor vehicles can be introduced into the atmosphere from exhaust, the crankcase vent, the carburetor and the fuel tank. The major components of the engine exhaust emissions which cause the air pollution are unburned hydrocarbons, carbon monoxide and oxides of nitrogen. Unburned hydrocarbons come out from the crankcase vents and fuel tank vents; and this is an evaporation loss of the fuel. A small amount of unburned hydrocarbons emitted in the engine exhaust due to incomplete combustion. Carbon monoxide and oxides of nitrogen are released in significant amounts from the engine exhaust.

Increasingly frequent instances of eye irritation, atmospheric occlusion

Table I : 1968 NATIONWIDE EMISSION ESTIMATES [1]*
(10^6 Tons/Year)

Category	HC	CO	NO_x
Gasoline engine	15.2	59.2	6.6
Diesel engine	0.4	0.2	0.6
Aircraft	0.3	2.4	0.0
Railroads	0.3	0.1	0.4

* *Numbers in brackets designate references*

and vegetation damage in the Los Angeles basin during the late 1940's were the
principal motivating factors for research which resulted in the proposal that a
photosynthesized chemical reaction was the responsible mechanism. This concept, generally credited to Haagen-Smit [2], established that unburned hydrocarbons and oxides of nitrogen in the presence of ultraviolet radiation were
mainly responsible for the irritation and damaging consequences of the so called
"smog" in Los Angeles.

Carbon monoxide concentration tends to follow directly the fuel air ratio.
Leaning of mixture results in proportionate reduction in carbon monoxide in all
of the studies which report on this phenomena. In principle, the concentration
of carbon monoxide contained in exhaust products should correspond to a chemical
equilibrium state represented by the water gas equation;

$$H_2O + CO = H_2 + CO_2.$$

At maximum flame temperatures this equilibrium yields significant quantities
of CO relative to CO_2, even for fuel lean mixture ratios. However as combustion
gases cool from peak flame temperatures to the much lower temperatures, this
equilibrium shifts in a direction favouring oxidation of CO to CO_2. Consequently for fuel lean or chemically correct mixture ratios, relatively small quantities of carbon monoxide ultimately appear in exhausted combustion products. For
fuel rich mixture ratios, however due to the simple insufficient concentrations
of carbon monoxide persist even in cool exhaust product.

The principal oxide of nitrogen formed in combustion process is nitric
oxide, NO. Nitric oxide is a high enthalpy specie relative to N_2 and O_2 from
the standpoint of basic thermodynamics. Therefore, its presence is favoured by
the existance of high temperature. In addition to being highly dependent upon
temperature, the rate of formation of nitric oxide depends upon the concentration of oxygen present in combustion products. Because of the thermodynamics
involved, the oxides of nitrogen produced at the engine exhaust port are almost
entirely nitric oxide, NO.   This is confirmed by measurements made with mass
spectrometer [3]. In the presence of oxygen and at low temperature the nitric
oxide becomes nitrogen dioxide (NO_2). This oxidation usually takes place in
the atmosphere, but it can also commence in the exhaust system, particularly if
air has been added to the gases at the exhaust value in an effort to partly
oxidize the unburned hydrocarbons.

2. LITERATURE REVIEW

The relationship between automobile exhaust emission and air pollution has been established largely as a result of studies of the air pollution problem in Los Angeles, California. In 1943 serious air pollution was detected in this area and as a result in 1948 the State Legislature passed a law permitting the formation of air pollution control districts with the power to curb emission sources [4].

A number of analyses of the state of automotive emission control technology have been made [5-7]. These contain estimates of emission control levels achievable by current and potential future power plants.

The first step in the analysis of the combustion process is to determine the composition and concentration of the products of combustion under different working conditions.

In 1951, Huff,et al.,[8] presented a report to NACA (National Advisory Committee for Aeronautics) introducing a general method for computation of equilibrium composition. This was the first triumph for calculation of exhaust emission. In 1964, a comprehensive report on chemical equilibrium study was prepared by a group of scientists leaded by R.D.Kopa [9]. The development of equation system defining the chemical equilibrium was based on the method presented by Huff,et al. The report showed that water injection into intake manifold and exhaust gas recirculation is effective in reducing the nitrogen oxides. In 1971, Gordon and McBridge compiled a computer program which can be used to calculate chemical equilibrium at assigned thermodynamic state. The program considers condensed as well as gaseous species. The solution technique minimizes the Gibbs free energy of all possible molecules that can be formed from given reactants. The two particular cases of constant pressure and constant volume combustion are special cases of assigned thermodynamic states in the NASA-Lewis Program [10].

"The idea of introducing water into an internal combustion engine is not new", Professor Hopkinson of Cambridge University wrote about sixtyeight years ago. In fact the concept of supplementing the fuel air mixture with water has accompanied the development of modern spark ignition engines. As engines have been developed the use of water addition has been repeatedly proposed as a solution to new problems.

An introduction of water vapor into the combustion chamber decreases the maximum cycle temperature because of its dilution effect. Intake manifold water injection has a further advantage in that its evaporation cools the gases in the induction system. As a result the charge becomes denser. The increased gas density results in higher mass flows through the engine and thereby provides higher maximum power levels.

The effects of adding water to gasoline (manifold injection) on engine efficiency and exhaust emission were examined by Peters and Stebar [11]. The results showed increased hydrocarbon emission and decreased vehicle driveability. However on the positive side it showed a higher octane rating and decrease nitric oxide emission.

Another method to control the CO and NO_x from gasoline engine which is presently undergoing extensive laboratory testing is exhaust gas recirculation (EGR). The procedure is to cool down a portion of exhaust and to introduce it below the carburetor, thereby deliberately diluting the fresh charge. This dilution has several effects. Obviously, since the exhaust contributes no energy to the combustion process the peak temperature reached with EGR is reduced. The formation of NO_x increases with increasing temperature, so the EGR works in the desired direction. Exhaust gas is preferred as a diluent over air, owing to the fact that CO_2 and H_2O being triatomic molecules have higher heat capacities than O_2 and are therefore more effective in achieving lower combustion temperatures.

Komiyama and Heywood[12] investigated the effect of EGR on NO_x concentration in the exhaust of gasoline engine. They also studied the effect on flame speed, ignition delay and cycle to cycle pressure variation.

A new concept simultaneously achieving low NO_x and improved fuel economy has been demonstrated by H.Kuroda, et al.[13]. It involves fast burning of the fuel and heavy amount of exhaust gas to be circulated. The fast burn being achieved by dual spark plug ignition in a sophisticated combustion chamber configuration. The idea has been practised at Nissan Motor Company and the experimental results have proved the fact that EGR is the best solution for the reduction of NO_x.

The present work is a computer simulation of the homogeneous combustion in a gasoline engine. It computes the composition and concentrations of the exhaust mixture at different combustion temperature and pressure. Effect of air-fuel ratio, oxygen-nitrogen ratio of the reacting mixture and carbon-hydrogen ratio of the fuel on the concentration of emission species has been studied and very useful conclusions have been drawn. The results have been compared with the data available in the literature. To reduce the quantities of CO, NO and NO_2 in the engine exhaust, effectiveness of the water injection and EGR techniques have been computed. Moreover, a comparison of methanol and gasoline by emission level of CO and NO_x is also given. The computer program used is a versatile one written in Fortran IV and run on IBM 158/370 + 3033 system. The program is very economical in respect of its execution time.

3. METHODOLOGY

For the calculation of the composition and concentration of the product species while using thermodynamic methods, a stringent condition of the chemical equilibrium is imposed. Iso-octane C_8H_{18} as a hydrocarbon fuel is chosen because its hydrogen-carbon ratio is close to that of commercial gasoline used in automotive transportation. The reaction equation of octane with air for stoichiometric combustion expressed in number of moles is

$$C_8H_{18} + 12.5\ O_2 + 12.5(3.76)N_2 = 8\ CO_2 + 9\ H_2O + 47.02\ N_2$$

As the combustion reaction takes place at quite a high temperature, dissociation takes place at large amount with the result that a large number of combustion products are formed. In order to find how plentiful a specie may be in the mixture, that specie must be included in the solution. In the present analysis it is assumed that the exhaust gas from the gasoline engine consists of 18 chemical species, they are

O, O_2, O_3, H, H_2, H_2O, OH, C, CO, CO_2, CH_4, N, N_2, NO, NO_2, NH_3, HNO_3, and HCN.

Some of these, eg., O_3, NO_2, HNO_3, HCN, CH_4, and NH_3 have negligible concentration but under extreme conditions the amounts of these substances is noticeable. When chemical equilibrium is established, the mole fractions of the above 18 species are related by equilibrium equations (eq.1-14 in the Appendix). The most reliable and latest values of equilibrium constants are available in JANAF Tables [14]. The non linear equations are reduced in number by considering some dominant species in the product mixture and transforming all the equations in terms of these species. The system of equation is simplified by linear approximation, using Taylor series and the solution is obtained by the Newton Method (eq.21-24).

The change in air-fuel ratio, oxygen-nitrogen ratio and carbon-hydrogen ratio of the reacting mixture will change the atomic ratios α, β, and γ defined in Appendix. So these are adjusted to take account of these effects. The combustion reaction used for methanol is

$$CH_3OH + 1.5\ O_2 + 1.5(3.76)N_2 = CO_2 + 2\ H_2O + 5.64\ N_2.$$

Then using the computed mole fractions of all the 18 species as well as the enthalpies of each product specie, the combustion temperature and pressure are determined. These results were compared with the available temperature-pressure data in the literature, Table V. The excellent agreement show that the strategy used is a correct one.

The next step was to determine the concentrations of the compounds of major concern, i.e., CO, NO, and NO_2 out of 18 species. This is done by interpolation. Effect of adding water and EGR on the concentration of CO and NO_x is computed based on the law of conservation of mass and internal energy in the reaction. The full description and the complete computer program are given in reference [15].

4. COMPUTATION OF EMISSION SPECIES

A computer program based on the equation system given in the Appendix was used to calculate the equilibrium composition of the exhaust gas. This step is very helpful in finding the effect of different combustion parameter on the combustion products. Figures 1 and 2 show the composition of the exhaust gas consisting 18 chemical species on different combustion temperatures. Temperature is the most effective combustion parameter which changes the concentration of every specie. The compounds of special interest CO, NO and NO_2 were found increasing with high temperature. This concludes that the reduction of combustion temperature is favourable in reducing the quantity of air pollutants. Figure 3 shows the effect of temperature on the above mentioned compounds.

Effect of air-fuel ratio of the reactant mixture upon the composition of the product mixture was also studied at a constant temperature of 3000°K. In case of lean mixture, the concentration of CO decreases but NO_x increases and for rich mixtures it is otherwise.

The effect of oxygen-nitrogen ratio of the reactant mixture upon the exhaust gas shown in figure suggests the solution of the problem that if nitrogen is used to dilute the reactant mixture, the concentration of both CO and NO_x decrease sharply. The exhaust gas which has a higher nitrogen to oxygen ratio can be used to mix with the reactant mixture for dilution. Moreover, CO_2 being triatomic has higher specific heat value than O_2 and hence will lower the temperature of exhaust gas. These results had been the cause of well known technique of EGR (Exhaust Gas Recirculation) for reducing the air pollutants in the engine exhaust. The effect of C/H ratio of the fuel is also computed and plotted in figure 6. The graph shows a fuel having low C/H ratio is more suitable as far as air pollution is concerned.

5. EMISSION CONTROL TECHNIQUES

The legal means to prevent and control air pollution have developed in the

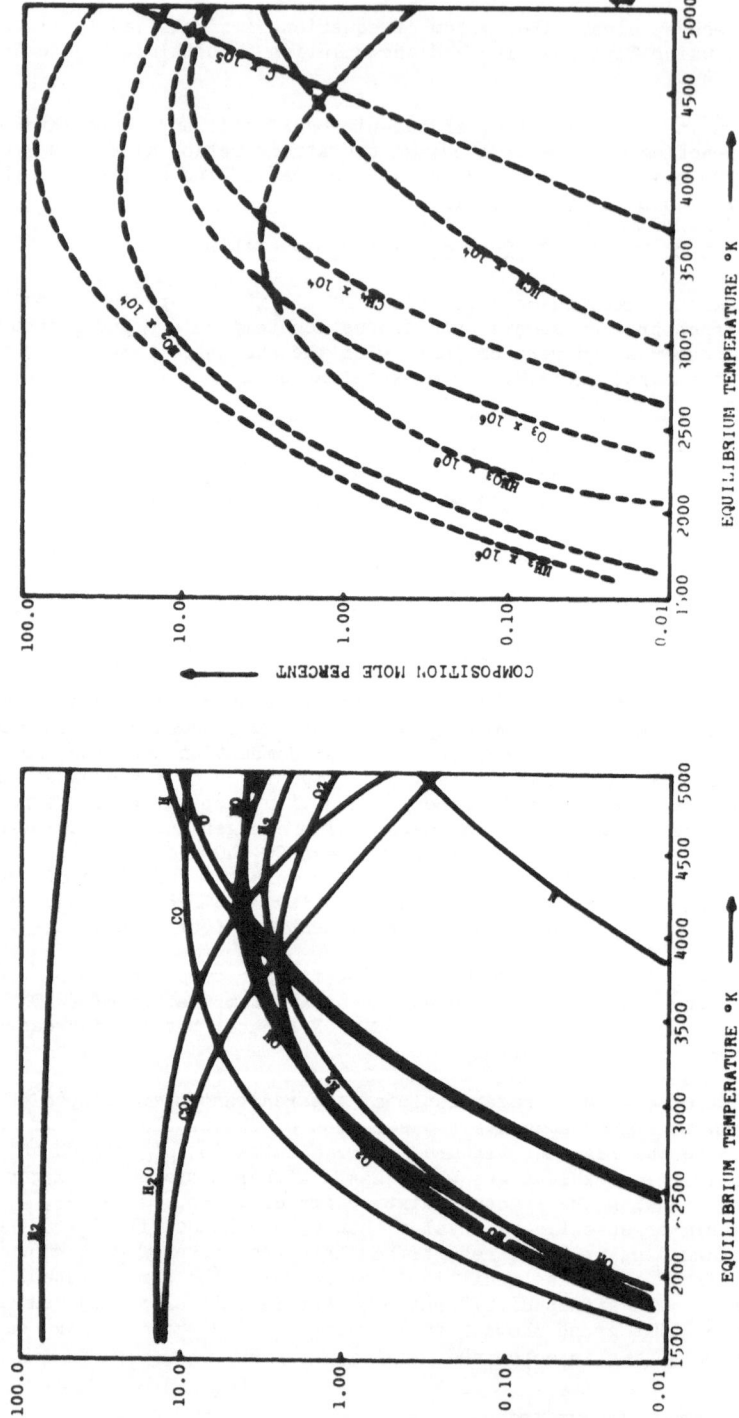

Fig. 1. Effect of Temperature on the Concentration of Emission Species.

Fig. 2. Effect of Temperature on the Concentration of Emission Species.

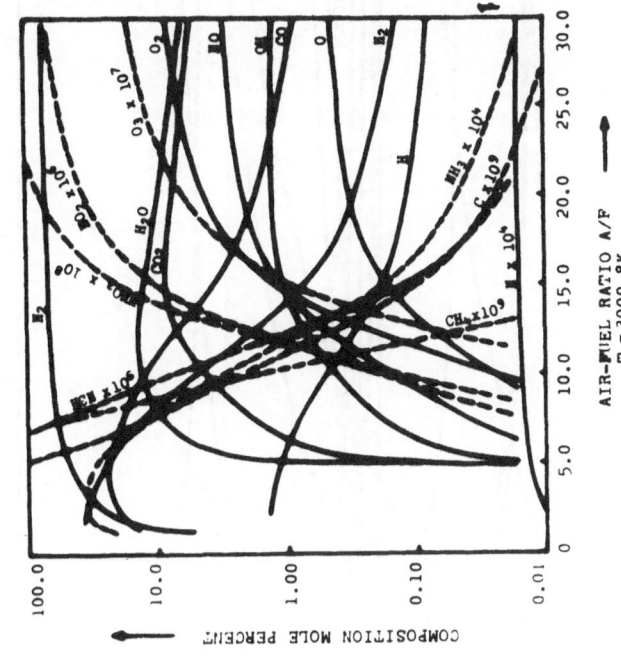

Fig. 4. Effect of Air-Fuel Ratio on the Concentration of the Emission Species.

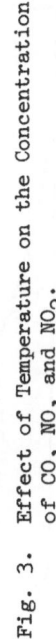

Fig. 3. Effect of Temperature on the Concentration of CO, NO, and NO$_2$.

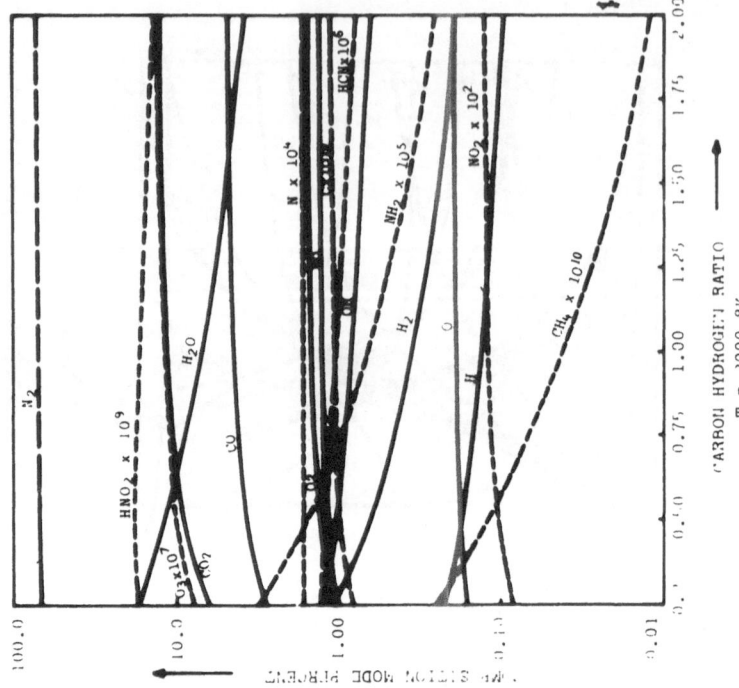

Fig. 6. Effect of Carbon-Hydrogen Ratio of the
Hydrocarbon Fuel on the Concentration
of Emission Species.

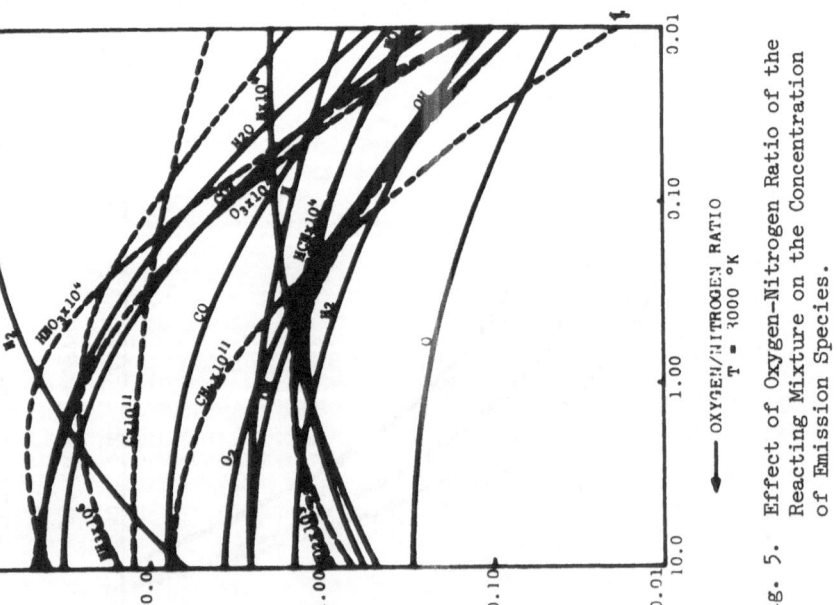

Fig. 5. Effect of Oxygen-Nitrogen Ratio of the
Reacting Mixture on the Concentration
of Emission Species.

U.S.A., Japan and other industrialized areas over a period of many years. Shortly after 1900, air pollution was the subject of ordinances in a number of U.S.municipalities. A country wide approach to air pollution control was taken by Los Angeles in 1947. NO_x control began in 1971 because it recognized the importance of NO_2 in the photochemical smog reaction as well as effects on health. The gradual decrease in Federal automotive exhaust emissions standards are shown in Table II.

Table II. FEDERAL AUTOMOTIVE EXHAUST EMISSIONS STANDARDS

Year	HC	CO	NO_x
1968	3.2 g/mi	3.3 g/mi	..
1970	2.2 g/mi	2.3 g/mi	..
1971	2.2 g/mi	2.3 g/mi	4.0 g/mi
1975	0.41 g/mi	3.4 g/mi	3.0 g/mi
1980	0.41 g/mi	3.4 g/mi	0.4 g/mi

5.1 Water Injection Through Manifold

The concept of supplementing the fuel air mixture with water has accompanied the development of modern spark ignition engines. As engine has been developed, the use of water addition has been repeatedly proposed as a solution to new problems.

The computer results (Table III, Figure 7) show a good effect of water injection on the emission species. There is substantial decrease in the concentration of CO, NO and NO_2. At 0.5 lb of water injected per lb of fuel there is 25 % reduction in CO and 24% reduction in NO and NO_2.

Table III. EFFECT OF INJECTING WATER INTO COMBUSTIBLE MIXTURE ON COMBUSTION TEMPERATURE AND CONCENTRATION OF CO AND NO_x

W/F lb of water per lb of fuel	Temperature °K	CO Mole %	NO Mole %	$NO_2 \times 10^4$ Mole %
0.000	2713.186	1.8925	0.5959	4.9802
0.200	2677.556	1.6496	0.5254	4.3027
0.400	2642.886	1.4788	0.4752	3.8484
0.600	2609.080	1.3292	0.4309	3.4523
0.800	2576.186	1.1554	0.3787	2.9762
1.000	2544.038	1.0338	0.3421	2.6556
1.200	2512.562	0.9269	0.3097	2.3757
1.400	2481.858	0.8014	0.2719	2.0513
1.600	2451.820	0.7147	0.2452	1.8273
1.800	2422.325	0.6387	0.2216	1.6315
2.000	2393.417	0.5517	0.1947	1.3996

5.2 Exhaust Gas Recirculation

One of the methods suggested for reducing the nitrogen oxides emitted from automobile exhaust and which is presently undergoing extensive laboratory testing is exhaust gas recirculation.

The computer results (Table IV) have been plotted in Figures 9 and 10 showing the effects of recirculating different amounts of exhaust gas per lb of fuel-air mixture. For 0.3 lb of EGR there is 70% reduction in CO and about 75% reduction in NO_x. Table III shows that the temperature drop due to 0.3 lb of EGR is 400°C.

Table IV. EFFECT OF EXHAUST GAS RECIRCULATION ON COMBUSTION TEMPERATURE AND CONCENTRATION OF CO AND NO_x

lb of EGR per lb of fresh charge	Temperature °K	CO Mole %	NO Mole %	$NO_2 \times 10^4$ Mole %
0.000	2713.186	1.8925	0.5959	4.9802
0.050	2638.229	1.5191	0.4787	3.8715
0.100	2566.482	1.2136	0.3823	2.9909
0.150	2497.604	0.9627	0.3041	2.3034
0.200	2431.673	0.7793	0.2465	1.8133
0.250	2368.218	0.6088	0.1939	1.3643
0.300	2306.963	0.4902	0.1567	1.0652
0.350	2249.327	0.3815	0.1347	0.8651
0.400	2193.620	0.2895	0.0935	0.5818
0.450	2140.817	0.2291	0.0721	0.4334
0.500	2089.723	0.1754	0.0578	0.3299

5.3 Alternative Hydrocarbon Fuels

Investigations of alternative fuels for internal combustion engines usually are aimed at one of the three objectives:

  i) to improve the efficiency and performance of engines;

 ii) to widen the availability of natural resources for fuel production
       and thus avoid fuel shortages; and,

iii) to reduce pollutants in engine exhaust gases.

Of the possible alternative automotive fuels, methanol takes a leading place. Due to its availability in abundance and its high octane quality, which may allow the use of higher compression ratio it has been studied by many investigators [16, 17] not only in laboratory engines but also in cars. Methanol is much more cheaper than gasoline and it can be made readily from a number of nonpetroleum resources. Methanol can be produced by passing compressed synthesis gas through a suitable catalyst. The source of synthesis gas can be coal, various natural wastes or even liquid and gaseous hydrocarbons. Methanol also can be manufactured by the destructive distillation of wood [17].

From technological aspects it is possible to use methnal as a gasoline extender or as methanol-gasoline blends for automobiles. Figure 11 and Figure 12 show a comparison between the emission levels of CO and NO produced by using gasoline and methanol. The reduction in CO is more than 90% and in NO about 55%. Moreover there is 400-800°C drop in maximum temperature achieved by methanol.

6. DISCUSSION

The results of the present work has been compared to the work of Steffensen, et al.[18] and Agrawal, et al.[19] in Table V. It is evident that the present results show an excellent agreement with those of other authors. A minute difference in the results is due to the fact that in present program the exact values of enthalpies, as reported by JANAF Tables, were read by the computer as an array of input.

Table V. COMPARISON OF PRESENT RESULTS WITH THOSE OF STEFFENSEN, ET AL. AND AGRAWAL, ET AL.

ψ Equivalance ratio	P_O Initial Pres. atm	T_i Initial Temp. °K	Present Work		Steffensen, et al.		Agrawal, et al.	
			P_O atm	T_O °K	P_O atm	T_O °K	P_O atm	T_O °K
0.80	1	298	8.589	2438.087	8.584	2436.0	8.5852	2434.460
0.80	10	298	87.048	2476.116	86.962	2473.0	86.9534	2471.460
1.00	1	298	9.529	2640.754	9.524	2638.0	9.5258	2637.457
1.00	10	298	98.267	2739.246	98.167	2736.0	98.1773	2735.026
1.25	1	298	9.885	2614.975	9.876	2612.0	9.8714	2609.214
1.25	10	298	100.115	2654.601	99.990	2651.0	99.9118	2646.956

The experimental values of the concentration of CO and NO_x as reported by some investigators are a bit higher than those of equilibrium composition [20]. The possibilities for this discrepencies are:

(a) Equilibrium calculations assume that sufficient time is available for the chemical reaction to maintain equilibrium. Recent theoretical and experimental investigations, however, have shown that this is not the case in typical engine operation.

The explanation for the presence of greater than equilibrium quantities of CO is that the reaction rate for conversion of CO to CO_2 is relatively slow one except at very high temperature. Thus the CO is 'frozen' at a value corresponding to very high temperature and very little piston travel after peak cylinder conditions.

Furthermore, equilibrium calculations predict that at temperature and pressures corresponding to exhaust gas conditions, NO should have mainly decomposed to N_2 and O_2. Measurements of actual concentration of NO in the exhaust gas

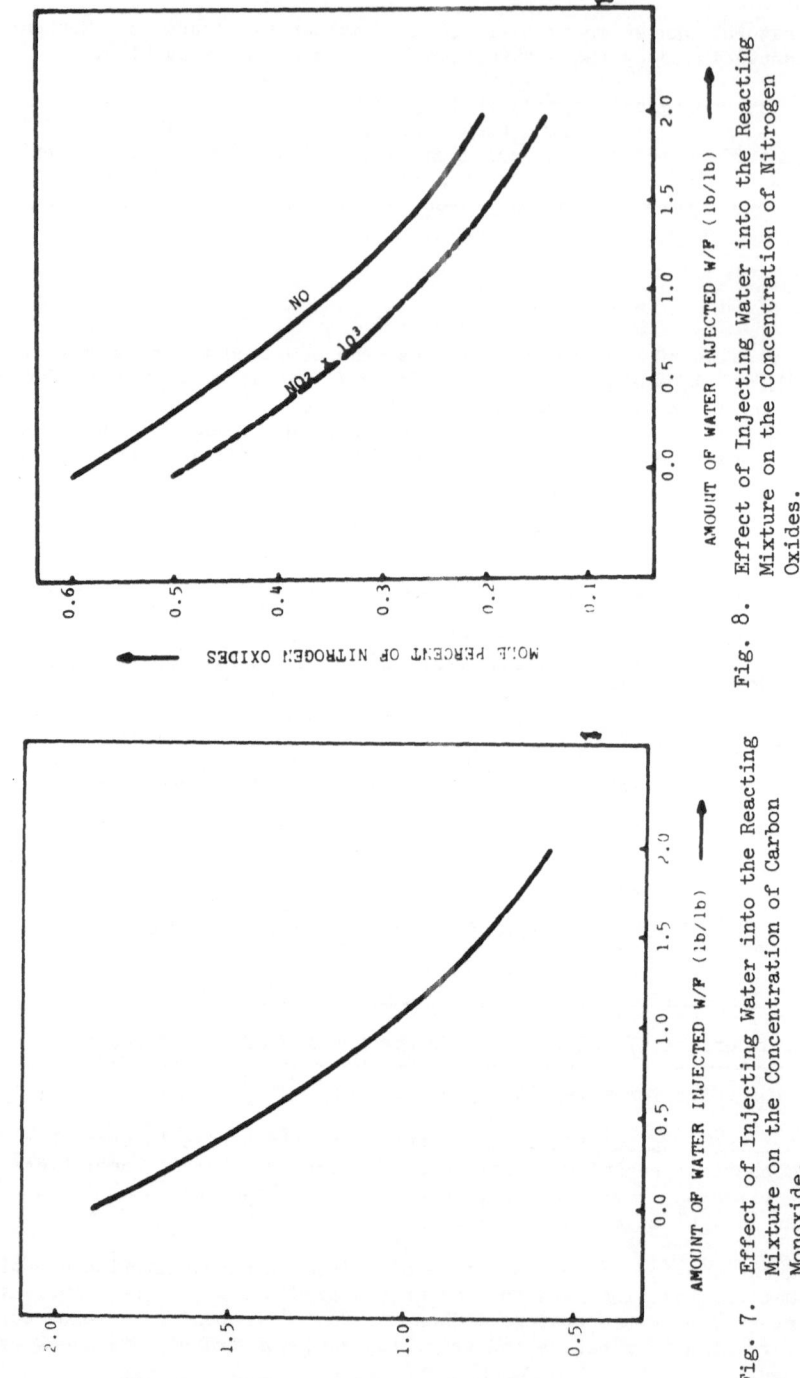

Fig. 8. Effect of Injecting Water into the Reacting Mixture on the Concentration of Nitrogen Oxides.

Fig. 7. Effect of Injecting Water into the Reacting Mixture on the Concentration of Carbon Monoxide.

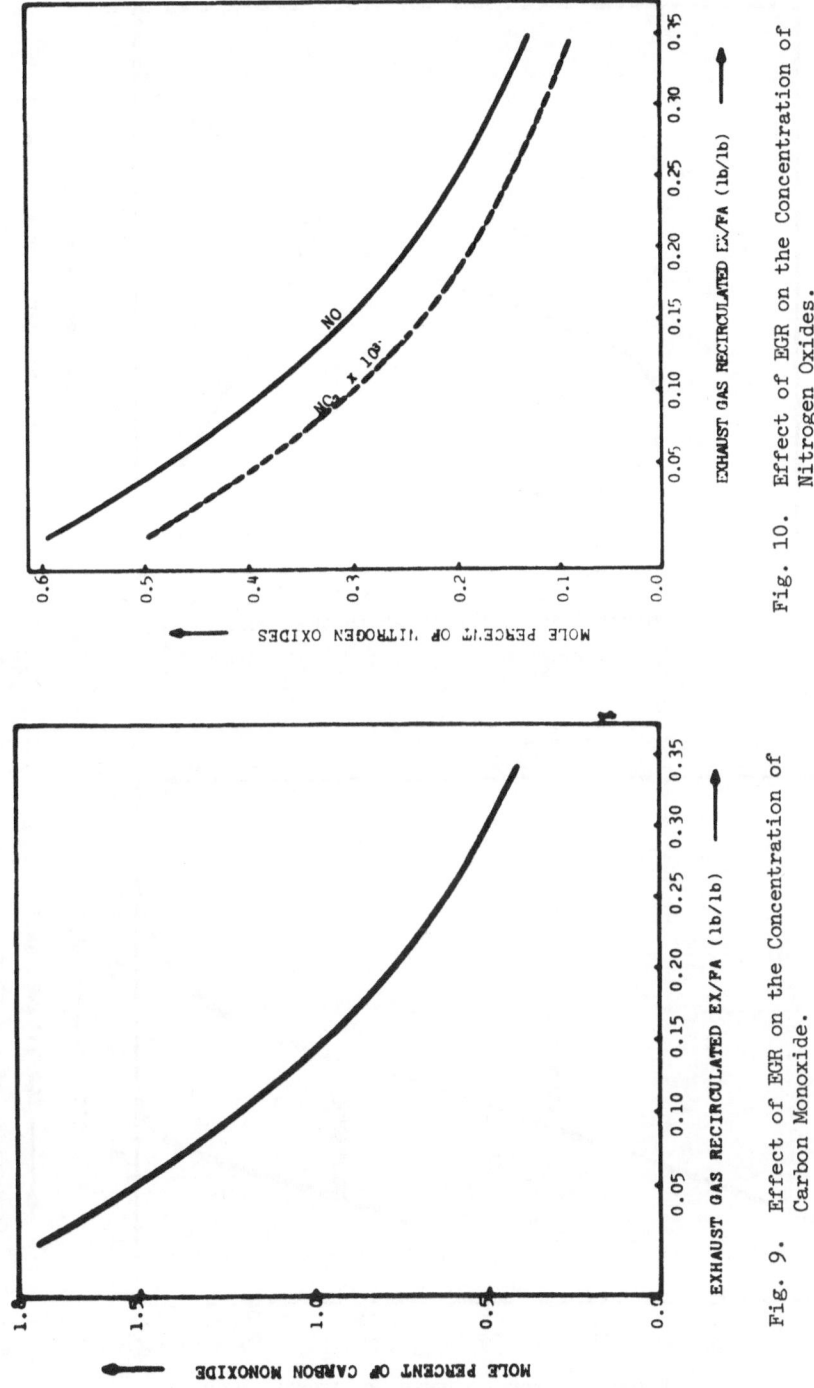

EXHAUST GAS RECIRCULATED EX/FA (lb/lb) ⟶

Fig. 10. Effect of EGR on the Concentration of Nitrogen Oxides.

EXHAUST GAS RECIRCULATED EX/FA (lb/lb) ⟶

Fig. 9. Effect of EGR on the Concentration of Carbon Monoxide.

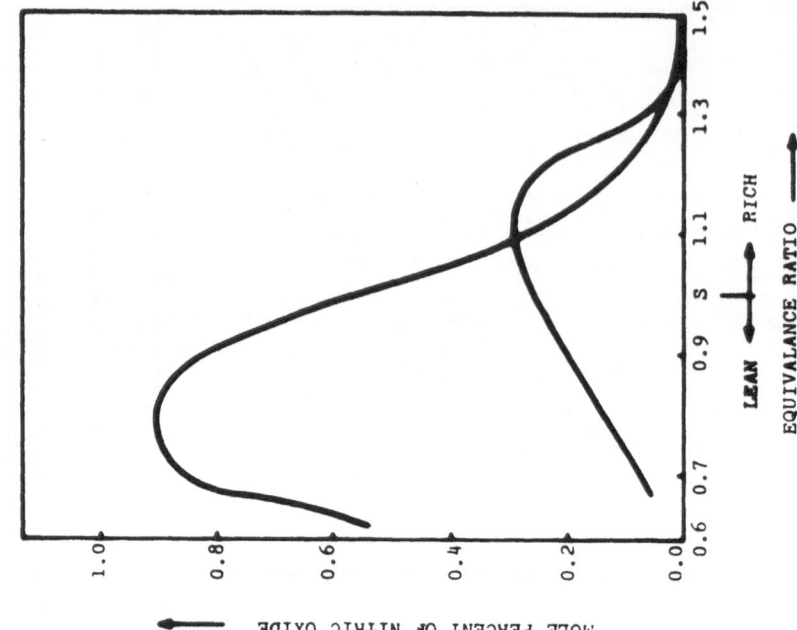

Fig. 12. Comparison of the NO Emission at Different Equivalence Ratio between Gasoline and Methanol.

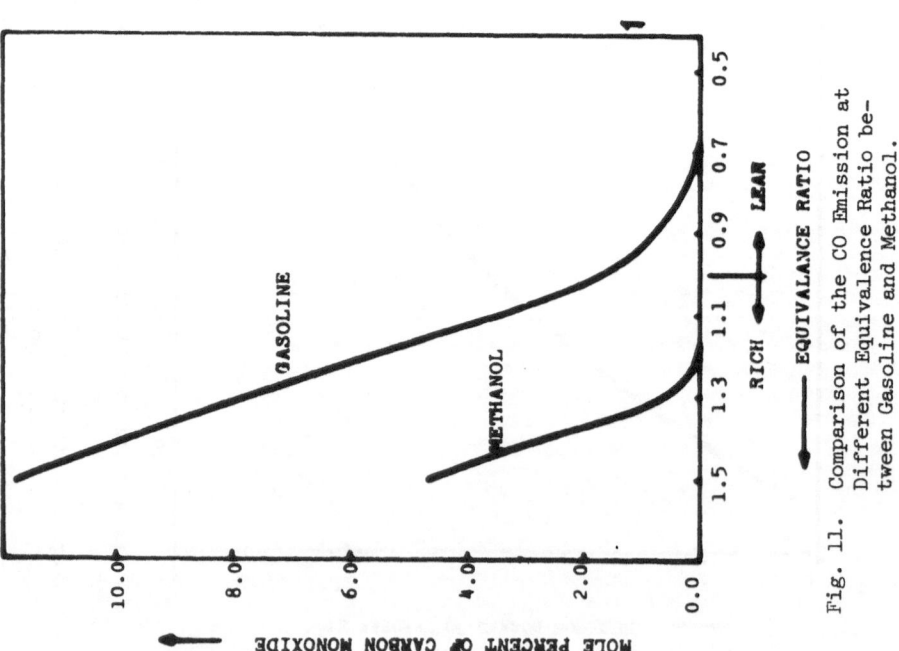

Fig. 11. Comparison of the CO Emission at Different Equivalence Ratio between Gasoline and Methanol.

show that this is not so, it is much higher than the equilibrium value at the exhaust pressure and temperature.

7. CONCLUSIONS

1. Intake manifold water addition reduces the concentration of CO and NO. About 25% reduction in CO and NO_x is obtained when 0.5 lb of water is injected per lb of gasoline.

2. Exhaust Gas Recirculation is more effective in reducing the air pollutants from gasoline engine. On recirculation of 0.3 lb of exhaust gas per lb of combustible mixture, there is 70% reduction in CO, 75% in NO_x and 406°C drop in peak combustion temperature.

3. When methanol is used in a spark ignition engine, CO emission reduces to negligible near stoichiometric air-fuel ratios and 55% reduction in NO_x is obtained. Moreover there is 400-800°C drop in peak temperature achieved by using methanol through an equivalance ratio of 1.5-0.8. Methanol seems to be fuel of automotive cars in future.

ACKNOWLEDGEMENTS

The authors acknowledge the full support of the University of Petroleum & Minerals, Dhahran, Saudi Arabia, in conducting this piece of research work. The cooperation of Dr.W.K.Rudloff and Data Processing Center's personnel is highly recognized. The authors appreciate the valuable help provided by Mr. Syed Ishrat Jameel of Mechanical Engineering Department at University of Petroleum & Minerals in typing the manuscript.

REFERENCES

1. National Air Pollution Control Administration. 1970. Determination of Air Pollutant Emissions from Gasoline Powered Motor Vehicles. U.S.DHEW, PHS, EHS, Durham, North Carolina.

2. Haagen Smit, A.J. 1952. Chemistry and Physiology of Los Angeles Smog. Ind. Eng. Chem. (44) pp.1342.

3. Starkman, E.S. 1969. Vehicular Emissions and Control. University of California, Berkeley.

4. Patterson, D.J. and Henein, N.A. 1972. Emission from Combustion Engines and Their Control. Ann Arbor Science Publishers, Inc., Ann Arbor, Michigan.

5. Morse, R.S. 1967. The Automobiles and Air Pollution: A Program for Progress. A Report by the Panel on Electrically Powered Vehicles, Department of Commerce Technical Advisory Board.

6. National Academy of Sciences, Washington, D.C. 1972. Semiannual Report by the Committee on Motor Vehicle Emissions of the National Academy of Sciences to the U.S.Environmental Protection Agency.

7. The Impact of Auto Emission Standards. 1973. Report of the staff of the Subcommittee on Air and Water Pollution to the Committee on Public Works, U.S.Senate, Serial # 93-11.

8. Huff, V.N. Gordon, G. and V.E.MOrell. 1951. General Method and Thermodyna-
    mic Tables for Computation of Equilibrium Composition and Temperature of
    Chemical Reactions. NACA Report 1037.

9. Kopa, R.D. Hollander, B.R. Hollander, F.H. and Kumura, H. 1964. Combustion
    Temperature, Pressure and Products at Chemical Equilibrium. SAE Progress in
    Technology, Vol.7.

10. Gordon, S. and McBridge, B.J. 1971. Computer Program for Calculation of
    Complex Chemical Equilibrium Composition, Rocket Performance, Incident and
    Reflected Shocks and Chapman-Jouguet Detonation. NASA Publication SP-273.

11. Peters, B.D. and Stebar, R.F. Water-Gasoline Fuels - Their Effect on Spark
    Ignition Engine Emissions and Performance. SAE Paper # 760547. 1971.

12. Komiyama, K. and Heywood, J.B. 1973. Predicting NO_x Emissions and Effects
    of Exhaust Gas Recirculation in Spark Ignition Engines. SAE Paper # 730479.

13. Kuroda, H. Nakajima, Y. Sugihara, K. Takagi, Y. and Muranaka, S. 1978. The
    Fast Burn with Heavy EGR, New Approach for Low NO_x and Improved Fuel Economy.
    Nissan Motor Company. SAE Paper # 780006.

14. JANAF Thermochemical Tables. Second Edition. The Dow Chemical Co, Midland,
    Michigan, 1971.

15. Mirza, A.S. 1981. Computer Simulation of Homogeneous Combustion Parameters
    and Emission Control in Gasoline Engines. M.S.Thesis. University of Petro-
    leum & Minerals, Dhahran, Saudi Arabia.

16. Lee. Wenpo and Geffers, W. 1977. Engine Performance and Exhaust Emissions
    Characteristics of Spark Ignition Engines Burning Methanol and Methanol-
    Gasoline Mixtures. Combustion and Reaction Kinetic Department, Wolfsheng,
    Germany.

17. Gallopoulos, N.E. 1977. Alternative Fuels for Reciprocating Internal Com-
    bustion Engines. General Motors Research Laboratories, Warren, Michigan.

18. Steffensen, R.J. Agnew, J.T. and Olsen, R.A. 1966. Combustion of Hydrocar-
    bons Property Tables. Bulletin # 122. Purdae University, Lafayette, Ind.

19. Agrawal, D.D. and Gupta, C.P. 1977. Computer Program for Constant Pres-
    sure or Constant Volume Combustion Calculations in Hydrocarbon-Air Systems.
    J.of Engg. for Power Transportations. Paper 76-DGP-2.

20. Watfa, A. and Daneshyar, H. 1976. Formation of Nitric Oxide Carbon Mono-
    xide and Unburnt Hydrocarbons in Spark Ignition Engines. Conference spon-
    sored by Automobile Division of I.M.E., Cranfield, England.

APPENDIX

EQUATION SYSTEM

In the present analysis it is assumed that the exhaust gas from the gasoline engine consists of 18 chemical species, they are:

CO_2, H_2O, CO, NO, N_2, O_2, H_2, OH, O, H, N, C, O_3, NO_2, CH_4, NH_3, HCN, HNO_3.

So at least 18 equations are required to compute the composition of each specie in the product mixture. The most common and simple chemical reactions involving the above mentioned species were selected based on the experience from the previous work on this problem. The equilibrium equations for those reactions are:

$$K_1 = \frac{X_O}{X_{O_2}^{\frac{1}{2}}} \, P^{\frac{1}{2}} \qquad \text{where X denote the mole fraction and P the combustion pressure.} \tag{1}$$

$$K_2 = \frac{X_{O_3}}{X_{O_2}^{3/2}} \, P^{-\frac{1}{2}} \tag{2}$$

$$K_3 = \frac{X_H}{X_{H_2}^{\frac{1}{2}}} \, P^{\frac{1}{2}} \tag{3}$$

$$K_4 = \frac{X_{H_2} \cdot X_{O_2}}{X_{H_2O}^2} \, P \tag{4}$$

$$K_5 = \frac{X_{OH}}{X_{O_2}^{\frac{1}{2}} \cdot X_{H_2}^{\frac{1}{2}}} \tag{5}$$

$$K_6 = \frac{X_C \cdot X_{O_2}}{X_{CO_2}} \, P \tag{6}$$

$$K_7 = \frac{X_C \cdot X_{O_2}^{\frac{1}{2}}}{X_{CO}} \, P^{\frac{1}{2}} \tag{7}$$

$$K_8 = \frac{X_C \cdot X_{H_2}^2}{X_{CH_4}} \, P^2 \tag{8}$$

$$K_9 = \frac{X_N}{X_{N_2}^{\frac{1}{2}}} \, P^{\frac{1}{2}} \tag{9}$$

$$K_{10} = \frac{X_N \cdot X_O}{X_{NO}} \, P \tag{10}$$

$$K_{11} = \frac{X_{NO} \cdot X_{O_2}^{\frac{1}{2}}}{X_{NO_2}} \, P^{\frac{1}{2}} \tag{11}$$

$$K_{12} = \frac{X_{N_2}^{\frac{1}{2}} \, X_{H_2}^{3/2}}{X_{NH_3}} \, P \tag{12}$$

$$K_{13} = \frac{X_{NO_2}^3 \cdot X_{H_2O}}{X_{HNO_3}^2 \, X_{NO}} \, P \tag{13}$$

$$K_{14} = \frac{X_C \cdot X_{H_2}^{\frac{1}{2}} \cdot X_{N_2}^{\frac{1}{2}}}{X_{HCN}} \; P \tag{14}$$

Let α, β, γ denote the atomic ratios between the elements of C-H-O-N

$$\alpha = \frac{\text{atoms of C}}{\text{atoms of O}}$$

$$\beta = \frac{\text{atoms of O}}{\text{atoms of N}}$$

$$\gamma = \frac{\text{atoms of C}}{\text{atoms of H}}$$

Then considering the product mixture

$$\alpha \left(2X_{CO_2} + X_{CO} + X_{H_2O} + 3X_{HNO_3} + 3X_{O_3} + 2X_{O_2} + X_O + X_{OH} + 2X_{NO_2} + X_{NO} \right)$$
$$= X_{CO_2} + X_{CO} + X_C + X_{CH_4} + X_{HCN} \; . \tag{15}$$

$$\beta \left(X_{NO_2} + X_N + 2X_{N_2} + X_{NO} + X_{NH_3} + X_{HNO_3} + X_{HCN} \right)$$
$$= 2X_{CO_2} + X_{CO} + X_{H_2O} + 3X_{HNO_3} + 3X_{O_3} + 2X_{O_2} + X_O + X_{OH} + 2X_{NO_2} + X_{NO}. \tag{16}$$

$$\gamma \left(X_H + 2X_{H_2} + X_{OH} + 2X_{H_2O} + 4X_{CH_4} + 3X_{NH_3} + X_{HNO_3} + X_{HCN} \right)$$
$$= X_C + X_{CO_2} + X_{CO_2} + X_{CH_4} + X_{HCN} \; . \tag{17}$$

and the summation of mole fractions equals unity ie.,Concentration Condition:

$$X_{CO_2} + X_{H_2O} + X_{N_2} + X_{CO} + X_{O_2} + X_{H_2} + X_{NO} + X_{OH} + X_O + X_H + X_N + X_C + X_{O_3} + X_{NO_2}$$
$$+ X_{CH_4} + X_{NH_3} + X_{HCN} + X_{HNO_3} = 1 \tag{18}$$

These 18 equations are sufficient to compute the composition of all the product species at given final pressure, P, and atomic ratios, α, β, γ, which depend upon the condition of reactant mixture.

To compute the flame temperature, the law of conservation of energy is applied to the combustion.

Denoting

Heat content of combustion products = H_O
Heat released by the reaction   = H_i

Then

$$H_i = H_O$$

where

$$H_i = n_f \left(h_{T_i} \right)_f + n_{O_2} \left(h_{T_i} \right)_{O_2} + n_{N_2} \left(h_{T_i} \right)_N \tag{19}$$

$$H_O = \sum n_i \left[\sum X_i \left(h_T \right)_i \right] \tag{20}$$

where summation on i in the second expression is for all the products of combustion.

Method of solving equations 1-18 is as follows:

The compounds H_2, H_2O, CO_2 and N_2 are best behaved so these are chosen as dominant species and hence the mole fractions of all other species are expressed in terms of these four, $X_{H_2} = A$, $X_{H_2O} = B$, $X_{CO_2} = C$, $X_{N_2} = D$, and the equilibrium constants K_1-K_{14}. Equations (15, 16, 17, 18) are rewritten in the form:

$F(A, B, C, D) = 0$

$G(A, B, C, D) = 0$

$H(A, B, C, D) = 0$

$J(A, B, C, D) = 0$

Linearizing the equations using Taylor series expansion

$$F = F_0 + F_A\, \delta A + F_B\, \delta B + F_C\, \delta C + F_D\, \delta D \tag{21}$$

$$G = G_0 + G_A\, \delta A + G_B\, \delta B + G_C\, \delta C + G_D\, \delta D \tag{22}$$

$$H = H_0 + H_A\, \delta A + H_B\, \delta B + H_C\, \delta C + H_D\, \delta D \tag{23}$$

$$J = J_0 + J_A\, \delta A + J_B\, \delta B + J_C\, \delta C + J_D\, \delta D \tag{24}$$

where $F_A = \delta F/\delta A$, etc., and $F_0 = F(A_0, B_0, C_0, D_0)$ etc.

Starting from assumed initial values A_0, B_0, C_0, and D_0 of A, B, C, and D, the set of above equations is solved by Newton Iteration method for δA, δB, δC, and δD which are then applied as corrections to A_0, B_0, C_0, and D_0 to yield new values of A, B, C, D. This process is repeated until the absolute values of the ratios $\delta A/A$, $\delta B/B$, etc., are each below an assigned minimum ε. The values of ε in this study was kept as 10^{-7} and the maximum number of iterations required to achieve this convergence was between 4 and 6. Once A, B, C, and D are computed, the composition of the equilibrium products is readily obtained.

Then, using the equilibrium composition of the products and the enthalpy data, the enthalpies of the products is calculated and compared with that of reactants. If the two corresponding energies do not tally, the enthalpies of products is calculated for the combustion temperature higher or lower than the previous value until two successive temperatures bracket the desired value. Further adjustments are then made by interpolation .

Combustion of Low Calorific-Value Gas Jets in a Cross-Flow

B. BHATTI, L.B. SAHGAL, and S.R. GOLLAHALLI
Combustion and Propulsion Laboratory
School of Aerospace, Mechanical and Nuclear Engineering
The University of Oklahoma
Norman, Oklahoma 73019, USA

ABSTRACT

The feasibility study of a method of enhancing the flame stability of low-calorific-value gas jets by subjecting them to cross-flows is the subject of this paper. An experimental investigation directed to examine the effects of the ratio of jet momentum flux to cross-flow momentum flux and the calorific-value of the gas fuels on stability, geometry, temperature profiles, radiation characteristics, and concentration profiles of pollutant species, in their diffusion flames has been presented. The results show that cross-flow decreases the lower-limit of the calorific-value of gases that can be burnt with stable jet flames and affects the flame-structure of conventional and low-calorific value gas fuel jets similarly.

1.   INTRODUCTION

For various reasons, it seems clear that the United States will rely heavily on coal and its derivatives as the source of energy in the near future. Gaseous fuels produced from coal are expected to become the substitutes for natural gas. However, the production of synthetic gases that have the same energy content and properties as the natural gases is not only expensive but also energy-intensive. Hence, the power gas or low-calorific-value gas with a HHV of 100-250 Btu/scf appears as an attractive fuel.

Low-calorific-value gases are also produced in a variety of other industrial and natural sources such as shale oil retorting, biomass fermentation, catalytic bed regeneration, solvent recovery, enhanced oil recovery by fire-flooding, volcanoes, and chemical-process by-products. Hence, the combustion of low-calorific-value gases appears as an attractive source of energy.

Because of the low volumetric energy content, the combustion of these gases poses several problems [1]. The difficulty in stabilizing the flames over a wide range of heat-release rates, because of low flame velocities and high gas velocities due to large volume flow rates required, is the most serious problem. The methods commonly proposed to combat this problem are: combustion with oxygen enriched air, mixing with a high calorific-value fuel, preheating the gases, and catalytic combustion. The first two methods are not economically attractive. Some studies [2] have been done to preheat the inlet gases using the energy in the exhaust and oxidizing them over catalysts [3]. These attempts have been limited to the burning of low calorific-value gases premixed with the required combustion air.

In practical situations, combustion in the form of diffusion flames has

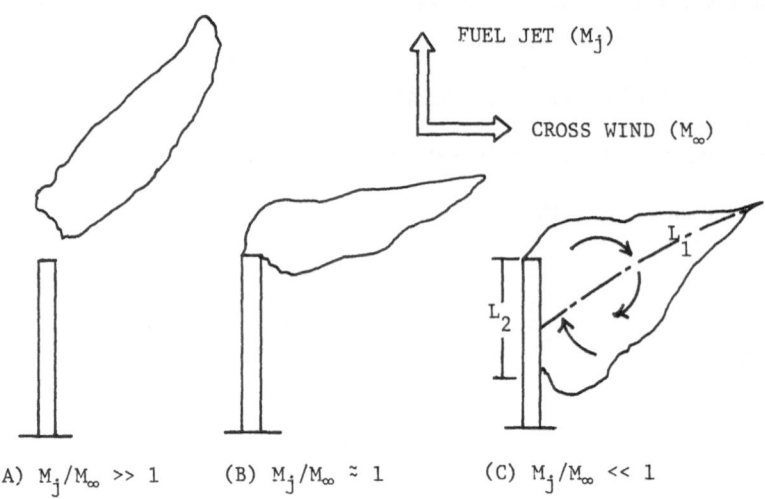

Fig. 1.   Effect of cross-flow on flame configuration.

been more attractive than premixed-gas combustion for the reasons of safety and
better control over the energy release rate which in the former is dependent
only on the fuel-feed rate. Further, in the case of low calorific-value gases,
the volume flow rate of the fuels itself is large because of their low energy
content, and premixing them with combustion air makes the problem of handling
them, and its consequence on the burner design doubly severe.

Secondly, the preheating of gases requires either an external heater
fueled by a high calorific-value gas or a complex large heat-exchanger to transfer energy from product gases to the inlet gases. Even in the case of cata-
lytic-combustion, preheating to a certain minimum light-off temperature, which
depends upon the composition of the incoming gases, is necessary [3]. Besides
these problems, the large quantities of catalysts needed will make that process
economically unattractive.

The method of burning low calorific-value gases described here circumvents
many of the drawbacks discussed above. In this feasibility study gas mixtures
with calorific values in the range of 4 - 12 MJ/m^3 were burnt in the form of diffusion flames in air. No external preheating or catalysts were used. This
investigation was motivated by an earlier study on the combustion of conventional gas jets in cross-flows [4]. Figure 1 shows the effects of cross-flow
on the flame configurations of gas jets emerging from cylindrical tubes as shown
in the photographs of that study. When the momentum-flux ratio (the ratio of
jet momentum flux $M_j = \rho_j U_j^2$ to cross-flow momentum flux $M_\infty = \rho_\infty U_\infty^2$) is high,
the flame lifts off the burner. The shear between fuel jet and cross-wind is
very high, which leads to intense mixing and short flames (Fig. 1A). When the
magnitudes of momentum-flux ratio are of the same order, the flame bends and
assumes the direction of the cross-wind. Except in the near-burner region, the
shear between fuel jet and cross-flow is not intense and consequently flame
length increases (Fig. 1B). At low values of M_j/M_∞ (high cross-winds), the
flame stabilizes in the wake of the burner tube itself (Fig. 1C). That can
result in three effects: (a) because of the strong recirculation vortex that
develops in the plane of the axis of the burner tube and gets attached to it,
the local gas velocity component in the flame-anchoring region becomes low,

(b) the flame gases licking the burner tube preheat the fuel gases flowing inside it, (c) the hot flame gas itself is brought back to the flame stabilization region, and hence, not only some thermal energy, but also some active chemical species, are added directly to the fresh fuel gases entering that region. It is reasonable to expect that these effects can enhance the stability of the flame, and thus decrease the low-end of the calorific-value of gases that can be burnt. However, as the aerodynamic and thermal pattern of the bent-over flames are markedly different from the straight flames, it is not possible to predict how the pollutant formation will be affected.

Hence, the objectives of the study presented in this paper were to determine the effects of cross-flow and calorific-value on (a) the flame stability and (b) the flame structure parameters such as flame size and configuration, temperature profiles, radiation emission, and emission of pollutants (CO, NO, and particulates).

2. EXPERIMENTAL DETAILS

The experiments of this study were carried out in an open-jet wind tunnel with a test section of size 68 cm x 42 cm x 137 cm and capable of producing cross-flow velocities up to 12 m/s. The burner was mounted such that the fuel jet emerged into the cross-flow at 34 cm above the tunnel floor. All the flames in this study were confined to the potential core of the wind-tunnel jet. Figure 2 shows the basic configuration of the nozzle and its various modifications employed in this study. The basic version of the burner consisted of a steel tube (ID = 9.5 mm and OD = 12.7 mm) jacketed with another steel tube (ID = 20.6 mm and OD = 25.4 mm). It was possible to change the exit diameter of the inner tube with an adapter and thus alter the ratio of the fuel-jet

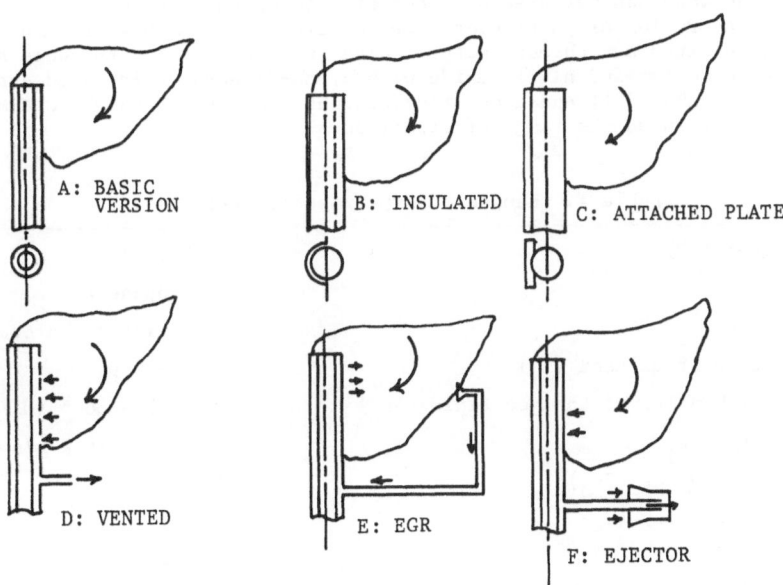

Fig. 2.   Burner configurations.

diameter to the outer diameter of the burner to which the recirculating vortex was attached. In order to maximize the fuel-heating and the effects of recirculating vortex, several modifications were effected to the basic version of the nozzle and flame stability tests were performed on them. These modifications included (i) insulating the upwind side of the burner (Fig. 2B) to decrease the cooling effect of cross-flow, (ii) the attachment of flat plates on the upwind side of the burner (Fig. 2C) to increase the thickness of the recirculating vortex, and (iii) allowing the flow of hot combustion products through the annular space to increase the preheating of fuel gases. The modifications of the third category included venting of gases through the jacket by simply making use of the static pressure differential between the wake of the burner and the free stream (Fig. 2D), collecting the exhaust gases near the end of the flame and passing through the jacket (Fig. 2E), and using a simple ejector mechanism that sucks the flame gases through the annular space (Fig. 2F).

Low calorific-value gases were simulated in this study by mixing propane with nitrogen or carbon dioxide. Technical grade bottled gases were used. Flow rates of gases were measured by means of calibrated rotameters. When diluents were added to the fuel gas, care was taken to adjust the flow rates so that the momentum-flux ratio remained constant. That was necessary to isolate the effects of calorific-value and fluid mechanics. Flames were photographed in color on panchromatic films at constant exposure conditions to compare the changes in appearance of different parts of the flame. Radiation emitted from the flames was measured by means of a water-cooled pyreheliometer (HYCAL Model 8410-B). This wide-angle radiometer was placed normal to the plane of the centerline of the flame at such a distance that the inverse-square relationship between the incident radiation flux and the distance was satisfied. Assuming complete combustion, the radiative fraction of heat release was calculated following [5]. The temperature profiles were determined by means of a chromel-alumel thermocouple and the readings were corrected for radiation losses from the beads and conduction losses along the wires. The attenuation of a helium-neon laser beam crossing the wake was measured by means of a laser power meter and the volumetric concentration of particulates was determined from those readings. Gas samples were withdrawn through water-cooled stainless probes and were analyzed for carbon monoxide and nitric oxide with a non-dispersive infrared analyzer (HORIBA Model MEXA 221) and a chemiluminescent analyzer (THERMOELECTRON model 10A). Table 1 shows the range of experimental variables.

Table 1. Ranges of Experimental Variables

Jet diameter (ID)	9.5 mm
Fuel	Propane
Diluent	Nitrogen, carbon dioxide
Mole fraction of diluent (X_j)	0 - 0.96
Lower heating value of the jet fluid (ΔH_B)	1.67 - 86.5 (MJ/m^3)
Jet velocity; U_j	0.17 - 14.9 m/s
Jet Reynolds number (Re_j)	360 - 10,800
Cross-wind velocity (U_∞)	2.1 - 4.9 m/s
Velocity ratio (U_j/U_∞)	0.035 - 7.1
Momentum-flux ratio (M_j/M_∞)	0.04 - 0.24
Froude number (Fr_j)	0.30 - 2380

3.   RESULTS AND DISCUSSION

3.1   Flame Appearance

As the momentum-flux ratio (M_j/M_∞) is decreased below 2, flame quickly bends over and stretches in the direction of cross-flow. A strong vortex with its axis perpendicular to the direction of cross-flow and burner tube appears in the downwind vicinity of the burner tube. A bluish flame region appears near the periphery of this vortex, but the core of the vortex does not appear to support combustion. The core of the vortex appears to be mainly the zone of fuel pyrolysis. The flame in the downstream region beyond the vortex appears yellow and the combustion of soot particles which have formed in the vortex appears to be dominant there.

When the cross-flow velocity is increased, the vortex grows and the recirculation increases. The extent of the blue flame near the vortex also increases. The appearance of flame in the downstream region, however, does not change perceptibly and the flame remains yellow. As the cross-flow velocity is increased, the increased vortex causes an enhancement of air entrained from the cross-flow into that region and results in the blue flame dominated by gas-phase oxidation. The flame in the far-downstream region assumes the direction of cross-flow and hence the relative shear between the flame jet and the cross-flow decreases. That probably retains soot burning as the dominant energy release mechanism and consequently yellow flame in that region.

As the calorific-value of the jet gases are lowered by the addition of

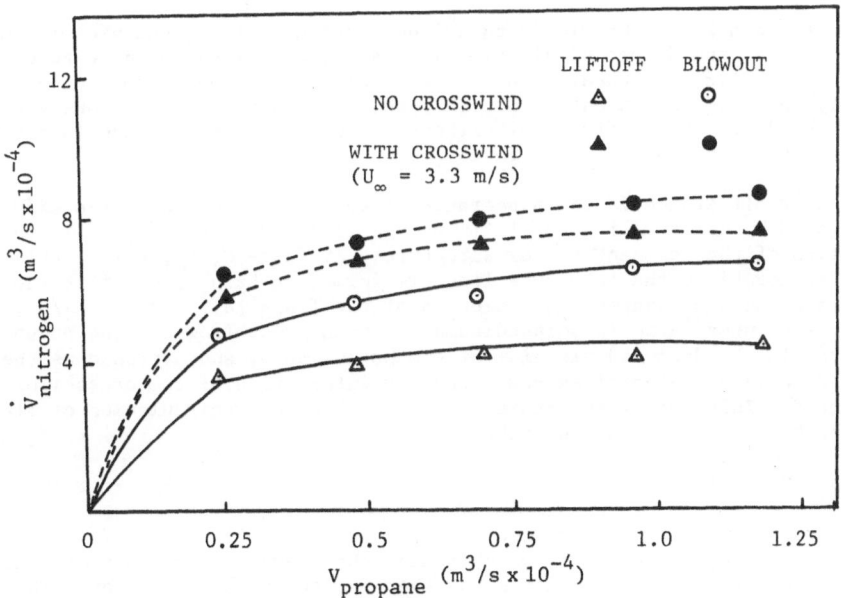

Fig. 3.   Effects of cross-flow on flame blow-out and lift-off limits of nitrogen-diluted propane jets.

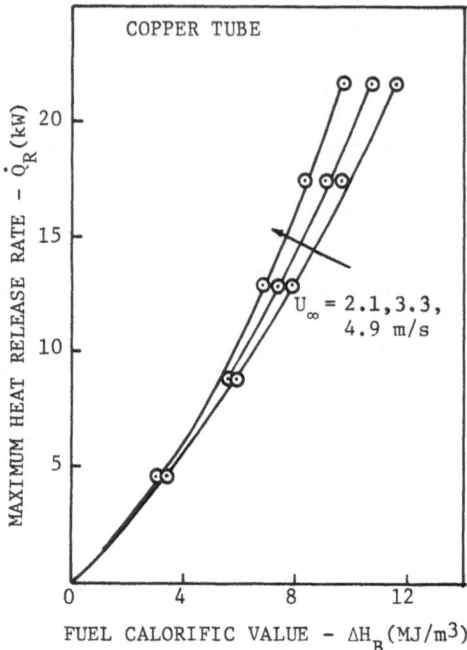

Fig. 4. Effect of cross-flow on the maximum heat
          release rate of low calorific-value gas jets.

diluents (N_2 and CO_2), the flames turn blue, particularly in the vicinity of
the vortex. As the degree of dilution increases, the extent of blue zone increases. The diluents appear to decrease pyrolysis and soot formation and consequently increase the extent of bluish zone, dominated by homogeneous gas-
phase reactions. These effects of diluents in bent-over flames are in conformity with the effects of diluents in co-flowing streams [6].

A major difference in the appearance between the flames at large M_j/M_∞
investigated by Gollahalli et al. [4] and the flames at low M_j/M_∞ is in the
orientation of the dominant vortex structure. At large M_j/M_∞, twin vortices in
the planes normal to the jet centerline are formed on both sides of it and
develop a horse-shoe shaped cross-section of the flame [4]. At low M_j/M_∞, a
single vortex whose axis is perpendicular to both cross-flow and the burner
tube forms in the downwind vicinity of the burner tube, and overshadows the
effect of Karman-type vortices caused by the interaction of the cross-flow and
burner tube. This vortex structure is responsible for recirculation of flame
gases and heating of the burner tube.

3.2 Flame Stability

Figure 3 compares the variation of the volumetric flow rate of nitrogen
that can be mixed with propane, which causes the flame to just lift off and
blow out, in quiescent surroundings and when subjected to cross-flow. These
measurements were obtained with basic version of the burner (Fig. 2A). It is
evident that both lift-off and blow-out limits of dilution are increased by
the cross-flow. Figure 4 shows the effects of cross-flow velocity in terms of

more useful parameters. The ordinate on Fig. 4 is the maximum heat-release rate ($\dot{Q}_R$) attained at blow-out limit and the abscissa is the calorific-value (ΔH_B) of the diluted fuel. This figure further shows that the maximum heat release rate for a given heating value of fuel can be increased by the cross-flow. Conversely, for a desired heat release rate, cross-flow allows the combustion of a lower calorific-value gas with a stable flame. For calorific-values below 4 MJ/m^3 the effect of cross-flow is negligible with the burner used in this case, and at 10 MJ/m^3, the increase of heat release rate is about 25% for roughly doubling the cross-flow velocity. This widening of the flame stability region and consequent increase of heat release rate is caused primarily by the effects of recirculating flame gases. Hence, in order to maximize those effects, the design of the burner was modified as described in Section 2.

Figure 5 shows the effect of changing the steel tube to copper tube to exploit the benefit of higher thermal conductivity of copper. The results, however, indicate that the improvement in $\dot{Q}_R$ is minimal. Also, the effect of insulating the upwind side of the burner to minimize the cooling effect of cross-flow is found to be insignificant. The effects of attaching flat plates on the upwind side of the burner are also shown in Fig. 5. Contrary to the expectations, the wider plate yields a lower $\dot{Q}_R$ in spite of the thicker wake and recirculation zone. However, a further examination reveals that the enhancement of the Karman-type vortices in the horizontal plane, caused by the wider plate increases the entrainment of surrounding air into the recirculation zone and dilutes it faster. That mitigates the heating effect of the vortex in the vertical plane and thus decreases the tolerance to fuel dilution.

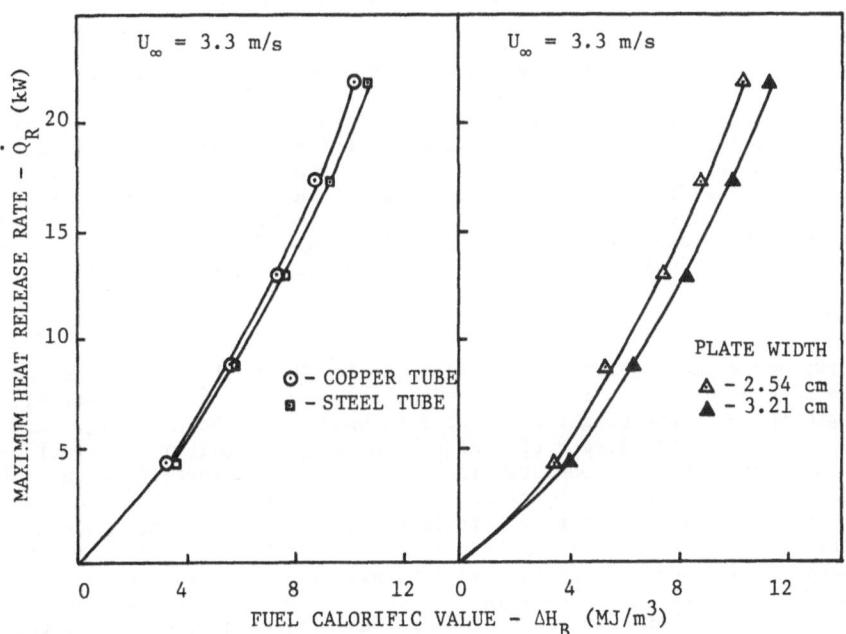

Fig. 5. Effect of burner design on the maximum heat release rate of low calorific-value gas flames in cross-flow.

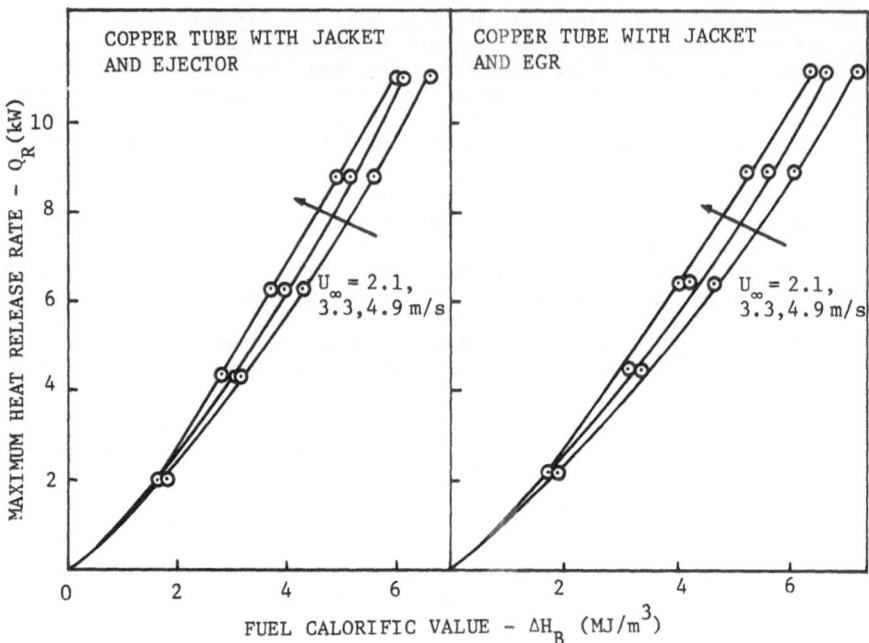

Fig. 6.   Effect of burner design on the maximum heat release rate
of   low calorific-value gas flames in cross-flow.

Figure 6 shows the variation of Q_R with $\Delta \dot{H}_B$ obtained with burner modifications shown in Figs. 2E and 2F.   In these cases the artificially induced flow
of product gases over the burner was provided in addition to the natural effects
of the recirculating vortex.   These tests were performed with low calorific-
value gases with $\Delta H_B \leq 7 MJ/m^3$.   It is noticed that these modifications improve
the effects of cross-flow on Q_R.   For instance, at $\Delta H_B = 4$ MJ/m^3, the increase
of $\dot{Q}_R$ is about 25 percent with modifications shown in Fig. 2F, whereas it was
negligible with the basic version of the nozzle.   This effect can be attributed
to the preheating of fuel and acceleration of the flame-induction reactions.

3.3   Flame Geometry

The two geometrical parameters of flame at low M_j/M_∞ that can be used as
indicators of the combustion rate and recirculation effects are the overall
flame length (L_1) and the height of flame attachment (L_2) shown in Fig. 1C.
Figures 7 and 8 show the effects of design parameters and dilution of fuel gases
(lowering of ΔH_B) at various momentum-flux ratios.   An examination of these figures reveals the following: (i) for both copper and steel tube burners without
outer jackets, L_1 increases up to a certain value of M_j/M_∞ and then decreases
and the peak value of L_1 occurs in the neighborhood of $M_j/M_\infty = 0.10$ for both
pure propane and diluted propane flames; (ii) with the jacketed tube or upwind-
side mounted plates, flame length (L_1) decreases by about 50 percent from its
value for bare tubes, but also increases monotonically with M_j/M_∞; (iii) the
effect of dilution on the peak flame length for similar burner designs appears
insignificant; (iv) the flame attachment length (L_2) decreases monotonically
with the increase of M_j/M_∞; and (v) the burner design parameters and the
calorific-value of fuel gas do not affect L_2 markedly.

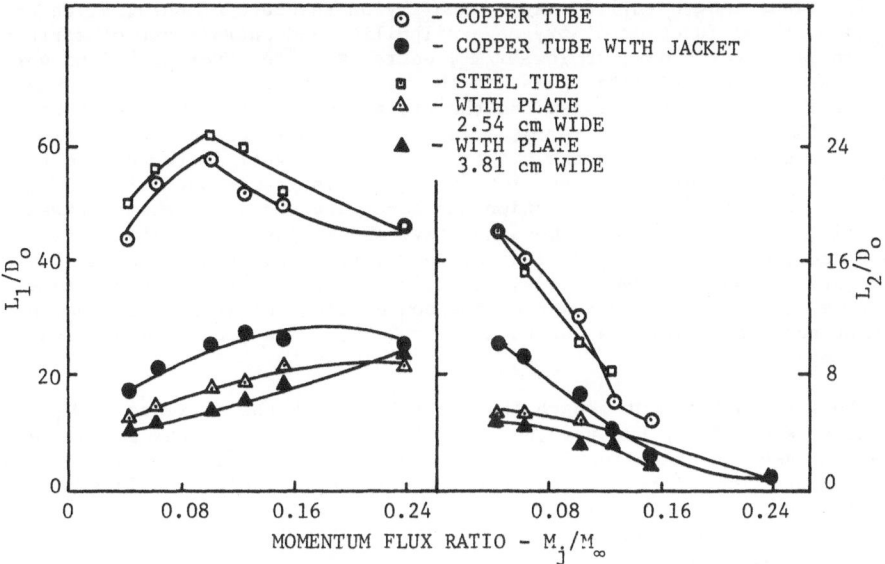

Fig. 7. Effects of design parameters on the visible flame length and
burner-attachment length of propane jets in cross-flow.

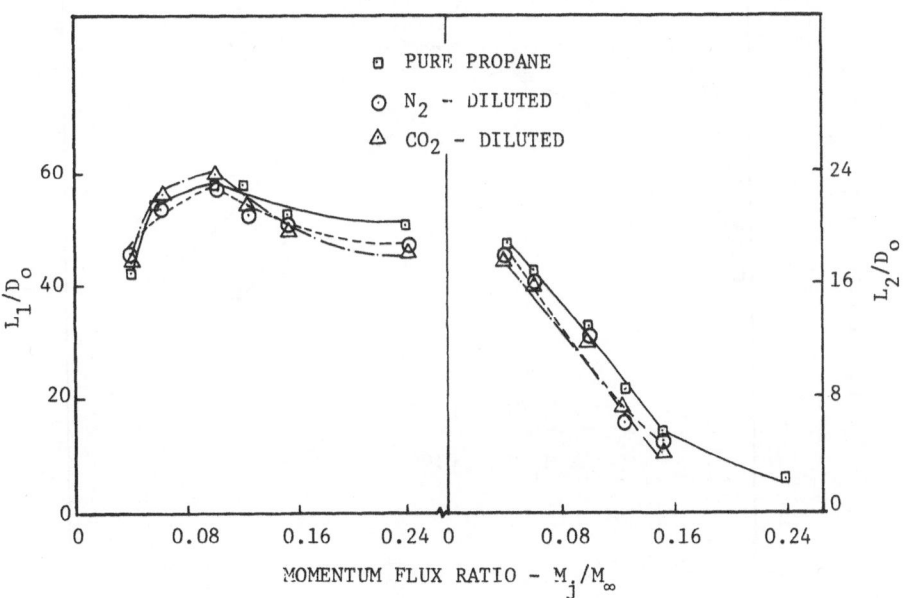

Fig. 8. Effects of dilution on visible flame length and burner-
attachment length of propane jets in cross-flow.

The flame length (L_1) depends primarily on the rate of burning, which in turn is a strong function of oxygen availability and entrainment of surrounding air. In bent-over flames at low M_j/M_∞, entrainment is governed by the vortex structures in the horizontal plane in the wake of the burner tube and the relative shear between the fuel jet and the cross-flow. For burners without outer jackets, the effects of the vortices in the horizontal plane become relatively small. This, combined with the higher flow-rate of fuel increases the flame length. Beyond a certain value of M_j/M_∞, (in this case about 0.1), the fuel jet bends towards vertical direction and hence increases the shear between cross-flow and the fuel jet. Consequently, the mixing and burning rate increase and the flame length decreases. However, in the case of burners with outer jackets and upstream plates, the effect of horizontal vortices remains so strong as not to be overshadowed by the consequences of increased shear between fuel-jet and cross-flow even up to $M_j/M_\infty = 0.24$. Hence, L_1 does not exhibit a decreasing trend.

The flame attachment length (L_2) increases with the decrease of M_j/M_∞ in all cases, essentially because the size of recirculation vortex in the vertical plane decreases as the effect of cross-flow increases. For low values of M_j/M_∞, L_2 of burners that yield thicker wakes is very small. Since L_2 is a direct measure of the effects of recirculating vortex governing the stability effects, this change of L_2 accounts for the lower flame stability of burners with plates mounted on upwind side.

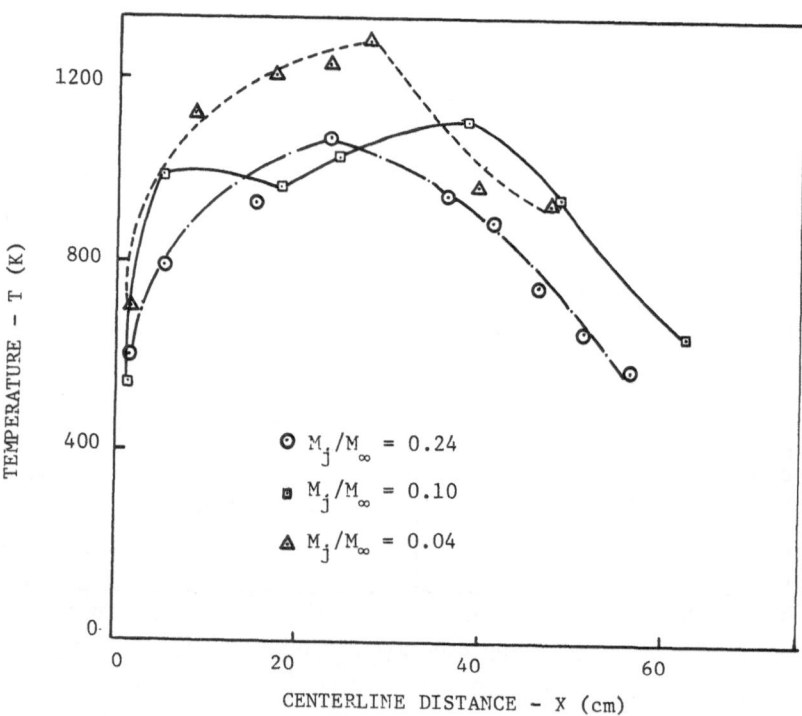

Fig. 9. Effect of momentum-flux ratio on centerline temperature profiles of propane jet flames in cross-flow.

3.3 Temperature Profiles

Figures 9 and 10 show the effects of momentum-flux ratio and dilution of
fuel gases on the centerline temperature profiles. The noteworthy results of
these figures are (i) the peak temperature increases as M_j/M_∞ decreases, (ii)
the near-nozzle temperature also increases as M_j/M_∞ decreases; (iii) the peak
temperature increases slightly with dilution; (iv) the near-nozzle temperature
decreases with dilution, (v) at M_j/M_∞ = 0.10 both pure propane and diluted propane jets exhibit a dip in the centerline temperature at X = 20 cm.

The increase of peak temperature value along the flame axis when M_j/M_∞ is
decreased is primarily caused by the decrease of entrainment into and dilution
of the far-nozzle region. As the wake effect of the burner tube increases (L_2
increases), the temperature in the near-nozzle region increases. However, with
dilution, the near-nozzle temperature decreases. Diluents appear to cause two
mutually competing effects. First, they curtail the endothermic fuel-pyrolysis
reactions which should result in a higher temperature. But, the diluent itself
absorbs its sensible heat and that should lower the temperature. The present
results indicate that the latter is more dominant. The effect of diluents on
the peak temperature is caused by the reduction of soot-particulate combustion.
That can lead to the dominance of relatively faster homogeneous gas-phase
reactions and lower heat losses. The combined effect of those factors can over-
shadow the effect of thermal dilution by the inert species in the fuel and thus

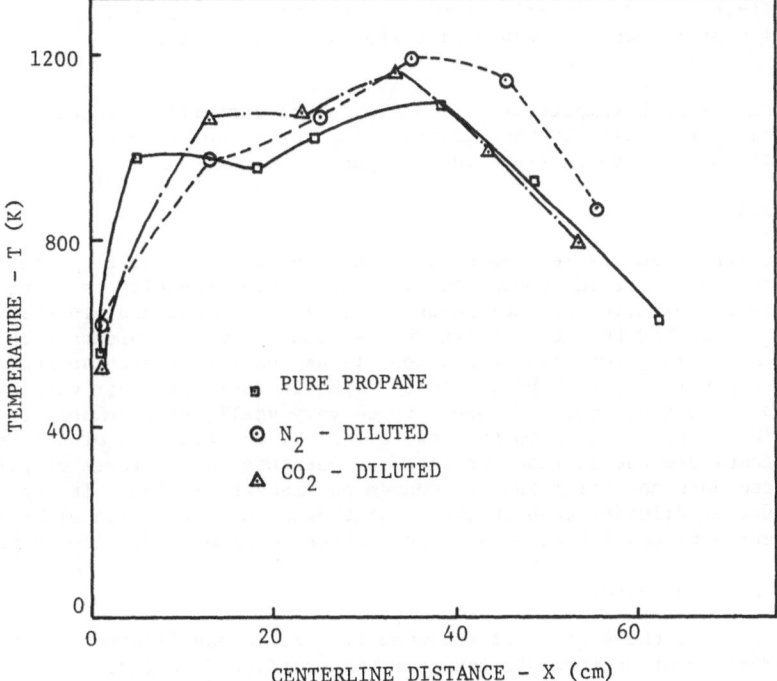

Fig. 10. Effects of dilution on centerline temperature profiles
of propane jet flames in cross-flow.

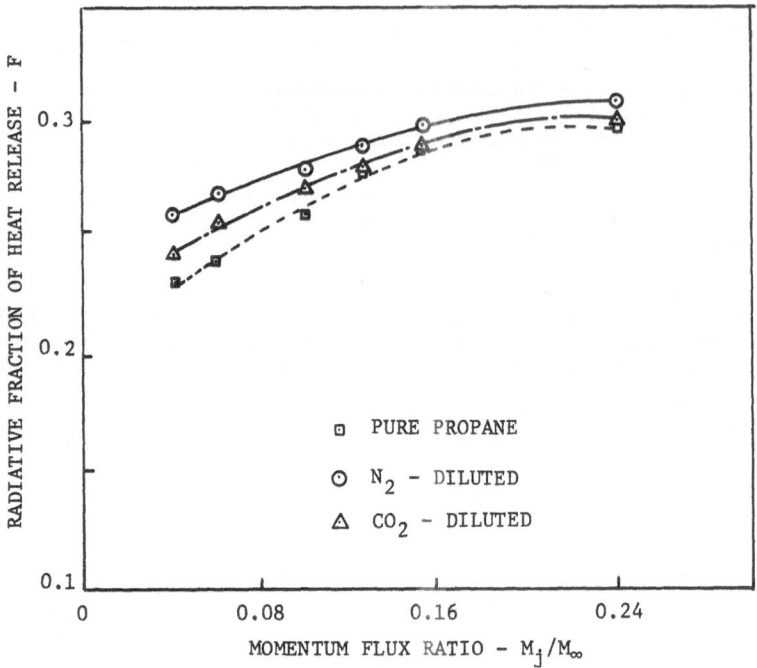

Fig. 11. Effects of momentum-flux ratio and dilution on radiative fraction
          of heat release of propane jet flames in cross-flow.

result in a higher peak temperature. The dip in the centerline-temperature profiles is caused essentially by the recirculating vortex whose core does not support combustion as noticed in flame photographs.

3.4 Flame Radiation

The fraction of energy release which is emitted in the form of radiation
is an important parameter in combustion systems. Hence, the effects of momentum flux ratio and dilution of fuel gases on it were examined and are shown in
Fig. 11. It is noticed that the radiative fraction of energy release of pure
propane flames at high cross-flow conditions is approximately same as that docu-
mented for propane flames at high values of M_j/M_∞ in co-flowing air streams [6].
The effects of dilution, however, appear to be very small, particularly at high
values of M_j/M_∞. The two effects that are caused by the addition of diluents
to the fuel gases are the increase of flame temperature and decrease of particulates. These have counteracting influences on flame radiation. It appears
with the amount of dilution used in the present case (mole fraction of N_2 = 0.28,
and mole fraction of CO_2 = 0.23), their net effect on flame radiation is minimal.

3.5 Particulate Concentration

Figure 12 shows the effects of momentum-flux ratio and dilution on the
centerline profiles of particulate concentration derived from laser-beam attenuation measurements. This figure reveals the following: (i) the particulate
concentration peaks at about 8 cm from the burner; (ii) the decrease of momentum
flux ratio increases the particulate concentration; (iii) the dilution decreases

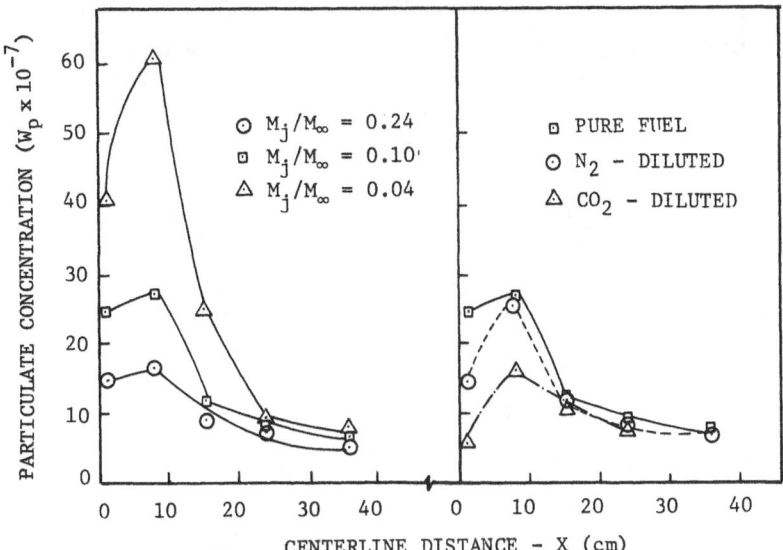

Fig. 12.   Effects of momentum-flux ratio and dilution on the particulate
concentration of propane jet flames in cross-flow.

particulate concentration; and (iv) the effects of dilution and momentum-flux
ratio on particulate concentration becomes negligible in the far-nozzle region.
It is interesting to note that the location of maximum W_p coincides with the
outer limit of the recirculation vortex. The soot particles nucleate and grow
in the near-nozzle vortex region. Beyond that region, the particulate concentration decreases because of soot-burning and dilution of flame gases by entrainment. The decrease of momentum-flux ratio increases the oxygen deficient
vortex in the near-nozzle region and consequently leads to higher W_p. The
dilution curtails the fuel pyrolysis and decreases formation of soot. These
results are in conformity with earlier studies on diluent effects in jet flames
in co-flowing air streams [6] and jet flames in cross-flows at high momentum-
flux ratios [7]. Also, these findings substantiate the results on the effects
of dilution on temperature and radiation profiles discussed earlier.

3.6 Carbon Monoxide Concentration

     Figure 13 presents the effects of momentum-flux ratio and dilution on the
centerline profiles of carbon monoxide concentration. This figure shows that
(i) the peak concentration of CO occurs at the end of the near-nozzle region
($X \approx 25$ cm) of the burner; (ii) the peak concentration of CO does not change
significantly when the momentum flux is changed from 0.24 to 0.10, but decreases
considerably when M_j/M_∞ is changed to 0.04; (iii) the dilution decreases the
peak concentration of CO and the amount of decrease is smaller when CO_2 is the
diluent.

     The production of CO in these flames can be traced to two sources: (a)
the partial oxidation of carbon, which is higher when soot concentration is
high and (b) the dissociation of CO_2. The small difference of CO concentration
between the flames at M_j/M_∞ equals 0.24 and 0.10 can be attributed to the small
changes in peak temperature, flame attachment length L_2, and particulate

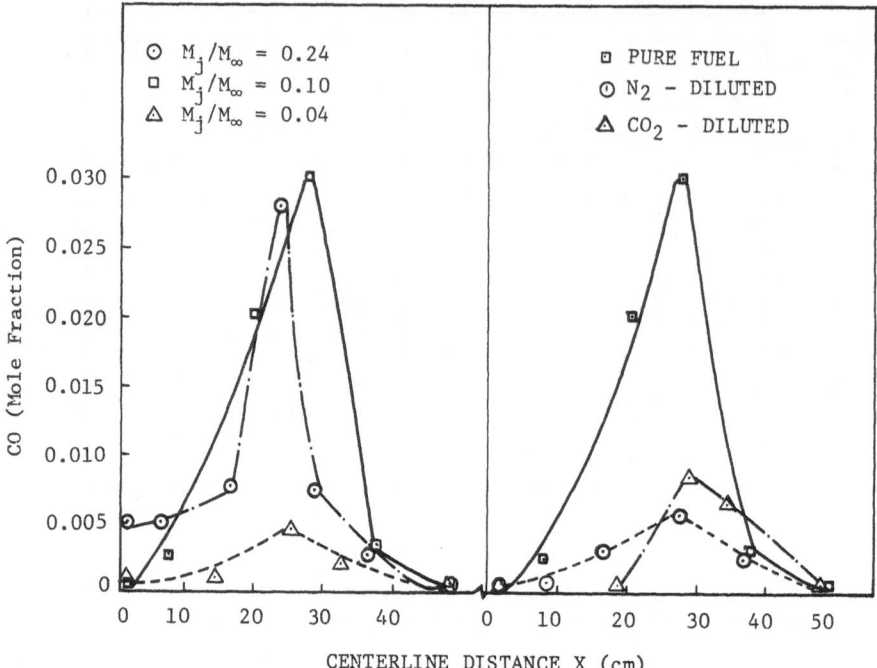

Fig. 13. Effects of momentum-flux ratio and dilution on carbon monoxide
concentration in propane jet flames in cross-flow.

concentration. The relatively large changes in these factors that occur when
M_j/M_∞ is changed to 0.04 accounts for the large change in CO concentration that
accompanies it. The decrease of CO concentration with dilution, however,
appears to be much larger than that suggested by the changes in temperatures
and soot concentration in the near-nozzle. The slightly higher CO concentration
in the flame where CO_2 was inert is probably the contribution of the dissociation of CO_2.

3.7 Nitric-Oxide Concentration

The effects of momentum-flux ratio and fuel dilution on the nitric oxide
concentration are shown in Fig. 14. It is seen that the peak NO concentration
(i) decreases when M_j/M_∞ is decreased from 0.24 to 0.10 but returns to its
original value when it is further decreased to 0.04 and (ii) decreases slightly
with dilution. As there is no chemically-bound nitrogen with the hydrocarbon
gas used, it is reasonable to expect most of NO is produced by the thermal
route and thus the variations of flame temperature and oxygen availability
should dictate the concentration of NO. The relatively invariant peak flame
temperature and decrease of air entrainment account for the lowering of NO that
accompanies the change of M_j/M_∞ from 0.24 to 0.10. Similarly, the higher temperature that is noticed in the flame at M_j/M_∞ = 0.04 brings up the NO concentration level. Furthermore, its concentration does not decrease sharply in the
far-nozzle region because of lower entrainment. Since the diluents cause slight
increase in flame temperature (Fig. 10) but decrease the availability of oxygen,
the NO concentration is not materially altered by them.

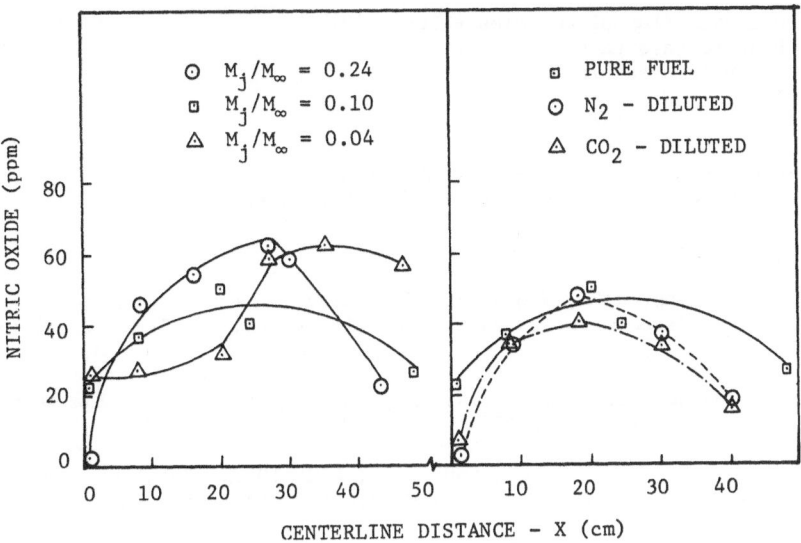

Fig. 14.   Effects of momentum-flux ratio and dilution on nitric-oxide
          concentration in propane jet flames in cross-flow.

4.   CONCLUDING REMARKS

This study was directed to examine the feasibility of burning low calorific-
value gases in the form of diffusion flames subjected to cross-flow. The results
have shown that the lower-limit of calorific-value of gases that can be burnt in
stable flames can be decreased by imposing cross-flow on the fuel jets, such
that the flames anchor to the burner tube in its wake. The ratio of jet momentum flux to cross-flow momentum flux under these conditions is less than unity.
The recirculation vortex that forms with its axis perpendicular to both cross-
flow and burner plays a vital role in enhancing the flame stability. The flame-
structure parameters such as temperature, radiation, and concentrations of pollutants are affected by the ratio of jet momentum flux to cross-wind momentum
flux and the calorific-value of fuel jet. The results of this study also suggest that further improvements in the flame stability range of low calorific
value gases can be achieved by using this technique in conjunction with catalytic combustion.

ACKNOWLEDGMENTS

The authors gratefully acknowledge the financial support of the Energy
Resources Center of the University of Oklahoma for this project.

NOMENCLATURE

D_o initial diameter of the jet
F radiative fraction of heat release
ΔH_B fuel calorific value per unit volume
L_1 flame length along the centerline
L_2 flame attachment length on burner tube
M_j momentum flux of the jet fluid

M_∞ momentum flux of the cross-flow
$\dot{Q}_R$ heat release rate
T flame temperature
U_j jet exit velocity
U_∞ cross-flow velocity
$\dot{V}$ volume flow rate
$\dot{W}_p$ volumetric particulate concentration
X distance along the centerline from the burner

REFERENCES

1. Syred, N., Dahmen, K.R., and Najim, S.A. 1977. A review of combustion
    problems associated with low calorific-value gases. J. Inst. Fuel, (50),
    p. 195.

2. Lloyd, A. and Weinberg, F.J. 1974. A burner for mixtures of low heat
    control. Nature, (251), p. 47.

3. Madagavkar, A.M., Vogel, R.G., and Swift, H.E. 1981. Catalytic combustion
    of low heat value gases. Ind. Eng. Chem. Prod. Res. Dev. (20), p. 628.

4. Gollahalli, S.R., Sullivan, H.F., Brzustowski, T.A. 1975. Characteristics
    of turbulent propane diffusion flame in a cross wind. Trans. C.S.M.E.,
    (30), p. 205.

5. Anon. 1969. Guide for pressure relief and depressurizing systems. RP-521,
    Am. Pet. Inst.

6. Gollahalli, S.R. 1977. Effects of diluents on the flame structure and
    radiation of propane jet flames in a concentric stream. Combustion Science
    and Technology, (15), p. 147.

7. Gollahalli, S.R. 1978. Aerodynamic and diluent effects on the emission of
    nitrogen oxides from hydrocarbon diffusion flames. The Can. J. Chem. Eng.,
    (56), p. 510.

Thermal Behaviour of Solid
Propellants Using a Shock Tube

R. RAMAPRABHU
Department of Mechanical Engineering
A.C. College of Technology
Perarignar Anna University of Technology
Madras 600 025, India

K.A. BHASKARAN
Department of Mechanical Engineering
Indian Institute of Technology
Madras 600 036, India

ABSTRACT

The efficient operation of a solid propellant rocket
encompasses a wide spectrum of scientific and engineering
desciplines ranging from fundamental laws of motion to the
extreme complexities of solid propellant combustion processes.
The thermal behaviour of solid propellants under varying
conditions of temperature and pressure is of great significance
in predicting overall performance of solid propellants. The
ignition of solid propellant is studied with reference to
variation in temperature (800K to 2500 K) and pressure (0.5 to
10 ata) and the results are compared with three prevailing
models of ignition. The ignition of propellants was achieved
by igniting them in a shock tube wherein very high temperatures
can be achieved instantaneously, which are not possible with
any conventional means of experimentation. Attempt is made to
independently establish the thermal role of thermodynamic,
transport and kinetic parameters on ignition mechanism. The
results clearly indicate strong dependance of ignition delay
on temperature and the effect was pronounced at higher
pressures (0.5 to 10 ata).

1. INTRODUCTION

One of the factors which strongly influence the thermal
behaviour of propellants is their ignition characteristics.
The solid propellant ignition is controlled basically by a
thermal induction interval during which a layer of propellant
material is raised to a temperature at which chemical reaction
rates are highly appreciable. This is followed by a runaway
chemical process in which chemical self heating quickly
becomes the dominant thermal source leading to ignition. The
thermal behaviour of solid propellant ignition is well
explained with the help of three theories of ignition.

2. THEORETICAL ANALYSIS

Three basic models of ignition are available to explain
the thermal behaviour of solid propellants namely solid phase

theory (1), gas phase theory (2) and solid gas interface
theory (3). Of these theories of ignition the gas phase,
solid-gas interface theories of ignition are based on the fact
whether the rate controlling reaction occur at the gas phase
or solid gas interface. But the solid phase theory deals with
transient thermal analysis with an exothermic chemical heating
in the solid with an exponential dependancy of rate on temperature
This theory assumes that the rate controlling reactions occur
at the solid surface. Detailed description and review of
these theories is well dealt by Price, E.W. et.al (4). The
analytical determination of ignition delay time involved
solution of nonlinear partial differential equations using a
finite difference method and employing a high speed IBM 365
system Computer. A typical plot of temperature and fuel distribution profiles at 50 and 250   seconds after shock wave
reflection is shown in Fig.1. As is evident from the figure,
as the distance from the end wall of the shock tube increases,
the fuel concentration decreases, emphasizing plausible explonation of ignition process by gas phase theory of ignition at
the gas zone near the propellant when environmental oxygen
concentration was high. Also heat transfer calculations indicate that depending upon the thermodynamic and transport pro-
porties of the polymer and the oxidizer, the surface temperatures of polymer and that of oxidizer reach different values
although both are subjected to same reflected shock heating.
Fig.2. This leads to analyse the role played by thermodynamic,
transport and kinetic properties on the thermal behaviour of
propellants. The analysis is exhaustive and carried out by
selecting suitable range of values for thermodynamic, transport
and kinetic parameters, and these values are fitted in the
solutions available for evaluation of ignition delay based on
the three theories of ignition, and the results compared to
ascertain role played by these parameters on the thermal
behaviour of solid propellants (5).

3. EXPERIMENTAL

     The propellant can be ignited in a shock tube to conform
to conditions of three modes of heat transfer, viz. conductive,
convective and conductive-convective heat transfers (Fig.3).
The experimental set up consisted of a stainless steel shock
tube with sophisticated instrumentation. Solid propellant
samples with Polyvinyl Chloride (PVC) as polymer and Ammonium
Perchlorate (AP) as oxidizer were mounted flush at the end
flange and was subjected to reflected shock heating (6).
Ignition was sensed by a photo tube kept at night angles to the
tube axis and shock velocity was computed using pressure transducers and digital counters. The propellants were tested for
varying conditions of temperature (800K-2500K) and pressures
0.5 to 10 ata with environmental oxygen concentration kept-
constant. The driver gas was Hydrogen-Nitrogen mixture and the
test gas was Oxygen. The shock tube was operated under tailored
conditions to have increased observation time (7).

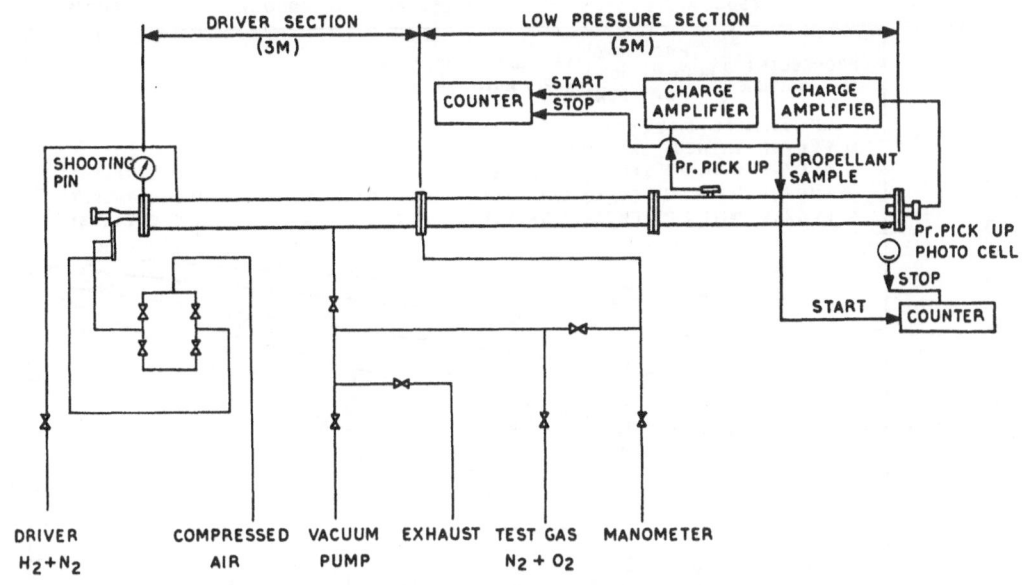

FIG.1. SCHEMATIC DIAGRAM OF EXPERIMENTAL SETUP

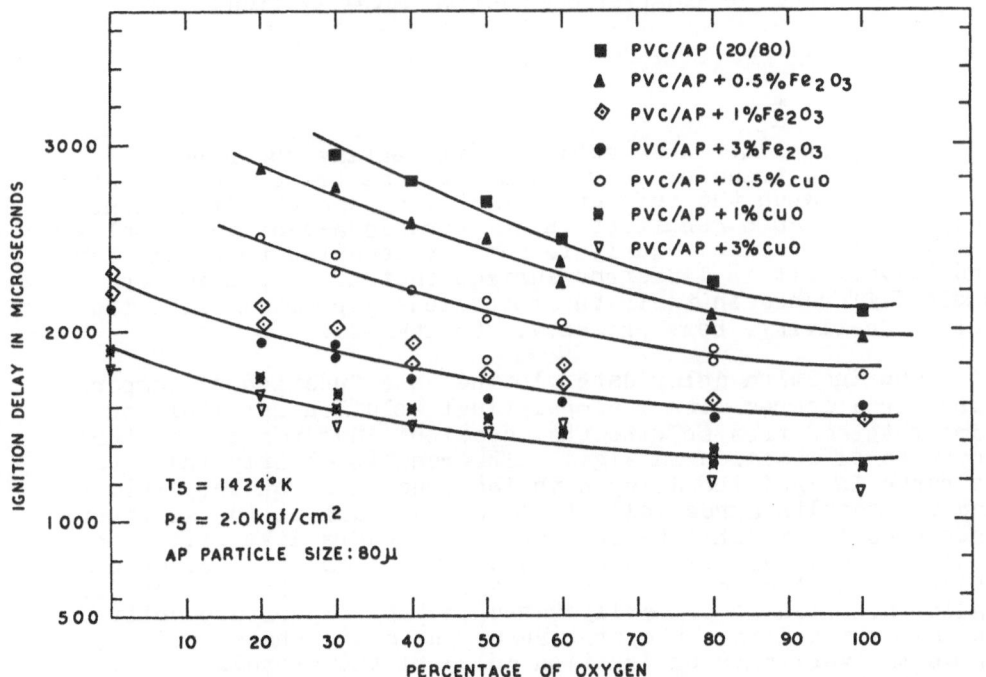

FIG.2. IGNITION DELAY VARIATION OF CATALYSED PROPELLANT WITH ENVIRONMENTAL OXYGEN CONTENT

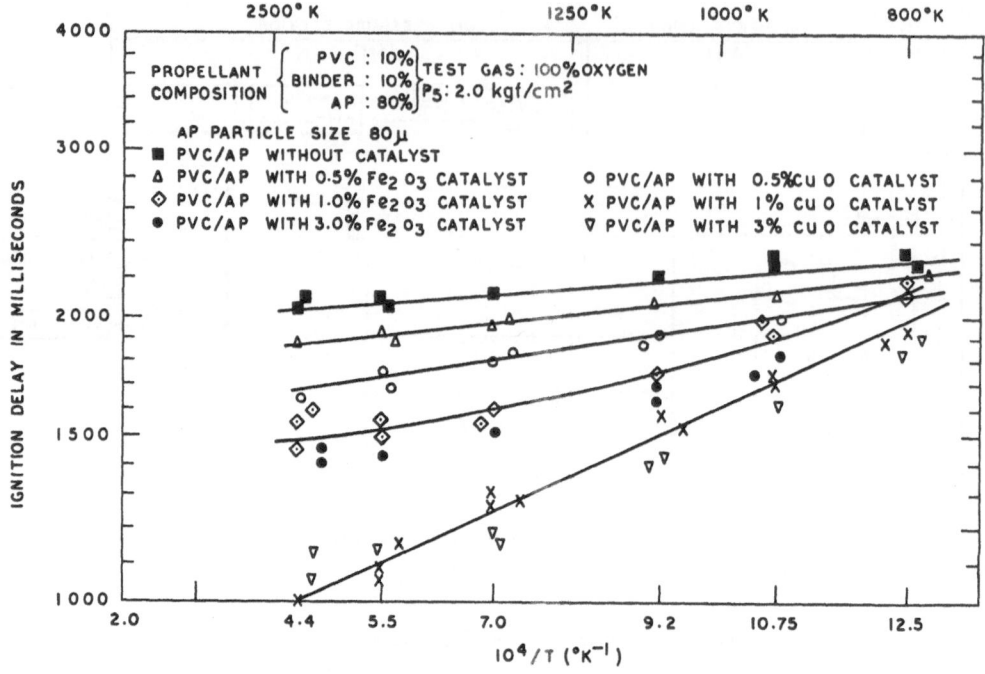

FIG.3. EFFECT OF CATALYST ON IGNITION DELAY OF PROPELLANT

4. RESULTS

The propellant when subjected to heating undergoes
unsteady heat transfer occuring due to the temperature gradient
existing between the reflected shock wave and the propellant
temperature, as a result of which there is a rise in propellant
surface temperature. Analysis further revealed that oxidizer
and polymer attain different surface thermal levels due to
their difference in their thermophysical properties like thermal conductivity, heat capacity, density etc.

The ignition delay data plotted as a function of temperature, is compared with the analytical solution for ignition
time obtained from solving the equations involved in the three
theories of ignition in Fig.4. The results clearly indicate a
decrease in ignition delay with increase in shock temperature
and the results agree well with gas phase and solid gas interface theories which take into account factors like diffusion,
consumption of propellant reactants due to chemical reaction
which are neglected in the treatment of solid phase theory of
ignition, which is primarily based on the temperature built up
to the self ignition temperature at the solid phase. Fig. 5
shows the variation of ignition delay of the propellant with
shock temperature for various pressure levels ranging from 0.5
to 10 atmospheres. There is an overall trend in the reduction
of ignition delay at higher temperatures and this effect is

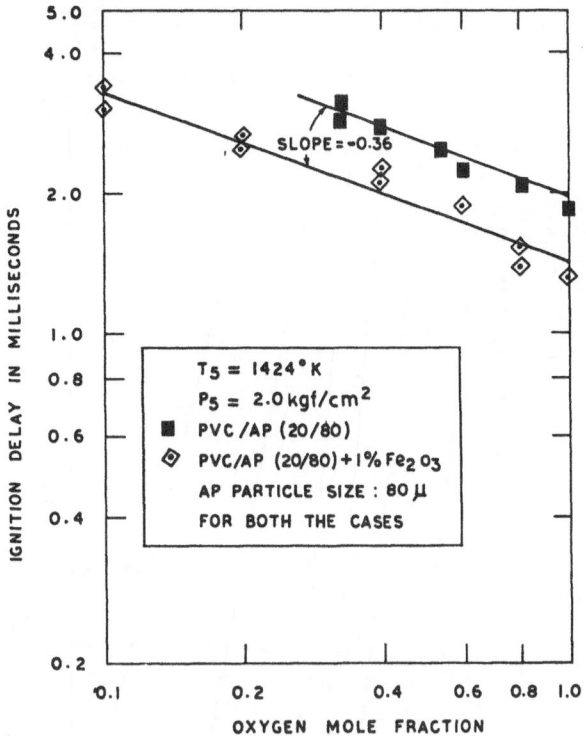

FIG.4. PLOT OF IGNITION DELAY VARIATION WITH OXYGEN
MOLE FRACTION

pronounced at higher pressures. The marked reduction in ignition delay observed at elevated temperatures is mainly due to
the increased magnitude of heat transfer from the hot stagnant
gas remaining behind the reflected shock wave. The heat transfer depends upon the temperature gradient existing between the
shocked gas and the propellant surface temperature, as well as
the thermophysical properties of the hot stagnant gas and the
propellant. Moreover at higher temperatures, the vaporization
rate of the propellant ingredients or the reaction rate
increases due to increased surface temperature of the propellant
as a result of heating, which results in changes in value of
pre exponential factor A figuring in the Arrhenius reaction
rate law, indicating for a first order reaction, increased
number of collisions among the molecules per unit time, which
also contributes to reduction in the delay time. These above
factors are sensitive to pressure variation also, especially
the thermophysical properties which result in marked reduction
of delay time at higher pressures and temperatures, since
higher temperatures and pressures result in increased surface
temperature which in turn result in increased vaporization
rate (gas phase theory) or increased reaction rate (gas-solid
interface theory) leading to shorter delay period and influen-
cing the overall thermal behaviour of the solid propellant.

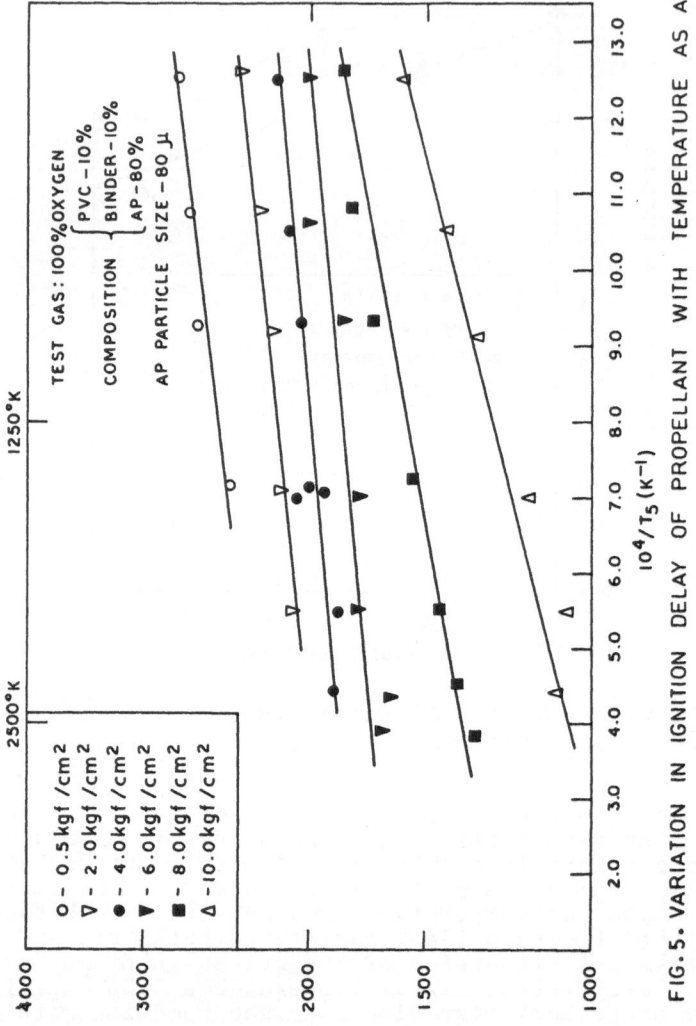

FIG. 5. VARIATION IN IGNITION DELAY OF PROPELLANT WITH TEMPERATURE AS A FUNCTION OF PRESSURE

A relationship for the induction time $= A \exp(E/RT)(O_2)^n$ was found to be applicable for the experimental condition of this study (Fig.6). The overall activation energy was found to be 7.8 Kcal/mole and the pre-exponential factor was found to be 2×10^{-5} with index

$$n = -0.49$$

$$\tau = 2 \times 10^{-5} \exp(7800/RT)(O_2)^{-0.49}$$

REFERENCES

1. Hicks, B.L. The theory of ignition considered as a thermal reaction. Journal of Chemical Physics, Vol.22, No.3, March 1954, p 414

2. McAlevy, R.F. III, The ignition mechanism of composite solid propellant. Ph.D thesis. June 1960, Princeton University

3. Anderson, R. Ignition theory of solid propellants. AIAA Journal, Vol.2, No.1, Jan 1964, p 179

4. Price, E.W. and Bradley, H.H. Theory of ignition of solid propellants, AIAAJ4, 1966, p 1153

5. Ramaprabhu, R. Studies on the ignition behaviour of composite solid propellants using a shock tube. Ph.D thesis, I.I.T. Madras, 1979

6. Ramaprabhu, R. and Bhaskaran, K.A. Shock tube study of the ignition characteristics of PVC-AP composite solid propellant. Presented at the V International Conference on Combustion Processes held at Poland, Sept 1977

7. Trass, O. and Makay, D. Procedure for contact surface tailoring. Technical note, AIAA Journal, Vol.1, June 1963, p 2151

Role of Catalyst on the Ignition Mechanism of Composite Solid Propellant

R. RAMAPRABHU
Department of Mechanical Engineering
A.C. College of Technology
Perarignar Anna University of Technology
Madras 600 025, India

K.A. BHASKARAN
Department of Mechanical Engineering
Indian Institute of Technology
Madras 600 036, India

ABSTRACT

Several metal oxides are known to promote ignition of composite solid propellants. Investigations have been done earlier
to study the effect of these metal oxides with reference to propellant combustion, but there has been not much of information
available with regards to the role of metal oxide catalysts on
the ignition behaviour of composite solid propellant. An
attempt has been made here to study the effect of catalyst addition on the ignition behaviour of Polyvinyl Chloride (PVC) -
Ammonium Perchlorate (AP) composite propellant which was ignited
using the reflected shock wave under varying conditions of temperature (800-2500K) and oxygen concentration Metal oxides such
as Cuprous Oxide (CuO), Ferric Oxide (Fe_2O_3) were used as
catalysts. The concentration of catalysts in the propellant was
varied from 0.5 to 3 percent without altering the stoichiometry
of the propellant. Ignition was achieved even in inert atmosphere when catalysts were added to the propellant. Higher concentrations beyond one percent had no appreciable effect on the
ignition delay. The catalyst CuO was found to be a better pro-
motor of ignition than the catalyst Fe_2O_3.

1. INTRODUCTION

Ignition characteristics of solid propellants are of great
relevance in the design of solid propellant grains as well as
pyrotechnic igniters. The shock tube, due to its versatality
makes it an ideal tool for the study of solid propellant ignition at any desired temperature environmental and oxygen
pressure. Ignition of the propellant was achieved by heating
behind the reflected shock wave. Ignition of the propellant is
due to a combination of chemical and thermal processes that
raise the propellant temperature and establish an environment
that can support combustion reactions. Propellant ignition can
be defined as an establishment of a runaway chemical reaction
accompanied by self heating of sufficient magnitude to accele-
rate the reaction. After ignition, steady state combustion
follows. Ignition research has emphasized the question of how long
it takes for the runaway reaction to build up, rather than the

approach to steady state. The time interval when the propellant surface is first exposed to the hot stagnant gas upto the instant when surface decomposition begins is called the chemical induction interval. The time to ignition is the sum of thermal and chemical induction intervals. Many metal oxides have been employed for increasing the rate of surface decomposition thereby reducing the ignition delay. It is understood that no real attempt has been made so far to analyse the catalystic role in propellant ignition. The objective of the present investigation is therefore to probe into the mechanism of the catalyst action on the propellant ignition.

2. EXPERIMENTAL

The experimental set up consists of a stainless steel shock tube of 71 mm ID with 3_m driver and 5_m test sections (1) . The test section of the shock tube is honed to a high degree of accuracy. The tube has a wall thickness of 14 mm. The schematic arrangement of the experimental set up is shown in Fig.1.

Catalysed propellant samples of 10 mm cube size was mounted in the end plange. The diaphragm, separating the high pressure side and the low pressure side was ruptured by a shooting pin operated by compressed air. Thin aluminium foils of 0.07 mm thickness were used as diaphragm material. The propellant samples were subjected to reflected shock heating under tailored conditions of the operation of shock tube (2) since the useful observation time of the present set up was about a milli second and ignition delays of composite propellants were generally of the order of one millisecond and above. The ignition delay was obtained by measuring the time interval between the arrival of the incident shock at the end flange (sensed by the piezeo-electric pressure transducer) and the emission of visible light due to ignition (sensed by a RCA 931A photomultiplier tube). The photo multiplier tube was kept at right angles to the tube axis. The shock velocity is calculated from the time taken for the incident shock to travel between two adjacent pressure transducers kept 35 cms apart, which trigger a digital electronic counter through charge amplifiers. A detailed description of the experimentation is described elsewhere (3) .

The catalysed composite propellants contained trace quantities of metal oxides such as Fe O , CuO in addition to the fuel and oxidizer. These catalysts were added taking care as not to alter the basic stoichiometry of the propellant. The composite solid propellant used in this investigation contained Polyvinyl Chloride (PVC) as the polymer and Ammonium Perchlorate (AP) as the oxidizer with Dybutyl Sebacate (DBS) as the liquid binder. The particle size of the Ammonium Perchlorate was maintained at 30 microns. Varying quantities (0.5, 1, 3 percent) of the catalysts (Fe_2O_3, CuO) were added to the propellants. Higher concentration of the catalysts have not been employed as this may alter the energy level of the propellant.

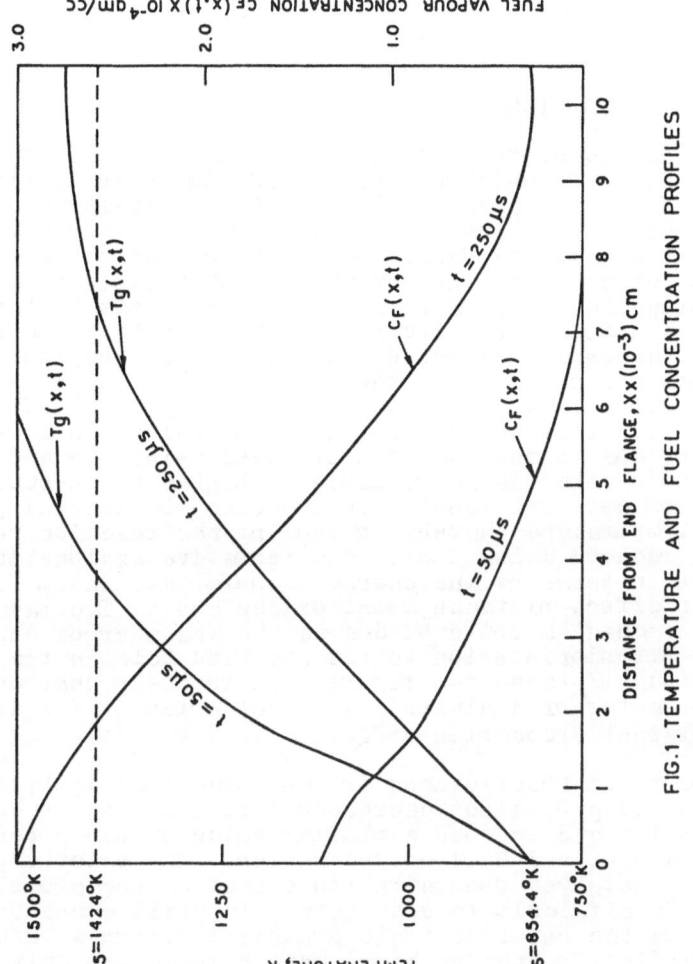

FIG.1. TEMPERATURE AND FUEL CONCENTRATION PROFILES

The propellants were tested for varying conditions of temperature ($800-2500^\circ$K) and environmental oxygen concentrations (zero to cent percent), by suitably selecting the filling pressures of driver and test sections and filling the test section of the shock tube with varying mole fractions of oxygen gas.

3. RESULTS AND DISCUSSION

The effect of environmental oxygen concentration on the ignition delay of catalysed propellant is shown in Fig.2. It is clear from the figure that propellant containing Cuprous Oxide (CuO) as catalyst yielded shorter delay times than the propellant containing Ferric Oxide (Fe_2O_3) as catalyst under identical conditions. Moreover both catalysts when added to PVC-AP propellant system helped the propellants to ignite even in neutral atmosphere. Fig.3 shows the variation in ignition delay of the catalysed propellant with shock temperature. These data are compared on the same plot both in Figs.2 and 3 with the delay data of uncatalysed PVC-AP propellant under same testing conditions to show the apparent reduction in the ignition delay due to the addition of these catalysts and the reduction in delay is appreciable at higher temperatures. The addition of catalyst results in lowering the initial propellant surface temperature thereby increasing the reaction rate yielding reduced delay time. The tentative explonations have been made in terms of the charge transfer mechanism for the catalytic effect of these metal oxides (4). The rate controlling mechanism could be due to the transfer of electron from the perchlorate ion to the positive hole in the oxides. From a study of these two figures, it is clear that CuO is a better promoter of ignition than Fe_2O_3 which is due to better thermophysical properties of CuO.

From these observations it was found that ignition delay of the catalysed propellant decreased with increase in catalyst concentration and reached a minimum value around 3 percent of catalyst (Fe_2O_3 and CuO) concentration. The exact explonation for such a catalyst concentration effect on the propellant ignition is difficult to speculate. In small concentration, the role of the catalyst is to promote ignition without significantly affecting the chemical heat release per unit mass of the propellant system. But when the catalyst concentration increases, it probably cuts into the weight of the actual propellant, thereby reducing the value of chemical heat release and altering the energy level of the propellant system. The ignition delay data of catalysed and uncatalysed propellant are plotted on log-log plot as shown in Fig.4. The slopes of the curves are same indicating the true nature of catalytic action of these catalysts. Catalysts are known to affect simultaneously the decomposition of the polymer, oxidizer and in turn the degradation of the propellant, leading to an increase in the overall reaction rate leading to quick ignition. They do not interfere with the chemical kinetics of the propellant system, thereby truly playing the role of catalyst.

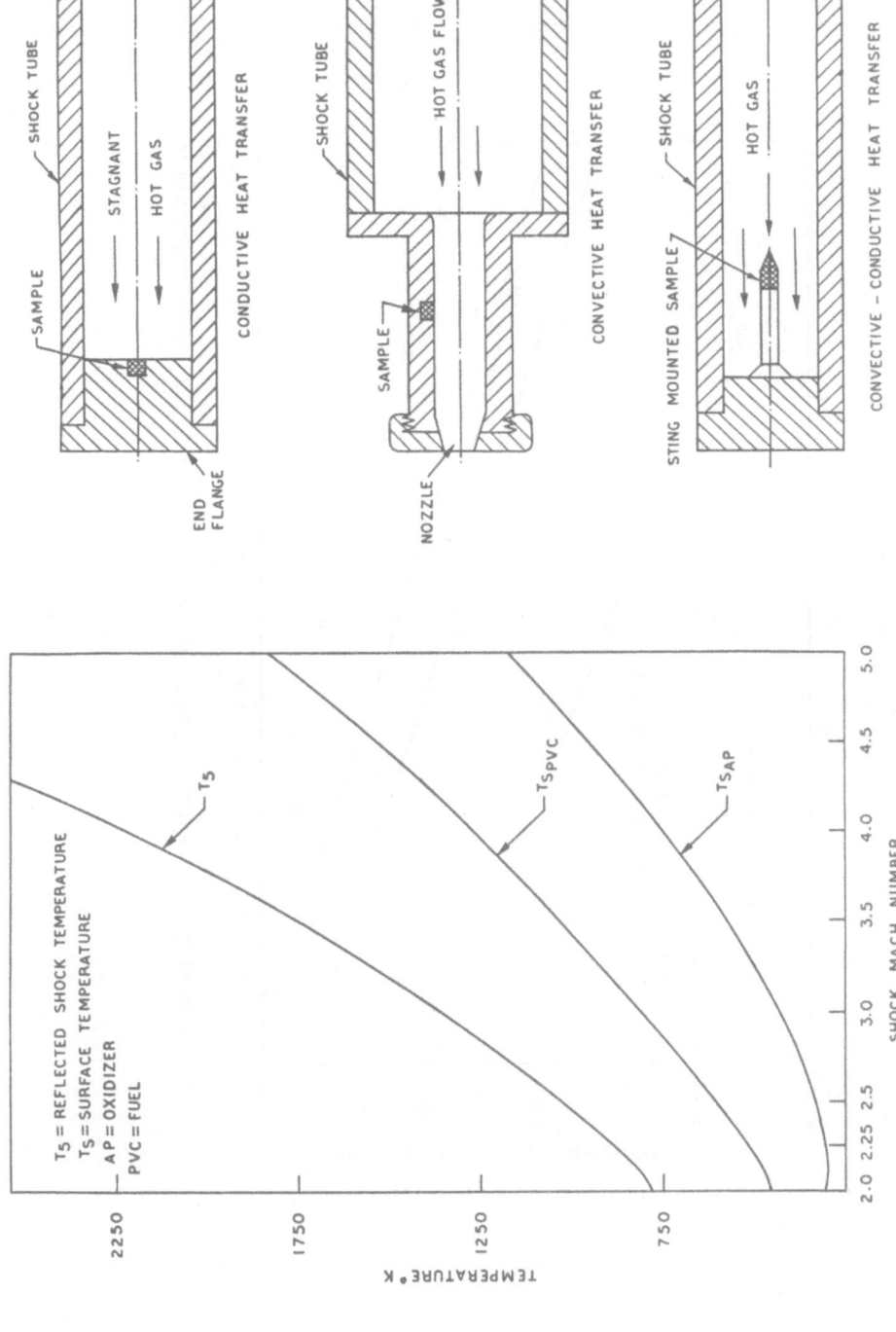

T5 = REFLECTED SHOCK TEMPERATURE
TS = SURFACE TEMPERATURE
AP = OXIDIZER
PVC = FUEL

FIG.2. SURFACE TEMPERATURE VARIATION FOR FUEL AND OXIDIZER

FIG.3. DIFFERENT MODES OF HEAT TRANSFER

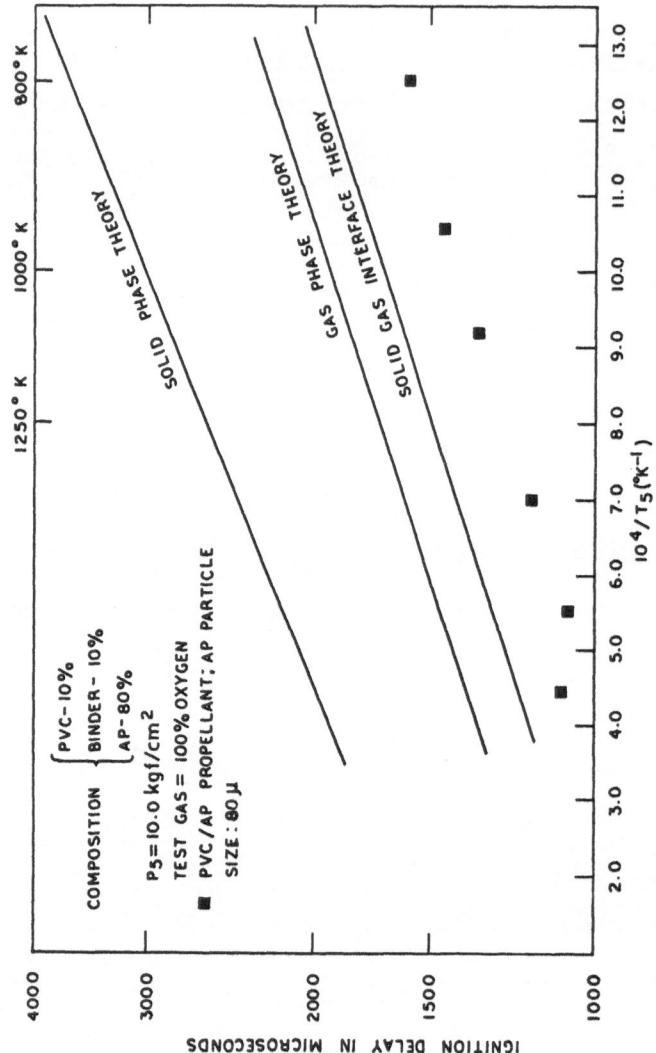

FIG. 4. VARIATION IN IGNITION DELAY OF PVC-AP PROPELLANT WITH TEMPERATURE

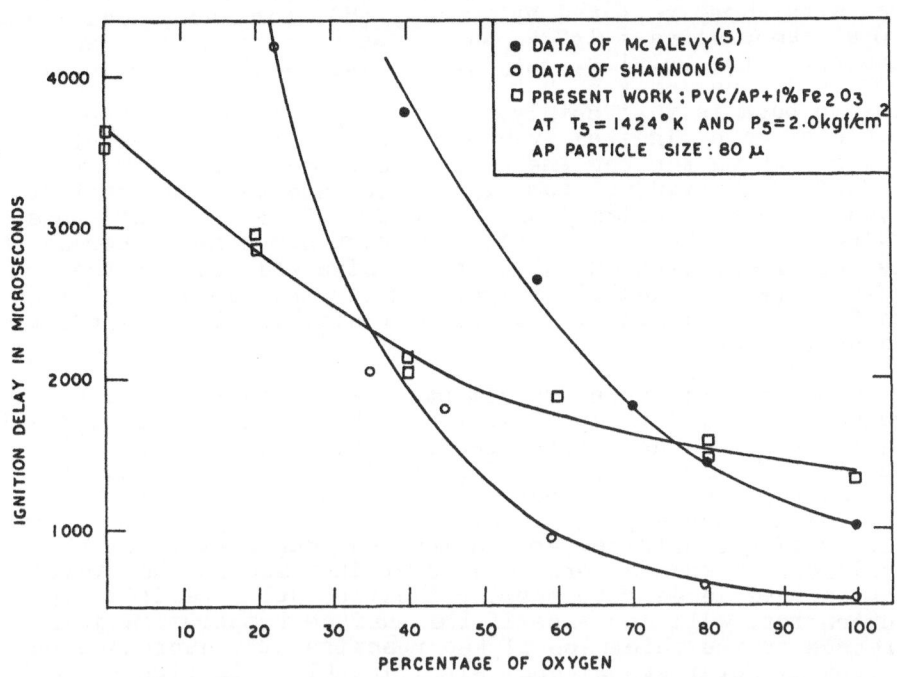

FIG.5. COMPARISON OF PRESENT WORK WITH PREVIOUS REPORTED DATA

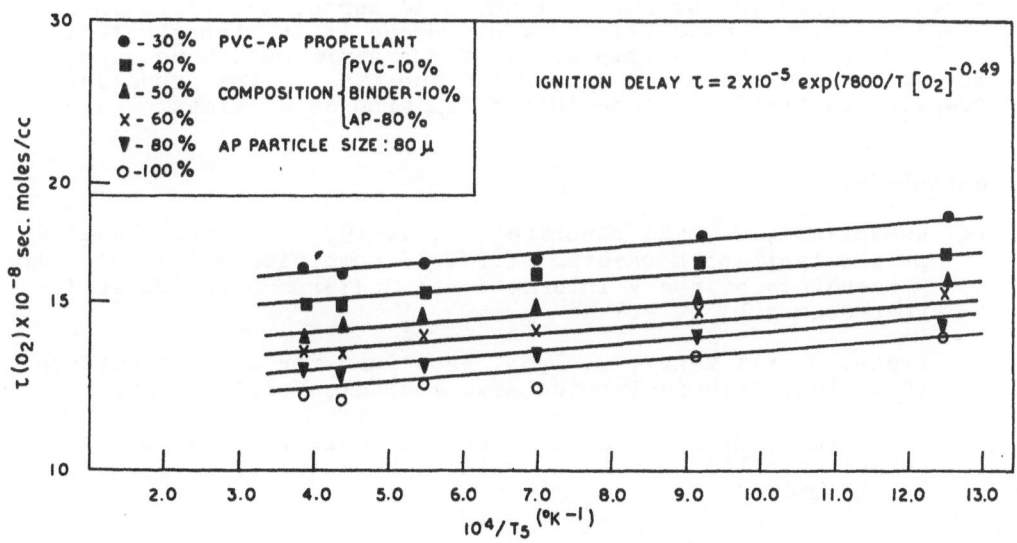

FIG.6. IGNITION DELAY VARIATION WITH TEMPERATURE AND OXYGEN CONCENTRATION

The ignition delay data of catalysed propellant are compared with those reported by McAlevy (5) for Epoxy fuel + 1.4%
Fe O at temperature = 1370°K and pressure = 51 kgf/cm^2 and
Shannon (6) for Polyiso-butylene + 0.25% Fe_2O_3 at temperature =
1250°K and pressure = 54.5 kgf/cm^2 in Fig.5. An interesting
point to note is that catalysed PVC/AP propellant system was
capable of being ignited even when the environment contained
inert gas which was not the case with those reported by McAlevy
and Shannon as evident from Fig.5. The addition of catalyst
(Fe_2O_3) to the propellant made with 80 AP particle size has
enabled achievement of ignition in inert atmosphere whereas the
same propellant with 80 AP particle size and without the
catalysts (Fe_2O_3) failed to ignite under identical conditions
when the environmental oxygen concentration fell below thirty
percent (Fig.2).

Based on the above experimental observations, it can be
concluded that the oxide additives such as CuO and Fe_2O_3 result
in a lowering of the initial decomposition temperature and
hence in acceleration of the decomposition process. Recent
evidence reported by Rastogi (7) has shown that catalysts
affect simultaneously the decomposition of the polymer, decomposition of the oxidizer and in turn the degradation of the
propellant. These factors lead to an increase in the overall
reaction rate leading to shorter ignition delay period. The
reaction rate will increase if the surface temperature jump
decreases or the thickness of the reaction zone decreases on
addition of catalysts without significantly affecting the calorific value of the propellant composition. Moreover it is
unlikely that the density, heat capacity, thermal diffusivity
would be affected by the addition of trace quantities of catalyst
It is felt that the initial propellant temperature, and the
adiabatic flame temperature which are primarily a function of
heats of formation of the fuel and oxidizer could not be changed
significantly by catalytic action. Hence it is strongly felt
that the propellant decomposition temperature must have been
lowered in order to account for the increase in the propellant
reaction rate and decrease in the ignition delay time.

REFERENCES

1. Ramaprabhu, R., and Bhaskaran, K.A., 1977. Shock tube study
    of the ignition mechanism of PVC-AP composite solid propellant
    Proceedings of the V International Conference on Combustion
    Processes held at Poland

2. Trass, O. and Makay, D. 1963. Procedure for contact surface
    tailoring, Technical note, AIAA Journal, Vol.1, p 2161

3. Ramaprabhu, R. 1978. Studies on the ignition behaviour of
    composite solid propellant using a shock tube, Ph.D thesis,
    I.I.T. Madras

4. Freeman, E.S. and Anderson, D.A. 1965. Effects of Radiation and ddoping on the catalytic activity of magnesium oxide on the thermal decomposition of potassium perchlorate, nature, 206, p 378

5. McAlevy, R.F.III. 1960. The ignition mechanism of composite solid propellant, Ph.D thesis. Princeton University

6. Shannon, L.J. 1966. Composite solid propellant ignition mechanism. UTC Report No.2138. Annual Scientific Report

7. Rastogi, R.P. Gurudup Singh. Ramrajsingh. 1977. Burning rate catalysts for composite solid propellants. Combustion and Flame, 30, p 117

SOLAR ENERGY COLLECTION AND STORAGE

General Solution of Collector Performance with Axial Conduction and End Effects

A.R. SHOUMAN
University of Petroleum and Minerals
Dhahran, Saudi Arabia

I.A. TAG
University of Qatar
Doha, Qatar

ABSTRACT

The Phillips solution of the flat-plate solar collector [6] is extended and utilized to examine the influence of the end losses on collector performance. The results of this study show that the influence of the end temperatures of the absorber plate is more significant than the losses due to the axial conductivity of an insulated end collector plate.

It will be shown that for an insulated end collector, the loss in the heat removal factor due to axial conductivity is negligible in the region of interest for flat-plate collectors. However, the end temperatures of the collector plate have more significant influence on the same factor, showing losses in some regions and improvements in other regions.

This study emphasizes the necessity of measuring the temperature of the absorber plate at both the fluid inlet and exit locations in order to determine accurately the collector performance parameters.

NOMENCLATURE

a	cross-sectional area
A_a	collector aperture area
A_r	collector receiver area
A, B, E	constants defined by Eqs. (13) and (14)
A_1, A_2, A_3	constants defined by Eq.(26)
B_1, B_2, B_3	constants defined by Eq.(27)
C_p	fluid constant pressure specific heat
C_1, C_2, C_3	constants defined in the text
D_1, D_2, D_3	constants defined in the text
F'	collector efficiency factor U_o/U_ℓ
F''	defined by Eq.(11)
F_k	correction factor
F_R	collector heat removal factor equation
k	thermal conductivity of the receiver
K	parameter defined by Eq.(3)

A. R. Shouman is on leave from New Mexico State University, Las Cruces, New Mexico 88003, USA.
I. A. Tag is on leave from Oak Ridge National Laboratory, Oak Ridge, Tennessee 37830, USA.

K'	parameter defined by Eq.(25)
L	axial length of collector
$\dot{m}$	rate of mass flow of the fluid
N	number of transfer units defined by Eq.(3)
N'	modified number of transfer units defined by Eq.(17)
N_x	dimensionless parameter defined by Eq.(3)
q_s	solar flux in the collector aperture
R	defined as θ_{fe}/θ_{fi}
r_1, r_2, r_3	constants defined by Eqs.(26) and (27)
S	slope of collector performance line
T_a	ambient air temperature
T_c	collector plate temperature
T_f	collector fluid temperature
	i, e -- subscripts for fluid inlet and outlet locations
U_g	overall heat transfer coefficient between receiver and the transfer fluid
U_ℓ	overall heat transfer coefficient between the receiver and the ambient
U_o	overall heat transfer coefficient between the transfer fluid and the ambient
x	axial coordinate
α_e	effective transmittance-absorptance product
β	defined as θ_{ce}/θ_{fi}
γ_2, γ_3	defined by Eq.(30)
η	collector thermal efficiency
η_o	collector thermal efficiency defined by Eq.(18)
θ	temperature defined by Eq.(3)
	a, c, f -- subscripts for ambient, collector, and fluid
	i, e -- subscripts for fluid inlet and outlet locations
Θ'	defined by Eq.(30)
$\lambda_1, \lambda_2, \lambda_3$	roots of the characteristic equation
μ	defined as θ_{ci}/θ_{fi}
Φ	parameter defined in the text.

1. INTRODUCTION

In a separate paper [1], we have discussed the existing literature on collector performance and its application to the analysis of existing collector test data. The application of the widely accepted solution of Hottel, Whillier, and Bliss (HWB)[2, 3, 4, 5] as well as the Phillips solution [6] to the test data produced inconsistent and impossible results. It was also shown that within the practical range of operating nonconcentrating flat-plate collectors, the HWB solution was as accurate as the Phillips solution. Based on those results, recommendations were made to modify the existing collector testing method by measuring the collector plate temperature at both fluid inlet and exit locations. It was shown that this would allow better determination of the collector performance parameters. It was noticed that the paper of Phillips considered axial conduction only for collectors with insulated ends or negligible end losses. The purpose of this paper is to extend the Phillips solution to include the end effects. This solution is applicable to concentrating collectors. It is also shown that the results can be presented in a very simple format.

2. MATHEMATICAL FORMULATION

The one-dimensional heat transfer equations for both the collector plate and fluid give

$$ak \frac{d^2 T_c}{dx^2} - \frac{A_r U_\ell}{L} (T_c - T_a) - \frac{A_r U_g}{L} (T_c - T_f) + \frac{eA_a q_s}{L} = 0 \tag{1}$$

and

$$\dot{m} C_p \frac{dT_f}{dx} + \frac{A_r U_g}{L} (T_f - T_c) = 0 \tag{2}$$

Solving Eq.(2) for T_c and combining it with Eq.(1) the following third order equation is obtained:

$$K \frac{d^3 \Theta_f}{dN_x^3} + K\left(1 + \frac{U_g}{U_\ell}\right) \frac{d^2 \Theta_f}{dN_x^2} - \frac{d\Theta_f}{dN_x} - \Theta_f = 0 \tag{3}$$

where

$$\frac{1}{U_o} = \frac{1}{U_\ell} + \frac{1}{U_g} ,$$

$$K = \frac{ak\ N^2}{A_r L (U_\ell + U_g)} ,$$

$$\Theta = \frac{\alpha_e A_a q_s}{A_r U_\ell} + T_a - T ,$$

$$N = \frac{A_r U_o}{\dot{m} C_p} , \quad \text{and}$$

$$N_x = N \frac{x}{L} ,$$

which is a dimensionless constant analogous to the heat exchanger NTU defined here as N but is distance dependent.

3. THE NONCONDUCTING COLLECTOR PLATE (HWB)

Considering the nonconducting collector (HWB) solution, namely $K = 0$, the solution to Eq.(3) is

$$\Theta_f = \Theta_{fi}\ e^{-N_x} \tag{4}$$

where Θ_{fi} is Θ_f at the fluid inlet condition. This gives the fluid exit condition Θ_{fe} as

$$\Theta_{fe} = \Theta_{fi}\ e^{-N} \tag{5}$$

Using the definition of the collector efficiency as

$$\eta = \frac{\dot{m} C_p\ (T_{fe} - T_{fi})}{A_a\ q_s} \tag{6}$$

and $\quad F' = \dfrac{U_o}{U_\ell} \tag{7}$

it can be shown that

$$\eta = F' \; \alpha_e \; \frac{\theta_{fi}}{\theta_a} \left(\frac{1-e^{-N}}{N} \right) \tag{8}$$

The familiar heat recovery factor F_R can be written in terms of the efficiency as

$$F_R = \frac{1}{\alpha_e} \frac{\theta_a}{\theta_{fi}} \; \eta$$

which gives

$$F_R = \frac{F'}{N} (1 - R)$$

where

$$R = \frac{\theta_{fe}}{\theta_{fi}}$$

We shall define a correction factor F_K given by $\dfrac{1-R}{1-R_0}$ where R_0 is R at K = 0 and is obtained from the HWB model as e^{-N}. Hence the following is obtained:

$$F_R = \frac{F'}{N} \; F_k \left(1 - e^{-N} \right)$$

and thereby F_K becomes a correction to the HWB model to account for the effect of conductivity.

For a heat exchanger with $\dot{m}C_p = \infty$, N = 0 and from Eq.(8):

$$\eta_0 = F' \; \alpha_e \; \frac{\theta_{fi}}{\theta_a} \tag{9}$$

thus giving

$$\frac{\eta}{\eta_0} = \frac{1 - e^{-N}}{N} \tag{10}$$

Fig.1 shows θ_f/θ_{fi} as a function of N_x and Fig.2 shows η/η_0 as a function of N.

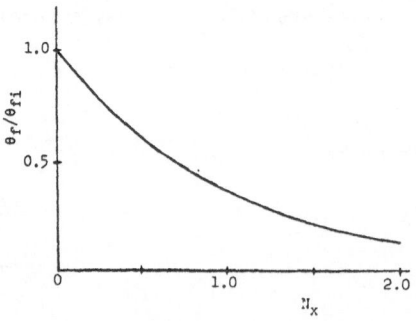

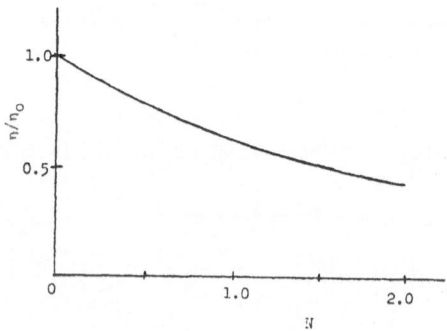

Fig. 1. θ_f/θ_{fi} as a function of N_x.   Fig. 2. η/η_0 as a function of N.

4. THE INFINITELY CONDUCTING COLLECTOR PLATE

Considering an infinitely conducting collector plate, $K = \infty$ and defining

$$F'' = 1 + \frac{U_g}{U_\ell} = \frac{1}{1 - F'} \tag{11}$$

Equation (3) becomes

$$\frac{d^3\Theta_f}{dN_x^3} + F''\frac{d^2\Theta_f}{dN_x^2} = 0 \tag{12}$$

whose general solution is

$$\Theta_f = A + BN_x + Ee^{-F''N_x} \tag{13}$$

and

$$\Theta_c = A + \frac{B}{F''} + E N_x \tag{14}$$

where A, B, and E are constants dependent on the boundary conditions.

It is interesting to notice that temperature distribution in an infinitely conducting collector plate is linear and that if the collector can be considered infinitely long, then the collector plate temperature becomes constant and the solution can be written as

$$\frac{\Theta_f - \Theta_c}{\Theta_{fi} - \Theta_c} = e^{-F''N_x} \tag{15}$$

which is similar to Eq.(4) except for the modification in Θ and N and this gives

$$\frac{\Theta_{fe} - \Theta_c}{\Theta_{fi} - \Theta_c} = e^{-F''N} \tag{16}$$

Using this result to determine the collector efficiency gives

$$\eta = \alpha_e \frac{\Theta_{fi} - \Theta_c}{F''\Theta_a} \left\{ \frac{1 - e^{-F''N'}}{N} \right\} \tag{17}$$

which produces

$$\eta_o = \frac{\Theta_{fi} - \Theta_c}{\Theta_a} \alpha_e \tag{18}$$

and

$$\frac{\eta}{\eta_o} = \frac{1 - e^{-F''N}}{F''N} = \frac{1 - e^{-N'}}{N'} \tag{19}$$

Replacing F''N by N', Eq.(19) becomes identical in form to Eq.(10) for the non-conducting collector plate case with N' replacing N.

5. END EFFECTS IN FINITE COLLECTORS

Considering a finite collector whose collector plate has an infinite conductivity $(K = \infty)$ and considering the three boundary conditions known to be the fluid temperature at inlet and the collector plate temperatures at both fluid inlet and exit locations, the solution is obtained as follows:

$$\theta_c = \theta_{ci} + (\theta_{ce} - \theta_{ci}) \frac{x}{L} \tag{20}$$

and

$$\theta_f = \theta_c - \frac{\theta_{ce} - \theta_{ci}}{F''N} + \left[\theta_{fi} - \theta_{ci} + \frac{\theta_{ce} - \theta_{ci}}{F''N}\right] e^{-F''N_x} \tag{21}$$

Equations (20) and (21) can be used to determine the exit fluid temperature which is given by

$$\frac{\theta_{fe} - \theta_{ce}}{\theta_{fi} - \theta_{ci}} = e^{-N'} - \frac{\theta_{ce} - \theta_{ci}}{\theta_{fi} - \theta_{ci}} \left[\frac{1 - e^{-N'}}{N'}\right] \tag{22}$$

Using Eq.(22) in Eq.(6) gives:

$$\eta = \frac{F'\alpha_e}{N} \left[\frac{\theta_{ce} - \theta_{ci}}{\theta_a} \left(1 - \frac{1 - e^{-N'}}{N'}\right) - \frac{\theta_{fi} - \theta_{ci}}{\theta_a} \left(1 - e^{-N'}\right)\right] \tag{23}$$

and

$$\eta_o = F'F''\alpha_e \left[\frac{\theta_{ce} - \theta_{ci}}{2\theta_a} - \frac{\theta_{fi}}{\theta_a}\right] \tag{24}$$

6. THE GENERAL PROBLEM

Substituting for K by $\frac{K'}{F''}$ where

$$K' = \frac{ak\ N^2}{A_r U_\ell L}$$

the third order differential equation can be written as

$$\frac{d^2\theta_f}{dN_x^3} + F'' \frac{d^2\theta_f}{dN_x^2} - \frac{F''}{K'} \frac{d\theta_f}{dN_x} - \frac{F''}{K'} \theta_f = 0 \tag{25}$$

Examining the characteristic equation for (25), we find that it has three real roots, one positive and two negative. The roots are given by

$$\lambda_1 = 2\sqrt{\frac{F''}{3} \left(\frac{F''}{3} + \frac{1}{K'}\right)} \cos \frac{\phi}{3} - \frac{F''}{3}$$

$$\lambda_2 = -2\sqrt{\frac{F''}{3} \left(\frac{F''}{3} + \frac{1}{K'}\right)} \cos \frac{\phi + \pi}{3} - \frac{F''}{3}$$

$$\lambda_3 = -2\sqrt{\frac{F''}{3} \left(\frac{F''}{3} + \frac{1}{K'}\right)} \cos \frac{\phi + \pi}{3} - \frac{F''}{3}$$

where

$$\phi = \cos^{-1} \frac{\frac{3}{2K'} - \frac{F''}{2K'} - \frac{1}{9} F''^2}{\sqrt{\frac{F''}{3} \left(\frac{F''}{3} + \frac{13}{K'}\right)}}$$

The general solution to Eq.(25) is

$$\theta_f = A_1 e^{\lambda_1 N_x} + A_2 e^{\lambda_2 N_x} + A_3 e^{\lambda_3 N_x} \tag{26}$$

and

$$\theta_c = B_1 e^{\lambda_1 N_x} + B_2 e^{\lambda_2 N_x} + B_3 e^{\lambda_3 N_x} \tag{27}$$

where A_1, A_2, and A_3 are constants determined from the boundary conditions and

$$\frac{B_1}{A_1} = 1 + \lambda_1 (1 - F') = r_1$$

$$\frac{B_2}{A_2} = 1 + \lambda_2 (1 - F') = r_2$$

$$\frac{B_3}{A_3} = 1 + \lambda_3 (1 - F') = r_3$$

Tables 1, 2, and 3 show the variation of λ_1, λ_2, and λ_3 as functions of F' and K'. It is important to notice that λ_2 is the smallest root in magnitude and is always negative. λ_1 is always positive and λ_3 is always negative and both are larger in magnitude than λ_2.

Table 1. FIRST ROOT AS A FUNCTION OF F' AND K'.

K' \ F'	0.050	0.100	0.200	0.400	0.600	0.800	0.900	0.950
0.005	14.485	14.855	15.694	17.944	21.655	29.747	40.545	54.601
0.010	10.236	10.490	11.066	12.605	15.124	20.535	27.577	36.421
0.050	4.567	4.669	4.897	5.498	6.451	8.372	10.615	13.037
0.100	3.224	3.291	3.440	3.827	4.426	5.576	6.818	8.026
0.500	1.436	1.458	1.508	1.631	1.805	2.087	2.322	2.492
1.000	1.013	1.026	1.056	1.126	1.222	1.363	1.467	1.534
5.000	0.451	0.454	0.462	0.479	0.500	0.525	0.541	0.549
10.000	0.318	0.320	0.324	0.333	0.344	0.356	0.363	0.366
50.000	0.142	0.142	0.143	0.145	0.147	0.149	0.151	0.151
100.000	0.100	0.100	0.101	0.102	0.103	0.104	0.105	0.105
1000.000	0.032	0.032	0.032	0.032	0.032	0.032	0.032	0.032

Table 2. SECOND ROOT AS A FUNCTION OF F' AND K'.

K' \ F'	0.050	0.100	0.200	0.400	0.600	0.800	0.900	0.950
0.005	-1.000	-0.999	-0.999	-0.998	-0.997	-0.996	-0.996	-0.995
0.010	-0.999	-0.999	-0.998	-0.996	-0.994	-0.992	-0.991	-0.991
0.050	-0.997	-0.995	-0.990	-0.980	-0.971	-0.963	-0.958	-0.956
0.100	-0.995	-0.989	-0.979	-0.961	-0.944	-0.930	-0.923	-0.919
0.500	-0.959	-0.929	-0.886	-0.828	-0.788	-0.757	-0.744	-0.738
1.000	-0.868	-0.825	-0.772	-0.710	-0.671	-0.641	-0.629	-0.623
5.000	-0.439	-0.431	-0.418	-0.398	-0.382	-0.369	-0.363	-0.361
10.000	-0.313	-0.309	-0.303	-0.293	-0.284	-0.277	-0.273	-0.272
100.000	-0.100	-0.099	-0.099	-0.098	-0.097	-0.096	-0.096	-0.095
1000.000	-0.032	-0.032	-0.032	-0.031	-0.031	-0.031	-0.031	-0.031

Table 3. THIRD ROOT AS A FUNCTION OF F' AND K'.

K' \ F'	0.050	0.100	0.200	0.400	0.600	0.800	0.900	0.950
0.005	−14.538	−14.967	−15.945	−18.613	−23.158	−33.751	−49.549	−73.605
0.010	−10.289	−10.602	−11.318	−13.275	−16.629	−24.543	−36.586	−55.430
0.050	−4.622	−4.785	−5.157	−6.185	−7.980	−12.409	−19.657	−32.080
0.100	−3.282	−3.413	−3.711	−4.533	−5.981	−9.646	−15.895	−27.106
0.500	−1.529	−1.640	−1.872	−2.470	−3.517	−6.330	−11.578	−21.754
1.000	−1.198	−1.313	−1.534	−2.083	−3.051	−5.721	−10.838	−20.911
5.000	−1.065	−1.134	−1.294	−1.748	−2.618	−5.156	−10.177	−20.188
10.000	−1.058	−1.122	−1.271	−1.707	−2.560	−5.079	−10.089	−20.095
50.000	−1.054	−1.113	−1.254	−1.675	−2.512	−5.016	−10.018	−20.019
100.000	−1.053	−1.112	−1.252	−1.671	−2.506	−5.008	−10.009	−20.009
1000.000	−1.053	−1.111	−1.250	−1.667	−2.501	−5.001	−10.001	−20.001

We shall now consider the different boundary conditions.

6.1 The Infinite Collector

For an infinite collector only the two negative roots λ_2 and λ_3 apply to the solution giving

$$\theta_f = A_2 e^{\lambda_2 N_x} + A_3 e^{\lambda_3 N_x} \tag{28}$$

and

$$\theta_c = A_2\{1 + \lambda_2(1 - F')\}e^{\lambda_2 N_x} + A_3\{1 + \lambda_3(1 - F')\}e^{\lambda_3 N_x} \tag{29}$$

Applying the boundary conditions that $\theta_f = \theta_{fi}$ and $\theta_c = \theta_{ci}$ at $N_x = 0$, we obtain

$$A_2 = \frac{\theta_{fi}\{1 + \lambda_3(1 - F')\} - \theta_{ci}}{(\lambda_3 - \lambda_2)(1 - F')}$$

and

$$A = \frac{\theta_{fi}\{1 + \lambda_2(1 - F')\} - \theta_{ci}}{(\lambda_2 - \lambda_3)(1 - F')}$$

This completes the solution, and it should be noticed that both θ_f and θ_c become zero at infinity.

Defining $\frac{\theta_{ci}}{\theta_{fi}} = \mu$, R can be determined from the two roots as:

$$R = \frac{(1 - \mu)\left(e^{\lambda_2 N} - e^{\lambda_3 N}\right) + (1 - F')\left(\lambda_3 e^{\lambda_2 N} - \lambda_2 e^{\lambda_3 N}\right)}{(\lambda_3 - \lambda_2)(1 - F')}$$

6.2 The Finite Collector

To complete the solution of a finite collector, three boundary conditions need to be specified. Phillips obtained his solution by specifying the inlet fluid temperature and assumed that the collector plate is insulated at both the inlet and exit fluid location.   We should mention at this point that θ in this paper is related to Phillips θ' by the following relationship:

$$\theta = \left[\frac{\alpha_e A_a q_s}{A_r U_\ell} + T_a - T_{fi}\right]\left(1 - \theta'\right) \tag{30}$$

Applying the boundary conditions $\theta_f = \theta_{fi}$ at $N_x = 0$ and $\dfrac{d\theta_c}{dN_x} = 0$ at $N_x = 0$ and $N_x = N$ to Eqs.(26) and (27) gives

$$A_1 = \frac{\theta_{fi}}{1 + \gamma_2 + \gamma_3}$$

where

$$\gamma_2 = \frac{A_2}{A_1} = \frac{\lambda_1 r_1}{\lambda_2 r_2} \; \frac{e^{\lambda_3 N} - e^{\lambda_1 N}}{e^{\lambda_2 N} - e^{\lambda_3 N}}$$

and

$$\gamma_3 = \frac{A_3}{A_1} = \frac{\lambda_1 r_1}{\lambda_2 r_2} \; \frac{e^{\lambda_2 N} - e^{\lambda_1 N}}{e^{\lambda_3 N} - e^{\lambda_2 N}}$$

Using these constants, the exit fluid temperature can be determined from

$$\theta_{fe} = \frac{\theta_{fi}}{1 + \gamma_2 + \gamma_3} \; e^{\lambda_1 N} + \gamma_2 e^{\lambda_2 N} + \gamma_3 e^{\lambda_3 N} \tag{31}$$

This gives

$$R = \frac{e^{\lambda_1 N} + \gamma_2 e^{\lambda_2 N} + \gamma_3 e^{\lambda_3 N}}{1 + \gamma_2 + \gamma_3}$$

Tables 4, 5, and 6 show the calculated values of F_k for $N = 0.1$, 1 and 2 as a function of both F' and K'. These values agree very well with Phillip's approximate solution. However, it is clearer from this presentation that F_k is almost unity except for very large K'.

Also the collector efficiency and all other desired collector parameters can be determined.

Table 4. F_k FOR AN INSULATED END COLLECTOR, $N = 0.1$

K' \ F'	0.050	0.100	0.200	0.400	0.600	0.800	0.900	0.950
0.005	1.000	1.000	1.000	1.000	0.999	0.998	0.998	0.997
0.010	1.000	1.000	1.000	1.000	0.999	0.998	0.996	0.995
0.050	1.000	1.000	1.000	0.999	0.999	0.997	0.994	0.990
0.100	1.000	1.000	1.000	0.999	0.999	0.997	0.994	0.988
0.500	1.000	1.000	1.000	0.999	0.999	0.997	0.993	0.987
1.000	1.000	1.000	1.000	0.999	0.999	0.997	0.993	0.986
5.000	1.000	1.000	1.000	0.999	0.999	0.997	0.993	0.986
10.000	1.000	1.000	1.000	0.999	0.999	0.997	0.993	0.986
50.000	1.000	1.000	1.000	0.999	0.999	0.997	0.993	0.986
100.000	1.000	1.000	1.000	0.999	0.999	0.997	0.993	0.986
1000.000	1.000	1.000	1.000	0.999	0.999	0.997	0.993	0.986

Table 5. F_k FOR AN INSULATED END COLLECTOR, N = 1.0

K' \ F'	0.050	0.100	0.200	0.400	0.600	0.800	0.900	0.950
0.005	1.000	1.000	0.999	0.999	0.998	0.998	0.998	0.997
0.010	1.000	1.000	0.999	0.998	0.997	0.996	0.995	0.995
0.050	0.999	0.998	0.996	0.992	0.988	0.982	0.979	0.977
0.100	0.999	0.997	0.995	0.988	0.980	0.969	0.963	0.959
0.500	0.998	0.995	0.990	0.976	0.956	0.923	0.901	0.889
1.000	0.998	0.995	0.989	0.973	0.948	0.905	0.876	0.860
5.000	0.997	0.995	0.988	0.970	0.939	0.883	0.843	0.823
10.000	0.997	0.995	0.988	0.969	0.938	0.879	0.838	0.817
50.000	0.997	0.995	0.988	0.969	0.937	0.876	0.834	0.813
100.000	0.997	0.994	0.988	0.969	0.937	0.876	0.833	0.812
1000.000	0.997	0.994	0.988	0.969	0.936	0.876	0.833	0.811

Table 6. F_k FOR AN INSULATED END COLLECTOR, N = 2.0

K' \ F'	0.050	0.100	0.200	0.400	0.600	0.800	0.900	0.950
0.005	1.000	1.000	1.000	0.999	0.999	0.999	0.999	0.999
0.010	1.000	1.000	0.999	0.999	0.998	0.998	0.997	0.997
0.050	0.999	0.999	0.997	0.995	0.992	0.989	0.987	0.987
0.100	0.999	0.998	0.995	0.990	0.985	0.980	0.977	0.975
0.500	0.997	0.993	0.986	0.970	0.952	0.932	0.921	0.915
1.000	0.996	0.991	0.981	0.959	0.933	0.903	0.887	0.878
5.000	0.994	0.988	0.975	0.942	0.899	0.850	0.825	0.813
10.000	0.994	0.988	0.974	0.939	0.893	0.839	0.812	0.800
50.000	0.994	0.987	0.973	0.936	0.887	0.829	0.801	0.787
100.000	0.994	0.987	0.972	0.935	0.886	0.827	0.799	0.786
1000.000	0.994	0.987	0.972	0.935	0.885	0.826	0.798	0.784

7.   END EFFECTS

For a noninsulated collector plate, the solution can be completed using two other boundary conditions together with the inlet fluid temperature. The other two most convenient boundary conditions are the collector plate temperatures at the fluid inlet and exit condition, using

$$\frac{\theta_{ce}}{\theta_{fi}} = \beta$$

and defining

$$C_1 = r_1 \, r_2 \left[e^{\lambda_1 N} - e^{\lambda_2 N} \right]$$

$$C_2 = r_2 \, r_3 \left[e^{\lambda_2 N} - e^{\lambda_3 N} \right]$$

$$C_3 = r_3 \, r_1 \left[e^{\lambda_3 N} - e^{\lambda_1 N} \right]$$

$$D_1 = r_1 \left[\mu e^{\lambda_1 N} - \beta \right]$$

$$D_2 = r_2 \left(\mu e^{\lambda_2 N} - \beta \right)$$

$$D_3 = r_3 \left(\mu e^{\lambda_3 N} - \beta \right)$$

which gives

$$\frac{A_1}{\theta_{fi}} = \frac{C_2 + D_3 - D_2}{C_1 + C_2 + C_3}$$

$$\frac{A_2}{\theta_{fi}} = \frac{C_3 + D_1 - D_3}{C_1 + C_2 + C_3}$$

$$\frac{A_2}{\theta_{fi}} = \frac{C_1 + D_2 - D_1}{C_1 + C_2 + C_3}$$

and

$$R = \frac{A_1}{\theta_{fi}} e^{\lambda_1 N} + \frac{A_2}{\theta_{fi}} e^{\lambda_2 N} + \frac{A_3}{\theta_{fi}} e^{\lambda_3 N}$$

Using $\mu = 0.95$ and $\beta = 0.94$, Tables 7, 8, and 9 show F_k as a function of F' and K' for N values of 0.1, 1 and 2. For this particular example, it can be seen that F_k has a higher variation due to end losses than thermal conductivity. Before finalizing such conclusion it is necessary to examine experimentally the practical values of both μ and β for actual collectors.

Table 7. F_k FOR COLLECTOR WITH END LOSSES, $N = 0.1$, $\mu = 0.95$, $\beta = 0.94$

K' \ F'	0.050	0.100	0.200	0.400	0.600	0.800	0.900	0.950
0.005	0.188	0.196	0.215	0.266	0.349	0.504	0.647	0.752
0.010	0.125	0.131	0.145	0.185	0.253	0.399	0.555	0.681
0.050	0.060	0.064	0.072	0.097	0.143	0.257	0.408	0.553
0.100	0.048	0.051	0.059	0.082	0.125	0.233	0.381	0.529
0.500	0.028	0.034	0.045	0.070	0.112	0.219	0.366	0.514
1.000	0.037	0.041	0.050	0.073	0.115	0.221	0.367	0.513
5.000	0.057	0.059	0.064	0.084	0.124	0.227	0.370	0.514
10.000	0.057	0.060	0.066	0.087	0.126	0.228	0.370	0.514
50.000	0.058	0.061	0.068	0.088	0.128	0.229	0.371	0.514
100.000	0.058	0.061	0.068	0.089	0.128	0.229	0.371	0.514
1000.000	0.058	0.061	0.068	0.089	0.128	0.229	0.371	0.514

8. CONCLUSION

The result of this study show that the end losses from the receiver plate of a collector could be more significant than the losses due to the axial conductivity. The study points out to the need for practical measurements to evaluate such losses.

Of particular emphasize is the necessity of measuring the temperature of the receiver plate at both the fluid inlet and exit locations in order to determine accurately the collector performance parameters.

Table 8. F_k FOR COLLECTOR WITH END LOSSES, $N=1.0$, $\mu=0.95$, $\beta=0.94$.

K' \ F'	0.050	0.100	0.200	0.400	0.600	0.800	0.900	0.950
0.005	0.457	0.456	0.456	0.455	0.458	0.471	0.494	0.529
0.010	0.492	0.491	0.490	0.489	0.492	0.508	0.538	0.582
0.050	0.633	0.630	0.625	0.617	0.616	0.637	0.679	0.734
0.100	0.717	0.714	0.707	0.695	0.690	0.708	0.749	0.800
0.500	0.884	0.881	0.874	0.862	0.852	0.857	0.879	0.901
1.000	0.920	0.918	0.913	0.903	0.894	0.894	0.907	0.920
5.000	0.954	0.953	0.950	0.943	0.936	0.932	0.934	0.936
10.000	0.959	0.957	0.955	0.949	0.942	0.937	0.937	0.938
50.000	0.963	0.961	0.959	0.953	0.947	0.941	0.940	0.940
100.000	0.963	0.962	0.960	0.954	0.947	0.942	0.941	0.940
1000.000	0.964	0.962	0.960	0.954	0.948	0.942	0.941	0.940

Table 9. F_k FOR COLLECTOR WITH END LOSSES, $N=2.0$, $\mu=0.95$, $\beta=0.94$

K' \ F'	0.050	0.100	0.200	0.400	0.600	0.800	0.900	0.950
0.005	0.209	0.210	0.212	0.219	0.231	0.260	0.301	0.356
0.010	0.237	0.238	0.242	0.251	0.267	0.305	0.358	0.427
0.050	0.348	0.350	0.355	0.370	0.398	0.462	0.543	0.634
0.100	0.423	0.425	0.430	0.446	0.478	0.550	0.636	0.723
0.500	0.668	0.667	0.667	0.675	0.700	0.760	0.821	0.869
1.000	0.769	0.768	0.767	0.770	0.786	0.830	0.871	0.901
5.000	0.903	0.902	0.900	0.898	0.901	0.913	0.924	0.932
10.000	0.926	0.925	0.923	0.920	0.921	0.927	0.932	0.936
50.000	0.945	0.944	0.943	0.940	0.938	0.938	0.939	0.939
100.000	0.948	0.947	0.945	0.942	0.940	0.940	0.940	0.940
1000.000	0.950	0.949	0.948	0.944	0.942	0.941	0.940	0.940

ACKNOWLEDGEMENT

The financial support from the University of Petroleum & Minerals, Dhahran, Saudi Arabia, for presentation of this work is acknowledged.

REFERENCES

1.   Shouman, A.R. and Tag, I.A. Method of Testing Solar Collectors for Determination of Heat Transfer Parameters. ASME Paper # 81-WA/Sol-2.

2.   Hottel, H.C. and Whillier, A. 1958. Evaluation of Flat-Plate Solar Heat Collectors. Transactions of Conference on the Use of Solar Energy, University of Arizona, Vol.II, pp.74-104.

3.   Bliss, R.W. 1959. The Derivation of Several Plate Efficiency Factors Useful in the Design of Flat-Plate Solar Heat Collectors. Solar Energy. 3. pp.55-64.

4.   Whillier, A. 1953. Solar Energy Collection and its Utilization for House

Heating.   Sc.D.Thesis.   Massachusetts Institute of Technology.

5.   Whillier, A.   1967.   Low Temperature Engineering Applications of Solar
     Energy.   ASHRAE, New York.

6.   Phillips, W.F.   1979.   The Effects of Axial Conduction on Collector Heat
     Removal Factor.   Solar Energy.   23.   pp.187-191.

The Optimum Flat Plate Solar Collector

KAMAL-ELDIN HASSAN
Mechanical Engineering Department
Tulane University
New Orleans, Louisiana 70118, USA

ABSTRACT

The solar fluid heater problem is formulated as an unsteady, two-dimensional conduction problem. Simplified to a steady, one-dimensional problem provides a direct formulation far more flexible than the formulation hitherto in use, without any loss of generality. This flexibility is used to determine the geometry of optimum collectors, and to determine the performance of fan-shaped ones.

An optimum collector would have a uniform effectiveness along the fluid path and, hence, effect a required fluid temperature rise with the least possible area. A fan-shaped collector of about the same geometrical proportions is shown to be nearly as effective as the corresponding optimum collector. The performance of either shape is determined for certain conditions. It shows that for this case a saving of some 6 to 13 percent could be obtained in comparison with the corresponding usual "parallel-tube" design.

1. INTRODUCTION

A direct formulation used earlier [1] is extended in the present work to the more general case of two-dimensional unsteady conditions. The direct formulation now in use [2,3,4] obscures the parameters that play important roles in the performance of solar fluid heaters [1]. Further, the direct formulation presented is far more flexible without any loss in generality. This flexibility allowed the solution, in the present work, for the case of the optimum solar fluid heater. This optimum collector heats a given fluid to the required temperature with the least surface area. It is found that such a collector is nearly fan-shaped.

For this reason, the performance of the fan-shaped collectors is also determined in this paper. It is shown that the performance of such collectors of a certain geometry approaches that of the corresponding optimum collector and uses considerably less area than the usual "parallel-tube" collector in common use.

2. GENERAL FORMULATION

The general mathematical model for the following discussions is shown in Fig. 1. It is assumed that along the "fin" of the collector, the temperature t is a function of position and time. It is also assumed that the temperature

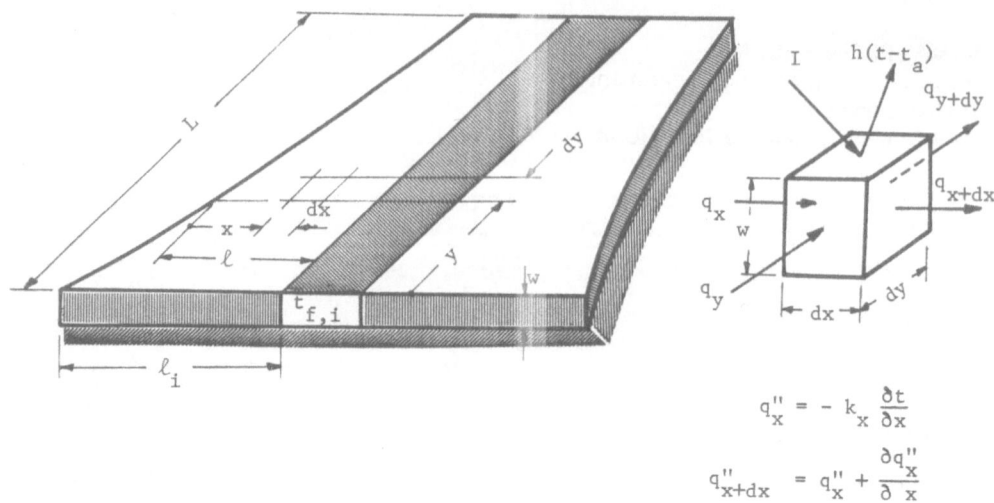

Fig. 1 Mathematical Model (Hatched areas are adiabatic)

is uniform over the thickness of the fin. This is not a serious restriction since fins are usually made quite thin (in the order of 1 mm) of a good conducting material. However, this thickness w may vary as a function of position. A further general assumption is that the thermal conductivity k_x in the x-direction is a constant but may be different from k_y in the y-direction.

Under the above conditions, the general differential equation that governs conduction in the fin would be:

$$\frac{\partial^2 t}{\partial x^2} + \frac{k_y}{k_x}\frac{\partial^2 t}{\partial y^2} - \frac{\rho c_f}{k_x}\frac{\partial t}{\partial \tau} = \frac{h}{k_x w}(t - t_a) - \frac{\alpha I}{k_x w} \tag{1}$$

In the above equation h is the heat transfer coefficient between the fin and its surroundings at temperature t_a.

To complete the formulation of the fluid heater problem, the change of the fluid temperature, $t_f = t_f(y,\tau)$, along its path in the y-direction could be obtained from:

$$\frac{\partial t_f}{\partial y} = \frac{UP}{\dot{m}c}(t_o - t_f) \tag{2}$$

The above two equations, solved simultaneously with appropriate boundary conditions, would give the temperature distribution over the fin and the fluid temperature along its conduit, both as functions of time.

2.1. Heat Transfer Coefficients

The heat transfer coefficients h and U merit a discussion here. The heat transfer coefficient h between the fin surface and surroundings should take into consideration the conductive "back" losses; however, it is essentially due to radiation and natural convection. This makes this heat transfer coefficient a function of the collector plate temperature. Considered as such, equation (1) would be non-linear and difficult to solve. In the following treatment it will be considered as a constant as usually assumed in similar analyses [1,2,3,4,5]. A reasonably good average value of h would not be difficult to obtain from previous experience. However, if the collector temperature range as obtained by calculations proves to be considerably different from that initially assumed, iteration could be made based on the new findings.

The heat transfer coefficient U between the fluid and its conduit is usually assumed constant in heat exchanger calculations. In the present case it is essentially an overall heat transfer coefficient that should take into consideration the geometry of the fluid conduit as related to the fin, the thermal contact resistance between the conduit and collector plate, etc. As such, the evaluation of U should not be difficult or, in most cases, critical. Indeed, it was shown that for the usual flat-plate collectors, the value of U is not critical in most cases [1] and its exact evaluation is not necessary as its value would not affect the final results considerably.

3. TWO-DIMENSIONAL STEADY-STATE CASE

The results of numerical experiment [5] indicate that the transient performance of flat plate solar collectors could be obtained, with an error of less than 5 percent, by considering its behavior as a succession of steady-state conditions. For this reason, the analysis presented here is limited to the steady-state case for which $\partial t / \partial \tau = 0$ and the inlet fluid temperature $t_{f,i}$ a constant.

Further, the assumptions usually made in heat exchanger analyses are upheld, namely:
- Thermal conductivity of the collector material in the fluid flow direction is negligible; i.e., $k_y = 0$.
- The properties and flow (heat and fluid) parameters are constants; these are: k ($=k_x$), h, U, I, α, w, P, ṁ, c and t_a.

The following dimensionless groups are also defined:

$$M = \sqrt{h/kw}\ \ell \tag{3}$$

$$\theta = (t_e - t)/(t_e - t_{f,i}) \tag{4}$$

where t_e is the equilibrium temperature defined by

$$t_e = t_a + (\alpha I/h) \tag{5}$$

$$N_E = U\,P\,/\,\sqrt{hkw} \tag{6}$$

$$X = x/\ell \tag{7}$$

and

$$Y = 2\,U\,P\,y\,/\,\dot{m}\,c \tag{8}$$

With these assumptions and definitions, equation (1) and its usual boundary conditions become [1]*.

$$\frac{\partial^2 \theta}{\partial x^2} - M^2 \theta = 0 \tag{9}$$

$$\frac{\partial \theta}{\partial X}\bigg|_{X=0} = 0 \tag{10}$$

$$\frac{\partial \theta}{\partial X}\bigg|_{X=1} = N_E M (\theta_f - \theta_o)/2 \tag{11}$$

Also, equation (2) and its boundary condition become

$$\frac{d\theta_f}{dY} = (\theta_o - \theta_f)/2$$

$$= -\theta_f / (2 + N_E \coth M) \tag{12}$$

and

$$\theta_f = \theta_{f,i} = 1 \quad \text{at} \quad Y = 0 \tag{13}$$

The general solution for equations (9, 10 and 11) was given before [1] as:

$$\frac{\theta}{\theta_f} = \frac{N_E}{N_E + 2 \tanh M} \frac{\cosh MX}{\cosh M} \tag{14}$$

From this, the reduced temperature θ_o at the fin root (at X=1, the conduit wall) was given as

$$\theta_o = N_E \theta_f / (N_E + 2 \tanh M) \tag{15}$$

In the above relation, θ_f is the local reduced fluid temperature.

The local flux per unit length in the fluid flow (y-) direction is given by

$$q' = U P (t_o - t_f)$$

$$= 2 U P \theta_f (t_e - t_{f,i})/(2 + N_E \coth M) \tag{16}$$

We define here a <u>local</u> collector effectiveness as:

---

* The definition of Y here is different from that used in [1].

$$\eta_\ell = q'/2 \ I \ \ell \tag{17}$$

From this, a "local reduced effectiveness E '' may be defined as:

$$E_\ell = \eta_\ell / N_I = N_E \ \theta_f/M \ (2 + N_E \ \coth M) \tag{18}$$

In this relation

$$N_I = h \ (t_e - t_{f,i})/I \tag{19}$$

$$= 1 \qquad for \qquad t_{f,i} = t_a \qquad and \qquad \alpha = 1$$

Again, we may define a "collector effectiveness η_c" for a length y of the collector as:

$$\eta_c = \dot{m} \ c \ (t_f - t_{f,i})/2 \ I \ {}_0\!\!\int^y \ell \ dy \tag{20}$$

A corresponding "reduced collector effectiveness E_c" would be:

$$E_c = \eta_c/N_I = N_E \ (1 - \theta_{f,Y})/ \ {}_0\!\!\int^Y M \ dY \tag{21}$$

The above relations, in particular equations (12, 16, 18 and 21) could be used to design a collector according to any given set of conditions. In this respect, it should be noticed that $M/M_1 = \ell/\ell_1$. Solutions were obtained for θ , M/M_1, E_ℓ and E_c as functions of the reduced distance Y along the flow direction.

Further, it may be noticed that equation (4) shows that θ_f has a maximum value of unity at the fluid inlet and decreases as t_f increases along the flow direction. For this reason, it is more convenient to have the reduced fluid ttemperature plotted as $T_f = 1 - \theta_f = (t_f - t_{f,i})/(t_e - t_{f,i})$ rather than θ_f itself.

The flexibility of the above formulation is used in the following parts for determining the geometry and/or performance of flat plate solar collectors for the following cases:

- The optimum collector of minimum area for a given temperature rise.

- Fan-shape.

4. OPTIMUM COLLECTOR

An optimum collector would have the least collecting area for a given service. Since the criterion for the utilization of the surface area is the effectiveness, an optimum collector should have a constant local effectiveness E_ℓ all along its length. This effectiveness is expressed by equation (18). If the value of E_ℓ has to be constant all along the collector including the inlet section at which $\theta_f = 1$ and $M = M_1$, then equation (18) gives

$$\frac{N_E\, \theta_f}{M(2 + N_E \coth M)} = \frac{N_E}{M_i(2 + N_E \coth M_i)}$$

or
$$\theta_f = \frac{2 + N_E \coth M}{2 + N_E \coth M_i}\, \frac{M}{M_i} \tag{22}$$

The fluid temperature change along the flow path is governed by equation (12) which was solved numerically by finite differences. The finite difference form used is as follows:

$$Y_{j+1} = Y_j + (1 - \frac{\theta_{f,j+1}}{\theta_{f,j}})(2 + N_E \coth M_j) \tag{23}$$

The solution to this problem as obtained from the above two equations for $N_E = 10$ is represented graphically in Fig. 2 with M_i as parameter. As could be seen from this figure, the pitch between two adjacent fluid conduits (2ℓ) drops rapidly in the Y-direction and finally becomes asymptotic to zero. Accompanied with that, the fluid temperature rises also rapidly but levels off at some value that increases with the value of M_i. An increase in M_i which, under given conditions, reflects an increase in ℓ_i, is accompanied by a decrease

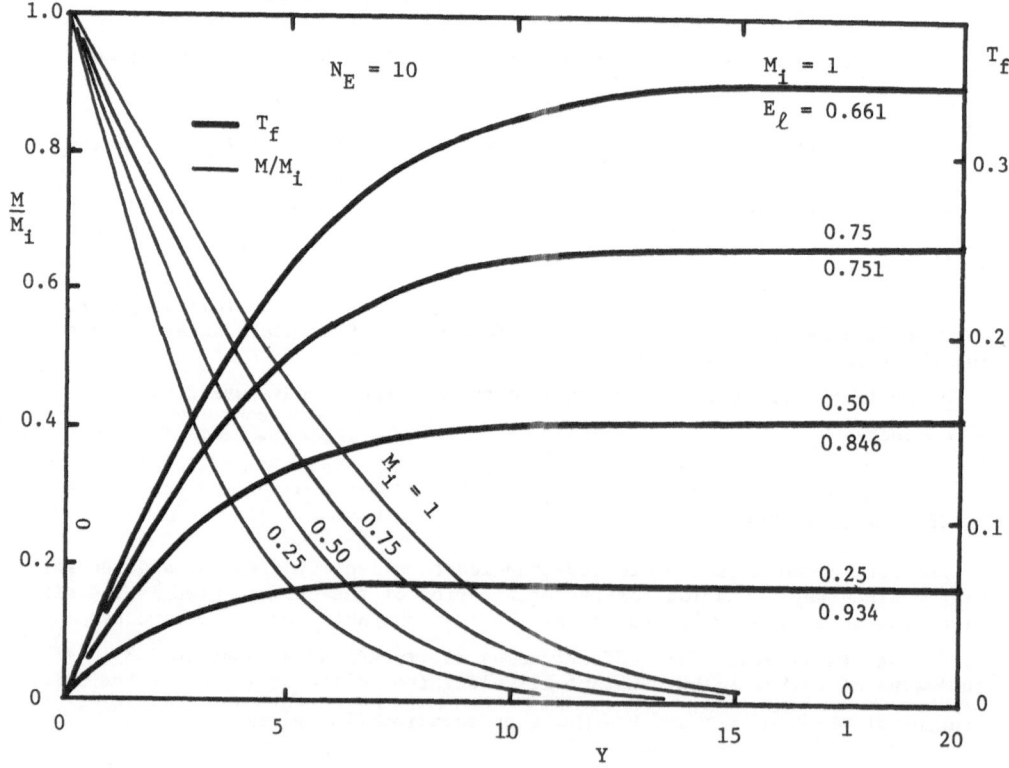

Fig. 2   Performance of Optimum Solar Collectors

in the reduced effectiveness E (in this case $E_c = E_\ell$). The decrease in the
fluid temperature rise that accompanies an increase in effectiveness might seem
paradoxical. However, an increase in the value of M_i indicates an increase in
the collector area which is not used as efficiently as a smaller one.

5. FAN-SHAPED COLLECTOR

It may be seen from Fig. 2 that the variation of M/M_i ($=\ell/\ell_i$) with Y is
nearly linear down to, say, $M/M_i \simeq 0.3$. This indicates that a fan-shaped collector would behave very nearly as the optimum collector while being far easier
to manufacture. For this reason this shape, the fan-shape, is studied in this
part.

A fan-shaped collector would physically be as shown in Fig. 3, and its
geometry would be represented mathematically by:

$$\ell_i - \ell = K y$$

In this relation, K represents the "gradient" of the conduits pitch as it decreases in the flow direction along the "fan radius". Using equations(3 and 7),
it could be stated in the following dimensionless form.

$$1 - (M/M_i) = Y/(2 N_K) \tag{24}$$

In this relation, the dimensionless group N_K is given by:

$$N_K = U P \ell_i / K \dot{m} c \tag{25}$$

Substitution from equation(24) into equation (12) gives

$$\frac{d\theta_f}{d M} = \frac{N_K}{M_i} \frac{2 \theta_f}{2 + N_E \coth M} \tag{26}$$

The boundary condition for this differential equation is

$$\theta_f = 1 \qquad \text{at} \qquad M = M_i \tag{27}$$

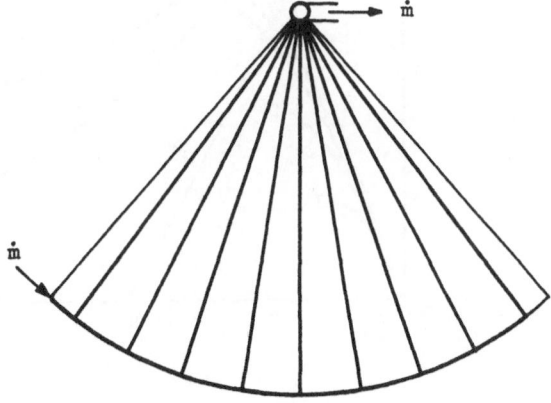

$\dot{m}$

$\dot{m}$

Fig. 3 Geometry of Fan-
        Shaped Collectors

The solution would then be

$$\theta_f = \left[\frac{2 \sinh M + N_E \cosh M}{2 \sinh M_i + N_E \cosh M_i}\right]^{\frac{2 N_E}{N^2 - 4} \frac{N_K}{M_i}} \exp\left[\frac{4 N_K}{N_E^2 - 4}\left(1 - \frac{M}{M_i}\right)\right] \quad (28)$$

The above relation is represented graphically in Fig. 4 for $N_E = 10$ and $M_i = 0.5$ and various values of the parameter N_K, including the case of parallel tubes where $K = 0$ and N_K, therefore, infinite. Except for the last case, all

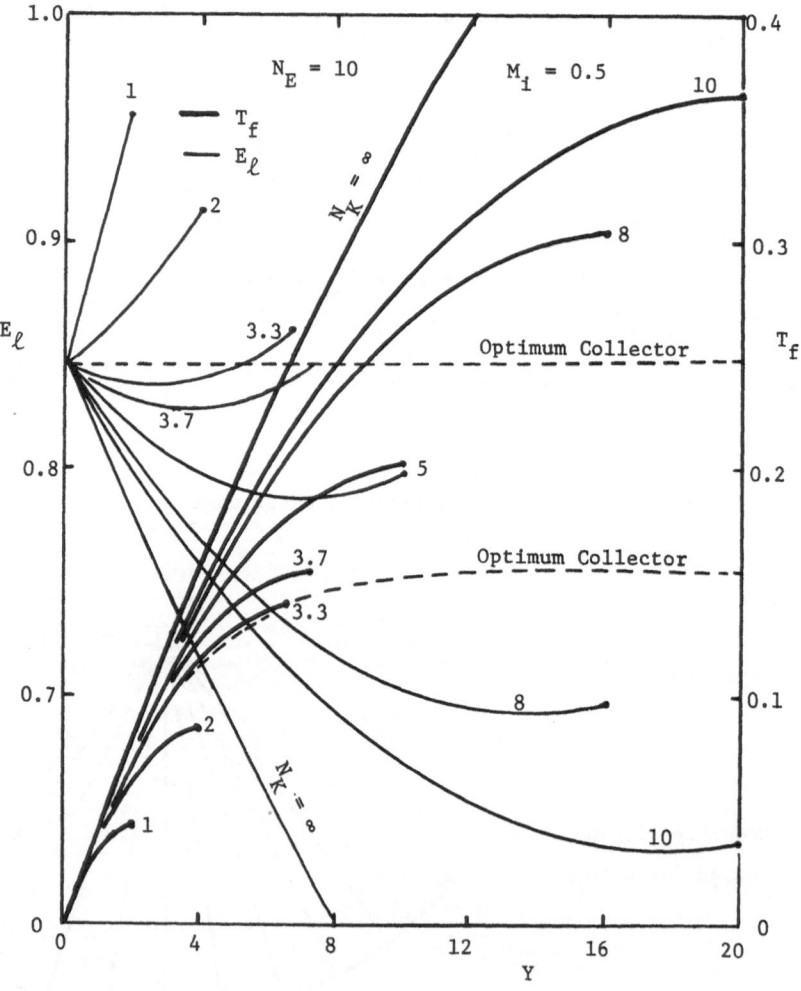

Fig. 4   Performance of Fan-Shaped Collectors

the collectors are restricted within a range of Y between 0 and $2N_K$ at the end
of which the fan-shape converges to a point. The figure shows also the variation
of the local reduced effectiveness E_ℓ with the reduced distance Y. Also plotted
as dashed lines are the temperature distribution and reduced effectiveness of
the corresponding optimum collector; the latter is a horizontal straight line
at E_ℓ = 0.846.

The curves for fan-shaped collectors could be divided into two general
groups according to the values of N_K. When compared with the optimum collector,
the fan-shaped collectors with low values of N_K have higher local effectiveness
values but lower end temperatures due to the inadequacy of area. The ones with
lower values of N_K have lower effectiveness values and attain higher end temperatures. Of particular interest are the collectors with N_K = 3.3 and 3.7 .
The former has its temperature line very near to that of the optimum collector
and, indeed, its average local effectiveness is close to the optimum. Also,
its slope K is the nearest to the straight-line part of the optimum collector
as shown in Fig. 2. Its end reduced temperature is, however, considerably
lower than that of the optimum collector. On the other hand, the fan-shaped
collector with N_K = 3.7 attains the end temperature of the optimum collector
but with considerably lower effectiveness.

The above discussion asserts the meaning of "optimum collector" as originally defined; it can heat the fluid to a required temperature with the least surface area and, hence, weight. This is illustrated clearly in Fig. 5 in which
the fluid reduced temperature rise T_f and the overall reduced effectiveness E_c
are plotted versus the reduced area A "scanned by the fluid". In this figure.
the reduced area A is evaluated by virtue of equation (24) as:

$$A = \int_0^Y M \, dY = M_i \, Y \left(1 - \frac{Y}{4N_K}\right) \tag{29}.$$

The value of the reduced effectiveness E_c is evaluated from equation (21) by
substituting for the value of the integral from equation (29); it gives

$$E_c = N_E (1 - \theta_{f,Y})/A \tag{30}.$$

The plots of Fig. 5 are similar to those of Fig. 4 and support the conclusions
obtained above. Indeed, for the particular case of N_E = 10 and M_i = 0.5, a
saving of area, as compared to the parallel-tube collector that gives the same
temperature rise, ranges between 6 percent for N_K = 5 and 13.5 percent for
N_K = 1.

It may have been noted also that the difference between the optimum collector and the optimum fan-shaped collector (for N_K in the range of 3.3 and 3.7
for the case presented) is very small and would easily be outweighed by the simplicity of the fan-shape geometry. The proportions of the optimum fan-shaped
collectors could be estimated from the geometry of the corresponding optimum
collectors of the same N_E and M_i as shown in Fig. 2. As mentioned earlier, the
curves of M/M_i versus Y in this figure are practically straight lines in the
region $0.3 < M/M_i < 1$. This value could be used to calculate the optimum

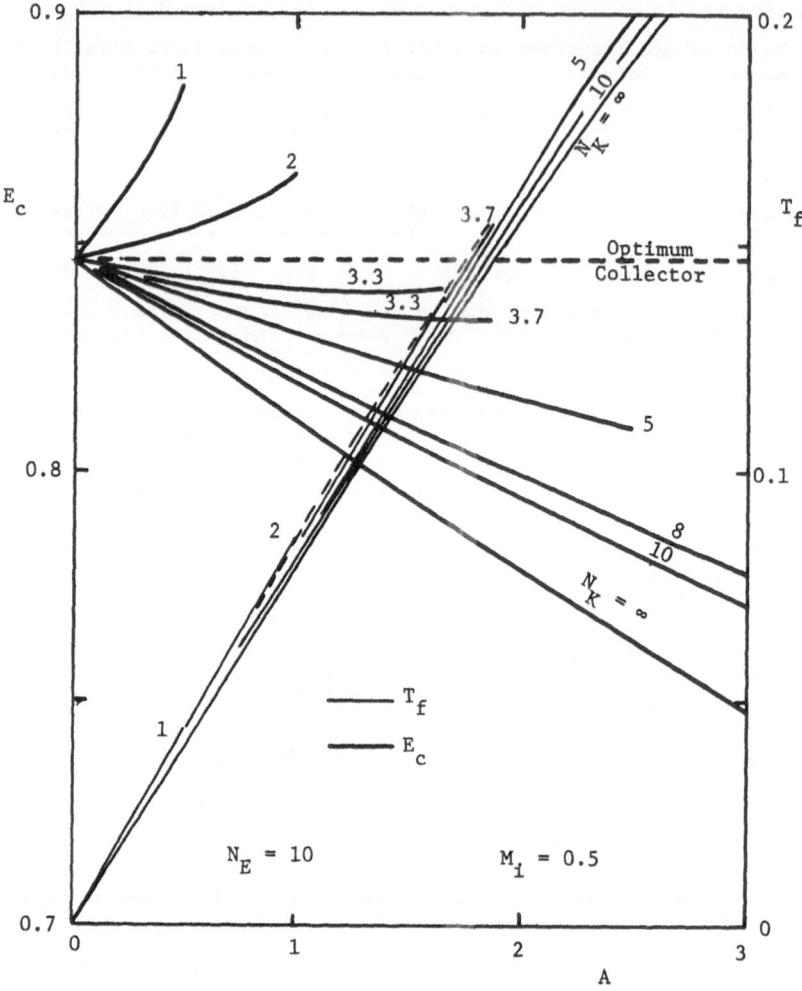

Fig. 5   Performance of Fan-Shaped Collectors

value of the parameter N_K for a given value of M_i which could be shown in the present case of $N_E = 10$ and M_i less than unity to be given by

$$N_{K,opt} = 1.8 + 3.1\ M_i \tag{31}.$$

6. CONCLUDING REMARKS

In the present work a new, direct and general formulation for the solar fluid heaters problem is simplified to the case of steady one-dimensional conduction and used to obtain the performance of collectors set to satisfy predetermined conditions. It was used to determine the form of the optimum solar collector that has the least surface area for a given fluid temperature rise. It was also shown that this optimum performance could be approached very closely by the simple fan-shape geometry. For a particular case, a saving of some 6 to 13 percent in the area could be effected through the use of the optimum shape, or the corresponding fan-shape. Indeed, this saving of area would be reflected practically in a saving of cost and space. The latter could be attained by proper positioning of the individual collectors in a battery by, say, positioning the "center" of one adjacent to the "rims" of adjacent ones.

Further, the new formulation allows for flexibility in design. Indeed, it could be used to determine the performance of collectors of almost any geometry. It could, therefore, be used for the "tailoring" of solar collectors to meet specific needs: fill a certain space, fit into an architectural theme, etc.

NOMENCLATURE

Symbols

A	reduced area defined by equation (29), dimensionless
c	specific heat, J/kg K
E	$= \eta/N_I$, reduced effectiveness, dimensionless
h	heat transfer coefficient between fin and ambient air, W/m^2 K
I	insolation, W/m^2
K	$= (\ell_1 - \ell)/y$
k	thermal conductivity, W/m K
L	length of collector, m
ℓ	fin width, m
M	dimensionless parameter defined by equation (3)
$\dot{m}$	rate of mass flow, kg/s
N_E	dimensionless parameter defined by equation (6)
N_I	dimensionless parameter defined by equation (19)
N_K	dimensionless parameter defined by equation (25)
P	perimeter, m
q	heat flux, W
T	$= 1 - \theta$
t	temperature, K
U	overall heat transfer coefficient between fluid and fin base, W/m^2 K
w	fin thickness, m
X	$= x/\ell$, dimensionless
x	distance from fin edge, m
Y	dimensionless distance defined by equation (8)
y	distance in flow direction, m
α	absorptivity, dimensionless

η effectiveness, dimensionless

θ reduced temperature, defined by equation (4)

ρ density of fin material, kg/m^3

τ time, s

Superscripts

'   per unit length

"   per unit area

Subscripts

a   ambient

c   collector

e   equilibrium

f   fluid

i   inlet

j   node indication

ℓ local

o   at fin root

opt   optimum

x   in x-direction

Y   at Y

y   in y-direction

REFERENCES

1. Hassan, K. and Abughres, S.M. 1978. Thermal performance of single-pass solar fluid heaters. Proceedings of the 2nd International Solar Forum, Hamburg, Vol. 1, pp. 569-580.

2. Hottel, H.C. and Whillier, A. 1958. Evaluation of flat-plate solar collectors performance. Trans. of Conference on the Use of Solar Energy, University of Arizona, Vol. 2, pp. 74-104.

3. Bliss Jr., R.W. 1959. The derivation of several plate efficiency factors useful in the design of flat-plate solar heat collectors. Solar Energy, Vol. 3, No. 4, pp. 55-64.

4. Parker, B.F. 1981. Derivation of efficiency and loss factors for solar air heaters. Solar Energy, Vol. 26, No. 1, pp. 27-32.

5. Klein, S.A., Duffie, J.A. and Beckman, W.A. 1974. Transient considerations of flat-plate solar collectors. J. of Engineering for Power, Trans. of the ASME, Series A, Vol. 96, pp. 109-113.

Effects of Radiation Losses on the Performance of Solar Collectors

ISMAIL A. TAG
Qatar University
Doha, State of Qatar

AHMAD R. SHOUMAN
University of Petroleum and Minerals
Dhahran, Saudi Arabia

ABSTRACT

As a result of the recent increase in the cost of conventional energy sources, a renewed interest in solar energy has led to the examination of the performance of solar collectors. The present manuscript focuses upon the influence of radiation-losses on the performance of solar collectors. A mathematical model, neglecting axial conductivity and normal resistance in the collector, is utilized. This resulted in a first order nonlinear ordinary differential equation in fluid temperature. A closed form solution for the equation is presented. A comparison of the magnitude of the error resulting from the linearization of the radiation losses is derived.

1. INTRODUCTION

As a result of the recent increase in the cost of conventional energy sources, a renewed interest in solar energy research has led to closer examination of the performance of solar collectors-for example, Phillips (1) has recently extended the Hottel-Whillier-Bliss (2,3,4) (HWB) model of flat-plate solar collectors to include the effect of axial conductivity of the absorber plate. He concluded that the error introduced by neglecting axial conduction was less than 30 percent for all collectors. More recently Shouman and Tag (5) showed that the influence of the end temperatures of the receiver plate of a collector is more significant than the losses due to the axial conductivity of an insulated end collector plate; however both axial conductivity and end effects can be neglected with minimum loss in accuracy. This renders greater importance to the HWB model.

Ismail A. Tag is on leave from Oak Ridge National Laboratory, Oak Ridge, TN 37830, USA. Ahmad R. Shouman is on leave from New Mexico State University, Las Cruces, NM 88003, USA.

In light of the above conclusion, the present manuscript focuses upon the influence of radiation on the performance of solar collectors, a phenomena that has more effect in the operation of concentrating collectors.

A one dimensional equation for the performance of solar collectors is presented and solved exactly with and without convection effect. The solution leads to an integral that is evaluated using the second order highly accurate Gaussian quadrature method.

2. MATHEMATICAL FORMULATION

The axial temperature distribution in a collector is expressed by the one-dimensional equation (energy balance)

$$\dot{m}C_p \frac{dT}{dx} + \frac{U_o A}{L} (T - T_a) + \frac{\sigma \varepsilon A_r}{L} (T^4 - T_a^4) = \frac{\alpha_e A_r q_s}{L} \tag{1}$$

Equation (1) is subject to the following initial condition

$$T = T_i \text{ at } x = 0$$

In Equation (1) it is assumed that the resistance is negligible in the transverse direction while thermal conductivity is zero in the axial direction.

Defining T_m by

$$q_s = \frac{U_o}{\alpha_e} (T_m - T_a) + \frac{\sigma \varepsilon}{\alpha_e} (T_m^4 - T_a^4) \tag{2}$$

if

$$\theta = \frac{T}{T_m} \tag{3}$$

Let

$$N = \frac{\sigma \varepsilon T_m^3 A_r}{\dot{m}C_p} \quad , \tag{4}$$

$$N_x = N \frac{X}{L} \quad , \tag{5}$$

and $r = \dfrac{U_o}{\sigma \varepsilon T_m^3}$ \hfill (6)

Thus, Equation (1) reduces to

$$\frac{d\theta}{dN_x} + r\ (\theta - 1) + (\theta^4 - 1) = 0 \tag{7}$$

and

$$\theta = \theta_i \quad \text{for } N_x = 0$$

3.   SOLUTION IN ABSENCE OF CONVECTION

In the absence of convection $(r = 0)$, Equation (7) can be integrated to give

$$N_x = \frac{1}{4}\ \ell n\ \frac{(1 + \theta)\ (1 - \theta_i)}{(1 - \theta)\ (1 + \theta_i)} + \frac{1}{2}\ (\tan^{-1}\theta\ - \tan^{-1}\theta_i) \tag{8}$$

Now let

$$\Phi(\theta) = \frac{1}{4}\ \ell n\ \frac{1 + \theta}{1 - \theta} + \frac{1}{2}\ \tan^{-1}\theta \tag{9}$$

The solution then becomes

$$N_x = \phi(\theta) - \phi(\theta_i) \tag{10}$$

with exit condition obtained from

$$N \doteq \phi(\theta_e) - \phi(\theta_i) \tag{11}$$

and the collector efficiency for this case can be calculated from

$$\eta = \frac{\theta_e - \theta_i}{N(1 - \theta_a^4)} \tag{12}$$

4.   THE GENERAL PROBLEM

Let

$$\phi(\theta,r) = \int_0^\theta \frac{d\theta}{(1 - \theta^4) + r(1 - \theta)}$$

$$= \frac{1}{4 + r}\ \int_0^\theta \frac{3 + 2\theta + \theta^2}{r+1+\theta+\theta^2+\theta^3}\ d\theta - \ell n\ (1 - \theta) \tag{13}$$

Then, the solution in presence of convection and radiation terms is given by

$$N_x = \phi(\theta,r) - \phi(\theta_i,r) \tag{14}$$

with the exit condition given by

$$N = \phi(\theta_e, r) - \phi(\theta_i, r) \tag{15}$$

and the collector efficiency is given by

$$\eta = \frac{\theta_e - \theta_i}{N\{1 - \theta_a^4 + r(1 - \theta_a)\}} \tag{16}$$

Once η is calculated, the familiar heat removal factor F_R is calculated from the following equation:

$$F_R = \eta\left[\frac{1 - \theta_a^4 + r(1 - \theta_a)}{1 - \theta_a^4 + r(1 - \theta_i)}\right] \tag{17}$$

For the sake of comparison, and to complete the solution, we shall examine the case of the absence of radiation. The solution can be written as

$$N_x = \frac{1}{r} \ln \frac{1 - \theta i}{1 - \theta} \tag{18}$$
$$= \frac{1}{r}\left[\ln \frac{1}{1 - \theta} - \ln \frac{1}{1 - \theta_i}\right]$$

5.   RESULTS AND DISCUSSION

Table 1 shows $\phi(\theta, r)$ for different values of θ and r. The same results are shown graphically in Figure 1. Superimposed on it is the function $\ln \frac{1}{1 - \theta}$ to handle the linear solution in absence of radiation or the linearized solution in the presence of radiation.

6.   EXAMPLE ON THE USE OF THE RESULTS

A flat-plate collector with a mass flow rate per unit area of 50 kg/hrm^2 has an inlet temperature of 20°C. The solar flux on the collector plate is 1,000 w/m^2. Assuming U_0 of 30 w/m^2 ^{o}K, $\alpha e = 0.9$, examine the influence of including radiation losses on the temperature leaving the collector for the cases of $\varepsilon = 0.9$ and 0.1. Ambient temperature is 20°C.

Table 2 gives the solution for the linear problem in the absence of radiation, linearized radiation, and the nonlinear radiation for emissivities $\varepsilon = 0.9$ and 0.1.

TABLE 1 φ AS A FUNCTION OF r AND θ

r \ θ	0.000	0.100	0.500	1.000	5.000	10.000	50.000	100.000
0.000	0.000	0.000	0.000	0.000	0.000	0.000	0.000	0.000
0.050	0.050	0.046	0.034	0.025	0.009	0.005	0.001	0.001
0.100	0.100	0.091	0.068	0.051	0.017	0.010	0.002	0.001
0.150	0.150	0.137	0.103	0.078	0.027	0.015	0.003	0.002
0.200	0.200	0.184	0.138	0.105	0.036	0.020	0.004	0.002
0.250	0.250	0.230	0.174	0.134	0.047	0.026	0.006	0.003
0.300	0.300	0.277	0.211	0.163	0.058	0.032	0.007	0.003
0.350	0.351	0.324	0.249	0.193	0.069	0.038	0.008	0.004
0.400	0.402	0.372	0.287	0.224	0.081	0.045	0.010	0.004
0.450	0.454	0.421	0.327	0.256	0.094	0.053	0.012	0.005
0.500	0.506	0.471	0.369	0.290	0.108	0.061	0.013	0.006
0.550	0.561	0.522	0.412	0.326	0.123	0.070	0.016	0.007
0.600	0.617	0.576	0.457	0.364	0.140	0.079	0.018	0.008
0.650	0.676	0.633	0.505	0.405	0.158	0.090	0.020	0.009
0.700	0.739	0.693	0.558	0.450	0.179	0.103	0.023	0.010
0.750	0.808	0.760	0.616	0.500	0.203	0.117	0.027	0.012
0.800	0.887	0.836	0.683	0.558	0.231	0.134	0.031	0.014
0.850	0.980	0.926	0.763	0.629	0.267	0.156	0.036	0.016
0.900	1.102	1.045	0.869	0.723	0.316	0.187	0.044	0.019
0.950	1.296	1.233	1.039	0.874	0.396	0.238	0.057	0.022
0.990	1.713	1.640	1.409	1.205	0.578	0.354	0.087	0.045

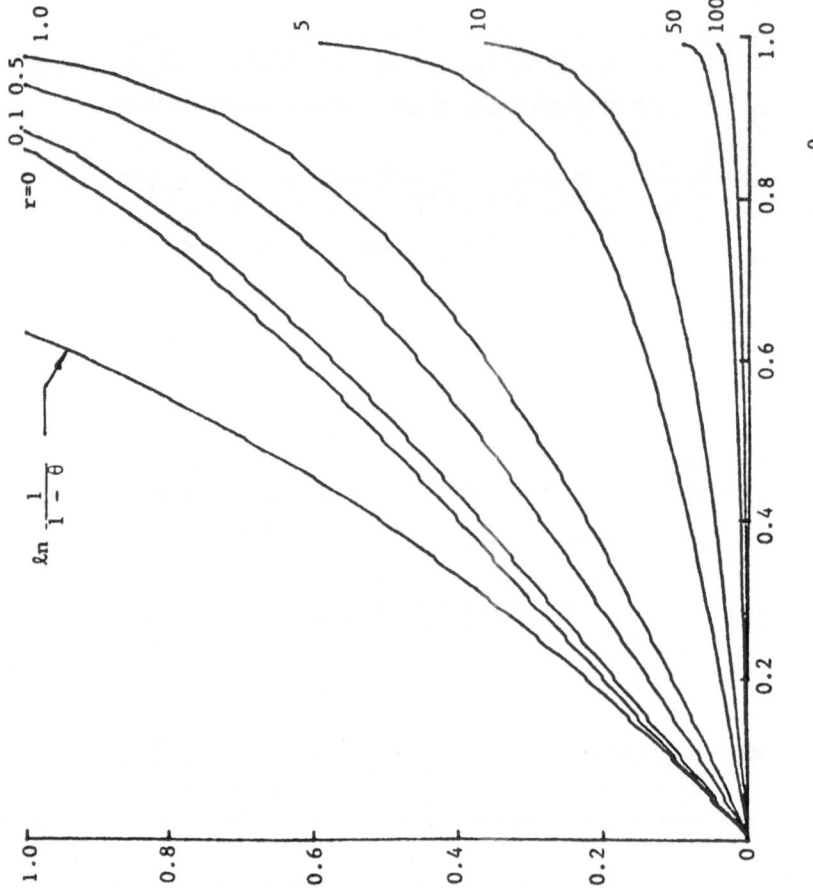

Figure 1. Φ versus θ for different r values

TABLE 2

	Linear	Linearized $\varepsilon = 0.1$	Exact	Linearized $\varepsilon = 0.9$	Exact
T_m $^\circ K$	323.	322.35	322.35	318.1	318.1
θ_i	0.907	0.9089	0.9089	0.9211	0.9211
θ_e	0.945	0.9463	0.9454	0.9574	0.9287
T_e $^\circ K$	305.09	305.04	304.76	304.55	295.42
T_e $^\circ C$	32.09	32.04	31.76	31.55	22.42

7. CONCLUSION

The exact solution for the one-dimensional heat transfer in solar collectors has been presented. The solution is analogous to the Hottel-Whiller-Bliss (HWB) model with consideration of the nonlinear radiation loss. Solution leads to a simply defined integral function that is evaluated and tabulated. The study shows that linearizing radiation does not fully account for radiation losses as determined from the exact solution, particularly for collectors with high emissivities. The exact solution can easily be utilized for determining collector performance. It should particularly be emphasized that the effect of radiation losses would be substantial for concentrating collectors and solar receivers operating at relatively high temperatures.

NOMENCLATURE

A_r collector receiver area
c_p fluid constant pressure specific heat
F_R heat removal factor
L axial length of collector
$\dot{m}$ rate of mass flow of the fluid
N parameter defined by Equation (4)
N_x parameter defined by Equation (5)
q_s solar flux in the collector aperture
r_s parameter defined by Equation (6)
T fluid temperature
T_a ambient temperature
T_e exit fluid temperature
T_i inlet fluid temperature
T_m parameter defined by Equation (2)
U_o overall heat transfer coefficient between the transfer fluid and the ambient
x axial coordinate
α_e effective transmittance-absorptance product
ε effective emissivity of collector plate
η collector thermal efficiency
θ parameter defined by Equation (3)
      i, e. -- subscripts for inlet and outlet locations
σ Stefan-Boltzman constant
Φ parameter defined by Equations (9) and (13)

REFERENCES

1. Phillips, W.F. 1979. The Effects of Axial Conduction on
    Collector Heat Removal Factor, Solar Energy, Vol. 23, pp.
    187-91.

2. Hottel, H.C. and Whillier, A. 1958. Evaluation of Flat-Plate
    Solar Heat Collectors, in Proc. Conf. on the Use of Solar
    Energy, Vol. 2, University of Arizona, p. 74.

3. Bliss, R.W. 1959. The Derivation of Several Plate Efficiency
    Factors Useful in the Design of Flat-Plate Solar Heat
    Collectors, Solar Energy, Vol. 3, p. 55.

4. Whillier, A. 1953. Solar Energy Collection and Its
    Utilization for House Heating, Sc. D. Thesis, Massachusetts
    Institute of Technology.

5. Shouman, A.R. and Tag, I.A. 19-24 April, 1982, General
    Solution of Collector Performance with Axial Conduction and
    End Effects, Proc. 16th Southeastern Seminar on Thermal
    Sciences, Clean Energy Research Institute, University of
    Miami.

6. Duffie, J.A. and Beckman, W.A. 1980. Solar Engineering of
    Thermal Processes, John Wiley and Sons.

7. Cristy, G.A. Private Communication.

Reversed Flat Plate Collector Having Two Absorbing Surfaces

PRIDA WIBULSWAS and SUKON AJLITHI
King Mongkuts Institute of Technology-Thonburi
Bangmod, Bangkok 14, Thailand

ABSTRACT

A reversed flat plate collector was developed from a flat plate collector having an absorbing area of 0.77 m^2 per surface. There were two transparent glass covers above the absorbing plate and one glass cover underneath. The absorbing plate consisted of copper tubes and fins painted matt black on both surfaces. A half cylindrical reflecting surface made of stainless steel reflected the solar radiation on the bottom surface of the absorbing plate. The collector faced to the south with a tilt angle of 14 degrees.

In order to carry out performance tests, a 40 litre storage tank was connected to the reversed flat plate collector. Circulation of the hot water in the system was achieved by natural convection. Short term test results show that the solar water heating system using the reversed flat plat collector had higher efficiency than those of the conventional single glazed and double glazed flat plate collectors.

The total first cost of the system was US $ 210 and the energy cost for the assumed life time of 15 years was about US ¢ 6/kWh at the rate of interest of 15 % per annum.

1. INTRODUCTION

Since the last few years, installation of flat plate solar collectors
for water heating has increased rapidly at a rate of over 100% per annum. It
is expected that in 1982, more than 10,000 sq.m. of flat plate collectors will
be installed mainly on houses, hotels and hospitals. As all locally produced
flat plate collectors have single-glazed covers and no selective coating, they
economically produce hot water at a temperature below 70 C. For applications
at higher temperatures such as solar refrigeration and air conditioning,
industrial process heat, etc., double-glazed collectors with selective coating
have proved to be more economical than single-glazed collectors by about 20%[1].
Reversed flat plate collectors developed in Russia and Japan[2] have also
indicated higher efficiencies than the single-glazed collectors. In this paper,
development of a reversed flat plate collector having two absorbing surfaces is
presented.

2. PRINCIPLE

The main objective of the development of reversed flat plate collectors in Russia and Japan is to reduce the thermal loss by reversing the absorbing surface downward. The top insulated surface of the absorbing plate facing
upward, loses very little heat to surrounding . A cylindrical surface is
provided for reflecting solar radiation to the reversed absorbing surface.

As the atmospheric temperature in Thailand is rather high, by replacing the top insulator with two transparent glass covers, the amount of solar
radiation absorbed by the top surface of the absorbing plate should be larger
than the heat lost to surrounding.

Useful heat gain from the collector may be written as

$$Q_u = A_c \left[I(\tau \rho \gamma \alpha)_e + I(\tau \alpha)_e - (U_b + U_t)(T_{pm} - T_a) \right] \quad \ldots \ldots (1)$$

where A_c = area of the absorbing plate,

= area of the transparent cover for the reflecting surface,

I = solar radiation intensity,

$(\tau \rho \gamma \alpha)_e$ = effective products of transmittance, reflectance, intercept
factor and absorbtance for the bottom surface of the
absorbing plate,

$(\tau \alpha)_e$ = effective transmittance absorbtance products for the top
surface of the plate,

Fig. 1a Reversed flat plate collector developed in Japan

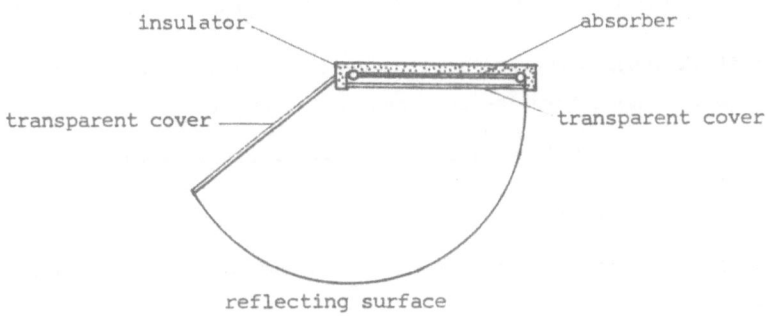

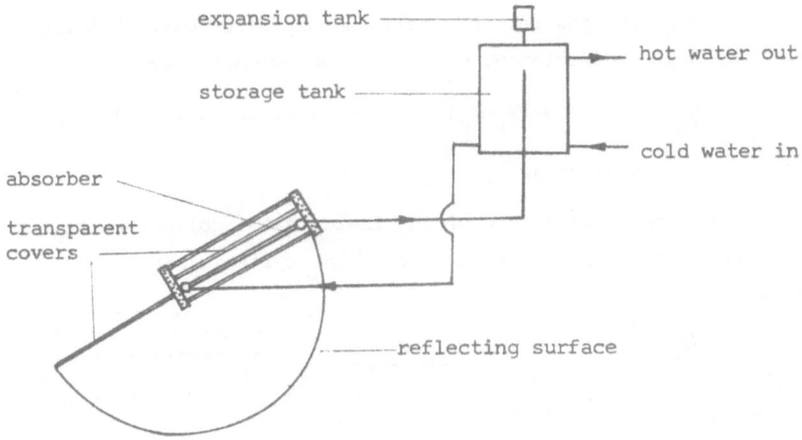

**Fig. 1b Diagram of solar water heating system using a
reversed flat plate collector having two
absorbing surfaces.**

$$U_b \quad = \quad \text{bottom heat loss coefficient,}$$

$$U_t \quad = \quad \text{top heat loss coefficient,}$$

$$T_p \quad = \quad \text{mean temperature of the absorbing plate,}$$

$$T_a \quad = \quad \text{ambient temperature.}$$

Collector efficiency is given by the equation

$$\eta_c \;=\; (\tau\rho\gamma\alpha)_e + (\tau\alpha)_e - (U_b{+}U_t)\,(T_p{-}T_a)/I \quad\cdots\cdots\cdots(2)$$

If it is assumed that optical properties and heat loss coefficients hardly change with temperature, the collector efficiency,

$$\eta_c \approx A + B\,(T_p{-}T_a)/I \qquad\cdots\cdots\cdots\cdots(3)$$

where A,B = constants.

The above equation is therefore in the form of Hottel–Whillier–Bliss equation[3].

A solar water heating system can be constructed by connecting the collector to a storage tank. If heat losses from the storage tank and conecting pipes are assumed constant and small in comparison with that from the collector, the system efficiency, η_s may be approximately written as

$$\eta_s \simeq C + D\,(T_p{-}T_a)/I \qquad\cdots\cdots\cdots\cdots(4)$$

where C,D = constants.

If the mean temperature of the fluid in the collector, t_{fm} is used instead of the mean plate temperature, t_p , the system efficiency may be expressed as

$$\eta_s \;=\; E + F\,(T_{fm}{-}T_a)/I \qquad\cdots\cdots\cdots\cdots(5)$$

where E,F = constants.

In practice, the useful heat gain, Q_u can be estimated from a number of measured quantities appearing in the following equation,

$$Q_u \;=\; \sum^{n} m{\cdot}C(T_{so}{-}T_{si}) + MC\,(T_{sf}{-}T_{si}) \qquad\cdots\cdots\cdots(6)$$

where m = mass of hot water drawn from the storage tank at a time,

  C = specific heat of water,

  n = number of times when hot water is drawn out during the test period,

T_{si}, T_{so}, T_{sf} =   inlet, outlet and final temperatures of hot water
in the storage tank,

M   =   final mass of water in the tank.

The system efficiency,   may also be estimated from

$$\eta_s \quad = \quad Q_u / 2A_c I \qquad \dots\dots\dots\dots\dots(7)$$

3. ECONOMIC ASSESSMENT

Economic analysis by the annual cost method can be employed to demonstrate the economic viability of the solar water heating system using the reversed flat plate collector.

Annual cost of a system[4]

= annual first cost + annual operating cost, AOC

- annual salvage value,

= P (CRF) + AOC + S.(SFF) $\dots\dots\dots\dots\dots(8)$

where P   = first cost of the system,

CRF = capital recovery factor,

= $i (1 + i)^n / \left[(1 + i)^n - 1\right]$

i   = rate of interest,

n   = useful life time of the system,

S   = salvage value,

SFF = sinking fund factor,

= $1 / \left[(1 + i)^n - 1\right]$.

4. TEST EQUIPMENT

The constructed reversed flat plate collector had a fin-and-tube absorbing plate of 1.1 x 0.7 sq.m. Ten copper tubes of 12.5 mm dia. were brazed to eleven copper fins being 0.15 m wide and 0.6 mm thick. Two transparent glass covers, each being 3 mm thick, were placed above the absorbing plate at distances of 25 and 50 mm from the plate. One transparent glass pane was also placed underneath the plate at a distance of 25 mm. The cylindrical reflecting surface was made of polished stainless steel. Another transparent glass pane was placed at the aperture of the reflecting surface.

To form a thermosyphon water heating system, the reversed flat plate collector was connected to a 40 litre storage tank. The glass fibre insulator around the tank was 25 mm. thick. Temperatures of the absorbing plate and of water inside and at the inlet and outlet of the storage tank were measured by means of thermocouples and a digital indicator. A solarimeter was used for recording solar radiation with an accuracy within 2.5% .

During the test runs, the flat plate collector was tilted at an angle of 14 degrees to the horizontal which was the lattitude of the test location.

5. TESTS AND RESULTS

5.1 Tests

Each day during the test period, an equal amount of hot water was drawn from the storage tank and replaced by cold water at every hour from 8.00 - 18.00 hr. While the test period lasted for several days, three hourly rates of the hot water drawn from the storage tank were used, namely 5, 10 and 20 litres per hour.

5.2 System Efficiency

Test results from figures 2 and 3 indicate that variations of the system efficiency with the plate and fluid temperature parameters follow the theoretical trends shown in equations (4) and (5). The system efficiency of the reversed flat plate collector is higher than that of the conventional double-glazed flat plate collector by about 10%. This is due to the fact that the bottom absorbing surface of the reversed flat plate collector faced the reflecting surface and lost little heat to surrounding.

5.3 Economic Analysis

First cost :	in US $
Reversed flat plate collector	165.-
Storage tank	29.-
Installation	11.-
Miscellaneous	5.-
Total first cost	210.-

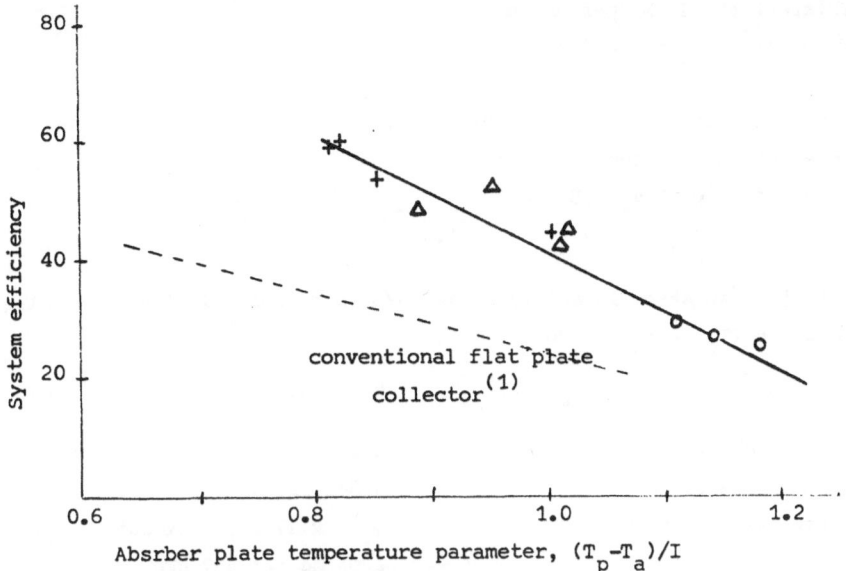

Fig.2 System efficiency based upon absorber plate temperature parameter

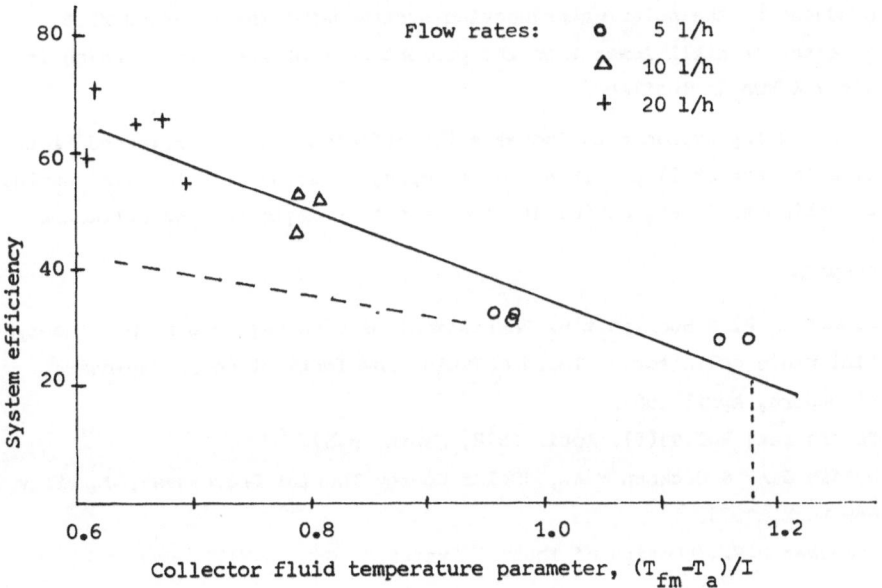

Fig.3 System efficiency based upon collector fluid temperature parameter

Useful life of the system   =   10 years

Rate of interest at 15 % per annum

Salvage value at 20 % of the first cost

$$= \text{US \$ } 42.-$$

Annual maintenance cost   = US \$ 5.-

Hence from equation (8), the annual cost

$$= 210 \times \frac{0.15 \times 1.15^{10}}{1.15^{10} - 1} + 5 - \frac{42}{1.15^{10} - 1} = \text{US \$ } 33.-$$

At the radiation intensity of 17 MJ/sq.m-d, which is the mean value for Thailand and $(T_{fm} - T_a)$ of 20 C,

$$(T_{fm} - T_a)/I \qquad = \qquad 1.17$$

From figure 3, the system efficiency   =   0.21

Total absorbing area, 2 A_c =   1.54 sq.m.

Useful energy gain per annum   =   0.21 x 17 x 1.54 x 365

$$= \quad 2005 \text{ MJ } = \text{ 557 kWh}$$

Cost of useful energy   = 33/557   =   US ¢ 5.9/kWh

From the above approximate economic analysis, the cost of useful energy produced by the solar water heating system using the reversed flat plate collector is still lower than the present cost of electricity which is about US ¢ 7.4/kWh in Thailand.

A few improvements to increase the efficiency of the reversed flat plate collector are still possible, for example, addition of selective coating on the absorbing surfaces, optimisation of the tilt angle for the collector.

6. REFERENCES

1. Wibulswas P. & Boonvanit P. "Assessment of double-glazed serpentine-type flat plate collectors", Third CEISEAN, The Institution of Engineers, Singapore, April 1981.

2. Technocrat, Vol.11(5), April 1978, Japan, p.57.

3. Duffie J.A. & Beckman W.A., "Solar Energy Thermal Processes", J. Wiley, New York, 1975.

4. Stoecker W.F., "Design of Thermal Systems", McGraw-Hill, New York, 1971.

The Effect of Parameters on the Performance of the Flat-Plate Solar Water Heating System

MILTON M. MUNROE
Federal University of Technology
P.M.B. 1526, Owerri
Imo State, Nigeria

ABSTRACT

The flat-plate solar water heating system is frequently used in pre-heating hot water for domestic and industrial purposes. To obtain the maximum benefit from solar energy systems, it is necessary to be able to predict the performance of the system in terms of the environmental conditions, the input radiation and the system parameters. This paper uses a simulation model of the flat-plate solar water heating system to investigate the response of the system to variations in parameters. The seven parameters of the system which are considered are the absorption-transmission product, heat loss coefficient and heat capacity of the collector, the mass flow rate of the circulating liquid, the heat loss coefficient and heat capacity of the tank and the water heating load. The simulation results indicate that the heat capacity of the collector has a negligible effect on performance and while absorption-transmission product and heat loss coefficient of the collector have the greatest effect, the other parameters also influence performance.

1. INTRODUCTION

The flat-plate solar water heating system is increasingly used in preheating water for domestic and industrial purposes, and much work has been done on performance studies of the system. In order to characterise the behaviour of the system, it is necessary to be able to predict the transient and steady-state performance of the system.

Steady-state analysis is based on the model by Hottel and Whillier [1], which assumes a negligible thermal capacity for the collector and expresses the output in terms of the incident radiation, the temperature difference between the collector and the ambient air and the collector loss coefficient.

Transient analyses have been done by Klein et al. [2], Close [3], Siebers and Viskanta [4], Wijeysundera [5], de Ron [6] and Munroe [7,8]. Klein et al. investigated a multi-node capacitance model of the collector, with nodes centred on the collector cover glass and the collector cover plate. Close considered a single-node capacitance model for the collector, with the node centred on the collector plate. Siebers and Viskanta analysed collector per-

formance using a two-dimensional heat transfer model in which collector cover plates and insulation are assumed to be quasi steady-
state and the collector plate and working fluid are considered to
be transient.

Wijeysundera characterised collector performance using the concept of collector response time and showed how transient behaviour
was influenced by the number of covers, the thicknesses of the cover
and absorber plates and the nature of the collector surface. De Ron
presented a detailed dynamic model, in which the collector was
divided into a finite series of equal segments. The transient behaviour of the cover glass and collector plate were described by
non-linear differential equations, while the fluid was described by
a non-linear difference equation.

The above analyses however have considered the dynamic behaviour
of the collector alone, in isolation from the rest of the system.
The collector, characterised by its absorption-transmission product,
heat loss coefficient and thermal capacity, is however only one component of the system and accounts for only three system parameters.
Other parameters of the system include the thermal capacity and heat
loss coefficient of the tank, the mass flow rate of the circulating
liquid and the water heating load. Analysis by Munroe [7] has shown
that while the dynamics of the collector is largely dominated by the
dynamics of the tank, the energy gained by the system is dependent
on all the system parameters. An earlier paper by Munroe [9] has
considered the effect on the performance of the system of the initial
and ambient temperatures of the system. The present paper extends
the analysis to consider the effect of the parameters on system performance.

2. THE SIMULATION MODEL

The usual configuration of a typical flat-plate solar water
heating system is shown in Fig. 1. An equivalent electrical circuit
or thermal network of the system, obtained by inspection from the
energy balance equations [7], is illustrated in Fig. 2.

The simulation model developed in [7] references the thermal
capacitance and heat loss coefficient of the collector to the mean
temperature of the collector fluid. The thermal capacity and heat
loss coefficient of the tank and the water heating load are referenc-
ed to the mean temperature of the storage tank. Heat losses in the
loop containing the circulation pump and heat exchangers are considered to be negligible.

If the solar radiation intensity on the collector is a constant
or step function, then the mean temperatures of the collector and
tank may be expressed at time t by

$$T_1 = A_o + A_1 \exp(-\delta t) + A_2 \exp(-\gamma t) \tag{1}$$

$$T_2 = B_o + B_1 \exp(-\delta t) + B_2 \exp(-\gamma t) \tag{2}$$

where T_1: the mean collector temperature,

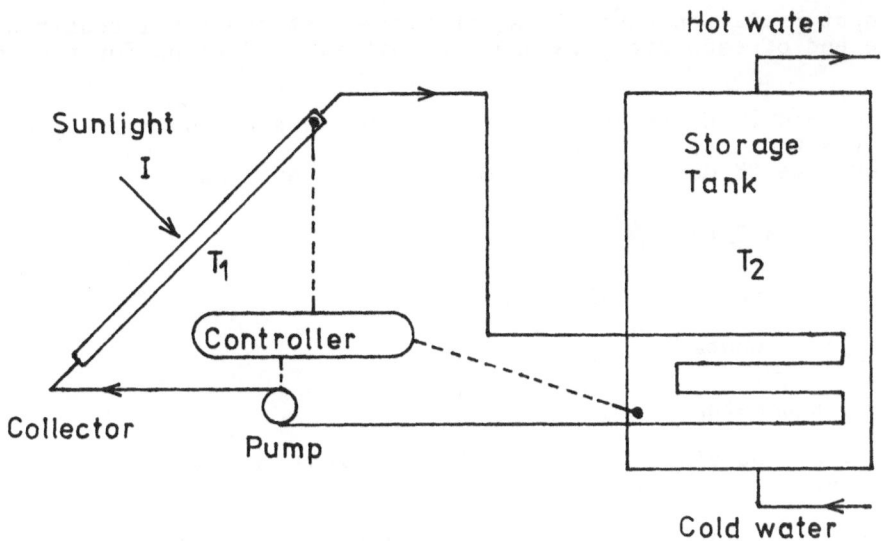

Fig.1. Flat-plate Solar Water Heater

T_2: the mean tank temperature.

A_0, A_1, A_2, B_0, B_1, B_2, δ, γ are constants given in the Appendix. These constants are functions of the initial and environmental conditions of the system and the incident radiation. Any solar radiation distribution in time may be approximated by a finite series of step functions; the shorter the duration of the step, the better the approximation. Equations (1) and (2) may therefore be used to estimate the mean temperatures of the collector and storage tank for

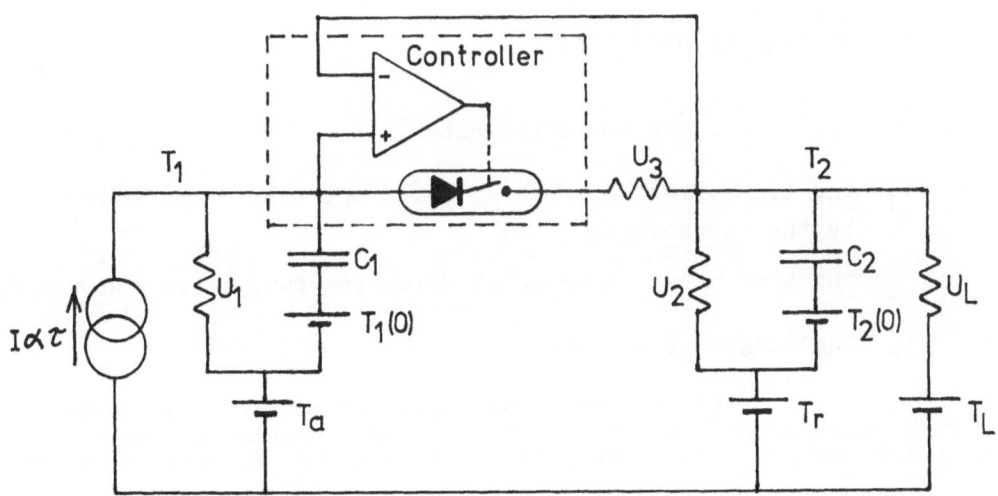

Fig.2. Equivalent Electrical Circuit

any input radiation conditions, provided that the final conditions at the end of each step are used as initial conditions for the next step.

The model for solar radiation is that described in [10], which estimates the intensity of solar radiation on a horizontal surface, from average UK meteorological data, by the expression

$$I = I_0 \exp(-\beta^2) \tag{3}$$

where I: the intensity of solar radiation,

  I_0: a constant,

  β: a constant.

I_0 and β are obtained from the meteorological data at the location of interest.

The instantaneous collector efficiency at time t is

$$\eta_1 = [I\alpha\tau - U_1(T_1-T_a) - C_1(dT_1/dt)]/I \tag{4}$$

where $\alpha\tau$: the absorption-transmission product for the collector,

  U_1: the collector heat loss coefficient,

  C_1: the collector heat capacity,

  T_a: the ambient collector temperature.

The mean collector efficiency over one solar day is

$$\bar{\eta}_1 = \int_{t_1}^{t_2} \eta_1 I dt / \int_0^{t_d} I dt \tag{5}$$

where $\bar{\eta}_1$: the mean collector efficiency,

  t_1: the time after daybreak at which the pump is switched on by the controller,

  t_2: the time after daybreak at which the pump is switched off,

  t_d: the length of the day.

The useful energy collected by the solar collector is considered as that stored in the storage tank or directly transferred to the water heating load. The rate at which energy is transferred to the tank is

$$I_2 = C_2 dT_2/dt \tag{6}$$

where C_2: the thermal capacity of the tank.

The rate at which energy is transferred to the load is

$$I_L = U_L(T_2-T_L) \tag{7}$$

where U_L: the water heating load,

T_L: the initial temperature of the load.

The fraction of the incident solar energy absorbed by the load is f,

$$f = \int_{t_1}^{t_2} I_L dt / \int_0^{t_d} I dt \tag{8}$$

The fraction of the water heating load which is supplied by solar energy is

$$F = \int_{t_1}^{t_2} I_L dt / \int_{t_3}^{t_4} U_L(T_F-T_L) dt \tag{9}$$

where T_F: the final temperature at which water is required,

t_3: the time after daybreak at which the load is on,

t_4: the time after daybreak at which the load is off.

3. RESULTS AND DISCUSSION

The performance of the system is simulated for a typical day in the summer, with a day length of 16.4 hours and a total solar energy input of 21 Mj/m2. The inclination of the collector is assumed to be 60 degrees to the horizontal and the solar radiation at normal incidence on the collector is approximated by the value on a horizontal surface.

The parameters of the system considered here are the absorption transmission product, heat loss coefficient and thermal capacity of the collector, the mass flow rate of the circulating liquid, the heat loss coefficient and thermal capacity of the storage tank and the water heating load. The performance of the system is considered in terms of the mean daily collector efficiency ($\bar{\eta}_1$), the fraction of the load supplied by solar energy (F) and the fraction of the incident solar energy absorbed by the load (f). Performance is determined for three values of each parameter (P) – for the reference value of the parameter,(Pr), 50 percent and 150 percent of the reference value. To determine the effect of each parameter independently, each is varied with the others held constant. The parameters of one system investigated have been taken as the reference values.

The variation of mean daily collector efficiency, $\bar{\eta}_1$, with

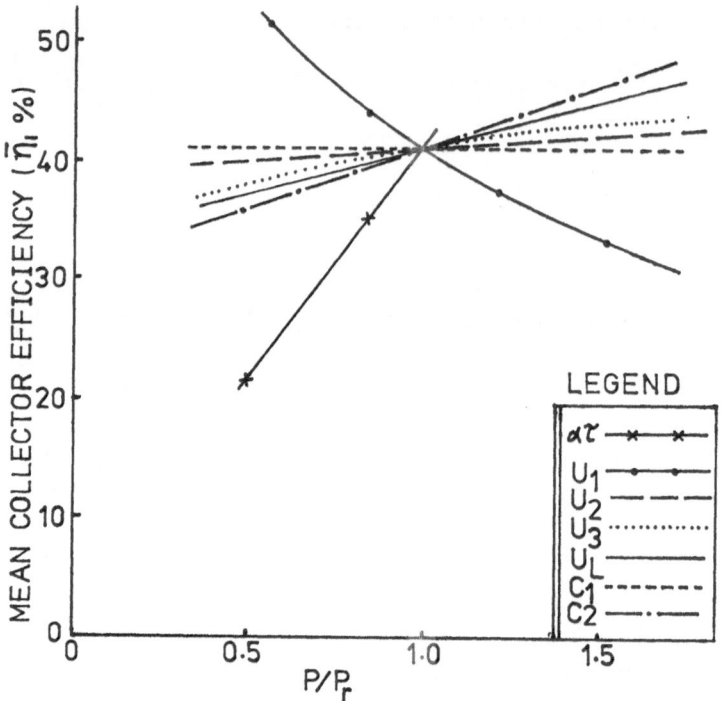

Fig.3. Variation of Mean Collector Efficiency with Parameters

parameters is illustrated in Fig. 3. The heat loss coefficient and
the absorption-transmission product have the most significant effect
on system performance. Increasing $\alpha\tau$ increases $\bar{\eta}_1$ by a corresponding
amount. Decreasing the size of U_1 does not reduce $\bar{\eta}_1$ as rapidly.
Collector thermal capacity has negligible effect on performance.
Increasing the rest of the parameters increases the mean collector
efficiency. It is significant to note that increasing the heat loss
coefficient of the tank (U_2) also increases the mean collector
efficiency. This suggests that collector efficiency is not a perfect
index of the performance of a system. In fact, any parameter change
which tends to lower collector operating temperature would likewise
increase collector efficiency and this is confirmed by an observation
of the effect of the other parameters U_3, C_2 and U_L.

The variation of the mean solar fraction F with parameters is
summarised in Fig. 4. As in the previous case, $\alpha\tau$ and U_1 have the
most significant effect on performance. Increasing C_2, U_L and U_2
decreases the value of F, while increasing flow rate U_3 increases F
rapidly at first, then more slowly. The thermal capacity of the
collector again has no effect on collector performance.

The variation of the fraction of the incident solar energy
absorbed by the load (f) is shown in Fig. 5. As before, U_1 and $\alpha\tau$
have the greatest effect on f. The value of f decreases with increasing C_2 and U_2 and increases with increasing U_3 and U_L. The
value of C_1 does not affect the fraction of solar energy absorbed
by the load.

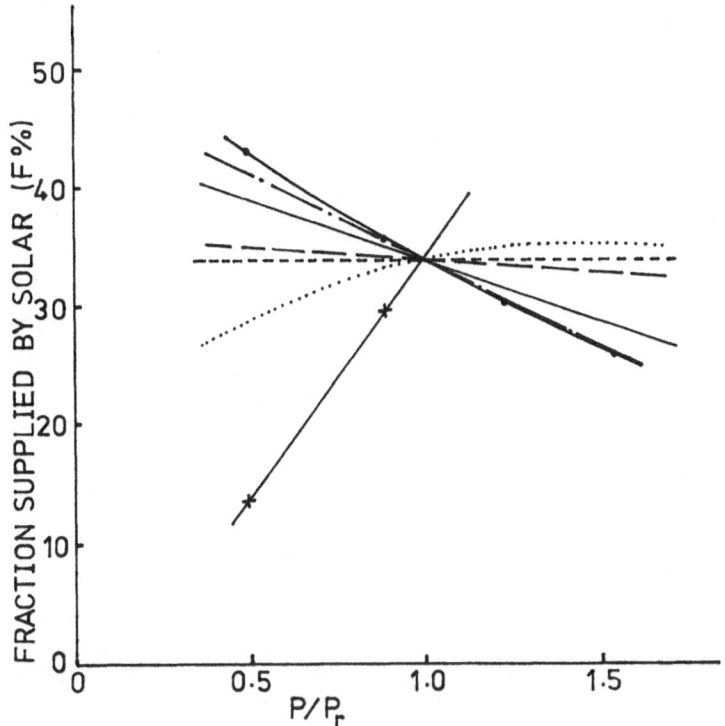

Fig.4. Variation of Fraction of Load Supplied by Solar Energy

4. CONCLUSIONS

The performance of a typical flat-plate solar collector system has been considered in terms of three different criteria, which are: 1) the mean efficiency of the collector, 2) the fraction of load supplied by solar energy and 3) the fraction of the incident energy absorbed by the load. An increase in collector absorption-transmission product and a decrease in collector heat loss coefficient clearly improve system performance from the viewpoint of the three criteria considered.

REFERENCES

1. Hottel, H.C. and Whillier, A. 1955. Evaluation of flat-plate solar collector performance. Trans. Conf. on the Use of Solar Energy. II. Thermal Processes. University of Arizona, pp. 74 -104.

2. Klein, S.A., Duffie, J.A. and Beckman, W.A. 1974. Transient considerations of flat-plate solar collectors. J. of Engineering for Power, Vol. 96A, pp. 109-113.

3. Close, D.J. 1967. A design approach for solar processes. J. of Solar Energy, Vol. 11, No. 2, pp. 112-122.

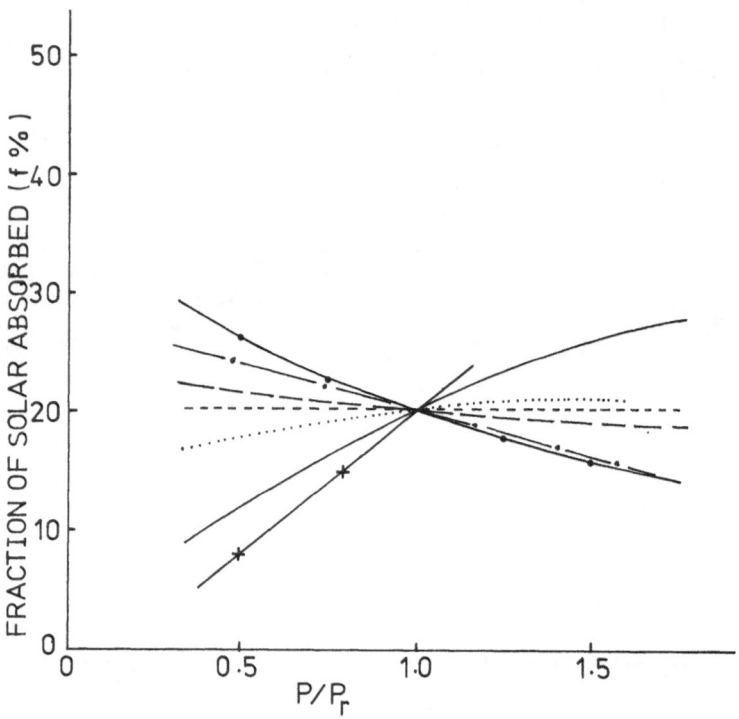

Fig.5. Variation of Fraction of Solar Energy Absorbed

4. Siebers, D.L. and Viskanta, R. 1978. Some aspects of the
    transient response of a flat-plate solar energy collector. J.
    of Energy Conversion, Vol. 18, pp. 135-139.

5. Wijeysundera, N.E. 1976. Response time of solar collectors. J.
    of Solar Energy, Vol. 18, No. 1. pp. 65-68.

6. de Ron, A.J. 1980. Dynamic modelling and verification of a flat-
    plate solar collector. J. of Solar Energy, Vol. 24, No. 2, pp.
    117-128.

7. Munroe, M.M. 1981. A generalised computer simulation model of
    the flat-plate solar water heating system. First International
    Conference on the Applications of Modelling and Simulation, Lyon,
    France, Vol. IV, Materials and Resources, pp. 238-241.

8. Munroe, M.M. 1980. Electric circuit simulation of the flat-plate
    solar water heating system. Proceedings of the Conference: Solar
    Energy in the 80's - Technical and Economic Viability, 12 Septem-
    ber, University of Birmingham, pp. 131-135.

9. Munroe, M.M. 1981. The influence of initial conditions on the
    performance of the flat-plate solar water heating systems. 4th
    Miami International Conference on Alternative Energy Sources.
    University of Miami, 14-16 December.

10. Munroe, M.M. 1980. Estimation of totals of irradiance on a horizontal surface from UK average meteorological data. J. of Solar Energy, Vol. 24, pp. 235-238.

APPENDIX

The constants in equations (1) and (2) are

$$A_o = X_o / \delta \gamma \tag{10}$$

$$A_1 = (X_o - \delta X_1 + \delta^2 X_2) / \{\delta(\delta - \gamma)\} \tag{11}$$

$$A_2 = -(X_o - \gamma X_1 + \gamma^2 X_2) / \gamma (\delta - \gamma) \tag{12}$$

$$X_o = \{(U_2 + U_3 + U_L)(I\alpha\tau + U_1 T_a) + U_3(U_2 T_r + U_L T_L)\} / C_1 C_2 \tag{13}$$

$$X_1 = \{C_1(T_1(o) + T_a)(U_2 + U_3 + U_L) + C_2(I\alpha\tau + U_1 T_a)$$

$$+ U_3(T_2(o) + T_r)\} / C_1 C_2 \tag{14}$$

$$X_2 = T_1(o) + T_a \tag{15}$$

$$B_o = Y_o / \delta \gamma \tag{16}$$

$$B_1 = (Y_o - \delta Y_1 + \delta^2 Y_2) / \{\delta(\delta - \gamma)\} \tag{17}$$

$$B_2 = -(Y_o - \gamma Y_1 + \gamma^2 Y_2) / \{\gamma(\delta - \gamma)\} \tag{18}$$

$$Y_o = \{(U_1 + U_3)(U_2 T_r + U_L T_L) + U_3(I\alpha\tau + U_1 T_a)\} / C_1 C_2 \tag{19}$$

$$Y_1 = [C_1\{U_3(T_1(o) + T_a) + (U_2 T_r + U_L T_L)\}$$

$$+ C_2(U_1 + U_3)(T_2(o) + T_r)] / C_1 C_2 \tag{20}$$

$$Y_2 = T_2(o) + T_r \tag{21}$$

$$\delta = \{a + (a^2 - 4b)^{\frac{1}{2}}\} / 2 \tag{22}$$

$$\gamma = \{a - (a^2 - 4b)^{\frac{1}{2}}\} / 2 \tag{23}$$

$$a = \left\{ C_1(U_2+U_3+U_L)+C_2(U_1+U_3) \right\}/C_1C_2 \qquad (24)$$

$$b = \left\{ U_1(U_2+U_3+U_L)+U_3(U_2+U_L) \right\}/C_1C_2 \qquad (25)$$

$T_1(o)$ and $T_2(o)$ are respectively the initial temperatures of the collector and storage tank above ambient. U_3 is the heat transfer coefficient between the collector and storage tank. Where heat losses from the circulation loop are small, U_3 is equal to the capacitance rate (mass flow rate x specific heat) of the circulating liquid. T_r is the ambient temperature of the tank.

Solar Energy Collection and Storage with Carbon-Gypsum Composites

ALLAN T. KIRKPATRICK
Mechanical Engineering Department
Colorado State University
Fort Collins, Colorado 80523, USA

ABSTRACT

Experiments were performed on carbon-gypsum test masses to measure the thermal performance as a function of the carbon-gypsum ratio, and the chemical state of the gypsum. The results indicated that addition of carbon increased the surface absorptivity of the gypsum.

1. INTRODUCTION

There has been a resurgence of interest in passive solar design in the past few years. One of the fundamental principles of passive solar design is that the whole building acts as the solar energy collector and the thermal energy storage, thus coupling the building very closely to the environment. Due to this coupling, proper design of the building is essential to insure thermal comfort. Two types of passive solar design that are receiving a great deal of attention are the indirect gain system, and the direct gain system, as shown in Figure 1. At present, the thermal energy elements in these designs are usually concrete, water, rocks, brick, or gypsum sheetrock [1]. A comparison of the thermal properties and economics of the above materials is shown in Table 1.

TABLE 1   THERMAL PROPERTIES COMPARISON (300°K)

MATERIAL	SPECIFIC HEAT KJ/Kg °K	DENSITY Kg/m^3	THERMAL CONDUCTIVITY W/m °K	HEAT CAPACITY KJ/m^3°K	THERMAL DIFFUSIVITY m^2/s x10^{-8}	COST $/KJ/°K
GYPSUM	1.0	2002	0.17	2002	8.49	0.24
CONCRETE	0.88	2300	1.4	2024	69.2	0.09
ROCK	0.84	2680	2.7	2251	120	0.06
WATER	4.18	1000	0.6	4180	14.4	0.05
CARBON	0.51	1950	1.6	995	161	--
BRICK	0.8	1920	0.72	1536	46.9	--

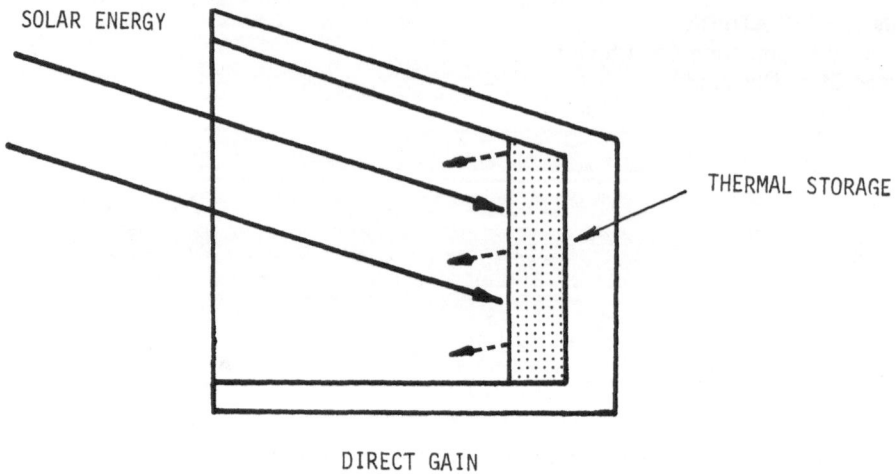

DIRECT GAIN

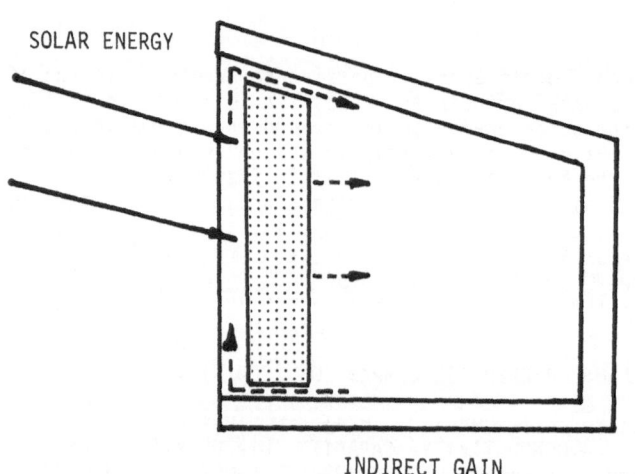

INDIRECT GAIN

------→ Heat transfer from storage to buildings

Figure 1. Passive Solar Building

Solar energy collection and storage in passive solar buildings is usually performed by the thermal storage mass, in contrast to active solar buildings, which have separate collection and storage elements. Therefore, it is important, in indirect gain systems, that the surface of the storage mass act as a selective surface, with high value of solar radiation absorbance (α) and a low value of longwave radiation emittance (ε). For a direct gain building, this is not the case, since a high value of ε is needed for subsequent radiant heating of the enclosure, and the building acts essentially as a blackbody cavity with a high value of α. Carbon selective surfaces have been developed using carbon microspheres with a diameter of approximately 0.5 microns [2].

Recently it has been proposed, and a patent application filed [3, 4], that gypsum is partially transparent to solar radiation in the near infrared. Thus, if an appropriate amount of carbon was added to the gypsum during mixing, the resultant composite would absorb solar radiation volumetrically, and therefore have a high absorptivity. This hypothesis is somewhat counter-intuitive, since crystalline structures such as gypsum usually act as Rayleigh scattering centers which attenuate infrared radiation. The purpose of this paper is to review the relevant gypsum chemistry and experimentally investigate the thermal performance of carbon-gypsum composites as solar energy collectors.

2. GYPSUM CHEMISTRY

There are three major types of hydrated calcium sulfate, as shown in Table 2. They are calcium sulfate dihydrate, commonly known as gypsum; calcium sulfate hemihydrate, commonly known as plaster of Paris or stucco; and calcium sulfate anhydrite [5, 6]. The anhydrite and dihydrate are found naturally. The hemihydrate is formed by dehydration at temperatures above 60°C, and the anhydrite can be formed by heating at temperatures about 200°C. The dehydration process is commonly called calcination. The reverse process of hydration is used in the construction industry to prepare wallboard, and facing material, i.e., plaster plus water yields gypsum. One important property exploited by the construction industry is its fire resistance. Properties which limit gypsum usage include plastic flow under load, strength loss in a humid atmosphere, and erosion in water. Thus gypsum is not normally used as a structural member, and in exposed exterior locations.

TABLE 2   CALCIUM SULFATE TYPES

NAME	FORMULA	MOLE. WT.	% H_2O
DIHYDRATE	$CaSO_4 \cdot 2H_2O$	172	20.9
HEMIHYDRATE	$CaSO_4 \cdot 1/2H_2O$	145	6.2
ANHYDRITE	$CaSO_4$	136	0

The dehydration process takes place by the removal of water molecules from the crystal lattice, which adjusts itself slightly, but remains virtually unchanged. The crystal lattice of the dehydrate is a layer lattice in which double layers of $CaSO_4$ alternate with double layers of water bonded by hydrogen bonds. Within the layers, the calcium and sulfate tetrahedra form parallel chains about three Angstroms apart.

The absorption spectra of a single crystal of dihydrate, also known as selenite; hemihydrate; and anhydrite have been measured by Coblentz [7], and his results are presented in Figure 2. The anhydrite of thickness 0.656 mm, curve a, is quite transparent in the near infrared, except for small absorption bonds at 1.9, 3.2, 5.7, 6.15, and 6.55 µ, and a large band at 4.55 µ, due to the SO_4 ion. The dihydrate of thickness 0.204 mm, curve b, is less transparent, and its absorption bonds correspond to the absorption bonds of water, which has large bonds at 1.5, 3, 4.75, and 6 µ. The hemihydrate, formed by heating of the dihydrate crystal, curve c, shows poor infrared transmissivity. Coblentz noted that the single dihydrate crystal became opaque upon heating, evidently due to the effect of the dehydration process creating additional crystal orientations. The hemihydrate was moistened with water and rehydrated into dihydrate, curve d, which shows the poorest transmissivity, except near 7 µ.

Comparison of curves c and d shows that hydrated calcium sulfate with a complex crystal structure can be made slightly more transparent in the 4-7 µ range by dehydration from the dihydrate to the hemihydrate state. However, the single crystal of dihydrate, and the anhydrite are much more transparent in the infrared region.

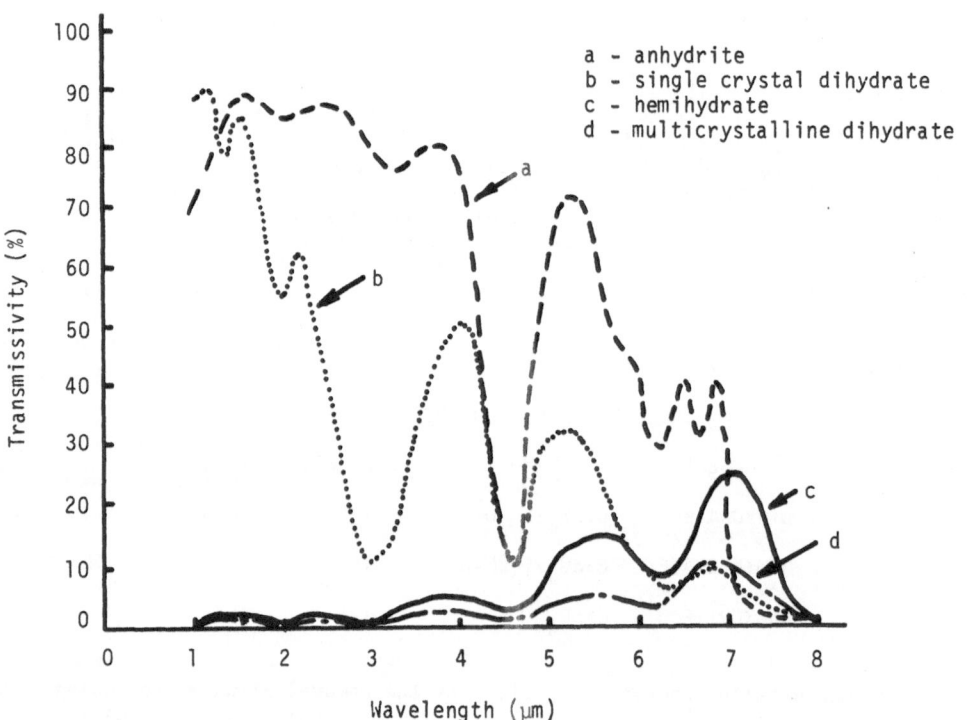

Fig. 2. Gypsum Transmission Curves

3. EXPERIMENTAL DESIGN

In order to assess the solar collection potential of carbon gypsum composites, an experimental program was designed. Small 80 g. masses instrumented with thermocouples were tested simultaneously to give relative performance comparisons. A schematic of a test mass is shown in Figure 3. The test masses had varying amounts of carbon added to the gypsum slurry, ranging from 0.625% to 2.5% by weight. The surface of the test mass had varying amounts of carbon, ranging from .625% to fully covered by carbon. The gypsum used is known commercially as HYDROSTONE; and the carbon used is known as B-COLORANT, a paint additive containing very small carbon particles, water, and ethylene glycol.

The test masses were placed six at a time into an isolated box with a glass cover, and the box was placed outside facing south. The temperatures of the carbon gypsum test masses were measured by a HP-85 digital data acquisition system at twenty minute intervals during a test run. Data was taken during October and November 1981, and again in March 1982. The first set of data was with dihydrate masses. The test masses were then heated at 75°C, and periodically weighed until a 16% weight loss was recorded, which indicated that dehydration from dihydrate to hemihydrate had occurred. The dehydration process was about three weeks long. The tests were then repeated with the hemihydrate test masses.

Due to the thermal diffusion time of gypsum, which is about 3 cm./hour, the axial temperature distribution will not be uniform, and thus the temperature measured by the thermocouple is only the midplane temperature.

A two dimensional finite difference program was written to predict the temperature profile through the gypsum test mass, and the results are shown in Figure 4. The analysis formulated energy balances for a glass cover plate and the gypsum surface with $\alpha=\epsilon=.9$, under solar radiation conditions typical for Colorado and applied the net heat flux as a boundary condition to the finite element nodes. The other gypsum surfaces were assumed to be adiabatic. Inspection of the curves suggests that for the 2.5 cm thickness used in the test mass, the midplane temperature is also approximately equal to the average temperature of the test mass, with a 15°C temperature swing from the front surface to the back surface. Therefore the change in the midplane temperature can be used as a measure of the net amount of solar radiation that has been absorbed by the carbon-gypsum test mass; and the test masses with the highest temperatures are assumed to have the highest absorptivity, and / or the lowest emissivity.

Figure 5 shows a schematic of the test arrangement. The test masses were arranged so that all in a given test were uniformly irradiated. The glass cover plate was 3.2 mm ordinary window glass. The test masses were thermally isolated from each other with insulation, and the test enclosure was insulated with 4 cm of styrofoam.

Finally, a typical solar spectrum is shown in Figure 6 indicating the relative proportions of the solar flux in the visible and in the infrared region. No attempt was made to measure the actual solar radiation passing through the glass cover plate, due to lack of suitable equipment, so the results presented in the next section are based solely on relative comparison of simultaneous measurements.

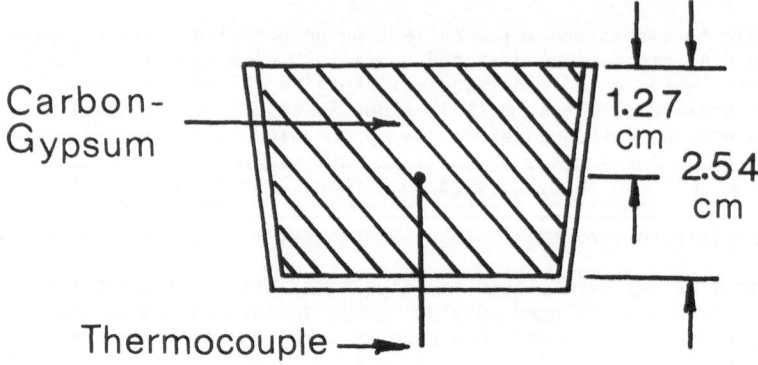

Fig. 3.   Carbon-Gypsum Test Mass

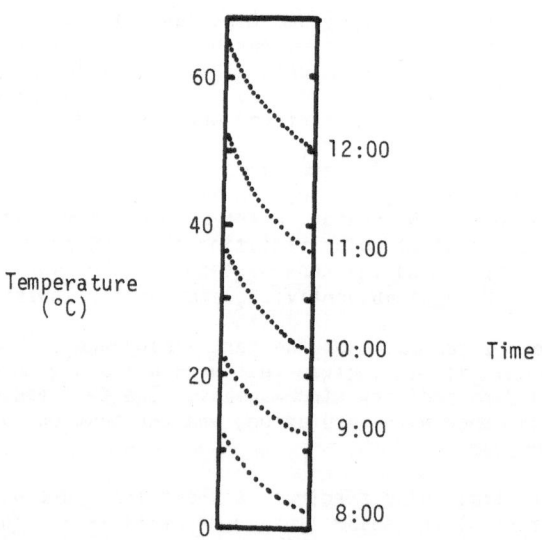

Fig. 4.   Predicted Temp. Profiles for 2.5 cm Thick Mass

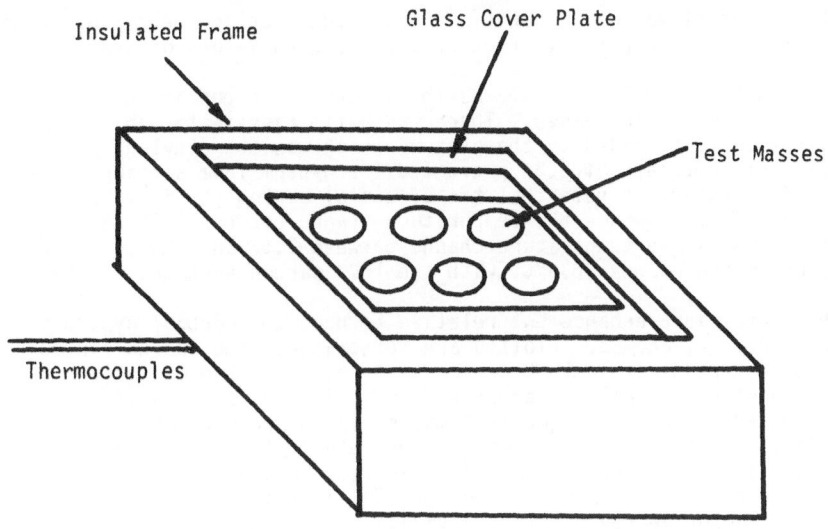

Fig. 5. Experimental Setup of Test Masses

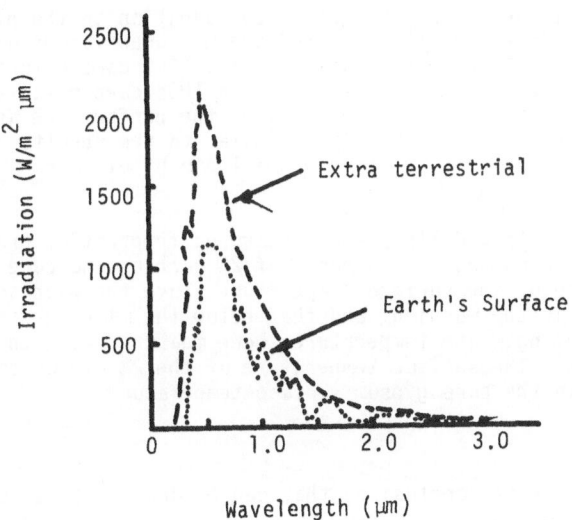

Fig. 6. Solar Spectral Distribution [8].

4. RESULTS

The effect of various amounts of carbon addition to the gypsum or dihydrate is shown in Figure 7. Plotted are measured values of temperature
versus time for five test masses: 0.625%, 0.83%, 1.25%, 2.5%, and 1.25% with a
fully carbon black 100% surface. As the amount of carbon is increased, the
measured temperature increases. There was a limiting factor in the amount of
carbon that could be added to the gypsum, which was the ethylene glycol. Amounts
of B-colorant in excess of 2.5% prevented the gypsum from setting properly. It
is quite interesting to note that the effect of carbon particles on the surface is logarithmic, since the temperature change between 1.25% and 2.5% is
about the same as the temperature change between 2.5% and 100%. The highest
temperature measured was 96.6°C, with the 100% carbon surface.

The performance enhancement relative to pure (0% carbon) gypsum is shown
in Figures 8a, 8b, and 8c. Plotted are temperature-time curves for pure
gypsum, pure gypsum with a 100% carbon surface, 2% carbon composite, and 2%
carbon composite with a 100% carbon surface. The curves show that the performance of the 100% carbon surface test masses is independent of the amount of
carbon in the composite, as expected. The relatively poor performance of the
pure gypsum is due to its higher reflectivity. Comparison of the curves also
shows that even a small amount of carbon (2%) increases the absorptivity of
the gypsum, and/or decreases the emissivity.

Comparison of the hemihydrate to the dihydrate is made in Figure 9 for a
2% carbon gypsum composite. The hemihydrate reaches slightly higher temperatures than the dihydrate. The slight increase can be explained by the slightly
superior infrared transmissivity of the hemihydrate relative to the dihydrate
as discussed earlier.

The influence of varying amounts of carbon addition to the plaster or
hemihydrate is shown in Figure 10. Plotted are measured values of temperature
versus time for pure plaster, pure plaster with a 100% carbon surface, 2%
carbon composite, and 2% carbon composite with a 100% carbon surface. The results indicate that with a 100% carbon surface, the performance is independent
of the amount of carbon in the composite, similar to the results for gypsum in
Figure 8. Also, addition of a relatively small amount of carbon increases the
plaster performance.

Finally, Figures 11a and 11b present measured temperature profiles in
9.0 cm thick dihydrate composites composed of 2% carbon, and pure gypsum. The
curves show the swing of the surface temperature above the midplane temperature during heating in the morning, and then below the midplane temperature
during afternoon cooling. The temperatures were measured with thermocouples
buried in the gypsum. The surface temperature of the 2% carbon composite is
about 30% higher than the pure gypsum surface temperature.

5. CONCLUSIONS

There are a number of conclusions that can be drawn from this work. First,
the solar energy collection and storage of gypsum and plaster can be enhanced
by about 12% with the addition of 1-2 percent carbon particles. The performance can be enhanced by about 20% by coating the surface entirely with carbon.
Second, the dihydrate and hemihydrate forms of gypsum are both essentially
opaque in the infrared region, and that radiation absorption and emission is
essentially a surface phenomenon. Finally, long term use of gypsum as a
thermal storage element, such as a facing for storage walls, will result in
calcination of the dihydrate to hemihydrate. Calcination, however, will not

affect the structural or thermal performance of gypsum.

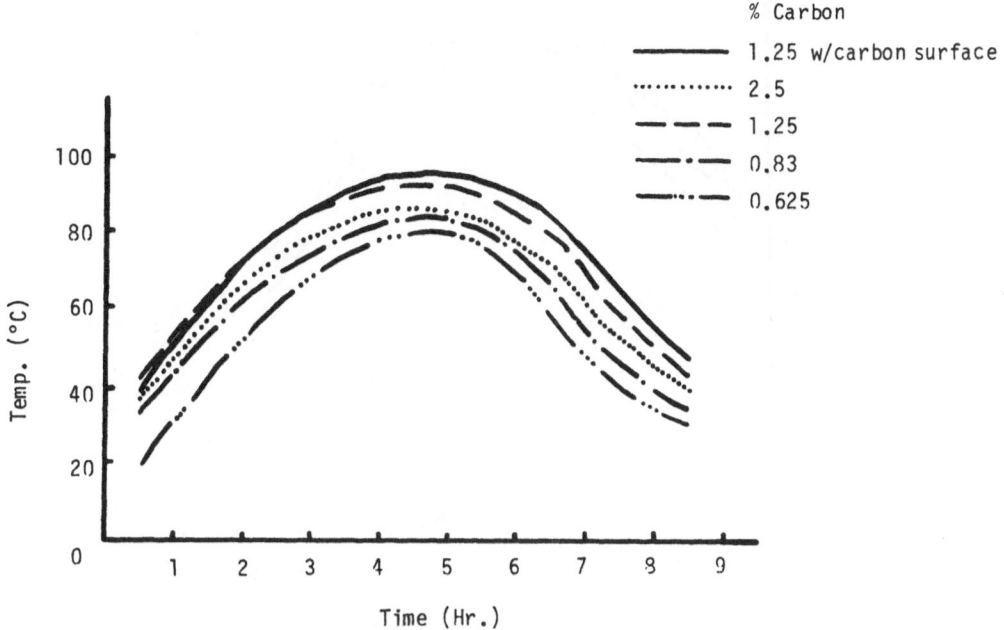

Fig. 7. Effect of Carbon Addition to Gypsum (Dihydrate)

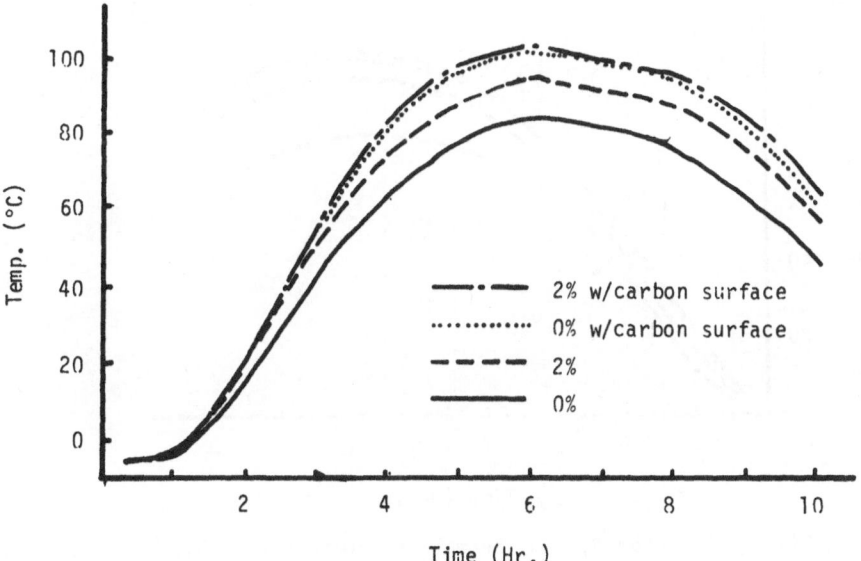

Fig. 8a. Carbon-Gypsum Thermal Performance (Day 1)

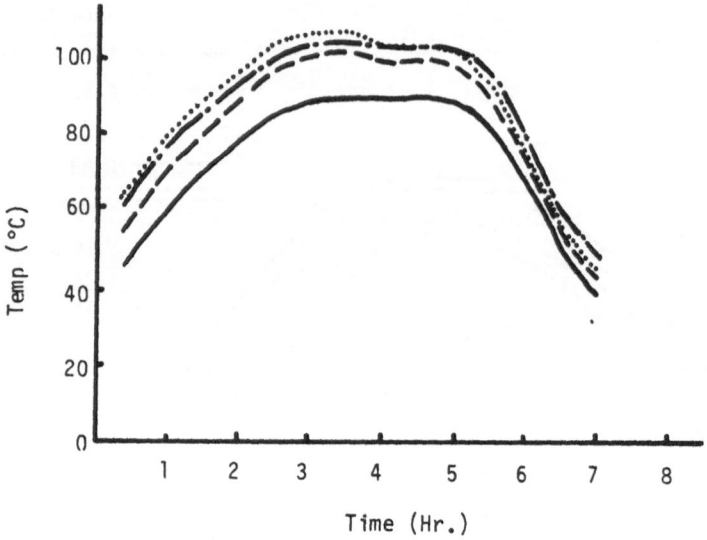

Fig. 8b. Carbon-Gypsum Thermal Performance (Day 2)

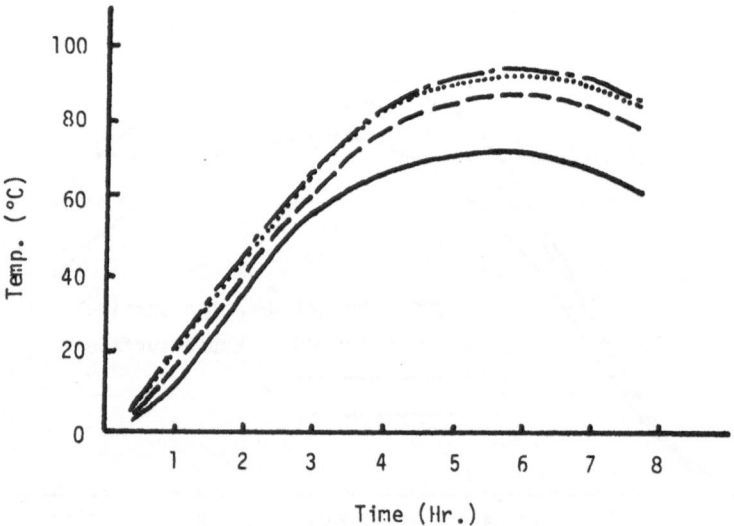

Fig. 8c. Carbon-Gypsum Thermal Performance (Day 3)

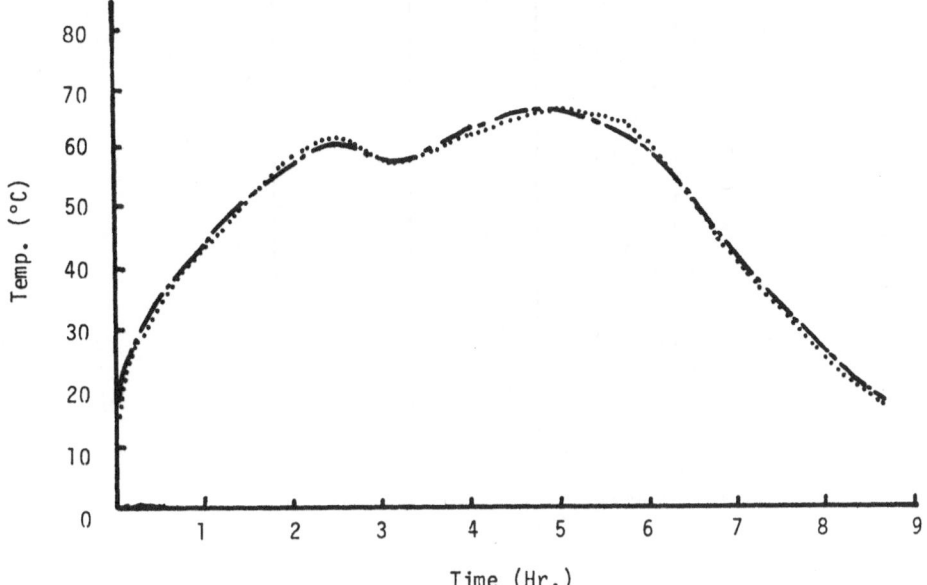

Fig. 9.   Comparison of Dihydrate (— - —) and Hemihydrate (....)
Performance with 2% Carbon.

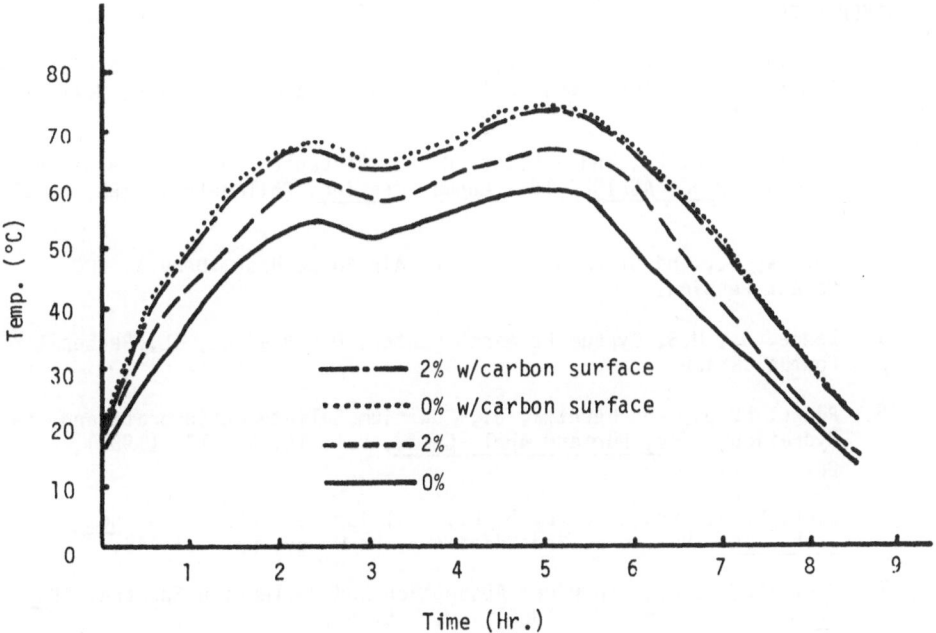

Fig. 10.   Effect of Carbon Addition to Plaster (Hemihydrate)

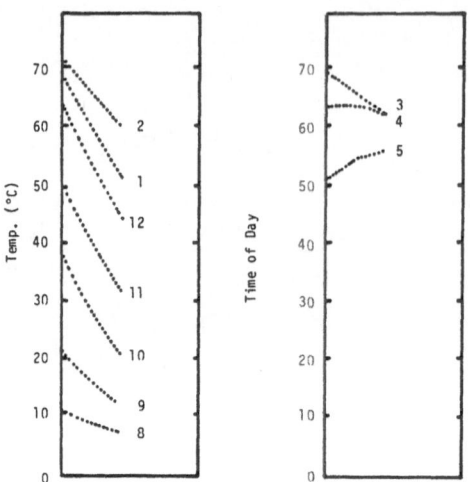

Fig. 11.   Temperature Profiles in 9 cm Thick 2%
          Carbon - Gypsum Block.

6. ACKNOWLEDGEMENTS

     This research was performed with the assistance of several undergraduates:
M. Lueker, R. Bernard, M. Coleman, K. Hopkins, C. Hurst, and P. McDaniel. An
equipment grant was provided by U.S. Gypsum Company.

7. REFERENCES

     1. Baylin, F., "Low Temperature Thermal Energy Storage: A State of the
         Art Survey," Golden, CO, Solar Energy Research Institute; July 1979,
         SERI/RR-54-184.

     2. Baldonado, O. C., and Schmitt, C. R., "Advanced Solar Concepts Using
         Carbon," Proc. AS/ISES 1981 Annual Meeting, Philadelphia, PA, p. 235-
         239.

     3. Bettis, J., and Clerman, R., Lowry Air Force Base, Denver, CO,
         Patent Pending.

     4. Lange, J., U.S. Gypsum Research Center, Des Plaines, IL, Personal
         Communication.

     5. Ridge, M. J., and Beretka, J., "Calcium Sulfate Hemihydrate and Its
         Hydration," Rev. Pure and Appl. Chem., Vol. 19, No. 17, (1969),
         pp. 19-43.

     6. Wenk, R. J., and Henkels, P. L., "Calcium Compounds," in Chemical
         Technology, Volume 4, 3rd Edition, John Wiley, pp. 437-448.

     7. Coblentz, W. W., "Infrared Absorption and Reflection Spectra," Phys.
         Rev., Vol. 23, No. 2, (1906), pp. 125-153.

     8. Incropera, F.P., and Dewitt, D.P., Fundamentals of Heat Transfer,
         John Wiley and Sons, New York, 1981, p. 601.

SOLAR ENERGY APPLICATIONS

Single and Multiple Stage Basin Type Solar Still

H.N. SHIRANI, I.A. TAG, and M.H. COBBLE
Mechanical Engineering Department
New Mexico State University
Las Cruces, New Mexico 88003, USA

ABSTRACT

A single stage still and a multiple stage basin type solar still were analyzed, and the output was numerically computed for a single day each month throughout the year, using climatological data for Las Cruces, New Mexico, and using actual measured solar input and compared to actual test data.

The multiple stage was clearly superior in output to the single stage still, and the analytical results compared favorably to the experimental results.

DIGITAL SIMULATION

Background Information for Numerical Solution

If a multiple-stage solar still is introduced, it results in an increase in distillate over what the solar still will normally produce during the course of a day. The cumulative daily distillate obtained by each system is a function of brine depth and, of course, the climatological variables and the design parameters of the solar still.

In this section, the direct use of solar energy as the only heating source in two basin-type solar stills is studied, through digital simulation, for Las Cruces, New Mexico. The city is located at $32^{\circ}17'$ N latitude, $106^{\circ}45'$ W longitude with 1188 meter ground elevation. In this study, hourly values of ambient air temperature (T_a), wind speed (V) and solar radiation (I_S) are used for Las Cruces for the year of 1980. The theoretical hourly values of solar radiation (direct normal, direct, diffuse and global) for clear days were estimated [14]. This is often justified for the sun-drenched city of Las Cruces, since the expected fraction of cloudy days compared to clear days is small. Results were obtained for the 21st of each month, with two or three more in addition to verify the theoretical and experimental results. Brine depths of 2.5 and 1.25 cm were used for single-and multiple-stage stills, respectively.

To calculate the still parameters (heat losses, temperatures, etc.) a computer program was developed in which the initial brine and glass cover temperatures are considered equal to the ambient air temperature at the sunrise hour (start of the operation).

Heat and Mass Transfer Relationships

The complexity of the energy transfer (heat and mass transfer) relationships governing the performance of a single-stage solar still are schematically
presented in Figure 1.   The same mechanism takes place in multiple-stage as
in single-stage solar stills with the exception that the latent heat of evaporation is utilized again instead of being dissipated to the ambient air through
the transparent cover.

The still is exposed to direct (short wave) and long wave atmospheric
solar radiation, the latter of which is all absorbed in the glass cover. The
major portion of the short wave radiation moves through the glass cover, and
is abosrbed by the brine mass representing the amount of solar radiation absorbed per unit still area.

In the basin-type stills, part of the solar intensity absorbed by the
brine liner is used in solar distillation, while the rest is lost to the surroundings (a) by conduction through the bottom and side insulation in a single-
stage solar still (the latter is negligible in a multiple-stage solar still
due to the short vapor path), and (b) by convection and radiation through the
glass cover.

A heat balance may be written for the glass cover and brine mass, assuming
that no vapor leakage is present:

    I.   The energy balance on the glass cover results in

$$q_{cw} + q_{ew} + q_{rw} + \alpha_g\, I_s = q_{ca} + q_{ra} + \dot{m}_e\, C_{pw}(T_g - T_a) + m_g\, C_g\, \frac{dT_g}{dt} \qquad (1)$$

    II.   Energy balance on the brine mass gives

$$\alpha_w\, I_s = q_{cw} + q_{ew} + q_{rw} + q_b + 2q_s - \dot{m}_e\, C_{pw}(T_{fw} - T_w) + M\, C_{pw}\frac{dT_w}{dt} \qquad (2)$$

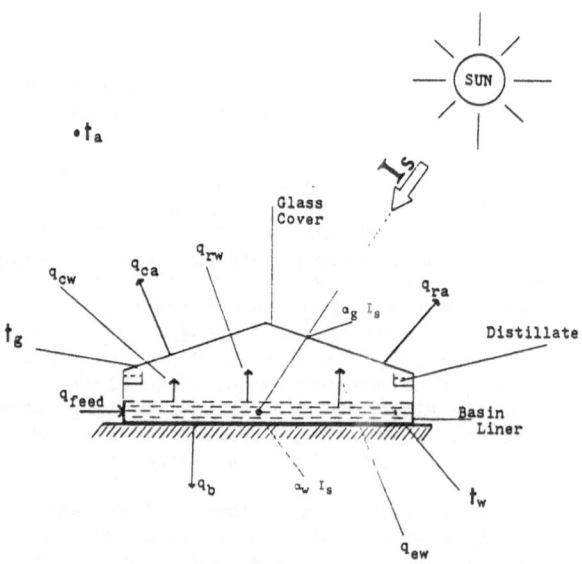

Figure 1.   Single-Stage Solar Still Schematic
Showing Various Heat Fluxes

Various energy flux terms listed in Equations (1) and (2) are defined below.
These are essentially the same as those developed by Dunkle (steady state) [15]
and with slight modifications by Cooper (Transient) [16].

$$q_{rw} = 0.9\sigma[(T_w + 273.15)^4 - (T_g + 273.15)^4] \tag{3}$$

$$q_{cw} = h_{cw}(T_w - T_g)$$

where

$$h_{cw} = C\left[(T_w - T_g) + \frac{P_w - P_{wg}}{2.65P_t - P_w} \cdot (T_w + 273.15)\right]^n$$
$$\cdot (T_w - T_g) \tag{4}$$

$$q_{ew} = \dot{m}_e\, h_{fg}$$
$$= 16.276 \times 10^{-3}\, q_{cw}\, \frac{P_w - P_{wg}}{T_w - T_g} \tag{5}$$

$$q_b = U_b(T_w - T_a) \tag{6}$$

$$q_a = U_s(T_w - T_a) \tag{7}$$

$$q_{ca} = h_{ca}(T_g - T_a) \tag{8}$$

$$q_{ra} = 0.9\sigma[(T_g + 273.15)^4 - (T_{sky} + 273.15)^4] \tag{9}$$

where

$$T_{sky} = T_a - 12 \tag{10}$$

$$q_{feed} = \dot{m}_e\, C_{pw}(T_{fw} - T_w) \tag{11}$$

Distillation rate is given by

$$\dot{m}_e = \frac{q_{ew}}{h_{fg}} \tag{12}$$

The more detailed of the terms used in Equations (1) through (12) along
with the modified terms for the distinction between the two systems (i.e.,
internal covective mode) are given in reference [14].

For a given time (T), Equations (1) and (2) were solved simultaneously
to determine water and glass temperatures, various heat fluxes and the instantaneous distillation rates. Cumulative distillates were also computed.

The brine and glass temperatures were iteratively estimated using succes-
sively improved brine and cover temperature values. The first estimate of the
brine temperature solution at the subsequent time interval was obtained (i.e.,
iteration m = 1) through Equation (2) by setting the loss terms (q's) to
zero. The first value of cover temperature, obtained from Equation (1) ignoring the losses, was used to estimate the brine temperature for the next
iteration (m = m + 1), and so on.

Brine Mass Initial Conditions and Digital
Simulation Considerations

For the distillate rate, a digital simulation program was developed which
studies the transient performance of the two basin-type solar stills. The
following assumptions were made in developing the computer program:

a) The temperature of all surfaces at the beginning of
   the day are the same as the ambient air temperature.

b) In the early sunrise hours, the solar radiation was

allowed to warm up the brine mass so that there
would be a temperature differential (i.e., $T_w-T_g>0$),
until then the distillate was assumed to be zero. In
reality the distillation starts immediately after
the insolation enters the brine mass. Initially,
it is through diffusion as in the very start of the
operation and then beyond that by free convection
(this differs from single-to multiple-stage solar
stills due to the range of Grashof number). However, the distillate produced in the first few
minutes is small and the error introduced by
letting it warm up a few degrees is negligible.

c) For the energy absorbed in the brine mass and
liner, the technique developed by Cooper [17]
has been used. The effective brine mass absorptance (α_w), which is the fraction of the incident solar flux absorbed by the brine mass, is
defined below:

α_w = (Cover Transmittance) · (Brine Absorp-
tance) + (Cover Transmittance) · (Brine
Transmittance) · (Brine Liner or Basin
Absorptance)

d) The reflected shortwave radiation is neglected
because it has no effect on vapor formation or
condensation.

e) The atmosphere inside the solar still is assumed
to be non-emitting, non-absorbing.

f) There is no temperature differential across the
glass cover and distillate film, so the distillate
is considered to leave the still at the cover
temperature.

g) Confines of the stills are air tight and no leakage of any kind is present.

For the multiple-stage still output, the various energy fluxes within the
solar still and those leaving the still boundaries were calculated using the
same approach as for a single-stage still. The energy balances on the covers
and brine masses of both solar stills were also the same.

Numbers of Case Studies for Digital Simulation

As mentioned earlier, brine depths of 2.5 cm and 1.25 cm were used for
single-and multiple-stage solar stills, respectively. Two different processes
were adopted to digitally stimulate the system parameters and distillate rates
for each still as follows:

a) The actual measured hourly ambient air
temperature and the corresponding calculated hourly solar intensity using
the earlier approach [14] , for the 21st
of each month (a total of 12 cases for
each system).

b) The actual hourly ambient air temperature
and solar radiation for the 21st of each
month (a total of 12 cases for each system)
are both measured.

The data for August 5th and 20th were employed in comparing the experimental and theoretical results.

EXPERIMENTAL INVESTIGATION

Construction of Single-and Multiple Stage
Solar Stills

Two different basin type solar desalination systems were constructed in mid-1980. The description of the systems is a follows:

a) The schematic of the single-stage solar stills is
shown in Figure 2. The sides were made of 2.5 cm
thick plywood sheets (concrete at the topmost part
in connection with glass cover) and for the bottom,
another 2.5 cm thick sheet of plywood was used.
The Grifolyn Fabric (nylon reinforced polyethylene)
sheet was used as a basin liner. A V-shaped stainless steel (22 gauge) trough was installed at a
slight incline to collect the distillate at one
end of the solar still. A sheet of 3.2 mm thick
A.F.G. tempered glass was utilized for the transparent cover. For the insulation under the bottom
and sides of the still, 8-3/4 cm of Fiberglass
(K = .046 W/m-°C) and 3-3/4 cm of Urethane foam
with aluminum foil backing (K = .024 W/m-°C) were
used, respectively. The piping, valves and con-
nectors for distillates header, line to tanks,
cleanout line, cleanout header, supply line and
header were chosen from 2.5 and 5.0 cm diameter
galvanized steel or polyvinylchloride (PVC). The
pipe connections to the basin liner were sealed
with P.V.C. fitting cement. The projected basin
area for the system was 1.45 m^2.

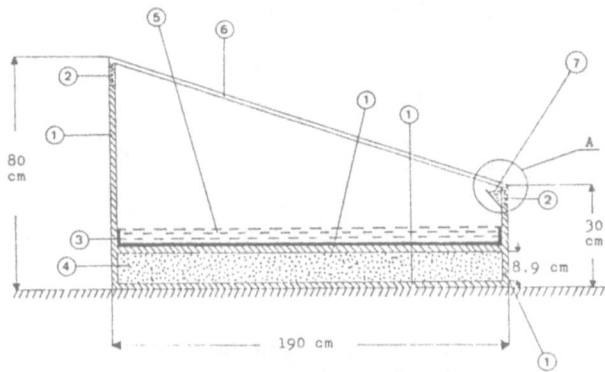

Figure 2.   Constructional Details of a Single-Stage
Solar Still

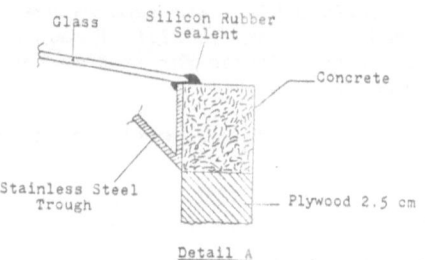

Detail A

NO	Description
1	Plywood 2.5 cm
2	Concrete
3	Grifolyn Fabric Liner
4	Fiberglass insulation 8.9 cm
5	Brine
6	Glass 3.2 mm
7	Stainless Steel Trough

*Figure 3. Constructional Details of a
Single-Stage Solar Still*

b) *The multiple-stage solar still is represented diagram-
matically in Figure 4. The system consisted of salt
water shelves and a collection trough for the distillate water. The same materials and procedures
used for the single-stage solar still were also
employed in the construction of the multiple-stage
solar still, except that no side insulation was
used due to the short vapor path. The projected
basin area for this system was 1.1 m^2.*

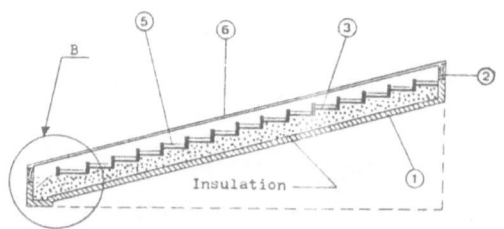

*Figure 4. Constructional Details of a
Multiple-Stage Solar Still*

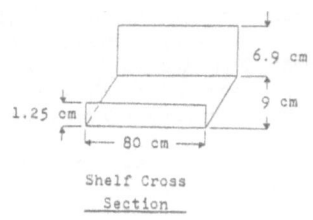

Shelf Cross
Section

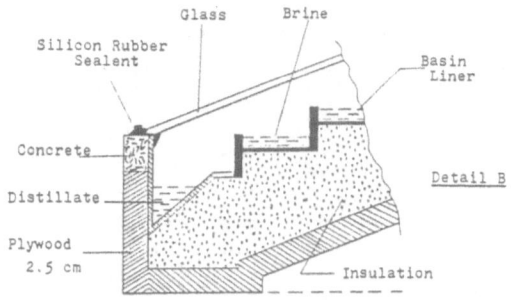

Figure 5.   Constructional Details of a
Multiple-Stage Solar Still

Experimental Set-Up

The stills were located at New Mexico State University, Las Cruces, New
Mexico. Las Cruces is located at 32°17' N latitude and 106°45' W longitude.
The ground elevation is 1183 meters above sea level. The experiments were
conducted in the months of July and August, 1980.

RESULTS AND DISCUSSION

Digitally Simulated Cases

The results from digitally simulated cases are summarized in Tables (1) to
(4.4) and are discussed below.

Discussion of Tables (1) and (2)

These tables list the cumulative daily distillates (m_e) for the following
cases:

    a)   Single-Stage Solar Still (brine depth of 2.5 cm)

    b)   Multiple-Stage Solar Still (brine depth of 1.25 cm)

In Table (1), the values of daily cumulative distillate are listed for the
single-stage solar still and the 21st of each month. These values were obtained
by using the actual values of hourly ambient air temperature and the calculated
values of hourly solar radiation for a clear day. In other columns, the sunrise hours, length of the day and total irradiation are also given. Table (2)
similarly summarizes the same type of information for the multiple-stage solar
still. The cumulative distillate listed in each table represents the sum of

the distillate during sunlit hours and the nightly distillate obtained after
sunset. The second component, i.e., nightly distillate after sunset and before
sunrise, is actually small compared to the sunlit distillate. It is seen from
Tables (1) and (2), that the daily cumulative distillate (m_e) increases from
January to June and then drops for the remainder of the year. This shows that
the cumulative distillate increases with increasing daily solar radiation and
ambient air temperature. It is also seen that the daily cumulative distillate
in the multiple-stage solar still is higher than that produced by the single-
stage solar still.

TABLE   1

Cumulative Daily Distillate (Kg/m^2) for Single-Stage Solar Still
(Theoretical Values of Solar Radiation)

Month 21st Day	Sunrise Hour	Length of the Day	Total Radiation (Kwh/m^2)	Cumulative Daily Distillate (Kg/m^2)
Jan	6:53	10.232	5.64	1.9
Feb	6:28	11.070	7.16	2.8
Mar	6:00	12.00	8.80	4.0
Apr	5:29	13.024	10.32	5.3
May	5:05	13.809	11.16	6.2
Jun	5:00	14.124	11.39	6.6
Jul	5:06	13.801	11.05	6.2
Aug	5:31	12.981	10.11	5.5
Sep	6:02	11.953	8.58	4.2
Oct	6:31	10.964	6.86	2.9
Nov	6:55	10.165	5.48	1.7
Dec	7:04	9.875	5.02	1.6

TABLE   2

Cumulative Daily Distillate (Kg/m^2) for Multiple-Stage Solar Still
(Theoretical Values of Solar Radiation)

Month 21st Day	Sunrise Hour	Length of the Day	Total Radiation (Kwh/m^2)	Cumulative Daily Distillate (Kg/m^2)
Jan	6:53	10.232	5.64	2.7
Feb	6:28	11.070	7.16	3.7
Mar	6:00	12.00	8.80	4.9
Apr	5:29	13.024	10.32	6.4
May	5:05	13.809	11.16	7.2
Jun	5:00	14.124	11.39	7.7
Jul	5:06	13.801	11.05	7.3
Aug	5:31	12.981	10.11	6.6
Sep	6:02	11.953	8.58	5.4
Oct	6:31	10.964	6.86	3.9
Nov	6:55	10.165	5.48	2.4
Dec	7:04	9.875	5.02	1.3

Discussion of Table (3)

In this table, the daily cumulative distillates for the two systems are
listed in the same way. The values of daily cumulative distillate were reached
by using the actual values of ambient air temperature and solar radiation.
Table (3) indicates that the daily cumulative distillate increases steadily
from January to its peak in June, but does not drop for the remainder of the
year in the same fashion discussed before. It occurs mainly because the hourly

*climatological values of those particular days do not follow those of the theo-
retical either closely or linearly. This could be explained by fluctuating
hourly values of solar radiation for cloudy days. For example, the months of
July and November show this more distinctively than the others.*

TABLE 3

Cumulative Daily Distillate (Kg/m^2) for Single-and Multiple-Stage
Solar Stills (Actual Values of Solar Radiation)

Month 21st Day	Total Radiation (Kwh/m^2)	Cumulative Daily Distillate (Kg/m^2)	
		Single Stage	Multiple Stage
Jan	3.21	1.1	1.7
Feb	4.51	2.0	2.8
Mar	5.19	2.9	3.6
Apr	7.70	5.0	6.1
May	7.56	5.3	6.2
Jun	7.57	6.1	7.1
Jul	6.53	4.5	5.5
Aug	7.44	5.3	6.3
Sep	6.34	4.1	5.2
Oct	5.20	2.9	3.9
Nov	1.66	0.4	0.6
Dec	3.20	1.2	1.9

Verification of the Numerical Model
Through Experimental Data

*The experimental values of the cumulative daily distillate for the two
systems considered herein are given in Table (4). Figure 6 is a typical plot
showing the cumulative distillate vs. total radiation (output vs. input) for
Table (4). It is seen from Figure 6 that the cumulative daily distillate of
the multiple-stage solar still is higher than that of the corresponding single-
stage solar still. This difference increases with increasing solar radiation.*

*The same procedures used in the digital simulation section for evaluation
of the cumulative daily distillate were also employed here for August 5th and
August 20th, and are presented in Tables (5) and (6). However, this was maily
done to check the validity of the methods used for digital simulation processes.
In Figures 7 to 17, typical plots for climatological variables, brine temperatures, cover temperatures, instantaneous distillation rate and cumulative distillate are shown as a function of time for the cases listed in Tables (5) and
(6). Figure 18 compares the actual and predicted variations of the cumulative
distillate for August 5th and August 20th. It is seen from Figure 18 that the
model predictions compared favorably with the experimental results. Nevertheless, the difference between the actual and the predicted outputs may be attributed to the assumptions that were made to calculate the hourly values of
solar radiation or to those assumptions made for the digital simulation model.*

*The predicted cumulative distillate values are higher than the actual
values. This is to be expected, because the digital simulation does not
account for the water vapor losses due to leakage, and condensate losses by
gravity from the glass cover into the brine mass. For solar stills with lower
brine depth, the distillate values are very sensitive to the brine and cover
temperatures because of the higher temperatures attained by the brine mass, or
alternatively, due to lower heat capacitance of the system. This is better
seen simply by comparing the temperatures and the instantaneous distillates
of the two systems.*

TABLE 4

List of Experiments Performed for Single-and Multiple-Stage Solar Stills

Date	Total Radiation (Kwh/m^2)	Cumulative Daily Distillate(Kg/m^2)		Increase %
		Single-Stage	Multiple-Stage	
Jul. 31, 1980	7.78	4.9	5.4	9.2
Aug. 4, 1980	5.83	2.9	3.5	17.1
Aug. 5, 1980	7.28	4.4	5.2	15.4
Aug. 6, 1980	6.03	3.9	4.7	17.0
Aug. 7, 1980	6.62	4.3	5.3	18.8
Aug. 8, 1980	7.45	3.7	4.6	19.5
Aug. 13, 1980	4.33	1.3	1.5	13.3
Aug. 14, 1980	5.07	1.8	2.2	18.2
Aug. 18, 1980	6.67	3.3	4.4	25.0
Aug. 19, 1980	4.78	1.9	2.1	9.5
Aug. 20, 1980	7.46	4.0	5.4	26.0

Overall
increase
17%

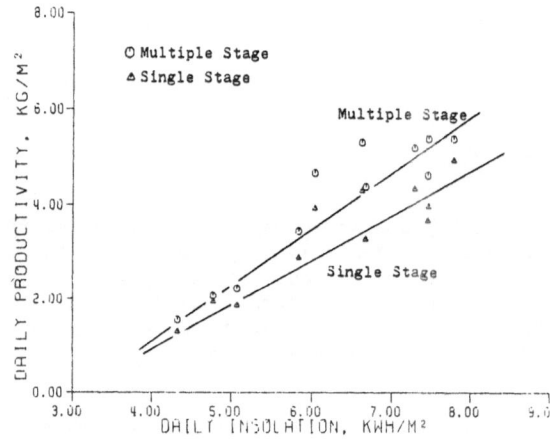

Figure 6. Comparison
in Performance of the
Single-and Multiple-
Stage Solar Stills
(Experimental Data)

TABLE 5

Cumulative Daily Distillate (Kg/m^2) for Single-and Multiple-Stage
Solar Stills (Theoretical Values of Solar Radiation)

Date	Sunrise Hour	Length of the Day	Total Radiation (Kwh/m^2)	Cumulative Daily Distillates (Kg/m^2)	
				Single-Stage	Multiple-Stage
Aug. 5, 1980	5:16	13.447	10.67	6.1	7.1
Aug. 20, 1980	5:30	13.012	10.15	5.5	6.5

TABLE 6

Cumulative Daily Distillate (Kg/m^2) for Single-and Multiple-Stage
Solar Stills (Actual Values of Solar Radiation)

Date	Total Radiation (Kwh/m^2)	Cumulative Daily Distillate (Kg/m^2)	
		Single-Stage	Multiple-Stage
Aug. 5, 1980	7.28	5.3	6.2
Aug. 20, 1980	7.46	5.3	6.3

The increase in the daily cumulative distillate values is 9% to 26%, as can be seen from Table (4). This is due to the reuse of the latent heat of vaporization by introducing steps into the basin to provide smaller volumes, i.e., a multiple-stage solar still. The latter, of course, contributes in lower time requirements for evaporation of the brine mass to begin and a larger quantity of distillate.

The reasons limiting the prediction accuracy of the digital simulation model for single-and multiple-stage solar stills presented in this study are as follows:

a) Water vapor losses due to leakage

b) Condensate losses from the glass cover, by gravity, into the brine mass.

The aforementioned may also be considered to improve the productivity of multiple-stage solar stills over single-stage solar stills, since the former has a shorter vapor path.

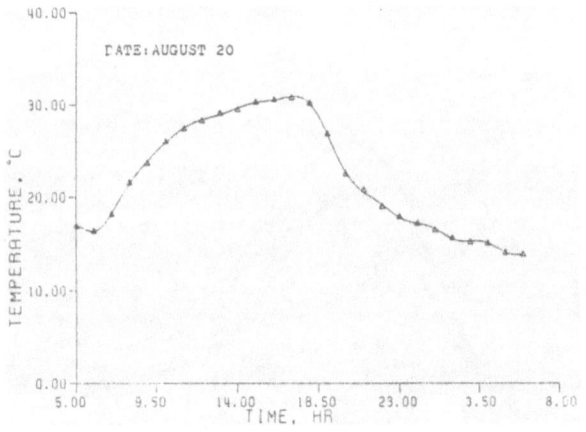

Figure 7. Ambient Air Temperature for Las Cruces, N.M., for the Year 1980

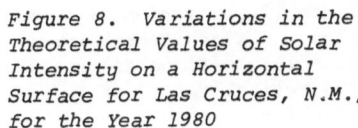

Figure 8. Variations in the Theoretical Values of Solar Intensity on a Horizontal Surface for Las Cruces, N.M., for the Year 1980

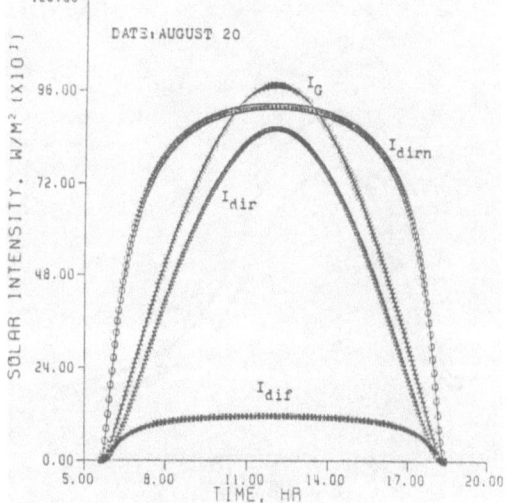

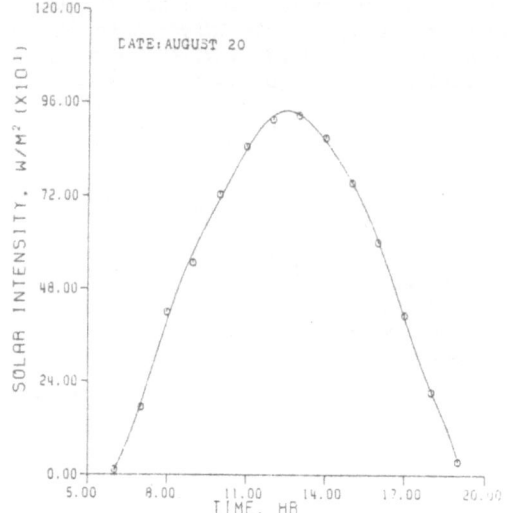

Figure 9. *Variation in the Actual Values of Solar Intensity on a Horizontal Surface for Las Cruces, N.M., for the Year 1980*

Figure 10. *Variation of Hourly Brine and Glass Temperatures for Single-Stage Solar Still (Theoretical Values of Solar Radiation)*

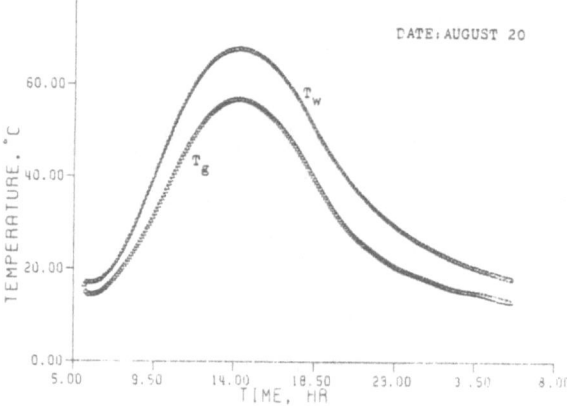

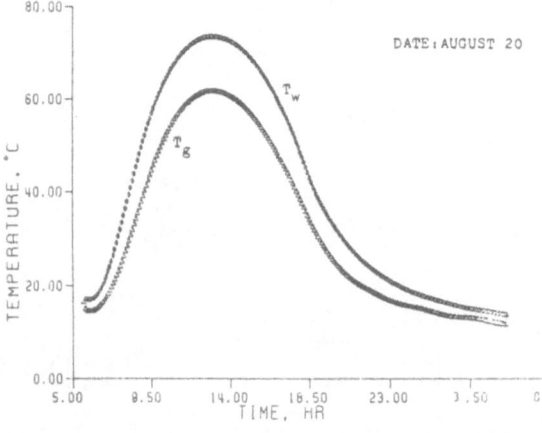

Figure 11. *Variation of Hourly Brine and Glass Temperatures for Multiple-Stage Solar Still (Theoretical Values of Solar Radiation)*

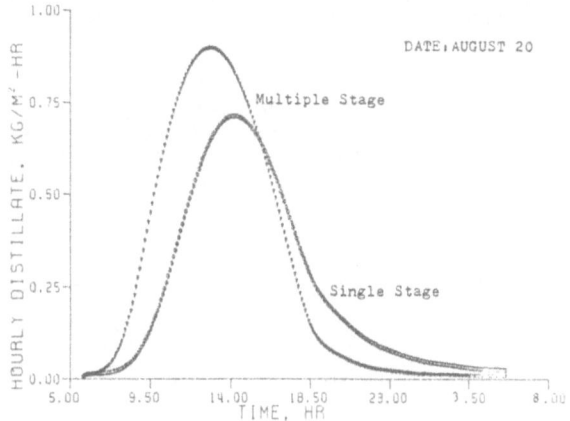

Figure 12. Variation of Hourly Distillates for Single-and Multiple-Stage Solar Stills (Theoretical Values of Solar Radiation)

Figure 13. Cumulative Distillates for Single-and Multiple-Stage Solar Stills (Theoretical Values of Solar Radiation)

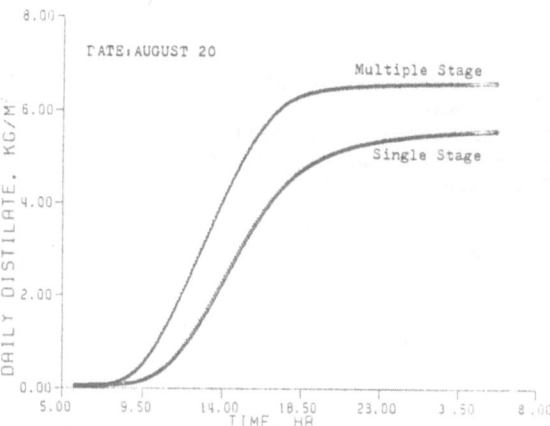

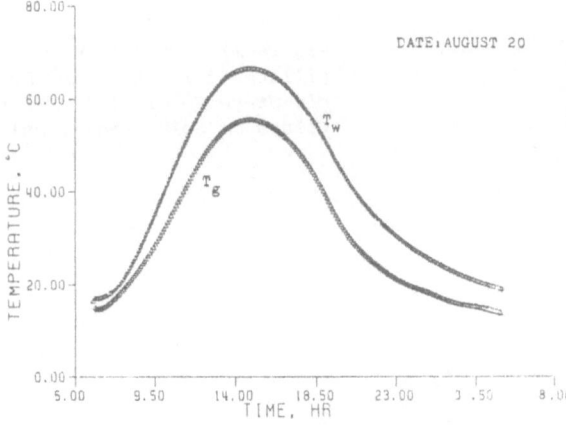

Figure 14. Variation of Hourly Brine and Glass Temperatures for Single-Stage Solar Still (Actual Values of Solar Radiation)

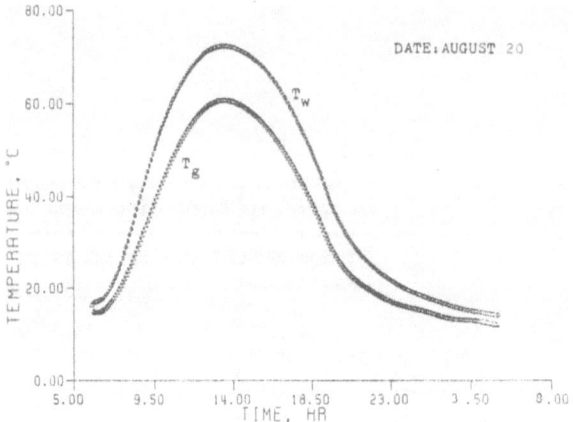

Figure 15. Variation of Hourly Brine and Glass Temperatures for Multiple-Stage Solar Still (Actual Values of Solar Radiation)

Figure 16. Variation of Hourly Distillates for Single-and Multiple-Stage Solar Stills (Actual Values of Solar Radiation)

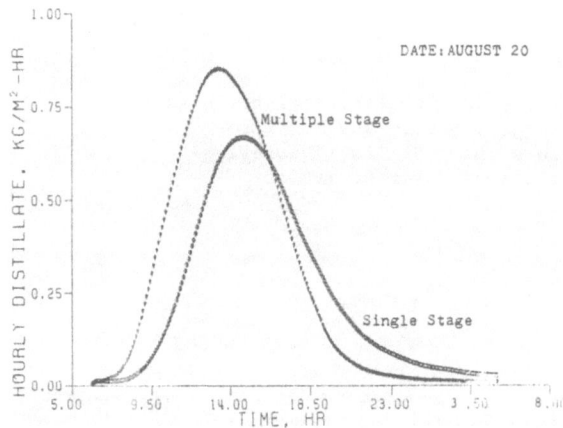

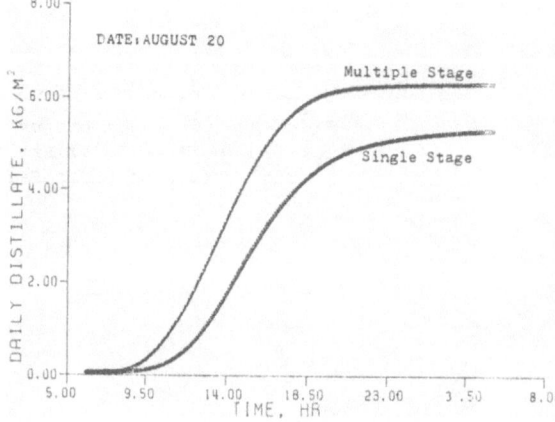

Figure 17. Cumulative Distillates for Single-and Multiple-Stage Solar Stills (Actual Values of Solar Radiation)

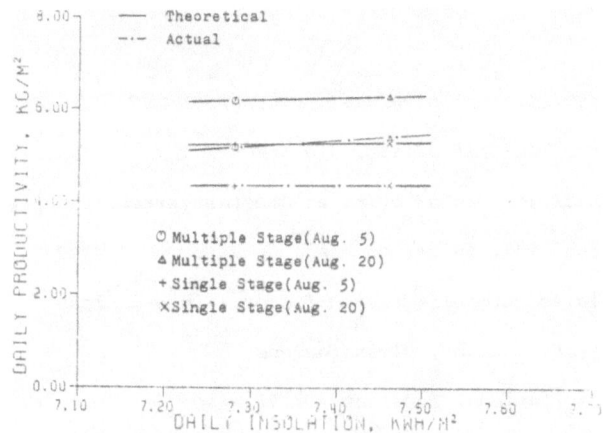

Figure 18. Comparison in Performance of the Single-
and Multiple-Stage Solar Stills for
August 5th & 20th

CONCLUSIONS

There is considerable evidence that the productivity of a solar still increases with decreased vapor path. The temperatures of the stills increase
with reduced thermal losses affected by insulation, since these are the major
sources of heat loss.

It is seen in this study that the output of a single-stage solar still
can be maximized by introducting more steps into the basin, i.e., by making it
a multiple-stage solar still. The above-mentioned have been proven in this
study, using climatological data for Las Cruces, New Mexico. The study, however, is flexible and can be readily adapted for any part of the world, by
using appropriate values of the climatological variables.

The higher productivity rate of the multiple-stage solar still constitutes
an attractive alternative to other types of solar stills; specifically, the
higher distillate production compensates for the increased cost of construction
and operational difficulties. As mentioned in the introduction, the use of
this study in creating an integrated food/water/power complex will tend to
make the whole system more economical. The concept becomes more suitable for
isolated communities which are far removed from power grids, have access only
to sea or brackish water, and where vegetables can not grow through outside
irrigation. There are a large number of such communities spread throughout
Asia and Africa where the availability of land area and solar radiation for
solar stills is not a problem.

NOMENCLATURE

Symbols

C_g	*specific heat of glass cover, J/Kg - $^\circ C$*
C_p	*specific heat of air, JKg - $^\circ C$*
C_{pw}	*specific heat of brine at constant pressure, J/Kg - $^\circ C$*
D_{AW}	*mass diffusivity, cm^2/sec*
F_s	*radiation configuration factor, dimensionless*
Gr	*Grashof number, dimensionless*
g	*gravitational constant = 9.81 m/sec^2*
h_{ca}	*convective heat transfer coefficient between cover and air, $watt/m^2 - ^\circ C$*
h'_{cw}	*convective heat transfer coefficient from brine to cover, $watt/m^2 - ^\circ C$*
h_{fg}	*latent heat of vaporization, J/Kg*
h_m	*mass transfer coefficient, m/sec*
I_{dif}	*diffuse solar radiation, $watt/m^2$*
I_{dir}	*direct solar radiation, $watt/m^2$*
I_{dirn}	*direct normal solar radiation, $watt/m^2$*
I_G	*global solar radiation, $watt/m^2$*
I_T	*total solar radiation, $J/m^2 - day$*
K	*thermal conductivity of air, $watt/m - ^\circ C$*
$\dot{m}$	*mass transfer rate, $Kg/m^2 - sec$*
m_e	*cumulative daily distillate, Kg/m^2*
m'_e	*instantaneous distillate, $Kg/m^2 - hr$*
M	*brine mass per unit basin area, Kg/m^2*
M_a	*molecular weight of air*
M_w	*molecular weight of water*
n	*constant*
Nu	*Nusselt number, dimensionless*
P_{Bm}	*logarithmic mean partial pressure, Pa*
Pr	*Prandlt number, dimensionless*

P_w partial pressure of water vapor at brine temperature, Pa

P_{wg} partial pressure of water vapor at glass temperature, Pa

P_t atmospheric pressure, Pa

q_b conduction heat loss from brine mass to ground, watt/m^2

q_{ca} convective heat loss from cover to air, watt/m^2

q_{cw} convection heat loss from brine mass to cover, watt/m^2

q_{ew} evaporative heat loss from brine mass to cover, watt/m^2

q_{ra} radiative heat transfer from cover to sky, watt/m^2

q_{rw} radiative heat transfer from brine mass to cover, watt/m^2

q_s conduction heat losses from sides, watt/m^2

R gas constant = 8.205 X 10^{-2} m^3 , atm/kg mol - $^\circ K$

Sc Schmidt number, dimensionless

Sh Sherwood number, dimensionless

t time, hour

T temperature, $^\circ C$

T_a ambient air temperature, $^\circ C$

T_{fw} feedwater temperature = T_a , $^\circ C$

T_g cover temperature, $^\circ C$

T_{sky} sky temperature = T_a - 12, $^\circ C$

T_w brine mass temperature, $^\circ C$

U_b & U_s overall heat transfer coefficients (bottom & sides), watt/m^2 - $^\circ C$

V wind speed, m/sec

X average distance between the brine surface and the glazing (vapor path), m

Greek Symbols

α_g absorptivity of glass, dimensionless

α_w effective brine mass absorptance, dimensionless

β coefficient of thermal expansion of the air-water mixture, 1/$^\circ C$

ε_g emissivity of glazing (cover), dimensionless

μ *absolute viscosity of air-water mixture, Kg/m - hr*

ν *kinematic viscosity of air-water mixture, m^2/sec*

ρ *density of air-water mixture, Kg/m^3*

τ *transmittance, dimensionless*

σ *Stefan-Boltzmann constant, $watt/m^2 - {}^\circ C^4$*

σ_{AB} *collision diameter, $\overset{O}{A}$ $(\overset{O}{A} = 10^{-10}m)$*

$\Omega_{D,AW}$ *collision integral or dimensionless function depending on temperature and intermolecular forces*

BIBLIOGRAPHY

1. Harding, J. "Apparatus for Solar Distillation." *Proceedings of Institute of Civil Engineers, Vol. 73, 1883, pp. 284-288.*

2. Telkes, M. "*Solar Distiller for Life Raft.*" *United State Office of Science, R & D Report No. 5225, P.B. 21120, 1945.*

3. McLeod, L. H. and McCracken, H. "*Performance of Greenhouse Solar Stills.*" *Sea Water Conversion program, University of California, Series 75, Issue 26, 1961, pp. 1-57.*

4. Sayigh, A.A.M. *Solar Energy Engineering. Academic Press, New York, 1977, pp. 431-464.*

5. Battelle Memorial Institute. "*Second Two Years Progress on Study and Field Evaluation of Solar Sea-Water Stills.*" *United States Department of Interior, Office of Saline Water, R & D Report No. 147, July 1965, pp. 1-86.*

6. Bloemer, J. W., Eibling, J. A., Irwin, J. R. and LoF, G.C.G. "*Development and Performance of Basin-Type Solar Stills.*" *Proceedings of Solar Energy Society Annual Meeting, Phoenix, Arizona. Solar Energy Journal, Vol. 9, No. 4, March 1965, pp. 197-200.*

7. LoF, G.C.G. "*Design and Operating Principles in Solar Distillation Basins.*" *Saline Water Conversion, Advances in Chemistry, Series No. 27, American Chemical Society, Washington, D.C., 1960, pp. 156-165.*

8. Bloemer, J. W., Gollins, R. A. and Eibling, J. A. "*Field Evaluation of Solar Sea-Water Stills.*" *Saline Water conversion, Advances in Chemistry, Series No. 27, American Chemical Society, Washington, D.C., 1960, pp. 166-177.*

9. Baum, V. A. and Bairamov, R. "*Prospects of Solar Stills in Turkmenia.*" *Solar Energy Journal, Vol. X, No. L, 1966, pp. 38-40.*

10. Garret, C. R. and Farber, E. "*Performance of a Solar Still.*" *ASME Publication, paper No. 61-SA-38, 1961, pp. 1-18.*

11. Hay, H. R. "V-Cover Solar Still." *Sun at Work,* 2nd Quarter, 1966, Vol. II, No. 2, pp. 6-9.

12. Khan, E. U. "Practical Devices for the Utilization of Solar Energy," *Solar Energy Journal,* Vol. VIII, No. 1, 1964, pp. 17-22.

13. Telkes, M. "Distillation With Solar Energy." Proceedings of World Symposium on Applied Solar Energy, 1955, Phoenix, Arizona, pp. 73-79.

14. Shirani, H. N. *Comparison In Performance of Single And Multiple Stage Basin Type Solar Stills*. Master's Thesis, New Mexico State University, December 1981.

15. Dunkle, R. V. "Solar Water Distillation: The Roof Type Still and a Multiple Effect Diffusion Still." Heat Transfer Conference, Part V, International Developments in Heat Transfer, University of Colorado, 1961, pp. 895-902.

16. Cooper, P. I. "Digital Simulation of Transient Solar Still Processes." *Solar Energy Journal,* Vol. 12, 1969, pp. 313-331.

17. Cooper, P. I. "The Maximum Efficiency of Single-Effect Solar Stills." *Solar Energy Journal,* Vol. 15, 1973, pp. 205-217.

An Approach for Cooling by Solar Energy

S.M. RABEIH, M.A. WAHHAB, and H.M. ASFOUR
Faculty of Engineering
Minia University
Minia, Egypt

Abstract:

The potential of solar energy and importance of its use in
Egypt have been discussed. Data for distribution of total insolation, temperature , wind velocity and relative humidity for each
day of the year at Minia governorate was collected. Calculation
of solar energy parameters for the whole year in Minia was made.

An experimental test rig was designed and erected. The apparatus consists of the collection unit and the refrigerator. The
measuring instrumentation used was explained.

A computer program covering calculations of all required parameters in the system was given. This program gives the quantity
of heat added to the generator of the refrigerator unit , efficiency of the used collection unit, angle of incidence and best angle
for inclination of the collection unit to give maximum efficiency.

The quantity of heat needed for the operation of the refrigerator at different temperatures and cooling loads during the whole
day-time was determined. It was proved that this quantity of heat
can be provided from the used collection unit. The required modifications on the generator of the refrigerator were made in order
to qaurantee the transfer of collected heat to the working fluid
in the refrigerator

Nomenclature

AL	:	Solar altitude angle, degree
B	:	Plate tilted angle , degree.
C	:	The value of measured amount of dust intensity, %.
Cp	:	Fluid specific heat, J/ Kg C.
D	:	Dirty factor,
DA	:	Declination angle , degree.
ET	:	Equation of time.
F_R	:	Heat removal factor,
F_c	:	Transmittance -Absorptance factor,

F   : Transmittance -Asorptance product,
F^e : Factor relating the surface angle to sky
G^{ss} : Mass flow-rate per unit area ,Kg/sec Cm^2
h_r : hour angle , degree
K   : Thermal conductivity, W/m ^oC.
N   : number of covers,
PA   : Plate angle, degree.
q_a : Solar energy absorbed , KJ/m^2
q_u : Useful heat collection , KJ/m^2
s   : Solar constant , 1377 W/m^2
S   : Shading factor,
SA   : Solar angle , degree
V   : Wind speed, m/s.
W   : Depth of precipitable water, Cm.
ZA   : Zenith angle , degree
θ : Incidence angle , degree
γ : Plate angle , degree
ϵ : Emissivity factor.

Introduction

     The proven conventional energy reserves suffer the ever-incr-
easing exhaustion rates by mankind. Actual threats are twoways enhanced; either by population rise and/or rapid technological progress that man realises . Such a position leads to hesitation as to
whether the present traditional energy conversion systems are reliable for long intervals of time , even with new discoveries. On
the other hand, dependence   on other renewable energy resources
as supports or supplements to the limited and exhaustible energy
fossils sounds as hopeful grounds. The role of the free, clean and
abundant solar energy may be economic and prominant in sharing to
solve the world energy crisis.

     Investigations of solar energy have covered many fields, such
as liquid and air heating (1), storage of heat (2), drying of agr-
icultural products, desalination ,and cooling. The absorption cycle of ammonia may be an appropriate cycle to use solar energy for
cooling because of the small amount of work required for its operation (3).

     It is the object of this work to investigate requirements to
change -over a household refrigerator from the use of a gas fuel
energy resource to the nonconventional solar energy. The heating
system is to be substituted by a two glass-cover designed solar
collector with an absorption area of 2 m^2 . Required changes in the
generator design are studied. Solar parameters at the location are
considered. Some variables such as incidence angle of light, fluid
nature and flow rate , temperature, cooling load and other relevant factors are discussed.

Mathematical considerations:

The mathematical procedures required for the thermal analysis of the used solar collector is given in this section. Factors governing the heat transfer processes are, the environmental factors such as ambient temperature, relative humidity, wind speed, cloud, dust and seasonal variations; the operating factors such as liquid flow rate, inlet fluid temperature and collector orientation; finally the collector design factors which include the collector dimensions, construction material and their properties.

As mentioned by SAKR (5), the direct solar energy incident is given by :

$$I_D = 1377 \ r \ \mathcal{Z}^m$$

The total transmissivity factor I is calculated (5) as:

$$\ln \mathcal{Z} = -0.29 - 0.012 \ W$$

The air mass is calculated from the following expression (7)

$$m = 1/ \cos (ZA)$$

where,

$$ZA = \cos^{-1}(\emptyset)$$

and,

$$Cos \ \emptyset = Cos \ (DA) \ Cos \ (AL) \ Cos \ SA + Sin(DA) \ Sin \ (AL).$$

The diffused radiation is evaluated as (6)

$$I_d = I_D \times Fss \times C$$

where,

$$Fss = 1 + Cos \ (PA) \quad / \ 2$$

and C is the value of the measured amount of dust intensity percent (6)
The total amount of incident energy at the considered lift angle is

$$I_t = I_D \ Cos \ \theta_t + I_d$$

where θ_t is obtained from Ref (7).

$$Cos \ \theta_t = Cos(DA) \ Cos \ (AL - PA) \ Cos(SA) + Sin \ (DA)$$
$$Sin \ (AL - PA)$$

The solar angle (SA) is determined as given in Ref. (4).

$$SA = \left[(hr - 12) \times 60 + ET + 4 \right] \ / \ 4$$

The transmittance of the used glass covers(7)is calculated from the following equation (7).

$$T = e^{-kL} \ (\ \frac{1- \ r}{1 + r})$$

where r is surface reflectance factor which equals 0.1188 if the difference between the reflection angle ($\emptyset$) and incidence angle($\emptyset$)

equals zero and

$$r = \tfrac{1}{2} \left[\frac{\sin^2(\theta - \theta')}{\sin^2(\theta + \theta')} + \frac{\tan^2(\theta - \theta')}{\tan^2(\theta + \theta')} \right]$$

The effect of dirt and shading factor is considered through the
following formula (7).

$$q_a = I_t F_e (1 - D)(1 - S)$$

where,

$$F_e = F_c + a_1(1 - e^{-K_1 L_1}) + a_2 T_2(1 - e^{-K_2 L_2})$$

$$F_c = 1.012 \, T_1 T_2 \alpha$$

$$= 1 - e^{(0.0255 - 6.683x \, 5.94 \, x^2 - 2.48 \, x^3)}$$

and $X = \cos \theta$

To obtain the useful heat collected the upper , edge and rear
losses must be determined. The upward heat transfer coefficient
(4).

$$U_{up_{45}} = \left[\frac{N}{(344/Tp)\left[(T_p - T_a)/(N + f)\right]^{0.31}} + \frac{1}{H_w} \right]^{-1} +$$

$$+ \frac{6(T_p - T_a)(T_p^2 - T_a^2)}{\left[\epsilon_p + 0.0425 \, N(1 - \epsilon_p)\right]^{-1} + \left[(2N + F - 1)/\epsilon_g\right] - N}$$

The wind heat transfer coefficient H_w is given as

$$H_w = 5.7 + 3.8 \, V$$

and the factor F can be determined from the following formula(4)

$$F = (1 - 0.4 \, H_w + 0.0005 \, H_w^2) \, 1.116$$

Considering the absorber plate angle, the upper heat loss coefficient becomes

$$U_{up_B} = U_{up_{45}} \left[1-(B -45)(0.00259-0.00144 \epsilon_p) \right]$$

$$U_{real} = \frac{K}{L}$$

$$U_{edge} = \frac{K \times Aedge}{L \times Acollector}$$

The solar energy absorbed by the collector is

$$Q_a = A_c F_R \left[q_a - U_T (T_i - T_a) \right]$$

where,

$$F_R = \frac{G Cp}{U_T} \left[1-e^{-(U_T F'/ G Cp)} \right]$$

and

$$F' = \text{efficiency factor} = \frac{1/U_T}{w \left[\frac{1}{U_T \left[D+(w-D)F \right]} + \frac{1}{C_b} + \frac{1}{\pi D_i h_f} \right]}$$

F is the fin efficiency as given in Ref (4,5)

$$F = \frac{\tanh \left[m (w - D)/2 \right]}{m (w - D)/2}$$

where W is the distance between the centere of the tubes and D is the outer diameter of the tube.

and C_b is the bond conductance

$$C_b = \frac{Kb}{\gamma}$$

The collector efficiency is expressed as

$$\eta_c = \frac{Q a}{A_c I_t}$$

It is evident that the upper , rear and edge heat transfer coefficients are functions of T_b, T_c, T_a, the incidence angle and collector orientation . A computer program was developed to evaluate the fluid temperature, heat absorbed, heat lost at bottom, upper and edge of collector and collector efficiency for any set of given design parameters and environmental conditions. The computer program is available on request (9).

Test rig and method

A schematic diagram of the experimental test rig is shown in Figures (1) and (2) . The flat plate solar collector has an area of $2m^2$ and its plane can be varied to different inclinations.

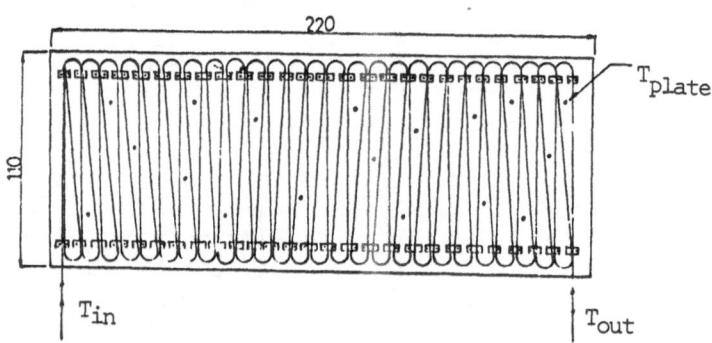

Figure 1.   Absorber

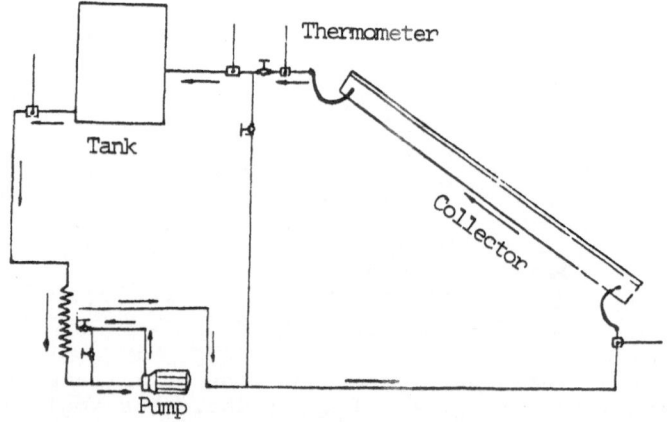

Figure 2.   Main Circulation

Measurement of different running parameters of the experimental set together with ambient conditions were taken each half
an hour every day beginning from 8 a.m. until 6 p.m. A sample of
these results is given in Fig. 3 . These data are processed also
for the theoretical analysis by the computer program mentioned
before.

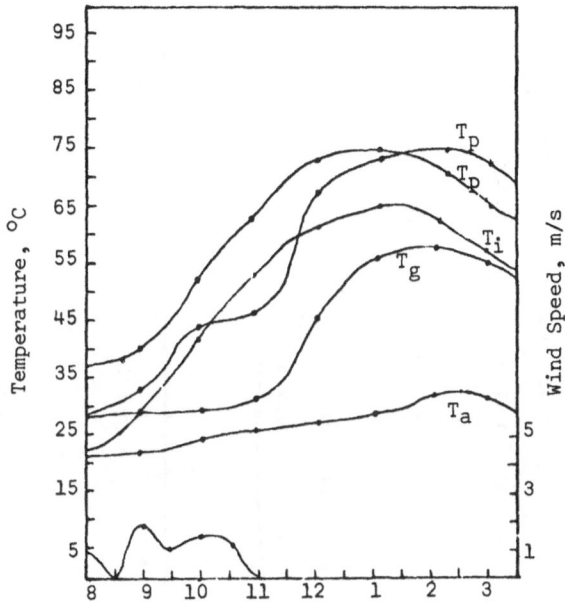

Figure 3. Temperature Distribution and Wind Speed on 19-11-1980

For determination of solar insolation, the measured air
relative humidity for the Minia governorate for the whole year
is given in table (1) . Moreover the sunrise and sunset were determined and represented in Fig. (4). They were part of the input
data to the computer.

The calculations were run every two hours from 6 a.m.until 6
p.m. The selected days are the 1$\underline{st}$, 8$\underline{th}$, 15$\underline{th}$and22 nd every month
of the year. Inclination of collector was varied from 0^0 to 98^0.

Based on Ref.(10)the analysis of the absorption refrigerator
system was made and given in the detailed study(9). For two cooling
loads $\frac{1}{12}$ and $\frac{1}{8}$ ton, the coefficient of performance , quantity
of heat input, heat rejection , maximum C.O.P . and the relative

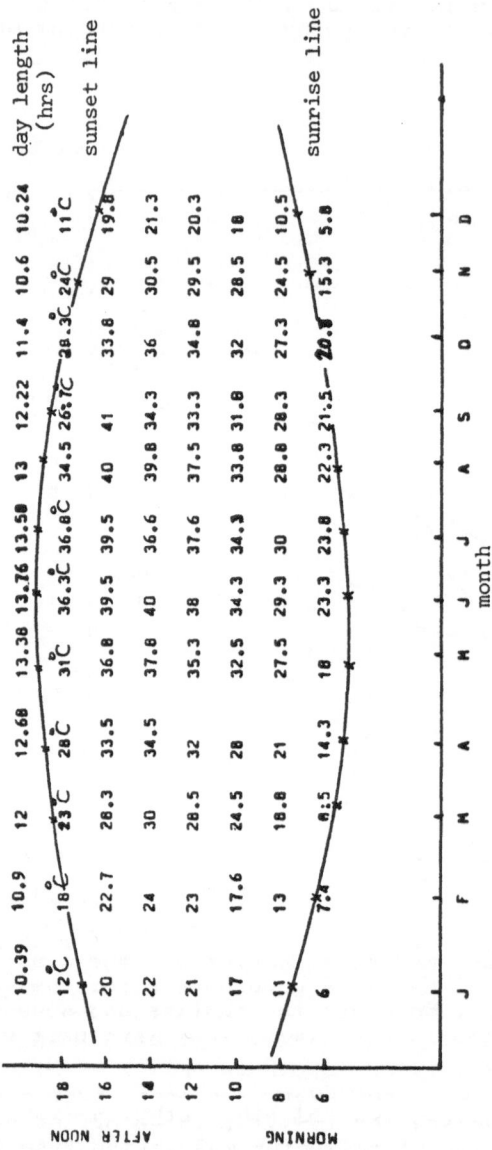

Figure 4. Mean Temperature in Minia, Sunrise-Sunset, and Day Length

TABLE-I. MEAN RELATIVE HUMIDITY IN MINIA THROUGHOUT THE YEAR.

Time, hr	J	F	M	A	M	J	J	A	S	O	N	D
18	45.5	40.3	34.3	28.8	38.8	30.5	30.5	29	36.5	38.8	63	64
16	40.5	3 1	29.5	27	29.8	28.8	29.3	24.3	39.8	29.5	49.8	53.5
14	40.0	35.7	30	38.5	29.8	33.8	35.5	29.3	36.5	32.3	47.8	49.5
12	39.5	36.7	32.5	39.8	36.3	40.8	41.8	41.8	43.3	41.5	53.8	52.8
10	58.0	57.6	43.8	47.5	46.3	56.8	55	54.3	49.8	56.3	58	58.8
8	79.3	74.5	72.5	77.5	61.3	71.8	76.3	76.8	68.8	76.3	82.3	83.5
6	89.6	80.7	90	94.5	91	88	96.8	98	99	94.3	96.8	96.6

Month

efficiency were evaluated. The generator temperature has been
varied from 85 °C to 115°C . Table 2 shows the results of both
cooling loads in a summarized form   Figure 5 shows   the required

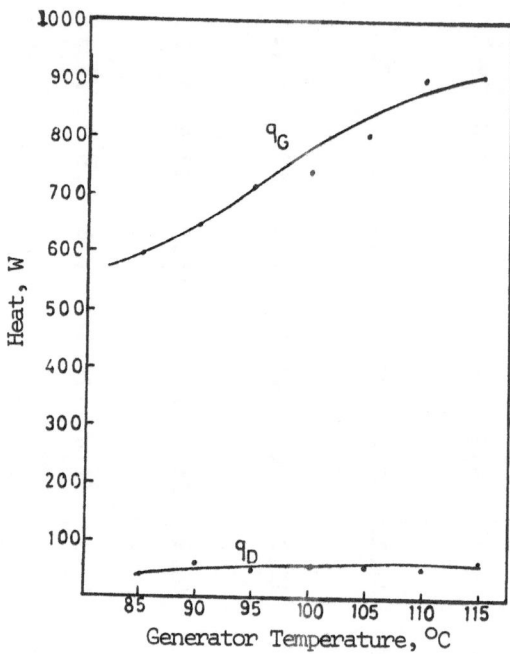

Figure 5.   Heat Required to Operate the Generator at 1/8 Ton Load

heat which must be supplied by the solar collection unit to
operate   the generator at 1/8 ton load.

TABLE 2. PERFORMANCE OF REFRIGERATOR

$T, °C$	$\dot{m}_7 = kg/hr$	$\dot{m}_6 = kg/hr$	$q_G = W$	$q_D = W$	C.O.P.	$C.O.P._{max}$	η_R	LOAD
85	1.38	3.8	600.8	40.358	.7317	1.033	0.71	
90	1.39	4.8	652.17	58.29	.674	1.1	0.61	
95	1.39	5.85	718.36	49.327	.612	1.175	0.52	
100	1.47	6.4	743.86	60.77	.59	1.15	0.51	1/8 TON
105	1.46	8.32	806.158	56.73	.545	1.21	0.45	
110	1.45	8.23	904.18	51.58	.486	1.27	0.38	
115	1.38	8.12	907.6	62.78	.484	1.4	0.34	
85	0.925	2.5	400.54	26.9	.732	1.033	0.7	
90	0.925	3.2	434.4	38.88	.675	1.1	0.61	
95	0.925	3.9	479.47	32.89	.61	1.175	0.52	1/12 TON
100	0.966	4.3	499.6	40.58	.587	1.15	0.51	
105	0.975	5.5	537.2	37.8	.546	1.21	0.45	
110	0.97	5.48	602.4	34.5	.485	1.27	0.38	
115	0.925	5.44	608.45	41.87	.482	1.339	0.36	

Results and discussion:

      The obtained results are given in diagram form, A sample of
these results was given in Fig.3. From this figure and others
given in the detailed studies (9), simpler relationships between inlet temperature, plate temperature and air temperature have
been deduced.

Time	Tinlet °C	Tplate °C
8 - 10	T_{air} + 20.32	T_{air} + 26.1
10 -12	T_{air} + 21.5	T_{air} + 29.5
12 -2	T_{air} + 27.4	T_{air} + 33.2
2 - 4	T_{air} + 25.85	T_{air} + 33.5

These relations are only valid for summer period. Similar

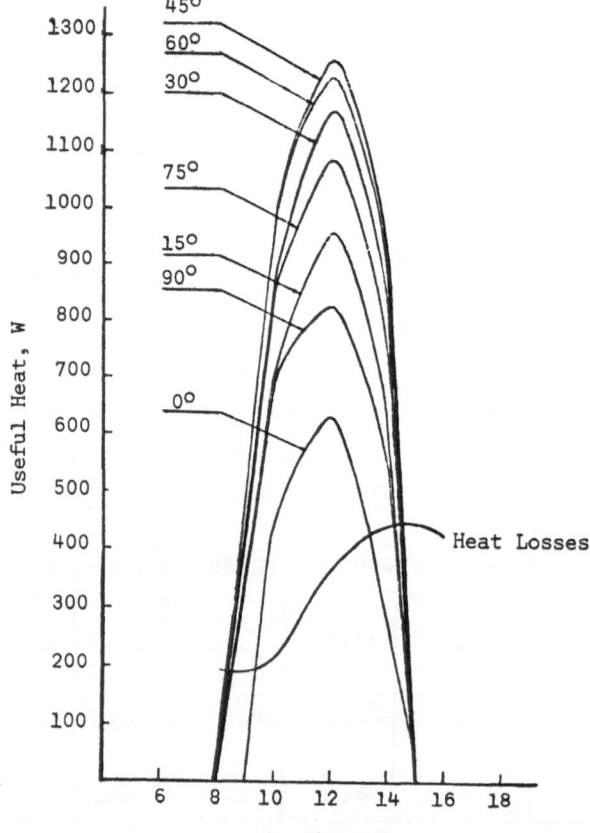

Figure 6. Distribution of Useful Solar Radiation in January, 1980

measurements covering the whole year were made and the deduced
relations are given in Ref (´9). These relations were used in the
computer program.

Figure 6 gives an example of the calculated amount of useful energy against time for January 1980. The results for the
whole months of the year are given elsewhere (9) . The area under
each curve represents the amount of useful heat for a complete
day at a certain angle of incidence.

The behaviour of thermal losses is shown also on the same
diagram. It is evident that the curves of thermal losses for the
different angles of incidence are approximatly coincident. Therefore only one curve for the thermal losses is plotted. Figure 7
illustrates different curves for the thermal losses of January
where it is obvious that the effect of the angle of incidence on
the thermal losses is very small and can be neglected.

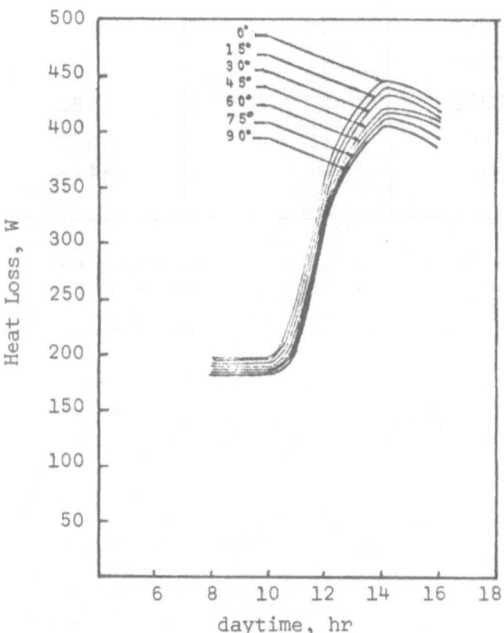

Figure 7.   Thermal Losses in January at
Different Angles of Incidence

following table (3) gives the maximum and minimum obtained useful
heat throughout the year.

TABLE 3.   MAXIMUM AND MINIMUM USEFUL HEAT FOR
THE USED COLLECTOR ($2m^2$ area)

Month	Useful heat KJ/day			
	maximum	Inclination angle	minimum	Inclination angle
Jan.	21600	45°	8730	0°
Feb.	25830	45	9360	90
March	27000	30	6660	90
April	29610	15	810	90
May	28980	15	3600	75
June	30150	0	0	90
July	31860	0	0	90
August	28350	15	0	90
Sep.	27810	15	4680	90
October	22320	30	8280	90
November	26100	45	7830	0
December	20860	45	7110	0

The average monthly value of useful heat for different angles
of incidence have been determined (9). It is noticed that the area

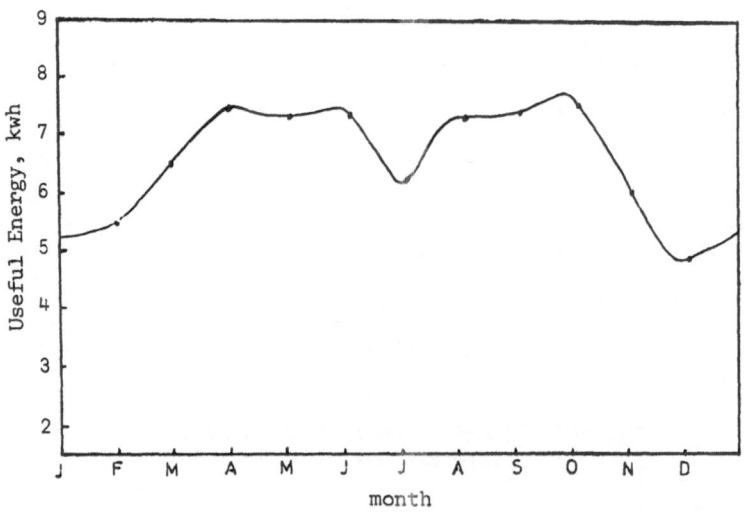

Figure 8.   Distribution of Solar Radiation at a
Plate Angle of 30°

under curve represents the total yearly amounts of useful heat col-
collected at a certain angle of incidence. Comparing these areas
it is evident that the greatest amount of total collected useful
heat (8518.5 MJ/year) for a whole year is reached with 30° angle
of incidence Fig. 8 . Therefore it is concluded that the best
angle of incidence at Minia is 30° .

Taking the best angle of incidence for every month individually and assuming the collector to be tilted at this angle for
every related month, a total amount of useful energy of 9666 MJ/
year can be reached . This means a gain of about 1148 MJ/ year
more than the collected heat from the collector with 30° incidence over the year can be reached as shown in Figure 9 . The amount of heat loss every month is shown in Fig. 10.

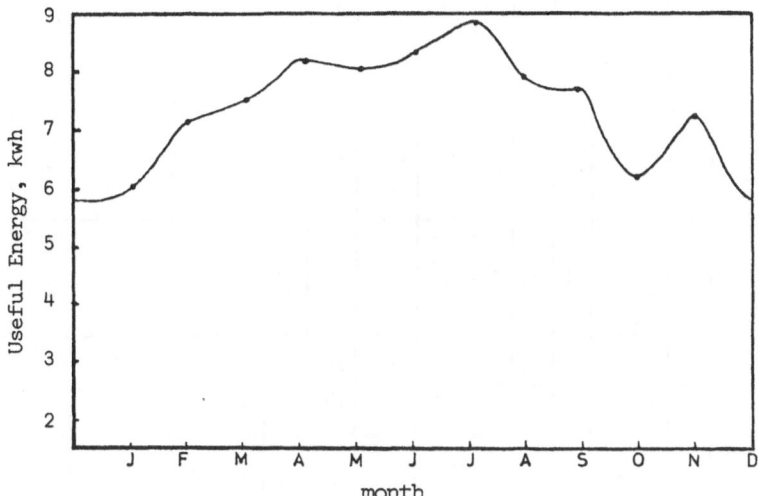

Figure 9. Distribution of Useful Energy at the Best Angle

It is seen that the difference is not too much. The total amount
of heat loss in the year is about 3169.8 MJ/year. The values of
useful heat every month at different angles of incidence,the yearly values is also given in table 4 . Table 4 reflects,again the
conclusion that the best angle at Minia over the year is 30°. It
is obvious that the average daily amount of total energy is greatest at 30°. Also the conclusions obtained before are summarised
in this table.

TABLE 4. MONTHLY, AVERAGE DAILY, AND YEARLY AMOUNTS OF USEFUL HEAT

MONTH	0°	15°	30°	45°	50°	75°	90°	THERMAL LOS-SES KW.hr.
JANUARY	2.425	4.325	5.500	6.000	5.825	5.225	3.600	2.725
FEBRUARY	3.375	4.625	6.550	7.175	6.825	5.350	3.725	2.600
MARCH	5.525	7.475	7.500	7.500	6.475	4.400	1.850	2.175
APRIL	8.000	8.225	7.325	6.925	5.225	2.150	2.250	2.300
MAY	6.600	8.050	7.375	5.200	2.750	1.000	0.000	2.325
JUNE	8.375	7.825	6.150	4.125	2.100	1.500	0.000	2.450
JULY	8.850	8.400	7.250	5.400	2.675	5.500	0.000	2.300
AUGUST	7.525	7.875	7.300	5.900	3.725	1.450	0.000	2.100
SEPTEMBER	6.600	7.725	7.625	7.050	5.275	3.625	1.300	2.125
OCTOBER	4.600	5.825	6.200	5.950	5.700	4.350	2.300	1.950
NOVEMBER	2.175	4.125	4.800	7.250	5.100	4.350	3.000	2.175
DECEMBER	1.975	1.750	5.275	5.800	5.700	5.125	4.300	2.225
TOTAL ENERGY EVERY YEAR KW/year	2017.5	2340	2366.25	2268.75	1768.5	1182	648	880.5
AVERAGE ENE-RGY KW.hr/day	5.604	6.500	6.5729	6.302	4.912	3.283	1.800	2.446

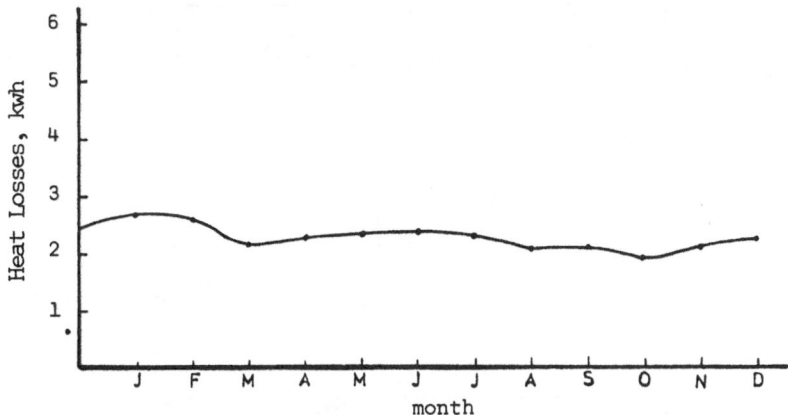

Figure 10. Distribution of Thermal Losses

        The calculated collector efficiency for January is given in
Fig 11. The other results for the whole year is given elsewhere (9) . From these diagrams it is seen that the highest efficiency is obtained at 12 O'clock in summer time and 11 O'clock in
the winter . Considering these diagrams , the best efficiency for
every separate month and the corresponding angle of incidence have
been deduced . The results are shown in table 5 together with the
efficiency obtained at the suitable angle of incidence for the
site . 30°.

                TABLE 5.   BEST EFFICIENCY AND ANGLE OF INCIDENCE

Month	J	F	M	A	M	J	J	A	S	O	N	D
efficiency	66	67	69	69	70	70	72	70	71	70	68	68
angle ,deg.	60	45	45	15	15	0	15	15	15	30	60	45
efficiency at 30° incid- ence.	64	66	68	68	65	65	69	69	69	70	67	65

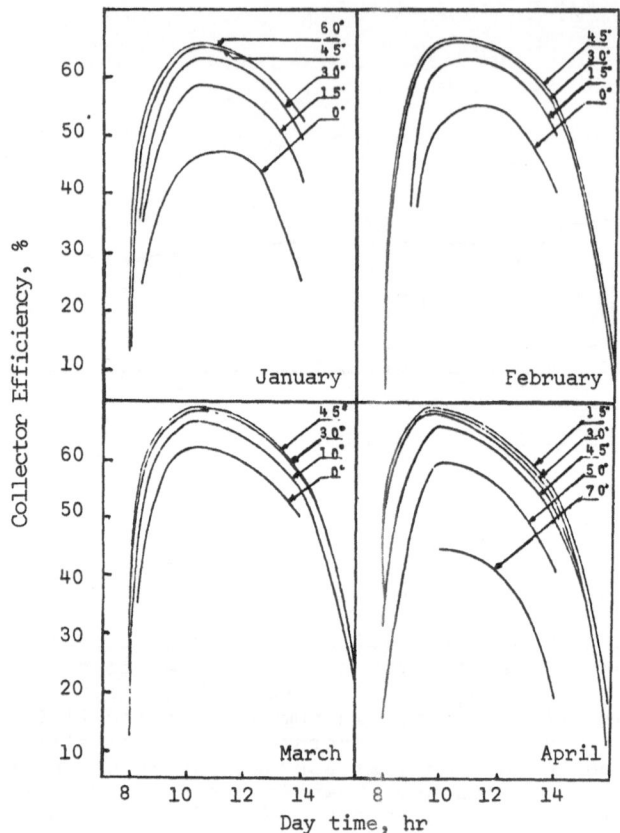

Figure 11. Collector Efficiency versus Day Time for Various Angles of Incidence

As previously deduced the suitable angle of incidence at Minia for the whole year has found to be 30° . For this angle the resulting efficiency every month is as mentioned also in table 5

Hence it is clear that the best efficiency of the collector can be obtained in october.

From the analysis made for the solar collector the useful heat collected have been evaluated. This evaluation was obtained at several operating conditions for the whole year . It is seen that at any state the quantity of heat given by the collector is not less than 907.6 W. On the other hand the analysis made for the refrigerator gives a wide variety of operation states of the refrigerator. The cooling load and also the required heat input to the generator of the unit have been evaluated . The analysis were made at different cooling loads and the required temperature difference and heat input was calculated. It is deduced that the greatest amount of heat required for the generator is less than the quantity of heat available by solar collector. Also the temperature difference is verified (9). That means that the designed solar collector covers with great satisfaction both of the heat required and temperature difference which are needed in the generator.

Conclusions:

      From the present research work, following conclusions can be drawn up:

   1- A complete survey of available calculation procedures for
      solar energy parameters with collector efficiency is given.
      These procedures were arranged in a computer program
      which facilitates the determination of direct, diffuse and
      total radition incident on a horizontal surface or that
      tilted to many angles , viz. 0,15,30 , 45, 60, 90°. Also
      the amount of heat transmitted from the two plate covers
      have been obtained.

      The best angle of incidence every month on which the
      collector must be inclined, the angle at which the coll-
      ector receives the sun rays and the collector efficiency
      were calculated.

   2- A special combination of a solar collector and refriger-
      ator system fitted with the measuring instrumentation was
      constructed and erected for experimentation.

   3- For Minia Governorate, the maximum and minimum collectable
      energy per each month were obtained. The corresponding
      angles of inclination were also calculated.

   4- The maximum amount of collectable energy amounted to
      2366.25 Kw.hr year for the used collection surface of 2m^2
      and it was reached at a 30° angle of incidence.

   5- The collector efficiency was calculated. For each month of
      the year   the maximum efficiency and the corresponding
      angle of incidence have been deduced.

   6- It was found that the best angle of inclination through
      the whole year is 30°. At this angle the collector effici-
      ency was calculated for each month.

   7- Taking the best angle of incidence for every month indiv-
      idually and letting the collector to be tilted at this
      angle for every related month which can be arranged beca-
      use of the special construction of the used collector, a
      total amount of useful energy of 9666 MJ/ year can be
      reached; this means a gain of about 1148 MJ/year more than
      the collected heat from the collector with 30° incidence
      over the year.

   8- Through experiments it was found that the best ratio of
      glycerol to water in the collector heating fluid is 1:3.
      The measured vaporization temperature was found to be 120°C.

This mixture was found the more suitable fluid for the cycle.

9- From the theoretical analysis of the used refrigerator ,it was found that the required heat input for the refrigerator in the range from 85 to 115°C is nearly 600 - 900 W at $\frac{1}{4}$ ton cooling load . At 1/12 ton cooling load the heat required for the same temperature range is about 400 to 600 W.

10-Comparing these data with the obtained values from the solar collector , it is obvious that the used collector can collect at the temperature range 85 - 115°C with great satisfaction the required heat for the operation of the refrigerator.

References:
----

(1)   Hassan, A.A.,Wahhab, M.A., Huzyien, A.S. : "Solar heating of liquids flowing across porous diatherminous solids, solar World Forum ISES-Brighton, August, 1981.

(2)   Asfour, A., Wahhab, M.A., Nassar, M., Wahid,S. :"New solids for solar heat storage", 4 th MIAMI International Conference on Alternative Energy Sources, Florida ,U.S.A, December, 1981.

(3)   Sayigh, A.A.M. :" Solar Energy Engineering " , 1 st Edition, Acadenic Press, P.P. 351-361, 1977.

(4)   Duffie , J.A. and Beckman, W. A.:" Solar Energy Thermal Processes ", Wiley Inter. Science P.P. 8-19 ,1974.

(5)   Sakr, I.A., Helwa, N. H.and Soliman , S.H. :"Methods for estimation of solar energy on normal surface", National Research Centre, Doki, Cairo, Egypt, 1978.

(6)   Sayigh, A.A.M.: "Estimation of total radiation intensities, a Universal formula", INGA/AMAP Jaint Assembly Conference. August 22 to Sept, 3, Seattle, U. S. A. 1977.

(7)   Richard C. Jordan . "Low-temperature Engineering Application of solar Energy ", New York, N. Y. 10017, PP. 1-2, 1977.

(8)   Kudert Seluck, " Fundamentals and Collectors of the Past and Present ", International Symposium-Workshop on Solar energy, 16 - 22 June Cairo, Egypt, 1978.

(9) Rabeih, S. M. : "Cooling by solar energy", M. Sc. Thesis, Mechanical Engineering Department , Faculty of Eng., Minia 1981.

(10) Jones, W.P. :"Air Conditioning Engineering", Edward Arnold (Publishers) Limited P.P.368-376,1973.

The Effect of Wind Gusts on Flow Conditions in a Solar Pond

HILLEL RUBIN and BARRY A. BENEDICT
Department of Civil Engineering
University of Florida
Gainesville, Florida 32611, USA

ABSTRACT

A solar pond is a shallow body of water in which a stabilizing salinity gradient prevents thermal convection. Proper operation of the solar pond depends on the ability to preserve its salinity gradient. Wind gusts lead to significant changes in transport phenomena through the pond's layers, and may neutralize its functions.

This study suggests a method of analysis by which wind effects can be predicted. The model refers to steady state conditions in a solar pond whose surface is subject to a constant shear stress generated by the wind gusts. By applying turbulent flow theories the density profile typical to this solar pond is predicted. The analysis refers to a density profile composed of three layers: (a) diffusion sublayer, (b) viscous sublayer, and (c) turbulent region.

It is shown that stratification can extend the thickness of the diffusion sublayer in which molecular diffusion takes place. Stratification may suppress turbulent vortices in the turbulent region and lead the transport process to be determined by molecular effects. The development of the molecular diffusion region in this case starts from the surface of the pond. Increasing the stratification causes the region of molecular diffusion to be extended towards the bottom of the pond. The last zone to be transferred to molecular diffusion is located at the top of the viscous sublayer.

The calculation refers to similar density profiles by applying the classical approach applying dimensionless parameters typical to the boundary layer theory. From these profiles the operation of the pond subject to wind shear stresses is predicted.

The model developed in this study is a completely diffusive simulator, simple to use and shows the effect of various dimensionless parameters typical to the shear pond system.

1. INTRODUCTION

A solar pond is a trap for solar radiation. It converts the solar radiation into thermal energy which is transferred to the deep water layers of

---

Hillel Rubin is on leave from Technion-Israel Institute of Technology, Haifa, Israel.

the pond. The major advantage of the solar pond stems from its long term
capacity to store thermal energy. Therefore various studies consider applications of solar ponds as an alternative for conventional energy sources (e.g.
[1], [2], [3]). Various alternatives for the design of solar ponds are suggested by different authors (e.g. [4], [5], [6]).

The fluid in the solar pond should be stratified. Then heating of the
deep layers of fluid does not induce convection currents in the solar pond.
Convection currents may lead to very high losses of thermal energy into the
atmosphere, associated with neutralization of the solar pond functions.

Wind gusts are extremely dangerous to the proper operation of the solar
pond. The wind exerts shear stresses on the surface of the pond, thereby
generating waves and circulatory currents which intensify transport processes
in the pond. Transport intensification leads to neutralization of the salinity gradient and thermohaline convection and losses of heat into the atmosphere.

The designer of solar ponds should have the capability to simulate wind
effects and therefore be able to design the pond surface protection.

The objective of this study is to develop a method of simulation which is
simple to perform and supplies quantitative information about the parameters
determining the wind effects.

The method of analysis refers to a complete diffusive model which considers the effect of the wind on the coefficients of diffusion through the fluid
layers.

The application of complete diffusive models was suggested by various investigators [7], [8], [9], [10] considering changes and variations in the
thermocline created by the absorption of solar radiation in lakes.

The advantage of the diffusive models stems from the ability to consider
the effect of major as well as minor variations in the coefficients leading to
heat and mass transfer in the flow field.

The disadvantage of the complete diffusive models stems from the neglect
of the explicit coupling between momentum, heat, and mass transfer processes.
According to this method momentum transfer affects other transport processes
implicitly through the variation of the coefficients of transport.

Some investigators (e.g. [11], [12], [13], [14]) suggested models that
take into account explicitly the coupling between momentum, heat, and mass
transfer in lakes. However, such ambitious models are very complicated to
use. As the numerical simulation of such models is subject to problems of
numerical stability and convergence certain simplifications in the coupling
models were used. These simplifications considered lake circulation to be
similar to cavity flow subject to laminar flow conditions. Therefore, the results have a certain lack of coherence with the reality.

This study applies the complete diffusive approach, extends, extrapolates
and modifies expressions and concepts in order to generate a modeling procedure that may simulate wind effects on solar ponds.

2. OPERATION OF THE POND UNDER NEGLIGIBLE SURFACIAL SHEAR STRESSES

Figure 1 represents a schematic description of the solar pond operation under steady state conditions. At the bottom of the pond salt is continuously added. The flowing thermal layer is almost saturated with salt. The temperature of this layer under normal operation is expected to be about 100°C. At the surface of the pond the temperature is almost identical to the average daily temperature of the environment. Due to the slow wash flow at the pond surface, the salinity gradient is kept constant.

The saturation concentrations of salts like NaCl and KCl are 270 kg/m^3, and 300 Kg/m^3, respectively. The density of freshwater at 25°C is 997 Kg/m^3, decreasing to 960 Kg/m^3 at 100°C. For this range of temperatures the volumetric thermal expansion of freshwater varies between 2.6×10^{-4} C^{-1} and 7×10^{-4} C^{-1}. Therefore, the concentration profile is the major contributor to the density profile shown in Figure 1.

We may assume that under normal operation at the bottom of the pond the fluid density is about 1.26 gr/cm^3. At the pond surface the density is about 1.03 gr/cm^3. The molecular diffusivity of salt in freshwater varies between 1.5×10^{-5} cm^2/sec and 4.5×10^{-5} cm^2/sec in the range of temperatures between 25°C and 100°C. However, such low diffusivities cannot be typical to a solar pond. Even extremely small wind gusts generate disturbances which increase significantly the diffusivities. The surfacial flow and the thermal layer flow also increase the effective diffusivities. Field measurements showed that at a surfacial velocity of 1 cm/sec thermal and mass diffusivities were as high as 0.10 cm^2/sec [7]. Such a value is typical of turbulent flows. Field measurements in naturally stratified lakes with surfacial velocities of 10-100 cm^2/sec showed diffusivities of 0.5-100 cm^2/sec [10], [9]. Even if the solar pond surface is protected against wind gusts, we may assume that at the pond surface the effective diffusivity is significantly larger than the molecular salt diffusivity. At the thermal layer region the effective diffusivity is also high, and its value is determined by the flow of the thermal layer, fluctuations, and small scale thermohaline convection.

The requirements for surfacial flow, supply of salt at the bottom of the pond, and the density profile under normal operation of the pond can be

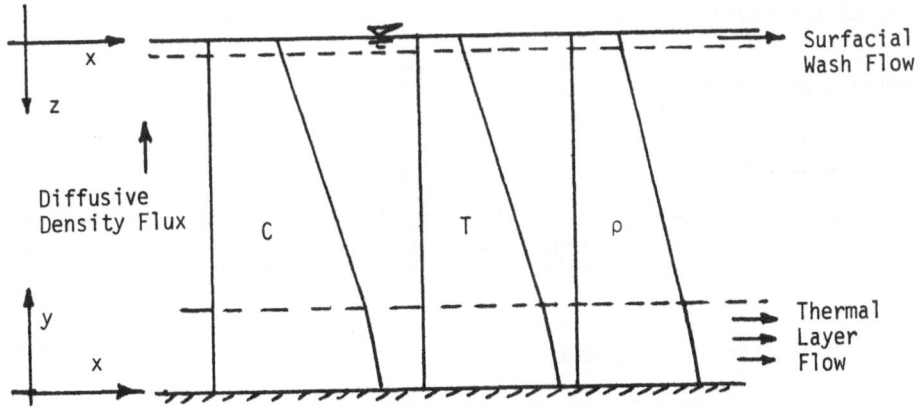

Fig. 1. Schematic Description of the Flow Field Subject at Steady State Conditions

determined by analyzing the "transport of density" in a vertical plane by applying the following equation

$$\frac{\partial \rho}{\partial t} = \frac{\partial \rho}{\partial z} \left(D_z \frac{\partial \rho}{\partial z} \right) \qquad\qquad (1)$$

where, ρ = fluid density; t = time; z = vertical coordinate; D_z = vertical effective diffusivity.

Eq. (1) neglects the effect of the "negative density sources" associated with the absorption of thermal radiation, as temperature effects are minor. This equation is subject to the following boundary conditions

$$\rho = \rho_B \quad \text{at} \quad z = H$$
$$\rho = \rho_T \quad \text{at} \quad z = 0 \qquad\qquad (2)$$

where, H = depth of the pond; ρ_B, ρ_T = density at the bottom and top of the pond, respectively.

Under normal operations steady state conditions prevail in the solar pond. Therefore the left hand side of eq. (1) vanishes. The density flux in the vertical direction is constant. However, solute concentration at the surface of the pond is kept constant due to the surfacial wash flow. The fluid entering the pond at its one end has a density ρ_f, of freshwater. When it leaves the pond at its other end its density is approximately ρ_T. Therefore, the volumetric discharge of the washflow per unit surface area of the pond is given as follows

$$V = D_z \frac{\partial \rho}{\partial z} / (\rho_T - \rho_f) \qquad\qquad (3)$$

As mentioned above the distribution of D_z is determined by the intensity of the thermal layer flow, the wash flow, arbitrary surfacial disturbances, and small scale convection motions due to thermal gradients at the bottom of the pond. Under quiescent conditions, it is possible to assume that the vertical diffusivity is of the order of magnitude of the molecular thermal diffusivity of the fluid.

3. MODELING WIND EFFECTS

3.1 Basic Concepts

Wind gusts exert shear stresses at the pond surface. There is a process of momentum transfer from the blowing wind into the pond. The shear stress is identical on both sides of the water-air interface represented as follows

$$\tau = (\rho u_*^2)_{air} = (\rho u_*^2)_{water} \qquad\qquad (4)$$

where, τ = surfacial shear stress; u_* = surfacial shear velocity.

The characteristic wind velocity is usually measured at a height of 10 m above the water surface. There are empirical relationships between the wind velocity at that elevation, u_{10}, and the surfacial air shear velocity [15].

$$(u_*)_{air} = 3 \cdot 10^{-2} \cdot u_{10} + 5.9 \cdot 10^{-6} \cdot u_{10}^2 \quad \text{(SI units)} \tag{5}$$

According to Eqs. (3) and (4) the shear stress at the pond surface varies between 0.5 dyne/cm^2 and 9 dyne/cm^2 for wind velocities between 20 Km/hr and 70 Km/hr. If the pond surface is protected by various means it is possible to reduce the effective shear stresses and the effect of the waves developed over the pond surface. We may assume that with surfacial protection the shear stress exerted on the surface of the pond obtains about one-third of its original value.

The surfacial shear stress leads to circulatory and drift flows leading to a significant increase of the effective mass diffusivity. As the wind blows over the pond, the turbulence in the upper layers of the pond is generated both by the mean shear and by breaking of waves. Although the momentum flux from the air to the waves is only a small fraction of the momentum flux transferred to the current, the energy flux to the waves is usually comparable to or greater than the energy flux to the current. Thus it is reasonable to suppose that the mechanical generation of turbulence in the solar pond can be characterized by the surface conditions and their interaction with the bottom of the pond without an explicit consideration of the current structure in the pond.

Even in quiescent conditions the solute diffusivity in the pond is probably much larger than the molecular diffusivity. Therefore, the basic conditions which are modified by wind effects are not those related to molecular diffusion. In order to consider the effect of the small scale thermal convection we assume that Schmidt number is of the order of magnitude of the Prandtl number, namely, smaller than 10. Such an approach is also consistent with the reference to density and density flux while ignoring the different effects of the temperature and the solute dispersion.

Considering flow and mass transfer in a smooth channel we follow the approach developed by Levich [16]. This approach concerns turbulent homogeneous flows. With modifications required by the stratification we intend to generate a model applicable for the description of mass transport in a solar pond. Encouragement for such an approach comes from a successful application of such concepts for the prediction of heat and mass transport in dilute polymeric solutions [17].

According to the model developed by Levich [16], the velocity profile in the channel consists of three different zones as follows

(a). <u>Molecular affected zone</u>, where momentum transfer as well as mass transfer are determined by molecular effects. In this zone

$$u^+ = y^+ \quad\quad\quad 0 < y^+ < 5 \tag{6}$$

$$\nu = \nu_0$$

where

$$u^+ = u/u_* \quad\quad y^+ = y \, u_*/\nu \tag{7}$$

Here, u = local velocity; y = vertical distance from the bottom of the channel; ν = kinematic viscosity; ν_0 = molecular kinematic viscosity.

(b). Molecular and turbulent affected zone, where momentum is determined by molecular effects as well as turbulent flucuations. The scale of the turbulent vortices, namely, the mixing length in this zone, is proportional to the square of the distance from the bottom of the channel. This region is characterized by

$$u^+ = 10 \arctan (0.1 \, y^+) + 1.2 \qquad 5 < y^+ < 30$$

$$\nu_{turb}/\nu_0 = 10^{-2} \, (y^+)^2 \qquad \nu = \nu_0 + \nu_{turb}$$

(8)

where, ν_{turb} = kinematic eddy viscosity.

(c). Turbulent region, where all transport processes are determined by the turbulent vortices. The mixing length in this region is proportional to the distance from the bottom of the channel. The turbulent region is characterized by

$$u^+ = 2.5 \ln (y^+) + 5.5 \qquad 30 < y^+$$

$$\nu_{turb}/\nu_0 = 0.4 \, y^+$$

(9)

We assume that the expressions representing the eddy viscosity by eqs. (8) and (9) can be applied even if these equations cannot be used for the description of the velocity distribution in the solar pond

Such an approach stems from the assumption that even if the velocity profile cannot be built-up due to the vertical walls of the pond and its stratification, still turbulent vortices leading to mass diffusion are developed. The scale of these vortices is correlated with the mixing length theory leading to eddy viscosities described by eqs. (8) and (9).

The stratification of the pond does not affect the horizontal scale of the vortices. Therefore, ν_x, representing the horizontal kinematic visocity, can be calculated according to eqs. (8) and (9).

The stratification of the pond reduces the vertical scale of the turbulent vortices leading to anisotropy of the turbulent viscosity. It does not affect the molecular viscosity.

The ratio between the horizontal and vertical eddy viscosities is generally represented as being dependent on the gradient Richardson number or Richardson number.

The gradient Richardson number is defined as follows

$$Ri_g = \frac{g(\partial\rho/\partial z)}{\rho(\partial u/\partial z)^2}$$

(10)

where, g = gravitational acceleration.

Richardson number is defined as follows

$$Ri = \frac{g \ y^2 \ (\partial \rho / \partial z)}{\rho u_*^2} \tag{11}$$

In the case of the solar pond it seems more appropriate to refer to the Richardson number expressed by eq. (11). The walls of the solar pond probably inhibit the initiation of flow in the horizontal direction in certain layers, but that does not indicate that turbulence does not diffuse into these layers.

A summary of various measurements [18] indicates that the ratio between the horizontal and vertical eddy viscosities in a turbulent stratified flow is given as follows

$$\nu_x / \nu_z = (1 + 10 \ Ri)^{0.5} \tag{12}$$

where, ν_x, ν_z = horizontal and vertical kinematic eddy viscosities.

According to Reynolds similarity the effective heat and solute diffusivities in the turbulent region of a homogeneous turbulent flow field are identical to the kinematic eddy viscosity. This statement implies for a stratified turbulent flow as follows

$$D_x = \kappa_x = \nu_{turb} \tag{13}$$

where, D_x, κ_x = horizontal turbulent solute and heat diffusivities.

A fluid particle advected in the vertical direction loses its momentum faster than its heat and solute concentration. Therefore the effect of the flow field stratification on the vertical solute and heat diffusivities is not represented by eq. (12).

Various studies ([17], [10]) supply scattered data concerning the ratio between horizontal and vertical heat diffusion in a stratified turbulent flow field. These data imply [20]

$$\kappa_x / \kappa_z = 1 + 1.6 \ Ri^{0.67} \tag{14}$$

where, κ_z = vertical heat diffusivity.

Data concerning solute diffusivities in a turbulent stratified flow are even more scattered than those concerning heat diffusivities. Different groups of data imply different correlations between Richardson number and the ratio between the horizontal and vertical solute diffusivities. Two of the possible correlations are suggested by Ref [20] as follows

$$D_x / D_z = 1 + 4 \ Ri \tag{15}$$

$$D_x / D_z = 1 + 2 \ Ri \tag{16}$$

where, D_z = vertical solute diffusivity.

Various studies (e.g. [7], [8]) consider that also for heat diffusivities linear expressions like eqs. (15) and (16) represent the anisotropy of the diffusivity tensor.

3.2 The Structure of the Density Profile

Levich [16] considers that the profile of solute concentration is also divided into three regions which are not exactly identical to those referring to momentum transfer, as follows:

(a). Diffusion sublayer, where mass transfer is determined by molecular diffusion. Therefore, we obtain

$$\rho^+ = \rho_B^+ - Sc\,y^+ \qquad\qquad 0 < y^+ < \delta^+ \tag{17}$$

where

$$\rho^+ = \rho u_*/J \qquad \rho_B^+ = \rho_B\, u_*/J \qquad Sc = \nu_0/D_0$$
$$\delta^+ = \delta u_*/\nu_0 \qquad\qquad D = D_0 \tag{18}$$

Here, J = vertical diffusive density flux; D_0 = molecular solute diffusivity; Sc = Schmidt number; δ = thickness of the diffusion sublayer; D = effective solute diffusivity.

Mass diffusion in the diffusion sublayer is not affected by the turbulence of the flow field.

Eq. (17) yields the maximal value of the density gradient in the flow field, namely

$$\max\left(-\frac{d\rho^+}{dy^+}\right) = Sc \tag{19}$$

Whenever referring to density gradient in this paper we consider its absolute value. Of course the density never increases with elevation.

(b). Viscous sublayer. In this region turbulence is the dominant factor determining the rate of density diffusion. Therefore the effective solute diffusivity is equal to the turbulent diffusivity. For the viscous sublayer Levich suggests

$$D = D_{turb} \qquad\qquad D_{turb}/\nu_0 = 10^{-3}\,(y^+)^4 \tag{20}$$
$$\delta^+ < y^+ < \delta_0^+$$

where, D_{turb} = the coefficient of turbulent diffusivity.

At the top of the viscous sublayer

$$\nu_{turb} \approx D_{turb} \tag{21}$$

Therefore Levich obtains

$$\delta_0^+ \approx 10 \qquad\qquad \delta_0^+ = \delta_0\, u_*/\nu_0 \tag{22}$$

where, δ_0 = thickness of the viscous sublayer. Note that eqs. (8) and (20) imply different expressions for ν_{turb} and D_{turb}. The thickness of δ_0^+ is different from the thickness of the molecular and turbulent affected zone referring to momentum transfer. These differences will be discussed later.

At the boundary between the diffusion and viscous sub-layers the turbulent diffusivity is almost identical to the molecular one. The constant value of the density flux in any horizontal cross section implies

$$-d\rho^+/dy^+ = \nu_0/D \tag{23}$$

Considering eqs. (20) and (23) for the top of the diffusion sublayer we obtain

$$\delta^+ = \delta_0^{+0.75} Sc^{-0.25} \tag{24}$$

(c). The turbulent region. In this region solute transport is determined by the turbulent effective diffusion. For this region

$$D_{turb} = \nu_{turb} \qquad\qquad \delta_0^+ < y^+ \tag{25}$$

where, D_{turb} = turbulent diffusivity.

In the following section we consider the effect of the flow field stratification on the various zones typical to the density distribution of the flow field.

3.3 Linear Model of Wind Effects

According to the linear model of wind effects we assume that the general relationship between the horizontal and vertical diffusivities is represented by the following expression, provided that $D_x = D_{turb}$

$$D_x/D_z = 1 + \sigma\, Ri \tag{26}$$

where, σ = constant for a wide range of Richardson numbers.

Applying the dimensionless variables of eqs. (7) and (18) we obtain the following expression for Richardson number

$$Ri = - \frac{g\nu}{u_*^3} \frac{(y^+)^2 \, (d\rho^+/dy^+)}{\rho^+} \tag{27}$$

The effect of the stratification on the turbulent diffusion depends on the scale of the turbulent vortices.

We consider the possible effect of the flow field stratification on each zone of the density profile as follows

(a). The diffusion sublayer. In this zone the stratification does not affect the diffusivity. Therefore eqs. (17) and (19) can be applied for the stratified flow. As implied in previous sections of this article the effective value of the Schmidt number expected in a solar pond is $Sc \approx 5$. Therefore an application of eq. (24) yields

$$\delta^+ \approx 3.72 \tag{28}$$

However if the flow field stratification suppresses completely the turbulent vortices, then the diffusion region is extended for the whole depth of the solar pond.

(b). The viscous sublayer. As represented by eq. (8), Levich [16] considers values of ν_{tur} for the molecular and turbulence affected zone of momentum transfer which are proportional to $(y^+)^2$. When he considers solute transfer in the viscous sublayer he refers according to eq. (20) to values of D_{turb} proportional to $(y^+)^4$. This difference between ν_{turb} and D_{turb} stems from the difference between scales of turbulent vortices considered for each one of the zones of velocity and density profiles. In the vicinity of the wall the scale of the turbulent vortices changes very rapidly with the distance from the wall. This phenomenon is associated with the decay of turbulence and energy dissipation. In the turbulent region the scale of the turbulent vortices is proportional to y^+ as indicated by eq. (9). Considering momentum transfer in the region of $5 < y^+ < 30$ it is reasonable to assume that the turbulent viscosity is proportional to $(y^+)^2$ as indicated by eq. (8). Referring to solute diffusion with no effect of small scale thermohaline convection, $Sc \approx 10^3$. In such a case according to eq. (24) $\delta^+ \approx 1$. Then for the region $\delta^+ < y^+ < \delta_0^+$ the assumption of turbulent solute diffusivity represented by eq. (20) is justified due to the small value of δ^+. However, in the solar pond, due to the small scale thermohaline convection probably the effective Schmidt number is much smaller than 10^3. Considering $Sc \approx 5$ we obtain the result represented by eq. (28). In such a case it seems that eq. (8) may represent the turbulent diffusivity in the viscous sublayer rather than eq. (20). By applying eq. (8) for the turbulent diffusivity we get the following expression for the thickness of the diffusion sublayer.

$$\delta^+ = \delta_o^+ \, Sc^{-0.5} \tag{29}$$

Introducing $Sc = 5$ into eq. (29) we obtain $\delta^+ = 4.47$. This value is not much different from the value indicated by eq. (28) which was obtained by applying eq. (24). We may summarize that in the solar pond an application of eq. (8) for the description of the turbulent diffusivity seems to be appropriate. Eq. (29) also yields $\delta^+ \to \delta_o^+$ as $Sc \to 1$.

Considering the density diffusion in the nonhomogeneous flow field, eq. (23) implies

$$- \, d\rho^+/dy^+ = (\nu_o/D_x)(D_x/D_z) \qquad D_x = D_{turb} \tag{30}$$

where, D_x, D_z = horizontal and vertical diffusivities.

Introducing eqs. (8), (26) and (27) into eq. (30) we obtain

$$- \frac{d\phi^+}{dy^+} \left(1 - \frac{\sigma \nu g}{u_*^3} \frac{10^2}{\rho^+} \right) = \frac{10^2}{(y^+)^2} \qquad \delta^+ \le y^+ \le \delta_o^+ \tag{31}$$

This equation is applicable where it yields positive density gradients smaller than Sc. As ρ^+ decreases with y^+, eq. (31) indicates that if in a certain point y_o^+ of the viscous sublayer the density gradient is equal to Sc then the diffusion sublayer is extended to y_o^+. For $y^+ > y_o^+$ the viscous sublayer is subject to turbulent effects.

Direct integration of eq. (31) yields

$$\rho^+ - \frac{\sigma \nu g}{u_*^3} 10^2 \ln \rho^+ - \frac{10^2}{y^+} = \rho_o^+ - \frac{\sigma \nu g}{u_*^3} 10^2 \ln \rho_o^+ - \frac{10^2}{y_o^+} \tag{32}$$

$$\delta^+ \le y_o^+ \qquad y_o^+ \le y^+ \le \delta_o^+$$

where

$$\rho_o^+ = \rho_B^+ - Sc \, y_o^+ \tag{33}$$

For small values of $\nu g/u_*^3$, namely large surfacial shear stresses, eq. (31) yields

$$- \frac{d\rho^+}{dy^+} \approx \frac{10^2}{(y^+)^2} \tag{34}$$

$$\rho^+ = \rho_B^+ - Sc \; \delta^+ - 10^2 \left(\frac{1}{\delta^+} - \frac{1}{y^+}\right) \tag{35}$$

These expressions refer to a homogeneous flow field with negligible effects of the stratification on the density profile in the viscous sublayer.

(c). <u>The turbulent region.</u> Introducing eqs. (9), (26) and (27) into eq. (30)

$$-\frac{d\rho^+}{dy^+} = \frac{2.5}{y^+}/\left(1 - \frac{\sigma v g}{u_*^3} \; \frac{2.5}{\rho^+} \; \frac{y^+}{\rho^+}\right) \qquad \delta^+ < y^+ \tag{36}$$

By an appropriate integration procedure of eq. (36) the density profile in the turbulent region can be determined. Eq. (36) can be used provided that it yields positive density gradients smaller than Sc. Because of this restriction eq. (36) cannot be used for large values of y^+/ρ^+ and vg/u_*^3. If the product of these two parameters is large then eq. (36) may indicate density gradients larger than Sc or even erronous positive density gradients. In such regions the turbulent vortices are completely suppressed and diffusion takes place by molecular effects leading to a density gradient equal to Sc. If in a certain location y_1^+ in the turbulent region the density gradient indicated by eq. (36) is equal to Sc then eq. (36) yields

$$\frac{2.5}{y_1^+} = Sc \left(1 - \frac{\sigma v g}{u_*^3} \; \frac{2.5 \; y_1^+}{\rho_1^+}\right) \tag{37}$$

where, ρ_1^+ = density at $y^+ = y_1^+$

It is indicated by eqs. (36) and (37) that for $y^+ > y_1^+$ the turbulent vortices are suppressed and molecular effects determine the diffusion of the density. If y_1^+ determined by eq. (37) is smaller than δ_0^+, then the turbulent vortices are suppressed in all the turbulent regions. The density at y_1^+ is, therefore, given as follows

$$\rho_1^+ = \rho_T^+ + Sc \; (H^+ - y^+) \tag{38}$$

where, ρ_T^+ = density at the pond's surface.

According to eq. (37) an increase of the shear velocity increases the value of y_1^+. It indicates that the effect of stratification is diminished by the shear stresses applied at the surface of the pond.

Eqs. (36) and (37) show that the effect of the stratification as a turbulent suppressor is less significant at the bottom of the pond than at those layers located above the thermal layer.

If the value of y_1^+ obtained from eq. (37) is much larger than δ_0^+ then probably small values of y^+/ρ^+ are obtained at the bottom of the pond. In these locations eq. (36) can be approximated as follows

$$- \frac{d\rho^+}{dy^+} = \frac{2.5}{y^+} \tag{39}$$

Direct integration of this equation yields

$$\rho^+ = \rho_{\delta_0}^+ - 2.5 \ln (y^+/\delta_0^+) \tag{40}$$

where, $\rho_{\delta_0}^+$ = density at the top of the viscous sublayer.

Eq. (40) represents the logarithmic density profile which is typical to turbulent homogeneous flow. It shows that turbulence in the flow field starts to develop at the top of the viscous sublayer. At that region the effect of the stratification is minimal.

Eq. (40) represents the density profile also when the parameter $\nu g/u_*^3$ obtains small values, namely, if the shear stress applied by the wind is large.

The discussion represented in the previous paragraphs indicates that an increasing shear stress of the wind increases the region of turbulent vortices in the pond. This region comprises values of y^+ from δ_0^+ towards y_1^+. For $y^+ > y_1^+$ a large gradient of the density develops. This region becomes smaller and smaller as the shear stress increases. This region can be named, "the separating layer of the pond", as it separates the turbulent region from the surface of the pond.

3.4 Nonlinear Model of Wind Effects

The linear model of wind effect supplies important information about the structure of the density profile developed in the solar pond. This information has direct implications with respect to the intensity of transport processes in the solar pond. However, certain studies showed that the ratio between the horizontal and vertical diffusivities is not necessarily linear, as implied by eqs. (12) and (14) - (16). The general expressions for nonlinear models are represented as follows

$$D_x/D_z = (1 + \alpha \; Ri)^\beta \tag{41}$$

or

$$D_x/D_z = 1 + \psi \, Ri^{\phi} \tag{42}$$

where, α, β, ψ, ϕ = constants.

The same approach applied for the linear model in the previous section can be used for the nonlinear models represented by eqs. (41) - (42). Series expansions can be used for the simplification of these models and obtaining analytical solutions for the density transport problem. However, the nonlinear models represented by eqs. (41) and (42) lead to complicated calculations which according to this study did not yield any specific contribution to the understanding of the physical phenomena. Under the current situation data concerning the ratio between D_x and D_z are very scattered and do not yield reliable values of the power coefficients appearing in eqs. (41) and (42). Therefore, models like eqs. (41) and (42) do not even supply quantitative information that has any superiority with respect to the linear model. Availability of more data may of course change this conclusion.

4. PARAMETRIC CONSIDERATION

According to eqs. (31) - (40) wind effects are determined by considering the relationships between the following dimensionless parameters

$$\rho_B^+ \, , \, y^+ \, , \quad g\nu/u_*^3 \, , \, Sc \tag{43}$$

The parameter ρ_T^+ directly indicates the quantities of fluid required for the wash flow

$$V^+ = 1/(\rho_T^+ - \rho_f^+) \tag{44}$$

This relationship is obtained from eq. (3). Therefore the wash flow determination is represented through the following dimensionless function

$$V^+ = f \, (\rho_B^+ \, , \, H^+ \, , \, g\nu/u_*^3 \, , \, Sc) \tag{45}$$

Although this dimensionless presentation of the wind effect is correct, it may be appropriate to correlate the relationships between the physical variables based on a different system of parameters.

We consider characteristic physical variables of the flow field. A characteristic density is represented by the density at the bottom of the pond which is constant as long as there is a sufficient supply of salt. A characteristic velocity, u, of the flow field can be considered as a surfacial velocity obtained from eq. (9). A characteristic length is the depth of the pond, H. The characteristic kinematic viscosity is ν_0. The gravity should be considered due to the effects of the buoyancy. The interaction between the buoyancy of the fluid, the viscous forces and the inertial forces is represented by a Reynolds number and the Grashof number expressed as follows

$$Re = UH/\nu_o \qquad Gr = g \, H^3 \, (1 - \rho_T/\rho_B)/\nu_o^2 \tag{46}$$

According to eq. (9) we define a Reynolds-dependent variable U^+ as follows

$$U^+ + 2.5 \ln U^+ = 2.5 \ln Re + 5.5 \tag{47}$$

This expression represents the interaction between Reynolds number and the shear stress applied by the wind on the surface of the pond.

Parameters listed in eq. (45) can be expressed as follows

$$\frac{u_*^3}{g\nu} = \frac{Re^3 \, (1 - \rho_T/\rho_B)}{Gr \, (U^+)^3} \qquad v^+ = \frac{VH}{\nu} \frac{U^+}{Re} = Re_w \, U^+/Re \tag{48}$$

where, Re_w = Reynolds number of the wash flow.

Therefore we may represent eq. (45) by the following dimensionless function

$$Re_w = F \, (\rho_T/\rho_B \, , \, Re \, , \, Gr \, , \, Sc) \tag{49}$$

According to this expression, the Reynolds number of the wash flow depends on the ratio between the top and bottom densities, Reynolds, Grashof and Schmidt numbers.

5. RESULTS AND DISCUSSION

5.1 Order of Magnitude of the Dimensionless Parameters

Prior to the presentation of the relationships between the dimensionless parameters we refer to their possible values in a realistic solar pond system.

The depth of the pond is probably less than 4 m. Therefore we may assume $1 < H < 4$ m. Maximum possible density of the fluid located at the bottom of the pond is 1.26 gr/cm^3. Minimum possible density at the surface of the pond is about 1.03 gr/cm^3. We refer to the Schmidt number whose value is 5 through all our calculation. However, the method can be applied for very wide values of Schmidt numbers as discussed later. The wind shear stress to be considered is in the range of 0.1 ÷ 10 dyne/cm^2.

Under conditions of molecular diffusion the density flux in a solar pond whose depth is 2 m is approximately 2.5 x 10^{-6} gr/(sec.cm^2). If the surfacial shear stress is 10 dy ne/cm^2 and molecular diffusion is hypothetically preserved then we obtain $\rho_B^+ = 1.6$ x 10^6. We may consider this value as the maximum possible magnitude of ρ_B^+ in a solar pond. While the pond surface is subject to shear stresses smaller then 10 dyne/cm^2 we obtain for a 2 m deep pond the following maximal values

$$Re \approx 2.10^6 \qquad U^+ \approx 33 \qquad Gr \approx 1.5 \times 10^{13} \qquad (50)$$

$$\rho_B/\rho_T \approx 1.22$$

Of course all these parameters cannot attain their maximal values simultaneously. The Grashof number and ρ_B/ρ_T always obtain their maximal values under quiescent conditions. The Reynolds number increases with increasing values of the wind shear stress. Our objective is to simulate the relationships between the dimensionless parameters, and to correlate our simulation with the physical variables typical to the solar pond operation.

5.2 Method of Simulation

The method of simulation adopted in this study stems from the reference to the dimensionless density profile being dependent on ρ_B^+, y^+ and $g\nu/u_*^3$. In order to obtain the density profiles we applied eqs. (17), (31), (32) and (36). Whenever necessary we applied fourth order Runge-Kutta numerical integration, and obtained density profiles for all reasonable values of ρ^+, y^+ and UF, where

$$UF = u_*^3/(\sigma g \, \nu_o) \qquad (51)$$

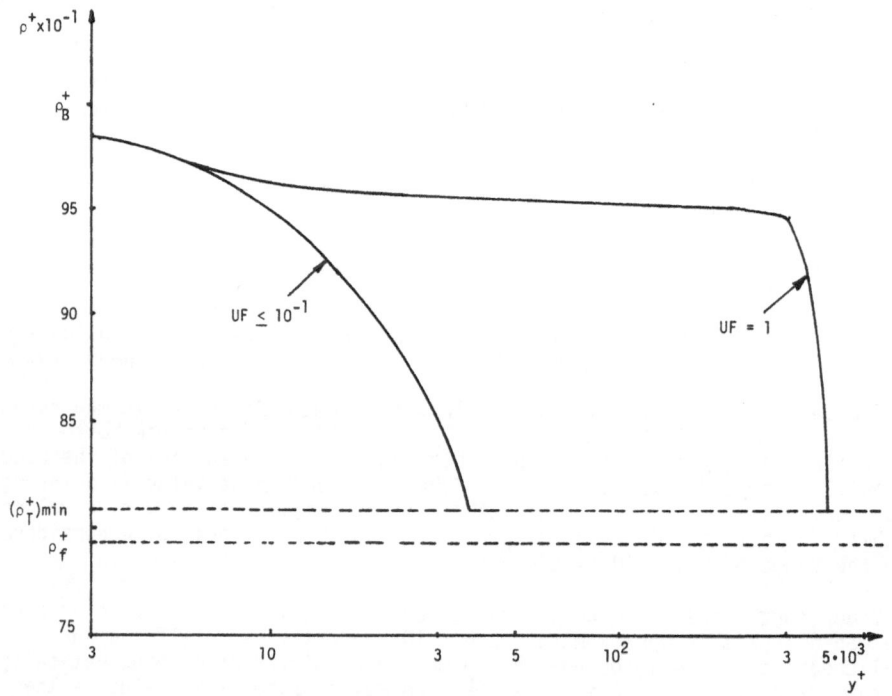

Fig. 2. Dimensionless Density Profiles for $\rho_B^+ = 10^3$

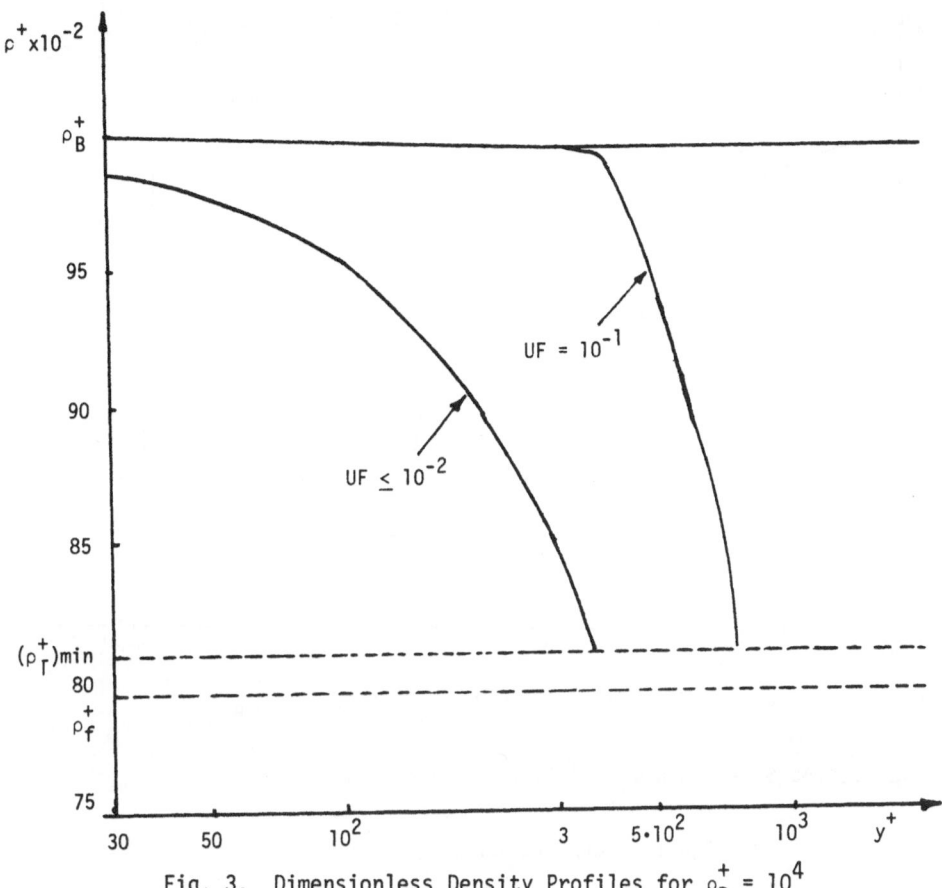

Fig. 3.   Dimensionless Density Profiles for $\rho_B^+ = 10^4$

Some of our results are shown in Figs. 2-4. Such results can be applied for obtaining various kinds of information as required. As an example consider that we are looking for an estimate of the phenomena typical to a pond whose depth is 2 m, being subject to surfacial shear velocity of 0.27 cm/sec. We may assume $\sigma = 2$. Therefore according to eq. (40) we get UF = 0.1, and according to eq. (7) $H^+ = 5400$. The curve representing UF = 0.1 intersects $y^+ = 5400$ for $\rho_B^+ = 10^5$ and $\rho^+ \approx 9.28 \cdot 10^4$, $\rho/\rho_B = 0.93$. Between $y^+ = 4000$ and $y^+ = 5400$ the diffusion is performed by molecular transfer. Therefore the density between the bottom of the pond and about 1.50 m is almost constant.

The large density gradient prevails in the top 50 cm of the pond. In this region located far from the black hot bottom of the pond probably the effective Schmidt number is larger than 5. On the other hand there is layer of fluid having an immediate contact with the atmosphere in which the density is almost constant. Our calculations do not consider this layer. Therefore an estimate of Schmidt number equal to 5 in the separating layer is not too far from reality.

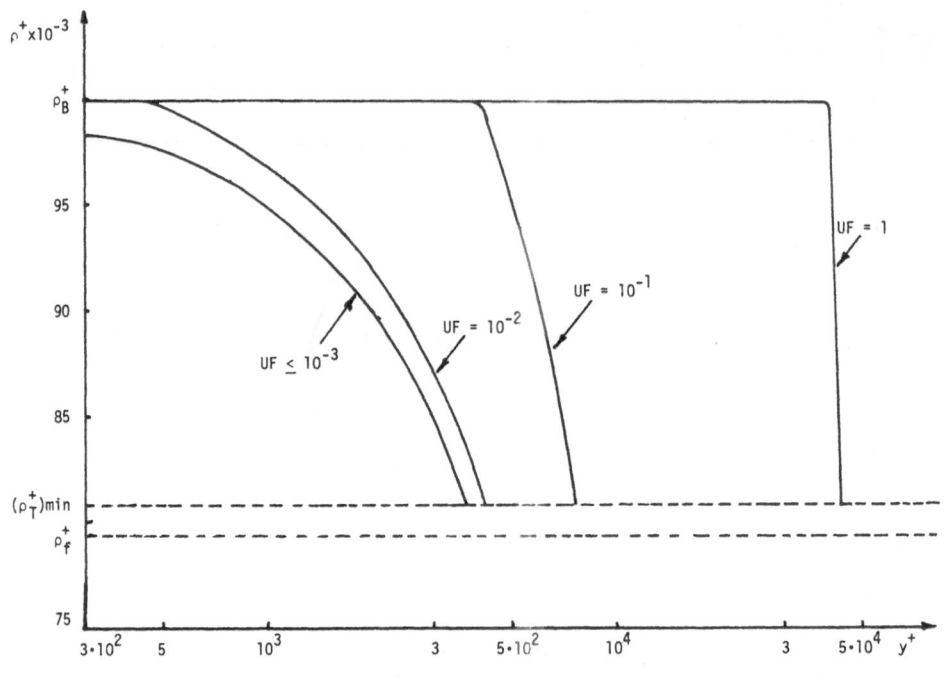

Fig. 4. Dimensionless Density Profiles for $\rho_B^+ = 10^5$

Considering $Sc = 5$ and $\rho_T \approx 1.20$ we obtain

$$J \approx 4 \times 10^{-6} \ gr/(sec.cm^2)$$

$$(52)$$

$$V \approx 2 \times 10^{-5} \ cm/sec$$

According to these expressions, for a pond whose surface area is 4,000 m^2 (2 acres) the wash flow is about 0.8 lit/sec, and quantities of about 160 gr/sec should be added to the pond.

The involvement of five different dimensionless parameters in the simulation of the physical phenomena as represented by eqs. (45) and (49), leads to a series of calculations whenever a realistic physical problem is considered, as examplified in this section.

5.3 Discussion

The method of simulation represented in the previous subsection is only one example of a possible approach towards the calculation of wind effects on a solar pond. This example had a major objective of representation of the basic effects of the wind gusts rather than suggesting a design procedure. Of course, design procedures should consider an appropriate representation of physical parameters determining the construction of the pond and their implication with respect to the performance of the pond under various environmental conditions. We have in mind various approaches for the achievement of such goals like application of charts, development of computer programs, etc. However this theme is beyond the scope of the present study.

Consideration of the physical phenomena themselves requires a certain discussion. As shown in Figs. 2-4, due to the turbulence generated by the shear stress a wide zone of almost constant density develops in the pond. Therefore in this region the stabilization against thermal convection is neutralized. Probably thermohaline convection develops and the thickness of the thermal layer increases significantly. This mixing action is probably associated with a reduction in the thermal layer temperature leading towards lower efficiency of the solar pond system. However, the major problem is associated with the tremendous increase in the quantities of wash flow and salt required in order to keep a certain layer at the surface of the pond being subject to molecular diffusion.

With respect to some basic aspects, this study shows that it is extremely difficult to completely avoid turbulent transport in the pond. In the vicinity of the bottom of the pond the stratification may affect only the extremely small size vortices, but vortices proportional to the distance from the bottom can hardly be suppressed at that region.

6. SUMMARY AND CONCLUSION

Considering the development of turbulence in a water body subject to surfacial shear stress, a model describing transport phenomena in a solar pond is developed.

The effect of the stratification in the solar pond is represented by the Richardson number effect on the coefficients of turbulent transport. The study suggests an adaptation of the model developed by Levich [16] for the description of transport phenomena in the solar pond. Modifications and consideration of the effect of stratification lead to certain changes in that model for the solar pond simulation.

The model predicts changes in the density profile typical to the solar pond due to the surfacial shear stress. These changes have significant implications with respect to the performance of the solar pond and its management and design.

The study refers to steady state conditions. However a certain time period is required for the establishment of steady state conditions in the solar pond. Therefore the predictions supplied by this study are conservative and give a certain factor of safety.

REFERENCES

1. Bryant, H.C. and Colbeck, J. 1977. A solar pond for London. Solar Energy, vol. 19, pp. 321-322.

2. Rabl, A. and Nielsen, C.E. 1975. Solar pond for space heating. Solar Energy, vol. 17, pp. 1-12.

3. Styris, D.L. et al. 1976. The nonconvecting solar pond applied to building and process heating. Solar Energy, vol. 18, pp. 245-251.

4. Assaf, G. 1976. The Dead Sea a scheme for a solar lake. Solar Energy, vol. 18, pp. 293-299.

5. Tabor, H. 1963. Solar ponds-large area solar collectors for power production. Solar Energy, vol. 7, pp. 189-194.

6. Tabor, H. Sep. 1964. Solar ponds. Electronic and Power, pp. 296-299.

7. Sundaram, T.R. and Rehm, R.G. 1971. Formation and maintenance of thermoclines in temperate lakes. AIAA Journal, vol. 9, No. 7, pp. 1322-1329.

8. Sundaram, T.R. and Rehm, R.G. 1972. Effects of thermal discharge on the stratification cycle of lakes. AIAA Journal, vol. 10, No. 2, pp. 204-210.

9. Powell, T. and Jassby, A. 1974. The estimation of vertical diffusivities below the thermocline in lakes. Water Resour. Res., vol. 10, No. 2, pp. 191-198.

10. Merritt, G. and Rudinger, G. 1970. Thermal and momentum diffusivity measurements in a turbulent stratified flow. AIAA Journal, vol. 11, No. 11, pp. 1465-1470.

11. Liggett, J.A. 1970. Cell method for computing lake circulation. ASCE J. Hydr. Div., vol. 96, No. HY3, pp. 725-743.

12. Liggett, J.A. and Lee, K.K. 1971. Properties of circulation in stratified lakes. ASCE J. Hydr. Div., vol. 97, No. HY1, pp. 15-29.

13. Young, F.D.L. and Liggett, J.A. 1977. Transient finite element shallow lake circulation. ASCE J. Hydr. Div., vol. 103, No. HY2, pp. 102-121.

14. Young, F.D.L., Liggett, J.A. and Gallagher, R.H. 1976. Steady stratified circulation in a cavity. ASCE J. Eng. Mech. Div., vol. 102, No. EM1, pp. 1-17.

15. Faley, H.T. 1974. Prediction of wind wave heights. ASCE J. Waterways Harbors and Coastal Eng. Div., vol. 100, No. WW1, pp. 1-11.

16. Levich, V.G. 1962. Physicochemical Hydrodynamics. Prentice-Hall, Englewood Cliffs, N.J.

17. Rubin, H. 1971. Scaling of heat transfer to dilute polymeric solutions. J. of Hydronautics, vol. 5, No. 4, pp. 148-150.

18. Mehta, U.H. and Lavan, Z. 1969. Flow in a two dimensional channel with a rectangular cavity. Trans. ASME J. Appl. Mech., vol. E36, pp. 897-901.

19. Ellison, T.H. and Turner, J.S. 1960. Mixing of dense fluid in a turbulent pipe flow. J. Fluid Mech., vol. 8, pp. 529-544.

20. Bachu, S. 1980. Analysis of transport and flow phenomena in solar ponds. D.Sc. Thesis, Dept. of Civil Engrg. Technion-Israel Inst. of Techn., Haifa, Israel.

COOLING AND DEHUMIDIFYING

Multi-staged Reversed Carnot Heat Pump Cycles

ANTHONY PETERS
New York City Technical College
The City University of New York
Brooklyn, New York 11201, USA

ROBERT PETERS
International Telecommunications Satellite Organization
Washington, D.C. 20024, USA

ABSTRACT

The multi-staged reversed Carnot cycle may have a performance exceeding twice that of the conventional reversed Carnot cycle. Two multi-staged cycles are analyzed. These are: (1) the heating of a receiver with a finite thermal capacity from an infinite thermal capacity heat source, and (2) the heating of a receiver with a finite thermal capacity heat source.

1. INTRODUCTION

The single staged reversed Carnot heat pump cycle is conventionally represented in the temperature-entropy diagram as in Figure 1. It is bound by isothermal lines T_1 and T_3 and isentropic lines S_1 and S_2.

The performance of the single staged reversed Carnot cycle is the highest attainable by any heat pump operating between a constant temperature heat source, T_1, and a constant temperature heat receiver, T_2.

In the heating mode, the performance is given by:

$$COP = \frac{T_3}{T_3 - T_1}$$

and in the cooling mode, the performance is given by:

$$COP = \frac{T_1}{T_3 - T_1}$$

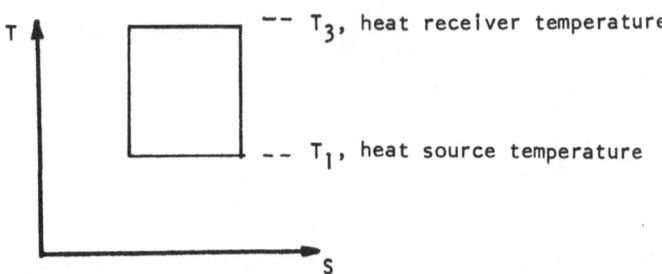

Figure 1. Temperature - Entropy Diagram of Single Staged Reversed Carnot Cycle.

A constant temperature heat source and heat receiver may be realized in practice by:

- processes involving a phase change, or

- processes using an infinite thermal capacity heat source and heat receiver. An infinite thermal capacity may be realized by a very.high flow rate.

Many processes involve the steady state heating of the receiver and/or the cooling of the source. The receiver and/or the source undergo significant temperature changes. Such processes may be idealized by multistaged reversed Carnot cycles. The performance of multistaged reversed Carnot cycles may exceed that of the single staged cycle.

This paper will analyze two such cycles and will develop relationships for their optimum performances.

2. CASE I.

The heat receiver has a steady state, finite mass flow rate. It is heated from T_2 to T_3 as it passes through the heat pump, while the heat source is maintained at constant temperature, T_1.

One example of such a process is the application of a heat pump to the generation of hot water (or hot air) for domestic home heating, while using a high flow rate of the heat source through the heat pump or using a heat source undergoing a phase change.

Such a process may be compared to a number of reversed Carnot cycle heat pumps, each receiving a quantity of heat from the same temperature source, T_1.

In the steady state, each Carnot heat pump transfers its heat to a receiver of constant mass and specific heat, whose temperature increases from T_2 to T_3. The thermodynamic properties of the working fluids may be represented in the temperature-entropy diagram as in Figure 2.

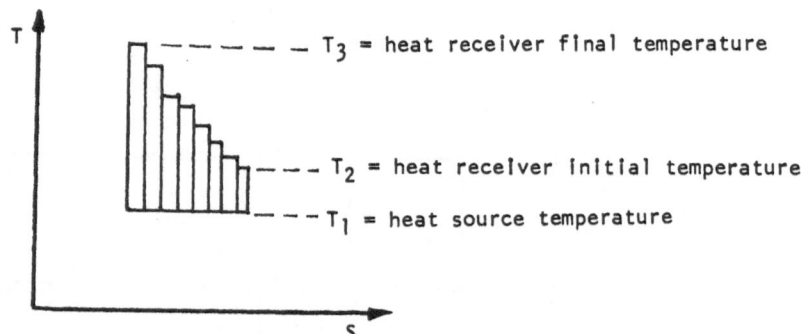

Figure 2.   Temperature - Entropy Diagram of the Working
             Fluids in a Multistaged Reverse Carnot Heat
             Pump System with Constant Source Temperature.

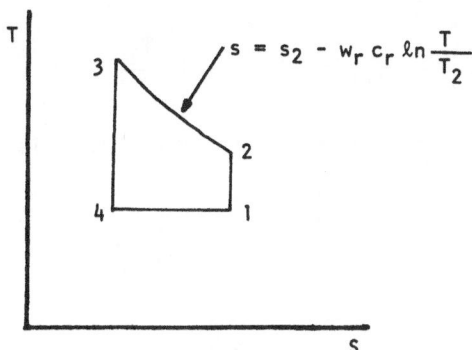

Figure 3. Temperature - Entropy Diagram of Working
Fluids of Infinitely Staged Reversed
Carnot Heat Pump System with Constant
Source Temperature.

If the number of reversed Carnot heat pumps is large, the properties of
the working fluids of the system may be approximated by smooth lines, as shown
in Figure 3.

The system may be analyzed by determining the heat output and heat input.
For constant flow rate w_r, and constant specific heat c_r of the receiver, the
heat into the receiver is:

$$Q_{out} = w_r c_r (T_3 - T_1)$$

For thermal equilibrium between the working fluid and the heat receiver,
and for reversible heat transfer, the combined entropy change of the working
fluid plus that of the receiver is zero:

$$(s_3 - s_2)_a + (s_3 - s_2)_r = 0 \, ,$$

and

$$(s_3 - s_2)_a = - w_r c_r \ln \frac{T_3}{T_2}$$

where the subscript "a" refers to the working fluid and "r" to the heat receiver.

The heat input at the source end is given by the definition of entropy:

$$Q_{in} = \int T \, ds$$

and for constant temperature, T_1:

$$Q_{in} = T_1 (s_1 - s_4)_s$$

where the subscript "s" refers to the source.

Because the system is bounded by isentropic lines,

$$(s_1 - s_4)s = (s_2 - s_3)a$$

yielding

$$Q_{in} = T_1 w_r c_r \ln \frac{T_3}{T_2} .$$

The performance of the system in the heating mode is given by:

$$COP = \frac{Q_{out}}{W_{in}} = \frac{1}{1 - \frac{Q_{in}}{Q_{out}}}$$

and by subsitution:

$$COP = \frac{1}{1 - T_1 \frac{\ln \frac{T_3}{T_2}}{(T_3 - T_2)}}$$

For the special case where the heat receiver is at substantially constant temperature, i.e., T_2 T_3, the performance of the system approaches that of the single staged reversed Carnot cycle. This may be verified by applying L'Hospitale's rule and evaluating the indeterminate form of

$$\lim_{T_2 \to T_3} \frac{\ln \frac{T_3}{T_2}}{(T_3 - T_2)} = \frac{\frac{d}{dT_2}\left(\ln \frac{T_3}{T_2}\right)}{\frac{d}{dT_2}(T_3 - T_2)} = \frac{-\frac{1}{T_2}}{-1} = \frac{1}{T_2}$$

By substitution,

$$\lim_{T_2 \to T_3} COP = \frac{1}{1 - \frac{T_1}{T_3}} ,$$

which is the single stage reversed Carnot cycle performance.

For the special case where the receiver is heated from source temperature T_1 to temperature T_3, as may occur in generating a hot water supply from feed-water at outdoor temperature, the performance may be determined by letting T_2 approach T_1.

$$COP = \frac{\frac{T_3}{T_1}}{\frac{T_3}{T_1} - 1 - \ln \frac{T_3}{T_1}}$$

This is single stage reversed Carnot performance modified by the loga-rithmic term. The effect of the logarithmic term is to increase the performance of the multi-staged system above that of the single-staged system. When T_2 approaches T_1, the ratio of the multi-staged system to that of the single staged system exceeds 2. This is shown in Table I.

TABLE I.   PERFORMANCE OF MULTI-STAGED REVERSED CARNOT
HEAT PUMP AS $T_1 \to T_2$ WITH CONSTANT TEMPERATURE
INPUT

T_3/T_2	1.1	1.2	1.3	1.4	1.5
COP Multi-Staged Reversed Carnot System	22.2	11	8	6.3	5.3
COP Single-Staged Reversed Carnot System	10.0	5	3.3	2.5	2
COP Multi-Staged/COP Single-Staged	2.2	2.2	2.4	2.5	2.6

The ideal performance of a heat pump generating hot water (or hot air)
which uses a substantially constant temperature heat source depends on the
ratio of the temperature of the hot water (or air) leaving to the temperature
of the incoming water (or air).

When the incoming water (or air) approaches the source (outdoor) temper
ature, the performance exceeds twice that of the single stage reversed Carno
cycle.

When the incoming water (or air) approaches the receiver temperature, t
performance approaches that of the single stage reversed Carnot cycle.

These results may be confirmed by expanding the logarithmic term into a
Taylor series using $T_3 = T_2 + \Delta T$:

$$\ln \frac{T_3}{T_2} = \ln \left(1 + \frac{\Delta T}{T_2} \right) = \frac{\Delta T}{2} - \frac{1}{2} \left(\frac{\Delta T}{T_2} \right)^2 + \frac{1}{3} \left(\frac{\Delta T}{T_2} \right)^3 - \frac{1}{4} \left(\frac{\Delta T}{T_2} \right)^4 \cdots$$

Discarding terms higher than the second order, letting T_1 approach T_2,
and substituting, an approximation for performance is given by

$$COP = \frac{1}{1 - [1 - \frac{1}{2} \left(\frac{T_3 - T_1}{T_1} \right)]} = \frac{2T_1}{T_3 - T_1}$$

Including terms higher than the second order decreases the denominator
and increases the performance.

The approximation shows that as the feedwater (or air) approaches the
outdoor temperature, the performance of the multistaged reversed Carnot heat
pump exceeds twice that of the single staged reversed Carnot system and con-
firms the results in Table I.

3.   CASE 2.

The heat receiver undergoes a significant temperature rise from T_2 to
T_3 as in Case 1, while hte heat sources also has a steady state, finite mass
flow rate and also undergoes a significant temperature change. The heat
source is cooled from T_1 to T_4 as it passes through the heat pump.

One example of such a process is the application of a heat pump to the
generation of hot water (or hot air) for domestic home heating, as in Case I,

while using a steady state, finite flow rate for the heat source as well as for the receiver.

Another example of such a process is the application of a heat pump for conventional air conditioning. Here, the source of heat is the room air. It enters the heat pump at approximately 80°F and is returned to the room at 56°F. The heat receiver is the outside air. It enters the heat pump at temperature T_2 and leaves at T_3.

Ideally, this may be compared to a system of an infinite number of reversed Carnot cycle heat pumps, each operating on a unit weight, w_a, of an ideal gas, each receiving an infinitesimal quantity of heat from an infinite thermal capacity heat source, the temperatureof each source changing in such a way to maintain the pressure of the ideal gas the same from stage to stage. The temperature of the sources increases from T_4 to T_1.

Each Carnot heat pump transfers its infinitesimal quantity of heat to a finite thermal capacity receiver whose temperature increases from T_2 to T_3.

Such a system of heat pumps would have a performance that would represent the limit of any system operating between a constant pressure heat source and a finite thermal capacity heat receiver, with no other heat input,

The thermodynamic properties of the working fluids may be represented in the temperature-entropy diagram as in Figure 4.

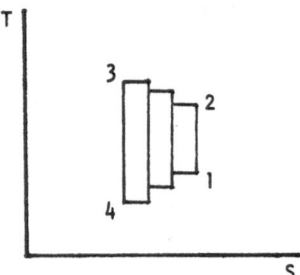

Figure 4.   Temperature-Entropy Diagram of the Working
            Fluids in a Multi-Staged Reversed Carnot Heat
            Pump System with Heat Input at Constant
            Working Fluid Pressure.

The system may be approximated with smooth lines if the humber of indi-vidual heat pumps is large, as shown in Figure 5 (next page).

This system may be analyzed by examining one of the heat pumps whose entropy is "s", whose receiver temperature is "T_h", and whose source tempera-ture is "T_c". Such a heat pump is shown in shaded lines in Figures 4 and 5.

From the start of the cycle (stroke 1-2), through the heat pump under examination, the entropy of the ideal gas at the source end is

$$(s - s_1)_{gas} = w_a \left[c_{pa} \ln \left(\frac{T_c}{T_1} \right) - \frac{R_a}{J} \ln \left(\frac{P}{P_1} \right) \right]$$

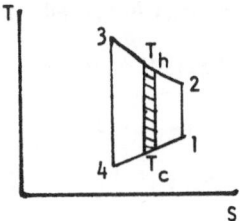

Figure 5.   Temperature - Entropy Diagram of Working
            Fluids of Infinitely Staged Reversed
            Carnot Heat Pump System with Heat Input
            at Constant Working Fluid Pressure.

For a constant pressure source, $P = P_1$, and

$$(s - s_1)_{gas} = w_a\, c_{pa} \ln\left(\frac{T_c}{T_1}\right)$$

where c_{pa} is the constant pressure specific heat of the ideal gas working fluid.

At the receiver end, heat has been removed from all the heat pumps from point 2 up to the heat pump under consideration. The heat went into the receiver to change the temperature of the receiver from T_2 to T_h. Because the system of the ideal gas and the receiver have no outside source of heat, and assuming reversibility, the combined entropy change is zero, and

$$(s - s_2)_a + (s - s_2)_{receiver} = 0$$

and

$$(s - s_2)_a + w_r\, c_r \ln\frac{T}{T_2} = 0$$

At the limit, when the system has reached T_3 and T_4:

$$(s_3 - s_2)_a + w_r\, c_r \ln\frac{T_3}{T_2} = 0$$

and

$$(s_4 - s_1) - w_a\, c_{pa} \ln\frac{T_4}{T_1} = 0 .$$

Because the system is bounded by reverse Carnot heat pumps, $s_4 = s_3$ and $s_2 = s_1$. Combining equations gives:

$$w_r\, c_r \ln\frac{T_3}{T_2} + w_a\, c_{pa} \ln\frac{T_4}{T_1} = 0 .$$

and

$$\frac{T_4}{T_1} = \left[\frac{T_2}{T_3}\right]^{w_r\, c_r\, /\, w_a\, c_{pa}} .$$

The heat input is along path 4-1, and is given by:

$$Q_{in} = Q_{4-1} = w_a c_{pa} (T_1 - T_4) .$$

The heat output is along path 2-3 and is given by:

$$\dot{Q}_{out} = Q_{2-3} = w_r c_r (T_3 - T_2) .$$

The work input is given by the conservation of energy,

$$W_{in} = Q_{out} - Q_{in} ,$$

and

$$W_{in} = w_r c_r (T_3 - T_2) - w_a c_{pa} T_1 \left[1 - \left(\frac{T_2}{T_3} \right)^{w_r c_r / w_a c_{pa}} \right] .$$

The coefficient of performance as a heat pump is given by:

$$COP = \cfrac{1}{1 - \cfrac{w_a c_{pa} T_1 [1 - (T_2 - T_3)^{w_r c_r / w_a c_{pa}}]}{w_r c_r (T_3 - T_2)}}$$

The coefficient of performance as a refrigerator is given by:

$$COP = \cfrac{1}{\cfrac{w_r c_r (T_3 - T_2)}{w_a c_{pa} T_1 \left[1 - \left(\frac{T_2}{T_3} \right)^{w_r c_r / w_a c_{pa}} \right]} - 1}$$

For the special case where the heat receiver is at substantially constant temperature, i.e. $T_2 \approx T_3$, the performance of the sytem approaches that of the single staged reversed Carnot cycle. This may be verified by applying L'Hospitale's rule and evaluating the indeterminant form of

$$\ell_{in} \frac{1 - \frac{T_2}{T_3}^{w_r c_r / w_a c_{pa}}}{w_w / w_a c_{pa} (T_3 - T_2)} = \frac{-w_r c_r / w_a c_{pa} \left(\frac{1}{T_3} \right)^{w_r c_r / w_a c_{pa}} (T_3)^{w_r c_r / w_a c_{pa} - 1}}{w_r c_r / w_a c_{pa}^{(-1)}} = \frac{1}{T_3}$$

and

$$COP \rightarrow \cfrac{1}{1 - \frac{T_1}{T_3}} ,$$

which is single stage reversed Carnot cycle performance.

For the special case where the receiver is heated from source temperature T_1 to temperature T_3, as many occur in generating a hot water supply from feed-water at outdoor temperature, the performance may be determined by letting T_1 approach T_2:

$$COP = \cfrac{1}{w_r c_r / w_a c_{pa}} \quad \cfrac{1 - \cfrac{T_2}{T_3}}{w_r c_r / w_a c_{pa} \left[\cfrac{T_3}{T_2} - 1\right]}$$

The performance depends on the fractions $w_r / w_a c_{pa}$ and T_3/T_2. This is shown in Table 2.

TABLE 2.   PERFORMANCE OF MULTI-STAGED CARNOT HEAT
           PUMP AS T_1 T_2 WITH CONSTANT PRESSURE
           HEAT INPUT

$w_r c_r / w_a c_{pa} = 0.5$					
T_3 / T_2	1.1	1.2	1.3	1.4	1.5
COP Multi-Staged Reversed Carnot	14.4	7.8	5.5	4.4	3.8
COP Single-Staged Reversed Carnot	11	6	4.3	3.5	3
COP Multi-Staged/COP Single-Staged	1.3	1.3	1.28	1.26	1.26

$w_r c_r / w_a c_{pa} = 1$					
T_3 / T_2	1.1	1.2	1.3	1.4	1.5
COP Multi-Staged Reversed Carnot	11	6	4.3	3.5	3
COP Single-Stage Reversed Carnot	11	6	4.3	3.5	3
COP Multi-Staged/COP Single-Staged	1	1	1	1	1

$w_r c_r / w_a c_{pa} = 1.5$					
T_3 / T_2	1.1	1.2	1.3	1.4	1.5
COP Multi-Staged Reversed Carnot	8.94	4.94	3.61	2.94	2.55
COP Single-Staged Reversed Carnot	11	6	4.3	3.5	3
COP Multi-Staged/COP Single Staged	.81	.82	.84	.84	.85

4.   THE MULTI-STAGED HEAT PUMP AND THE REVERSED CARNOT CYCLE

The literature gives reversed Carnot COP:

$$COP = \frac{T_3}{T_3 - T_1}$$

and points to this as the highest performance attainable by a heat pump operating between a heat source at temperature T_1, and a heat receiver at T_3. This
seems to contradict the results of paragraph 4. However, a close examination
of the literature shows the results of paragraph 4 are entirely consistent
with the theory.

Professor Said Carnot[1] wrote:

"We take, for example, one body kept at a temperature of 100°, and
another body kept at a temperature of 0°. . . [and] imagine two bodies
A and B kept each at a constant temperature . . . to which we can give,
or from which we can remove heat without causing the temperature to
vary. . . ." (emphasis added).

The authors of this analysis suggest that Professor Carnot was writing
about cycles operating between constant source and receiver temperatures.
Processes where the heat source or the heat receiver do not remain at constant
temperature are not necessarily bound by the restrictions of the reversed
Carnot performance, in the opinion of the authors.

In the case of heating and cooling, as in the applications for the heat
pump, the water being heated (or the air being cooled) undergoes a significant temperature change. The heat pump transmits heat to the receiver while
the temperature of the receiver and the owrking medium (the air) is increasing from T_2 to T_3. The temperature at which heat is transferred to the receiver is the lowest possible temperature of the receiver and the working
medium. The lowest possible grades of energy are used. The savings in work
are shown in the temperature-entropy diagrams in Figures 5 and 6.

When the temperature change $(T_3 - T_2)$ is large, as occurs in hot water
generation, the effect is a substantial decrease in the owrk input. The
savings is labeled "Work Saved".

When w_w gets to be large, it has the effect of decreasing the exhaust

temperature, $T_4 = T_1 \left[\dfrac{T_2}{T_3} \right]^{w_r c_r / w_a c_{pa}}$. This forces the heat transfer

from the environment into the air, along path 4-1, to occur at a lower temperature, thereby in creasing the work input. The added work input is label-
led "Work Added".

A low w_w and a high $(T_3 - T_2)$ yield a high COP. A high w_w and a small $(T_3 - T_2)$
yield a lower COP.

5.   TEMPERATURE - ENTROPY DIAGRAMS

[1] Reflections on the Motive Power of Fire, and on Machines Fitted to
Develop that Power. Sadi Carnot. 1824. Publisher: Chez Bachelier,
Libraire, Paris, France.

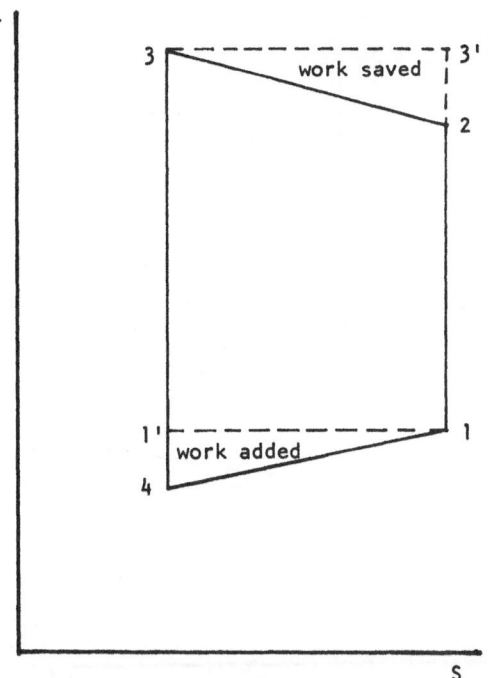

Figure 6.   Temperature - Entropy Diagrams
in the Heating Mode

Legend:

Figure 1-2-3-4 is the multi-staged reversed Carnot cycle with heat
input at constant pressure.

Figure 1-3'-3-1' is the single stage reversed Carnot cycle.

Area 2-3'-3 is the work saved by the multi-staged reversed Carnot
cycle with heat input at constant pressure over the
the single-staged reversed Carnot cycle.

Area 1-1'-4 is the work saved by the single-staged reversed Carnot
cycle over the multi-staged reversed Carnot cycle with
heat input at constant pressure.

T_1 is the outdoor (heat source) temperature.

T_2 is the initial receiver temperature.

T_3 is the final heat receiver temperature.

T_4 is the temperature of the working fluid after expansion to
atmospheric pressure.

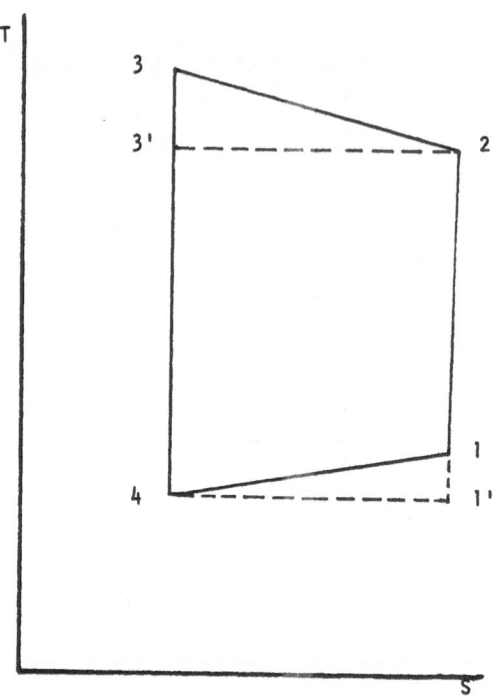

Figure 7.   Temperature - Etropy Diagrams
in the Cooling Mode

Legend:

Figure 1-2-3-4 is the multi-staged reversed Carnot cycle with heat
input at constant pressure.

Figure 1'-2'-3'-4 is the single staged reversed Carnot cycle.

Area 1-4-1' is the work saved by the multi-staged reversed Carnot
cycle with heat input at constant pressure over the
reversed Carnot cycle.

Area 2-3-3' is the work saved by hte single-staged reversed Carnot
cycle over the multi-staged reversed Carnot cycle with
heat input at constant pressure.

T_1 is the room (heat source) temperature.

T_2 is the initial heat receiver temperature.

T_3 is the final heat receiver temperature.

T_4 is the refrigerated air temperature.

Hybrid Desiccant Vapor Compression Cooling System

WILLIAM M. WOREK
Illinois Institute of Technology
Department of Mechanical Engineering
Chicago, Illinois 60616, USA

ABSTRACT

The concept of an integrated hybrid, desiccant, vapor compression cooling system is presented and the improvement in performance over standard vapor compression cooling systems is estimated. In such a system, which uses the sorbent material manufactured in the form of a desiccant sheet, significant temperature gradients can occur in the desiccant material. A sorption model which includes this effect is presented.

1. INTRODUCTION

Significant peak electric loads occur in summer months due to air-conditioning demand. In vapor compression air-conditioning units, the heat removed from the residence, as well as the electrical energy supplied to the unit, is rejected from the condenser as waste heat. The proposed hybrid cooling system uses this waste heat to activate a dehumidification cycle which uses a solid desiccant, silica gel, as the sorbent.

Such hybrid systems have received little attention and only recently has the merit of such systems been realized. Limited analyses and conceptual designs of hybrid systems can be found in the recent literature. These concepts exclusively couple existing cooling machines with a separate dehumidifier [1]. The major problem with such an approach is the large size and the complexity of the system.

2. SYSTEM CONFIGURATION

The integrated hybrid desiccant vapor compression air-conditioning system consists of two condensers, which are integrated with a sorbent, and one evaporator. In the typical operation of the system, heat rejected from the cooling system will be used to reactivate one of the condensers, while the other condenser will be used as a dehumidifier upstream of the evaporator. When the desiccant in the condenser, which is providing dehumidification, is no longer effective (i.e., the rate of moisture sorbed is low) it will be switched with the desiccant integrated condenser which was reactivated by waste heat rejected from the cooling system.

This switching will coincide with the typical cycling of the compressor in the cooling machine.

Figure 1 shows, schematically, the air flow in the hybrid system. Outdoor enters and is directed by the first bidirectional damper to the desiccant integrated condenser. In the condenser, the sorbent is desorbed by waste heat rejected from the cooling machine. The warm, moist air at the exit of the condenser is then returned to the outdoors. The room air, which is to be cooled and dehumidified, enters on the opposite side of the first bidirectional damper and is directed to the dehumidifier where the air is dried. The heat generated in the sorption process is removed by a secondary stream which rejects the heat to the outdoors. The dry air at the exit of the dehumidifier then enters the evaporator where it is cooled. This conditioned air then is returned to the room.

The refrigerant flow schematic, shown in figure 2, gives the method which the refrigerant will be switched between the two components which are integrated with the desiccant material.

3.   ANALYSIS

The desiccant integrated components (dehumidifier/condenser) uses a sorbent in the form of sheets. The manufacture of such desiccant sheets was developed at IIT [2] and is being patented. These sheets, which use 0.009 mm regular density silica gel particles held in a teflon web, are typically 2 mm thick. The sheets are adhered to the fins of a finned tube heat exchanger. Two identical heat exchangers are used to form the condenser/dehumidifier components.

The desiccant integrated heat exchangers have two streams. The usage of these two streams which depends if the process is dehumidification or desorption. In the dehumidification cycle, a process air stream is dried and a cooling stream is used to remove the heat generated during the process of sorption. In the desorption mode a process air stream removes the moisture liberated from the sorbent and a refrigerant provides the heat input.

Silica gel has a low thermal conductivity. Therefore, the temperature distribution in the desiccant material must be accurately modeled to evaluate the process of sorption in the desiccant. The geometry used for the analysis is shown in figure 3, where stream 1 is the process air stream and stream 2 is either the cooling fluid (dehumidification) or the refrigerant (desorption) (Note: In the actual dehumidifier/condensers these will be distinct streams). Figure 4 shows an enlargement of the desiccant thickness.

The mass balance on the process stream and wall yields,

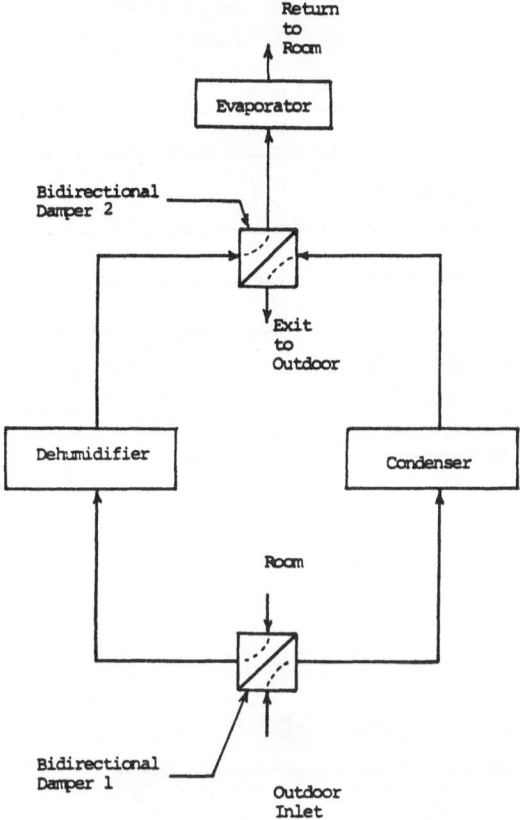

Figure 1. Hybrid System Air Flow Schematic

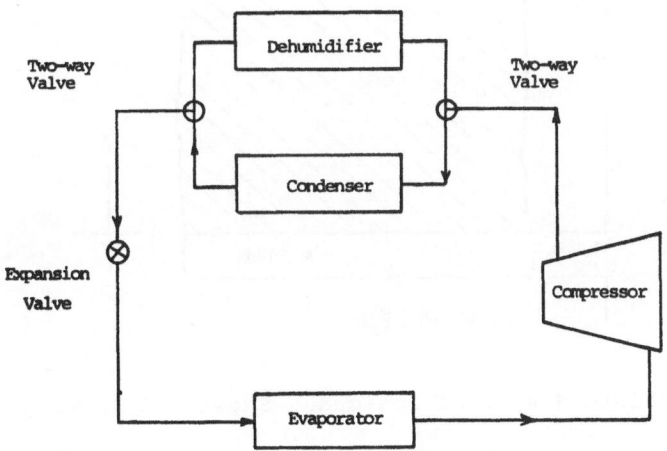

Figure 2. Hybrid System Refrigerant Flow Schematic

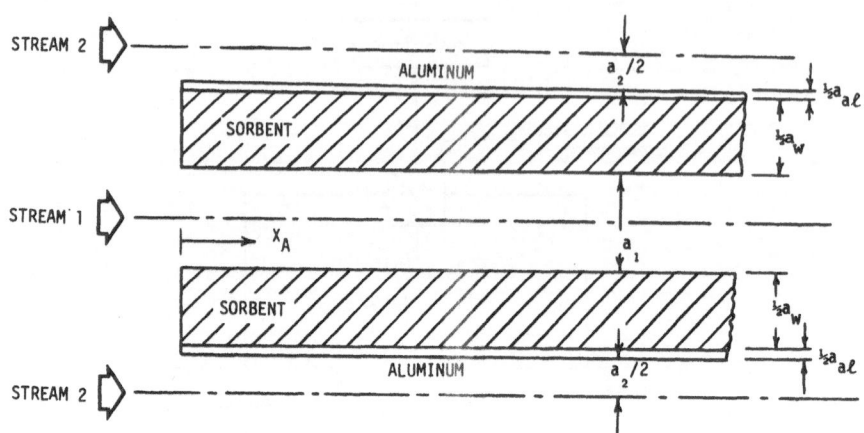

Figure 3.   Schematic Showing the Dimensions of a Typical Subsection
of a Dehumidifier Channel.

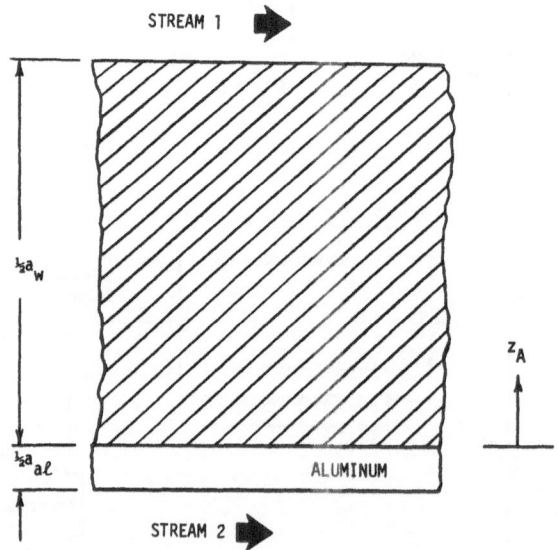

Figure 4.   Enlarged View of Desiccant Sheet and Aluminum Wall
Assembly.

SORPTION

Mass Balance

Process Stream

$$\frac{\dot{m}_1}{ny_{AF}} \frac{\partial Y}{\partial x_A} = 2K_y (Y_w - Y) \tag{1}$$

Wall (Sorbent)

$$(1-g) \rho_w a_w \frac{\partial w}{\partial t_A} = 2K_y (Y - Y_w) \tag{2}$$

An energy balance yields,

Energy Balance

Process Stream

$$\frac{\dot{m}_1 c_1}{ny_{AF}} \frac{\partial T_1}{\partial x_A} = 2h_1 (T_w - T_1) \tag{3}$$

Secondary Stream
  (dehumidification)

$$\frac{\dot{m}_2 c_2}{ny_{AF}} \frac{\partial T_2}{\partial x_A} = 2h_2 (T_w - T_2) \tag{4}$$

(desorption)

$$\frac{\dot{m}_2}{ny_{AF}} \frac{\partial H_2}{\partial x_A} = 2h_2 (T_w - T_2) \tag{5}$$

Wall

$$\rho_w \left(\frac{\partial H_w}{\partial t_A}\right) - k_w \frac{\partial^2 T_w}{\partial z_A^2} = \frac{2K_y (Y - Y_w)}{a_w} \frac{\partial H_1}{\partial Y} \tag{6}$$

In the desiccant material

$$H_w = H_w (T_w, w) \tag{7}$$

therefore

$$\rho_w \left(\frac{\partial H_w}{\partial T_w} \frac{\partial T_w}{\partial t_A} + \frac{\partial H_w}{\partial w} \frac{\partial w}{\partial t_A} \right) - k_w \frac{\partial^2 T_w}{\partial z_A^2} = \frac{2K_y (Y-Y_w)}{a_w} \frac{\partial H_1}{\partial Y} \qquad (8)$$

defining the heat of sorption, Q

$$Q = \frac{\partial H_1}{\partial Y} - \frac{1}{1-g} \frac{\partial H_w}{\partial w} \qquad (9)$$

we obtain from equations (2) and (9)

$$\frac{\partial H_w}{\partial w} \frac{\partial w}{\partial t_A} = \frac{2K_y (Y-Y_w)}{\rho_w a_w} \left(\frac{\partial H_1}{\partial Y} - Q \right) \qquad (10)$$

also defining

$$\frac{\partial H_w}{\partial T_w} = \lambda_4 \, C_{wR} \qquad (11)$$

where

$$\lambda_4 = \lambda_4 (T_w, w) \qquad (12)$$

we obtain for the wall

$$\rho_w C_{wR} \, \lambda_4 \frac{\partial T_w}{\partial t_A} - k_w \frac{\partial^2 T_w}{\partial z_A^2} = \frac{2K_y (Y-Y_w)}{a_w} \qquad (13)$$

therefore, in non-dimensional form we have

$$\frac{\partial Y}{\partial x} = \lambda_3 (Y_w - Y) \qquad (14)$$

$$\frac{\partial w}{\partial t} = \lambda_2 (Y-Y_w) \qquad (15)$$

$$\frac{\partial T_1}{\partial x} = T_w - T_1 \qquad (16)$$

$$C \frac{\partial T_2}{\partial x} = T_w - T_2 \tag{17}$$

$$\frac{C}{c_2} \frac{\partial H_2}{\partial x} = T_w - T_2 \tag{18}$$

$$\lambda_4 \frac{\partial T_w}{\partial t} - \lambda_6 \lambda_9 \frac{\partial^2 T_w}{\partial z^2} = \lambda_5 \lambda_9 (Y - Y_w) \tag{19}$$

Sorbent boundary conditions

$$\frac{\partial T_w}{\partial z} = \frac{1}{\lambda_6} (T_1 - T_w)$$

at z=1

$$\frac{\partial T_w}{\partial z} = \frac{b}{\lambda_6} (T_2 - T_{a\ell}) + \frac{1}{\lambda_6 \lambda_{10}} \frac{\partial T_{a\ell}}{\partial t}$$

$$T_w(o,t) = T_{a\ell}(t) \qquad \text{at z=0}$$

System Inlet Conditions

　　　Dehumidification　　　　　　　　　　　Desorption

$$T_1(t,o) = T_{1i}(t) \qquad\qquad\qquad T_1(t,o) = T_{1i}(t)$$
$$Y(t,o) = Y_{1i}(t) \qquad\qquad\qquad\quad Y(t,o) = Y_{1i}(t)$$
$$T_2(t,o) = T_{2i}(t) \qquad\qquad\qquad H_2(t,o) = H_2(t)$$

The initial conditions are taken from the previous sorption process.

This system is numerically solved using a finite difference formulation. The results are obtained after periodic, steady state operation is achieved.

4.　PERFORMANCE RESULTS

A schematic representation showing a comparison between a standard vapor compression cooling system and the hybrid system is given on the psychrometric chart in Figure 5.

Preliminary results of the performance of the hybrid cooling

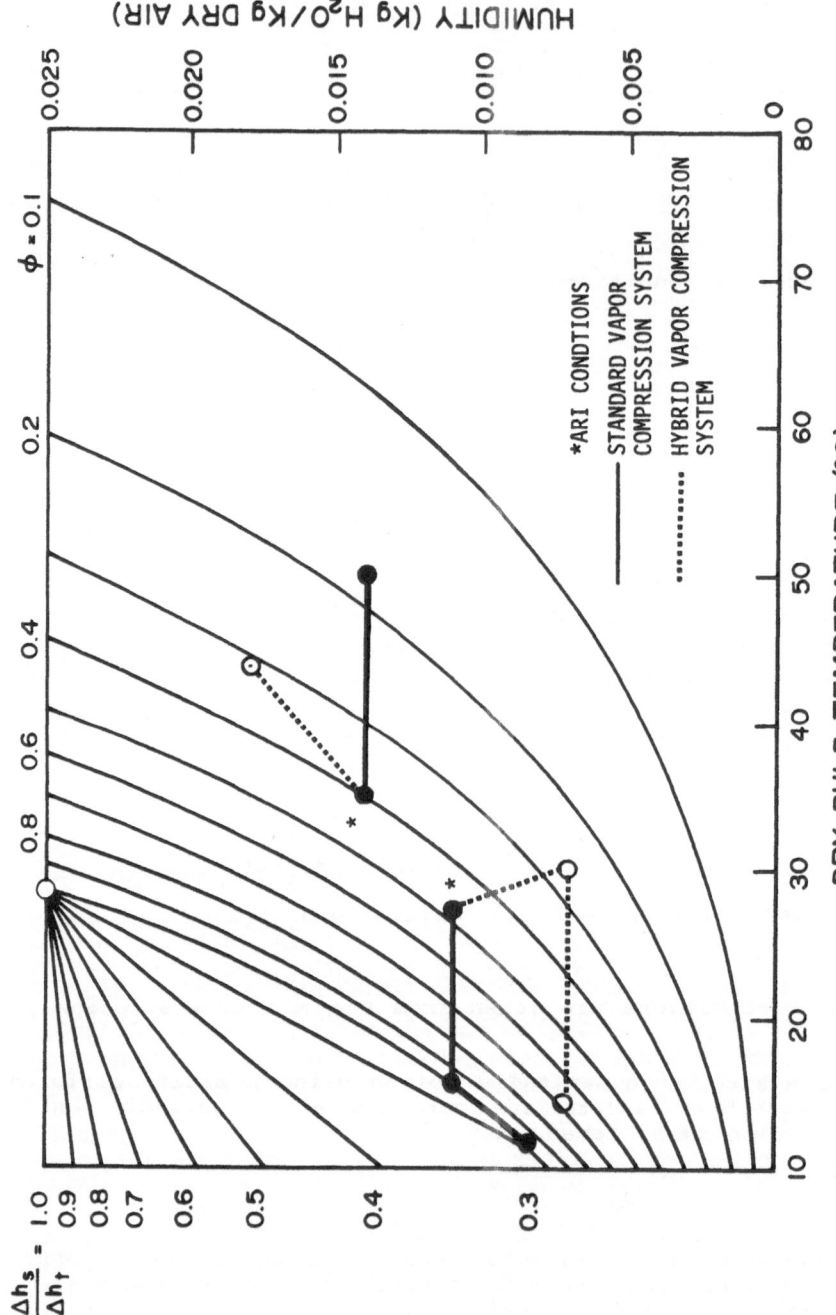

Figure 5. Performance Comparison of a Standard Vapor Compression System with a Hybrid Desiccant Vapor Compression System.

system shows that the hybrid system operates at higher evaporator and lower condenser temperatures. By utilizing the waste heat rejected in the condenser of the cooling system, the latent load is satisfied by the dehumidifier no significant level of energy input from the cooling system. This enables approximately a 30% improvement in the systems electrical coefficient of performance.

5. CONCLUSIONS

The integrated hybrid desiccant vapor compression cooling system utilizes heat reject in the condenser to activate dehumidification process which uses a solid desiccant. The sorbent material, silica gel, is used to supply the latent load and vapor compression cooling system satisfies the sensible load. The silica gel, which is manufactured in the form of a sheet, is integrated into the two components which can be switched to operate either as a condenser or a dehumidifier. In the condenser the desiccant is regenerated while in the dehumidifier the desiccant satisfies the latent cooling load.

Integration of the desiccant into the vapor compression system allows the evaporator and condenser to operate at higher and lower temperatures respectively. Preliminary results show the overall improvement in COP is 30% with little additional size and complexity over standard vapor compression cooling systems.

6. NOMENCLATURE

a_1	process channel width
a_2	cooling channel width
a_{al}	twice the aluminum thickness
a_w	twice the sorbent thickness
b	h_2/h_1
c	$(m_2 c_2)/(m_1 c_1 b)$
C_{wr}	reference wall specific heat
c_1	process stream specific heat
c_2	secondary stream specific heat (dehumidification)
g	fraction of teflon in sorbent
H_w	enthalpy of wall
H_1	enthalpy of stream 1
H_2	enthalpy of stream 2
h_1	process stream heat transfer coefficient
h_2	secondary stream heat transfer coefficient

K_y mass transfer coefficient

k_w sorbent thermal conductivity

m_1 total mass flow of process stream

m_2 total mass flow of secondary stream

n   number of channels

Q   heat of sorption

T_w wall temperature

T_1 stream 1 temperature

T_2 stream 2 temperature

t_A actual time

t   nondimensional time,

x_A actual process channel location

x_{AF} actual process channel length

x   nondimensional process channel length,

y_{AF} actual cooling channel length

z_A actual desiccant thickness

z   nondimensional thickness

λ_2 $K_y C_{wR} \; \lambda_9 / (h_1 (1-g))$

λ_3 $K_y C_1 / h_1$

λ_4 $\dfrac{\partial H_w}{\partial T_w} \, C_{wR}^{-1}$

λ_5 $K_y Q / h_1$

λ_6 $2k_w / (a_w h_1)$

λ_9 $1 + (\rho_{al} a_{al}) / (\rho_w a_w)$

λ_{10} $1 + (\rho_w a_w) / (\rho_{al} a_{al})$

7. REFERENCES

[1] Curran, H.M., Personal Communication

[2] Gidaspow, D., Lavan, Z., Onischak, M. and Perkari, S., "Development of a Solar Desiccant Dehumidifier," _Proceedings of the 13th Intersociety Energy Engineering Conference_, 1978, pp. 1623-1627.

Effect of Sorption Properties on Desiccant Dehumidifier Performance

S. DINI
Western New England College
Springfield, Massachusetts 01119, USA

Z. LAVAN, W.M. WOREK, and **D. GIDASPOW**
Illinois Institute of Technology
Chicago, Illinois 60616, USA

ABSTRACT

The sorption isotherms for silica gel manufactured in the form of sheets were experimentally determined for high temperatures and humidities. Using this data the heat of sorption and the integral heat of wetting are calculated using a Clausius-Clapeyron type equation. These relationships are then used to study the nonlinear coupling between heat and mass transfer processes in a cross-cooled dehumidifier which is the principal component of a solar powered desiccant cooling system.

The effect of the shape of the equilibrium isotherm and the heat of sorption on the cooling capacity, the thermal coefficient of performance and the moisture cycled was studied.

1. INTRODUCTION

A number of different desiccants have received attention for solar cooling, the most popular being natural zeolite, molecular sieve, and silica gel. The sorption capacity of silica gel strongly depends on the vapor pressure. This dependence results in silica gel having a good adsorption capacity at low temperatures (20-25°C) and enables desorption at moderate temperatures (60-65°C). Hence, silica gel dehumidifiers can utilize the air delivered by a simple flat plate collectors for regeneration of the desiccant.

Natural zeolite and molecular sieve, on the other hand, show much less water vapor pressure dependence and have good adsorption characteristics at low relative humidities or high temperatures. However, the required regeneration temperatures for these materials are usually high (150-170°C).

The principal component of the IIT cross-cooled system is the silica gel dehumidifier. This is the unit which removes moisture from the air to be conditioned. Since the system of equations governing the dynamics of sorption has to be solved along with the equilibrium relations of the silica gel sheets, it is apparent that the adsorption characteristics of the desiccant would influence the cooling capacity and the COP of a system. Thus the thermodynamics of the desiccant system as well as properties of the silica gel sheet were studied.

To study the equilibrium sorption capacity of silica gel
sheets at high temperatures and high humidities, a TGA (Thermo-
gravimetric Analysis) test setup was designed and constructed.
The sorption isotherms for silica gel was determined at various
temperatures. In most cases, a small hysteresis effect was noted.
Using the equilibrium vapor pressure data, heat of sorption and
hence heat of wetting was obtained. These results were then used
in the analysis of solar powered desiccant cooling systems developed at IIT.

2. THERMOGRAVIMETRIC ANALYSIS

A TGA test system was constructed to study the equilibrium
adsorption capacity of desiccant materials at high temperatures
and high humidities. The test setup consists of two sections, one
section is used for regeneration and the other section is used for
adsorption/desorption measurements. The two parts of the setup
may be used simultaneously.

Figure 1 shows the schematic diagram of the regeneration
section of the TGA test setup. The glass sample holders are
equipped with caps. The caps permit air tight sealing of the
sample during weighing. A sample of desiccant material is placed
in a bottle of approximately 2 cm in diameter. The sample is
dried by inserting the bottle into the setup shown in figure 1 and
passing dry air at $110^{\circ}C$ through the sample until the sample
weight becomes constant with time. Heating tapes are used in
conjunction with SCR voltage controllers to control the temperature of the air and the sample. In addition, the drying section
is covered with a low thermal conductivity insulating material to
reduce heat losses. Two thermocouples are used to monitor the
temperatures of inlet and outgoing air, and another thermocouple
is used to monitor the temperature of the sample. The inlet and
outlet temperatures of the heated dry air, as well as the sample
temperature are measured by a digital thermometer. The dew point
of the inlet and outgoing air are measured by a hygrometer. When
equilibrium is reached the sample holder is removed from the setup
and the weight is recorded using a single pan analytical balance.

When drying is complete, the sample is placed into the test
setup as shown in figure 2. For adsorption measurements in the
neighborhood of room temperature, a simple bubbler provides
saturated air which is mixed with dry air to obtain the desired
inlet moisture conditions. For adsorption measurements well
above room temperature, a steam generator provides steam which is
mixed with dry air to produce the inlet moisture conditions.
Heating tapes and insulating material are wrapped around the
connecting tubes to keep the tubes wall temperature above the dew
point to avoid condensation. The hygrometer sensor is also heated
in a heating box at least $5^{\circ}C$ above the dew point to avoid condensation.

Desorption runs are made in the same setup as in the adsorption operation. During the desorption process, dry air and steam
are mixed, heated, and then passed through the sample. The temperature and moisture content of each run were set at the begin-

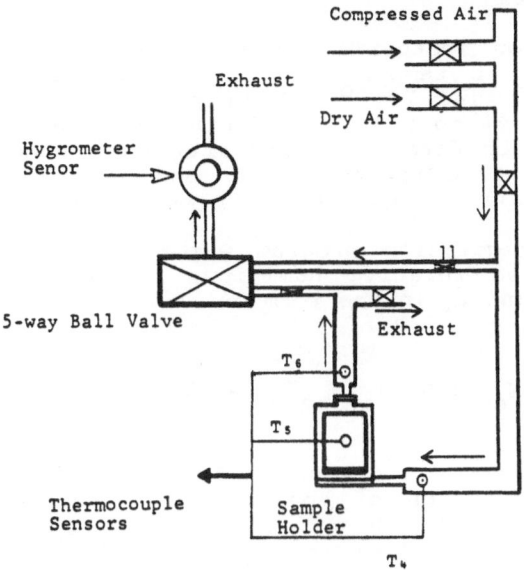

Figure 1. Schematic Diagram of the Regeneration Section of the
TGA Test Set-up

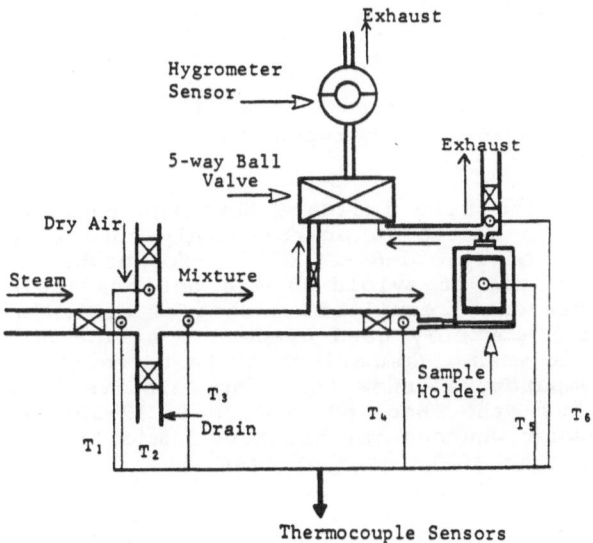

Figure 2. Schematic Diagram of Adsorption/Desorption Section of
the TGA Test Set-up

ning and were maintained throughout the duration of the test.

2.1 Test Results

The sorption isotherms for moist air and silica gel sheet were determined at temperature $22.4^{\circ}C$, $37.8^{\circ}C$, $45.2^{\circ}C$, $52.4^{\circ}C$, $60.1^{\circ}C$, $71.1^{\circ}C$, and $79.9^{\circ}C$ which covers a range of relative humidities from 4% to 95%. The isotherms obtained are presented in figure 3. Each data point for silica gel sheet was determined at least twice, once for adsorption and once for desorption. In most cases a small hysteresis effect was noted.

A linear regression program was used to obtain functional relations between relative humidity, , equilibrium moistures content, W_{eq}, and T_w in the form of

$$\phi = A_1(T_w)W_{eq} + A_2(T_w)W_{eq}^2 + A_3(T_w^2)W_{eq}^2 + \text{----} \tag{1}$$

For variable acceptance criteria of 0.03 and 0.05, the relative humidity as a function of equilibrium temperature and moisture content is given by

$$\phi = A_1 + A_2T_w^2 + A_3W_{eq} + A_4W_{eq}^2 + A_5W_{eq}^3 + A_6T_w^3W_{eq} \tag{2}$$

where

$$
\begin{aligned}
A_1 &= -9.310771 & A_2 &= 0.00171765 \\
A_3 &= 478.0868 & A_4 &= -1417.118 \\
A_5 &= 2094.818 & A_6 &= 9.183715(10^{-5})
\end{aligned}
$$

The plot of this polynomial fit is shown in figure 4. The results indicate a close agreement between the polynomial fit and the experimental data.

The test results also indicate that the equilibrium adsorption capacity of the sheets is approximately 15-35% less than the equilibrium values of syloid-63 at $22^{\circ}C$. For higher temperatures, the equilibrium values of syloid-63 are not available. However, Hubard [5] has presented equilibrium data of silica gel which are applicable to such commonly used grades of silica gel as 3-8, 6-12, 6-16 and 14-20 mesh. Comparison of test results with Hubard's data for high temperatures show that for relative humidities in a range of 20 to 75%, the sheet adsorption is 12-30% less than that of silica gel powder whereas for relative humidities less than 20% and above 75% , it is 10-15 percent less.

3.   HEAT OF SORPTION AND INTEGRAL HEAT OF WETTING

The adsorption of a gas on a solid is a spontaneous, exothermic process. In physical adsorption, the formation of an adsorbed layer is similar in many respects to normal condensation of a layer. When a vapor is adsorbed on silica gel or a similar adsorbent, heat is liberated equivalent to the latent heat of evapora-

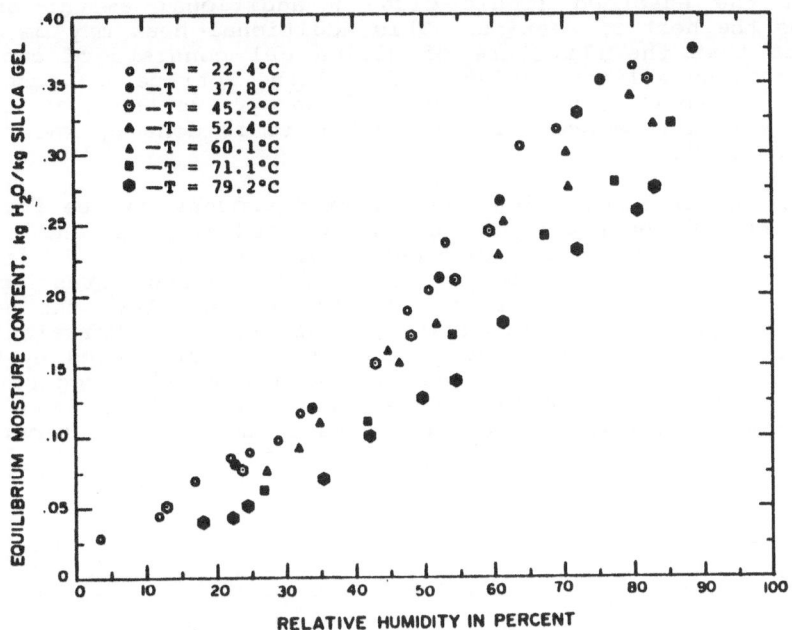

Figure 3.   Equilibrium Characteristics of Silica Gel Sheet

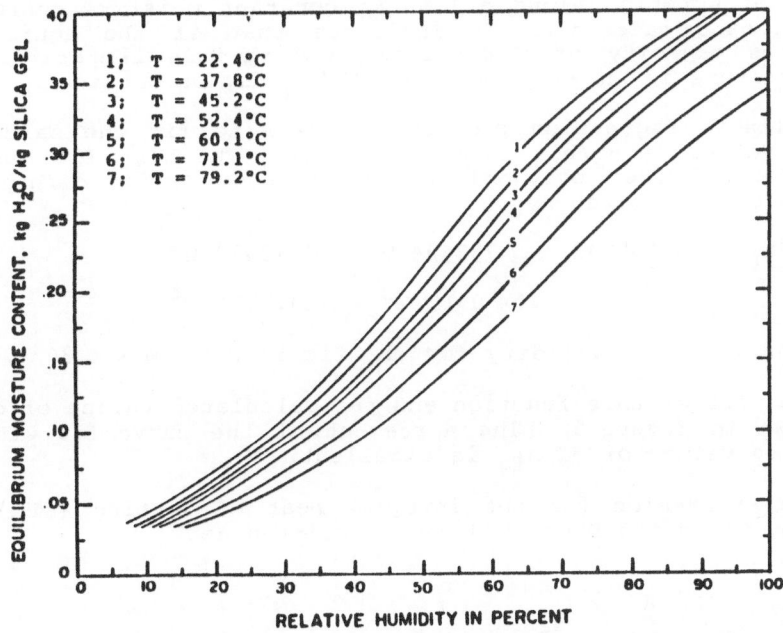

Figure 4.   Polynomial Fit of Equilibrium Characteristics of Silica
Gel Sheet for a Confidence Level of 0.05

tion of the adsorbed liquid plus an additional amount of heat known as the heat of wetting. This additional heat may be due to the fact that the structure of silica gel consists of capillary pores creating a large surface. This large surface has associated with it a great amount of surface energy, destruction of this surface will give a heat effect whether it is done by water vapor or water liquid.

For sorbents in which the sorption process can be reversed, the Clausius-Clapeyron equation can be used to link the heat of sorption to the vapor pressure data. In other words, if isotherm measurements are available at neighboring temperatures down to relatively low pressure, it is possible to calculate the heat of adsorption and also the integral heat of wetting. Correlation of vapor pressure and latent heat data has been suggested by Othmer [7]. Close and Banks [2] have extended Othmer's suggestion to obtain a relation similiar to equation (3), to relate vapor pressure-concentration-temperature data to heat of sorption.

$$\left(\frac{\partial \ln P_V}{\partial \ln P_g}\right)_W = \frac{q_s}{h_{fg}} \qquad (3)$$

Equation (3) has been obtained with the assumption of perfect gas behavior for the gas mixture and negligible surface effects for the desiccants. This equation correlates the vapor pressure data to heat of sorption along a line of constant moisture content of the desiccant material. It indicates that if the equilibrium adsorption capacity of a desiccant for various temperature and vapor pressure is known, the ratio q_s/h_{fg} may be obtained.

A linear regression program was used to fit the calculated average values of q_s/h_{fg} to a polynomial. The best fit obtained was a fourth order polynomial. The expression for q_s/h_{fg} as a function of W is given by

$$q_s/h_{fg} = 1.545942 - 7.733934 \, W + 52.70972 \, W^2$$
$$- 154.0512 \, W^3 + 161.5251 \, W^4 \qquad (4)$$

where the range of validity for the fit is .02 < W < 0.36.

The plot of this function and the calculated values of q_s/h_{fg} are shown in figure 5. The agreement of the curve fit with the calculated values of q_s/h_{fg} is excellent.

The expression for the integral heat of wetting (ΔH_w) suggested by Close and Banks [3] may be written as

$$\Delta H_w = h_{fg} \int_0^W \left(1 - \frac{q_s}{h_{fg}}\right) dW \qquad (5)$$

Substituting for q_s/h_{fg} from equation (4) in equation (5), the polynomial for the integral heat of wetting for silica gel sheet is obtained. For W < 0.36 the relation is given by

$$H_w = h_{fg}[-0.545942\ W + 3.866967\ W^2 - 17.569907\ W^3$$
$$+ 38.5128\ W^4 - 32.20502\ W^5] \tag{6}$$

where $h_{fg} = 2504.4 - 2.4425\ T_w$

4. CROSS-COOLED DESICCANT COOLING SYSTEM

The complete IIT cross-cooled desiccant cooling system consists of two cross-flow dehumidifiers, two sensible heat exchangers and three evaporative coolers. The major components of this system is the cross-flow silica gel dehumidifier. The advantage of a cross-flow dehumidifier has been explained extensively by Roy and Gidaspow [8]. They have shown that the cross-flow dehumidifiers can be smaller and require less power than adiabatic dehumidifiers. The section of the cross-flow dehumidifier considered in the present work is shown in figure 6. The side of the wall facing stream 1, the process stream, is made of silica gel sheet adhered to the sides of the aluminum channels. The silica gel is capable of exchanging water with the process stream. The cooling stream, flowing in the cross-flow channel 2, exchanges heat only with the solid wall separating the channels.

The differential equations describing the simultaneous heat and mass transfer in a cross-flow regenerator were formulated by Roy and Gidaspow [8] using Schumann's model. They have used a generalized-equilibrium relationship at the gas-solid interface to obtain nonlinear coupling between the heat and mass transfer processes. Mathiprakasam [6] developed a similar system of nonlinear coupled heat and mass transfer equations without neglecting the resistance in the sheet and obtained solutions for periodic steady state operation using a finite difference scheme.

This system of equations and the method of solution described in [6] is used in the present work and will not be presented here. However, the computer program was modified and the present analysis differs in several ways: 1) The equilibrium data of the silica gel sheet was obtained experimentally and a linear regression program was used to obtain the best fit to these data. 2) Functional relations between the heat of adsorption and moisture content, W, were formed and used in the analysis. 3) The effect of the heat of wetting is considered for calculating the specific heat of the wall.

4.1 Thermodynamic Analysis of the Desiccant Cooling System

Figure 7 shows the arrangement of the components of a cooling system, with the dehumidifiers in adsorption, desorption and purging modes of operation respectively. In the adsorption mode, the air leaving the room at State 1 is first sent to the dehumidifier, where the silica gel removes moisture from stream. The heat of adsorption (the energy released when water is converted from a gaseous state to an adsorbed state) increases the air temperature at State 2. Cross-cooling helps to remove part of this heat. The

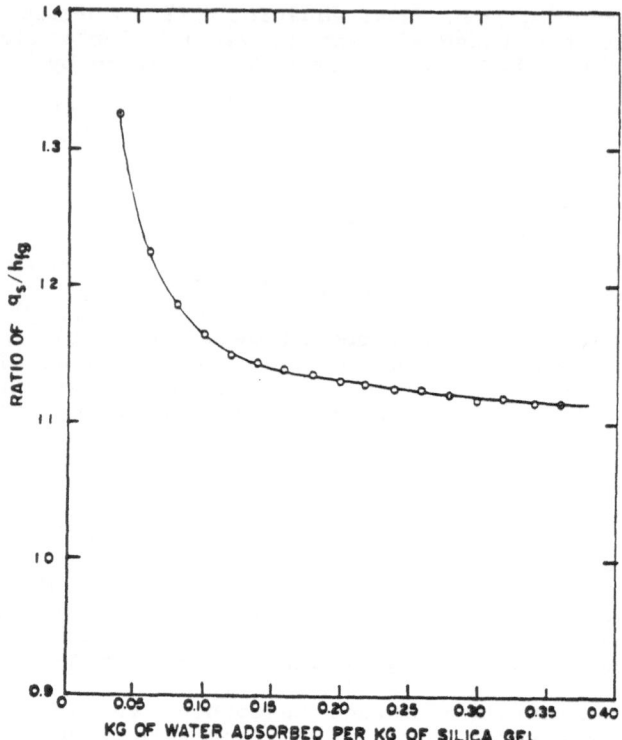

Figure 5. Ratio of Heat of Sorption to Latent Heat of Vaporization for the Silica Gel Sheet

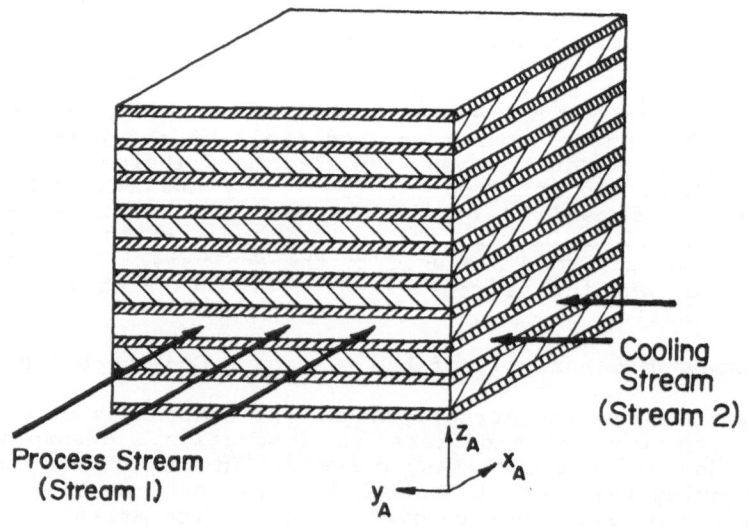

Figure 6. Schematic of a Cross Flow Desiccant Dehumidifier

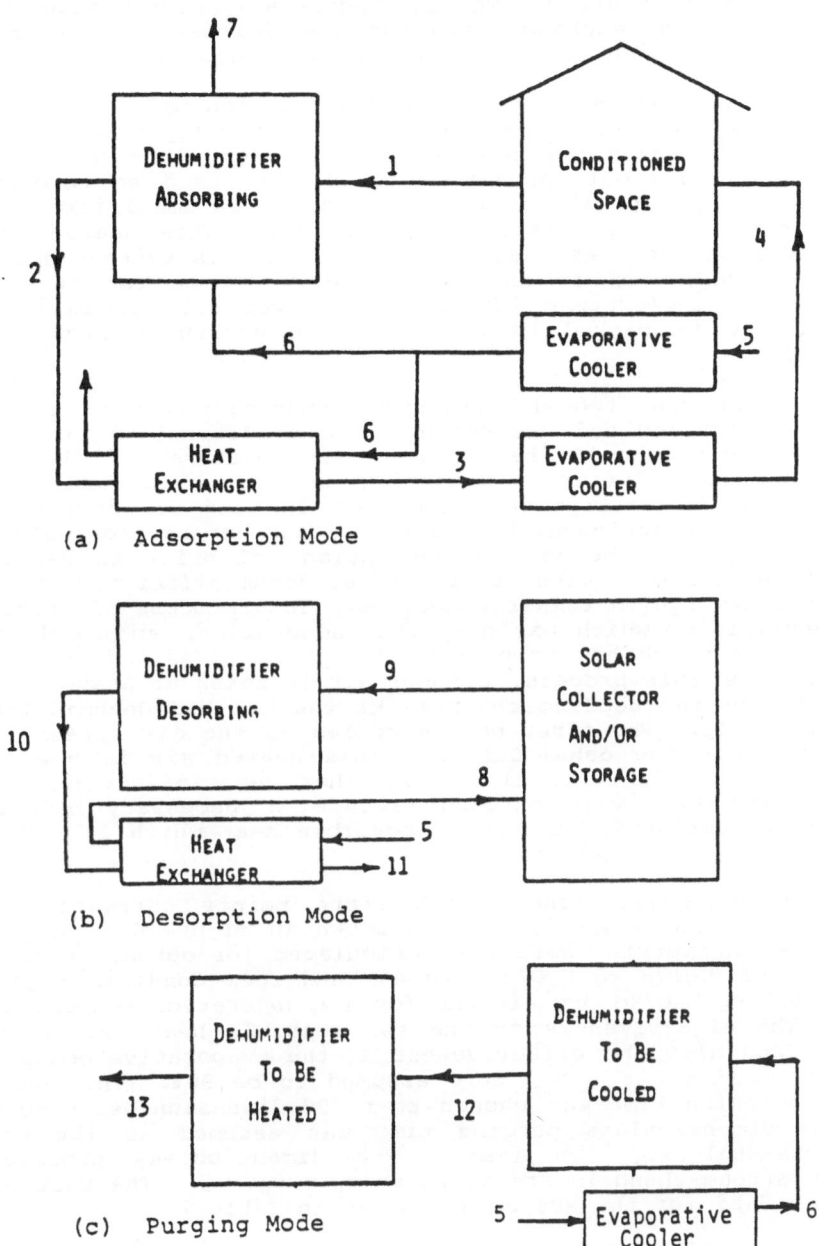

Figure 7. Details of the Cross-Cooled Desiccant Cooling System,
with Dehumidifiers in Adsorption, Desorption and Purging Modes
of Operation

process stream is then sensibly cooled to State 3, and is evaporatively cooled to State 4. The cooling stream which passed through the sensible heat exchanger and the dehumidifier, is the outdoor air at State 5 which has been evaporatively cooled.

When the dehumidifier is in the desorption mode, the moisture in the bed is removed by passing hot air typically 60-65°C, from a solar air collector, a rock storage or any other heat sources through the bed. Outdoor air at State 5 is used to recover the sensible heat, from the stream leaving the dehumidifier at State 10, by means of a sensible heat exchanger. This heated outdoor air leaves the heat exchanger at State 8 and is further heated in the solar collector or any other energy source to the desired regeneration temperature (State 9). The wet air stream from the dehumidifier is passed through the heat exchanger and is then discarded.

Each of the two dehumidifiers undergoes cyclic switching between adsorption and desorption at a predetermined cycle time. When the first dehumidifier is adsorbing moisture, simultaneously the second dehumidifier is desorbing moisture which was adsorbed in the previous adsorption cycle. At the end of adsorption and desorption, the desiccant beds are cool and hot respectively. The desiccant bed at the end of adsorption and prior to desorption needs to be heated. Likewise the other dehumidifier bed which has completed desorption requires cooling. This process of preheating the dehumidifier which has completed adsorption, and cooling the dehumidifier which has completed desorption is called the purging process. For this process, outdoor air is taken at State 5 and is sent through the cooling channels of the hot bed dehumidifier to reach State 12. The first bed is cooled as the air stream purges the heat from the dehumidifier. This heated air is now passed through the cooling channel of the other dehumidifier to preheat it to State 13. This purging process is necessary because it recovers a substantial amount of sensible heat which is stored in the hot bed dehumidifier.

A psychrometric chart with state points corresponding to locations on the block diagram is shown in figure 8. The values on the psychrometric chart were calculated for outdoor conditions of 35°C and 0.0178 kg H_2O/kg dry air and room conditions of 26°C and 0.0126 kg H_2O/kg dry air and for a regeneration temperature of 65°C. The effectiveness of the two sensible heat exchangers η_1 and η_2 and also the effectiveness of the evaporative cooler used for the cooling air, η_3, were assumed to be 90%. The nondimensional sorption time was chosen to be 20 dimensionless time units and the dimensionless purging time was assumed as 16% of the nondimensional sorption time. The dimensionless process and cooling stream channels are taken as $x_F = y_F = 3$. The rest of the physical data for the system are shown in Table 1.

The amount of cooling delivered to the building is the difference between the enthalpy of air leaving the building and the enthalpy of air entering the building.

$$Q_c = \dot{m}_1 H_1 - \dot{m}_4 H_4 \qquad\qquad\qquad (7)$$

Table 1. Input and Physical Data for Cross-Cooled Desiccant Cooling System

C_s	specific heat of silica gel	0.092093 kJ/kg°C
C_f	specific heat of Teflon	1.17209 kJ/kg°C
C_a	specific heat of aluminum	0.90418 kJ/kg°C
C_w	specific heat of liquid water	4.18603 kJ/kg°C
C_1	specific heat of process air	1.033 kJ/kg°C
C_2	specific heat of cooling air	1.033 kJ/kg°C
f	mass fraction of silica gel	0.5 kg/kg total mass
g	mass fraction of Teflon	0.025 kg/kg total mass
η_1, η_2	effectiveness of sensible heat exchangers	0.9
η_3	effectiveness of evaporative cooler	0.9
x_F	dimensionless length of process channel	3
y_F	dimensionless length of cooling channel	3
t_F	dimensionless sorption time	20
t_{FP}	dimensionless purging time	3.2
$\dfrac{a_1}{a_2}$	ratio of process channel width	1.0
$\dfrac{\dot{m}_1}{\dot{m}_2}$	ratio of mass flowrate of process air to that of regenerating air	1.0
T_o	input temperature of sorbing air	26°C
Y_o	inlet absolute humidity of	0.0126 kg H_2O/kg dry air
T_A	ambient air temperature	35°C
Y_A	ambient absolute humidity	0.0178 kg H_2O/kg dry air

where Q_c is the cooling capacity and m_1, m_4, H_1 and H_4 represents the mass flowrates and the enthalpies of process air kJ/kg leaving and entering the conditioned space respectively. Assuming the mass flowrate out of the building equals the mass flowrate into the building, equation (7) may be written as

$$Q_c = \dot{m}_1 (H_1 - H_4) \tag{8}$$

From conservation principals, for an adiabatic humidification process 3-4 enthalpy essentially remains constant, therefore

$$Q_c = \dot{m}_1 (H_1 - H_3) = \dot{m}_1 [H_1 - H_2) + (H_2 - H_3)] \tag{9}$$

The quantity $(H_1 - H_2)$ represents the change of enthalpy for the dehumidification process and $(H_2 - H_3)$ represents the change of enthalpy for the sensible cooling process.

The thermal coefficient of performance is defined as

$$COP = \frac{\text{Cooling Capacity}}{\text{Heat Input}} = \frac{Q_c}{Q_h} \tag{10}$$

The heat input, Q_h, during desorption is by hot air from the solar collectors or any other source of energy. Referring to figure 7, air enters the solar collector at State 8 and leaves at State 9. Thus the heat supplied to the desorbing air is

$$Q_h = \dot{m}_8 (H_9 - H_8) \tag{11}$$

where $\dot{m}_8$ is the mass flowrate of desorbing air. Substituting equations (9) and (11) in equation (10) gives

$$COP = \frac{\dot{m}_1 (H_1 - H_3)}{\dot{m}_8 (H_9 - H_8)} = \frac{\dot{m}_1 [(H_1 - H_2) + (H_2 - H_3)]}{\dot{m}_8 (H_9 - H_8)} \tag{12}$$

In the analysis we have chosen $\dot{m}_1 = \dot{m}_8$. Hence,

$$COP = \frac{H_1 - H_3}{H_9 - H_8} = \frac{(H_1 - H_2) + (H_2 - H_3)}{H_9 - H_8} \tag{13}$$

Equation (9) and (13) along with the psychrometric chart of figure 8 give us insight into the desirable characteristics of the system.

4.2 Numerical Results

The cooling capacity with and without the heat exchangers and the coefficient of performance were obtained using the input and

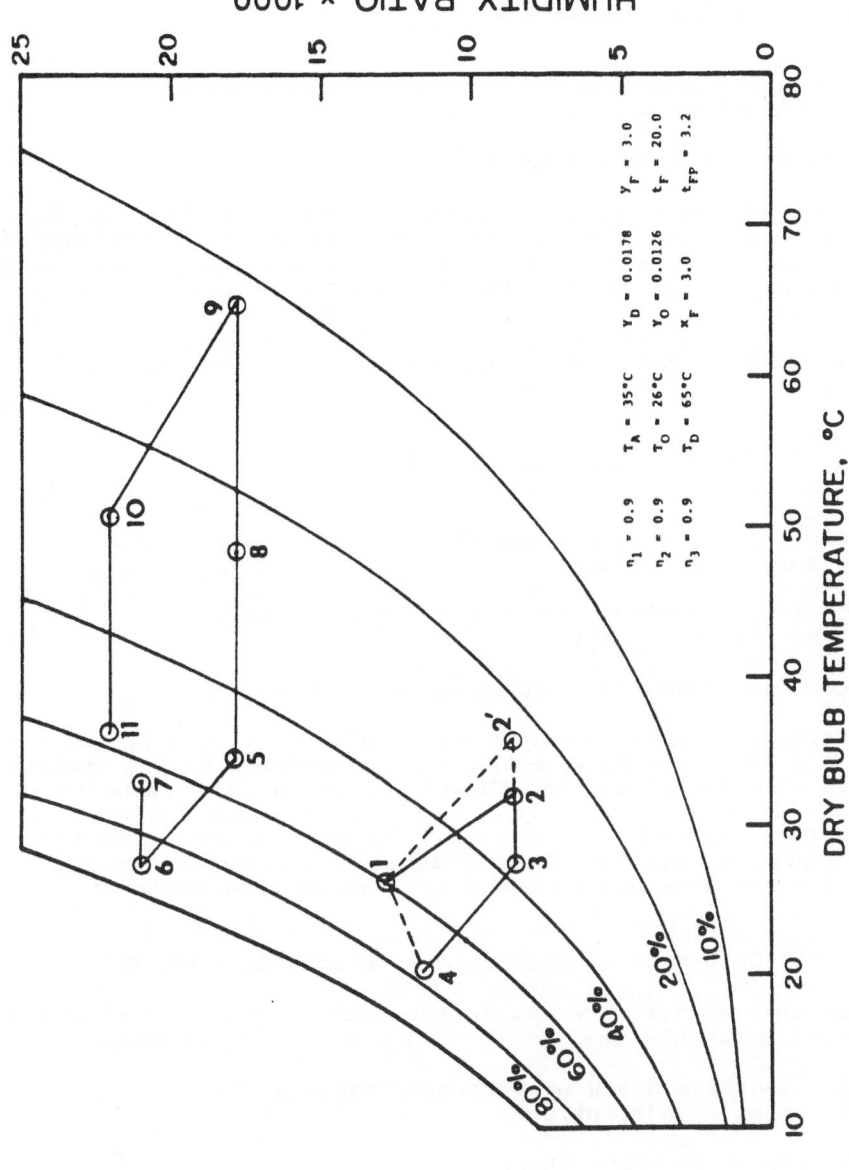

Figure 8. Psychrometric Chart Showing Adsorption and Desorption Processes

physical data given in Table 1.   The results are as follows:

> Cooling capacity without sensible heat exchanger =
> 3.93 kJ/kg process air

> Cooling capacity with sensible heat exchanger =
> 9.08 kJ/kg process air

> COP with sensible heat exchanger = 0.54

These results were compared with the solution obtained from the same governing equation without considering the functional relation between the heat of adsorption and the equilibrium moisture content, W_{eq}, and neglecting the effects of the heat of wetting for calculating the specific heat of the wall.   The comparison indicates a lowering in COP and cooling capacity when the functional relation for heat of sorption and W_{eq} is considered.   For the case of q_s = 2700 kJ/kg and

$$\frac{\partial \Delta H_w}{\partial T_w} = 0 \text{ the result are as follows:}$$

> Cooling capacity without sensible heat exchanger =
> 4.6 kJ/kg dry air

> Cooling capacity with sensible heat exchager =
> 9.7 kJ/kg dry air

> COP with sensible heat exchanger = 0.585

Referring to the psychrometric chart of figure 8, the line 1-2' is a line of constant enthalpy, while the line 1-2 depicts the process line for silica gel sheet. As shown, the enthalpy of air at State 2 is less than that at State 1 which is partly due to latent heat removed, and the rest is due to cross-cooling. Had the adsorption process been adiabatic instead of cooled, then State 2 would have had an enthalpy greater than State 2'.

5.   EFFECT OF DESICCANT MATERIAL PROPERTIES ON PERFORMANCE

In this section, we study the cooling system performance of desiccant materials having the following different properties.

1. Isotherm Shape of Desiccant Material
2. Heat of Adsorption

5.1   Effect of Isotherm Shape

The effect of isotherm shape on the performance was investigated for four different types of isotherms.   Three of the isotherms are at 20°C and the fourth is for a temperature of 50°C. Isotherms 1 and 3 may be categorized to represent the Type 1 and Type III of Van der Waals adsorption isotherms resepectively, and isotherms 2 and 4 have a linear variation with W_{eq}. Figure 9 shows the shape of the four isotherms. Since the heat of sorption is related to the equilibrium vapor pressure by equation (3) the complete set of equilibrium vapor pressure data can be obtained by prescribing the heat of sorption. A value of $q_s = 1.2 \, h_{fg}$ was chosen for the present analysis, since for most desiccants in use today, with the exception of natural zeolites, the heat of adsorption is normally 5-30% greater than the latent heat of vaporization.

Having chosen an isotherm shape at one temperature and the heat of sorption equal to $1.2 \, h_{fg}$, the set of equilibrium data can now be calculated and the shape of isotherms for 40, 50, 60, and 80°C are obtained from equation (14). Integrating equation (3) along a constant W line yields

$$\ln P_v = \frac{q_s}{h_{fg}} \ln P_g + C \tag{14}$$

If P_{g1}, P_{g2}, P_{g3}, P_{g4} and P_{g5} represent the saturation vapor pressure of water at 20, 40, 50, 60 and 80°C respectively, then

$$\ln P_{v2} = \ln P_{v1} + \frac{q_s}{h_{fg}} (\ln P_{g2} - \ln P_{g1}) \tag{15}$$

$$\ln P_{v3} = \ln P_{v1} + \frac{q_s}{h_{fg}} (\ln P_{g3} - \ln P_{g1}) \tag{16}$$

$$\ln P_{v4} = \ln P_{v1} + \frac{q_s}{h_{fg}} (\ln P_{g4} - \ln P_{g1}) \tag{17}$$

$$\ln P_{v5} = \ln P_{v1} + \frac{q_s}{h_{fg}} (\ln P_{g5} - \ln P_{g1}) \tag{18}$$

Using the above relations, the values of P_{v2}, P_{v3}, P_{v4} and P_{v5} were calculated and their corresponding relative humidity were obtained.

A linear regression program was used to obtain the functional relation between relative humidity, T_w and W_{eq}. For a variable acceptance criterion of 0.05, the relative humidity, as a function of equilibrium temperature and moisture content was obtained. These relationships are

$$\phi = 161.4523 \, W_{eq} - 1394.472 \, W_{eq}^2 + 3677.958 \, W^3$$
$$- 4.880806 \, W_{eq}^2 T_w \tag{19}$$

for Isotherm I

$$\phi = 191.0246 \, W_{eq} + 2.788867 \, T_w W_{eq}$$
$$+ 8.462908 \, 10^{-3} \, T_w^2 W_{eq} - 2.339123 \times 10^{-5} \, T_w^3 W_{eq} \tag{20}$$

for Isotherm II

$$\phi = 1.300757 + 486.8825\ W_{eq} - 947.8368\ W_{eq}^2$$
$$+ 1963.815\ W_{eq}^3 + 14.32028\ T_w\ W_{eq}$$
$$- 77.41191\ T_w\ W_{eq}^2 - 0.17287449\ T_w^2 W_{eq}^2 \qquad (21)$$
$$+ 2.973282\ T_w^2 W_{eq}^3$$

for Isotherm III
and

$$\phi = 133.5375\ W_{eq} + 2.237173\ T_w\ W_{eq} \qquad (22)$$
$$- 1.841987\ 10^{-3}\ T_w^2 W_{eq}$$

for Isotherm IV

The above polynomials were used in the computer program of the cross-cooled desiccant cooling system and the results obtained are tabulated in Table 2.

The isotherm shape is probably the most important property of desiccant materials. Comparison of the first three isotherms indicates that for the same value of heat of sorption, type I isotherm produces a higher cooling capacity per kg of air than the other two isotherms. It is 43.2 percent higher than the type III isotherm and 2.6 percent higher than the linear isotherm.

These results indicate that the shapes of isotherms would effect the system performance significantly. The comparison shows that the overall performance of type 4 isotherm is the best among the 4 types and is followed by type 1 and type 2 isotherms.

5.2 Effect of Heat of Adsorption

The cooling capacity and the COP of the unit decrease by increasing the heat of sorption. This is due to the decrease in temperature during regeneration, because additional energy is required to convert the water from the adsorbed phase to the vapor phase.

The moisture cycled per kg of desiccant is less sensitive to changes in the heat of sorption. The moisture cycled is reduced as the heat of sorption is increased, but not as fast as the reduction in COP and cooling capacity. These effects are shown in figure 10.

6. CONCLUSIONS

The sorption isotherm for moist air and silica gel sheet at various temperatures was determined. The experimental results for adsorption capacity of slica gel in the form of sheets were compared to the data available for silica gel powder. It was determined that the equilibrium sorption capacity of the sheet is 12 to

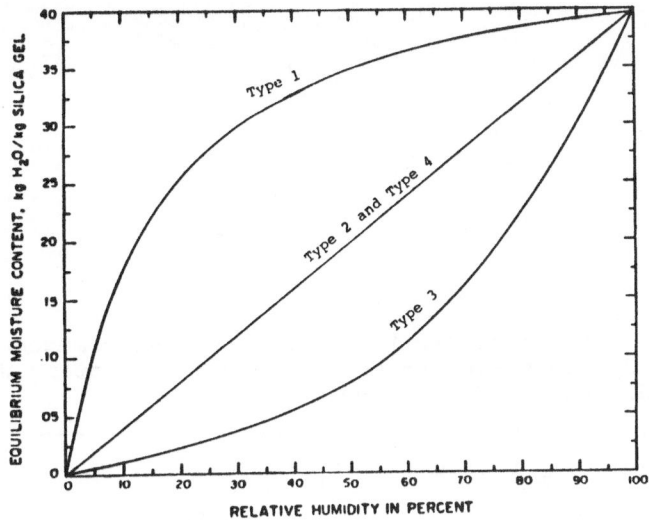

Figure 9. Assumed Adsorption Isotherms for Type 1, Type 2 and
Type 3 Desiccant Materials at T=20°C and Type 4 Desiccant
Material at T=50°C

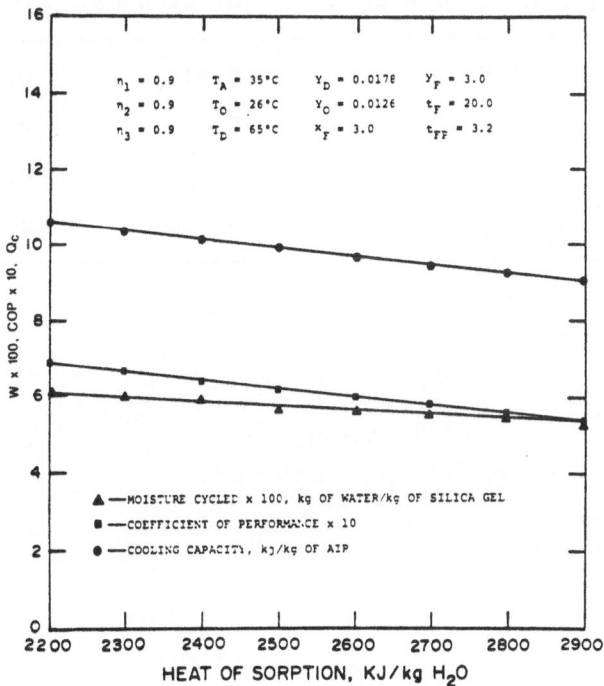

Figure 10. Effect of Heat of Sorption on COP, Cooling Capacity
and Moisture Cycled.

Table 2. Effect of Isotherm Shape on Performance of
  Cross-Cooled Desiccant Cooling System Operating Under
  Periodic Steady-State Condition

$\eta_1 = 0.9$	$T_A = 35$ C	$Y_D = 0.0178$	$y_F = 3.0$
$\eta_2 = 0.9$	$T_O = 26$ C	$Y_O = 0.0126$	$t_F = 20.0$
$\eta_3 = 0.9$	$T_D = 65$ C	$x_F = 3.0$	$t_{FP} = 3.2$

Isotherm Number	Cooling Capacity per kg of Desiccant	Coefficient of Performance	Moisture Cycled
1	9.218	0.514	0.0548
2	8.981	0.528	0.0535
3	5.238	0.406	0.0334
4	9.550	0.550	0.0564

30% lower than the silica gel powder. The heat of sorption and the integral heat of wetting were obtained from sorption isotherm measurements. The results show that the heat of adsorption of the sheet is 10 to 30% greater than the heat of condensation/vaporization of water, depending on the amount of moisture adsorbed. The calculated results for the integral heat of wetting of the sheet were compared to the results obtained by Ewing, et.al.[5]. This comparison indicated that the heat of adsorption of the sheet is 5 to 15% higher than the results obtained for the silica gel powder.

The present data were subsequently used in a computer program to predict the performance of cross-flow silica gel desiccant cooling system. For the operating conditions given in Table 1, the following results were obtained:

Cooling Capacity = 9.087 kJ/kg dry air
Coefficient of Performance = 0.542
Total Moisture Cycled = 0.0539 kg H_2O/kg silica gel

The above results were compared with the solution obtained from the same governing equations without including the functional relationship between the heat of adsorption and the equilibrium moisture content and also neglecting the effect of the heat of wetting. The present results show a reduction in COP by 7.4%, cooling capacity by 6.8% and moisture cycled by 5.3%.

The effect of the properties of desiccant material used in the cross-cooled desiccant cooling system are very significant. The desirable properties are:

(1) Low heat of adsorption. A high heat of sorption reduces the cooling capacity and the coefficient of performance significantly, however, the moisture cycled is slightly reduced.

(2) High adsorption capacity at low temperatures and low relative humidities and high desorption capacity at higher temperatures. This means that the isotherms corresponding to the adsorption and desorption processes should be farthest apart. This produces a double positive effect on COP because of increase in cooling capacity and decrease in heat input.

7. NOMENCLATURE

a_1, a_2 process and cooling channel width respectively, m

C_a, C_f, C_s, C_w specific heat of aluminum, teflon, silica gel and liquid water respectively, kJ/(kg^{o}C)

C_1, C_2 specific heat of process and cooling air respectively, kJ/(kg^{o}C)

COP   coefficient of performance

f   mass fraction of silica gel

g   mass fraction of teflon

$H_1, H_2, \cdots$ enthalpy of air kJ/kg dry air

h_{fg} latent heat of vaporization of water, kJ/kg

ΔH_w integral heat of wetting, kJ/kg desiccant

P_v, P_g partial vapor pressure and saturation vapor pressure H_2O, Pa

Q_c cooling capacity, kJ/kg dry air

Q_h heat input, kJ/kg dry air

q_s heat of sorption kJ/kg H_2O

T_w temperature the desiccant wall $^{\circ}C$

T_D, T_O inlet temperatures of desorbing and adsorbing air respectively, $^{\circ}C$

T_A ambient air temperature, $^{\circ}C$

t, t_F dimensionless time variable, total dimensionless sorption time

t_{FP} dimensionless purging time

W average moisture concentration in the silica gel during any sorption process, kg H_2O/kg desiccant

W_{eq} equilibrium moisture content of silica gel, kg H_2O/kg desiccant

x, x_F nondimensional space variable in the direction of process air flow, nondimensional process channel length

Y humidity ratio of process air, kg H_2O/kg dry air

Y_O, Y_A, Y_D inlet humidity ratio of process air, adsorbing air and desorbing air respectively, kg H_2O/kg dry air

Y_w humidity ratio of the air in equilibrium with the wall during any sorption process, kg H_2O/kg dry air

y_F nondimensional space variable in the direction of cooling air flow, nondimensional length of the cooling channel

η_1, η_2, η_3 effectiveness of desorbing stream heat exchanger, adsorbing stream heat exchanger and evaporative cooler, respectively

8. REFERENCES

1. Brunauer, S., Deming, L.D., Deming, W.E. and Teller, E. 1940. On a Theory of the Vander Waals Adsorption of Gases, _Journal of American Chemical Society_, 62: 1723-1733.

2. Close, D.J. and Banks, P.J. 1972. Coupled Equilibrium Heat and Single Adsorbate Transfer in Fluid Through a Porous Medium-II Predictions for a Silica Gel Air Drier Using Characteristic Chart, _Chemical Engineering Science_, 27: 1157-1169.

3. Dehler, F.C., May 1940. Silica Gel Adsorption, _Chemical and Metallurgical Engineering_, 47: 307-310.

4. Ewing, D.T. and Bauer, G.T. 1937. The Heat of Wetting of Activated Silica Gel. _J. of American Chemical Society_, 59: 1548-1553.

5. Hubard, S.S. 1954. Equilibrium Data for Silica Gel and Water Vapor. _Indiana Engineering Chemical_, 46: 356-358.

6. Mathiprakasam, B. 1979. _Performance Predictions of Silica Gel Desiccant System_. Ph.D. Thesis, Illinois Institute of Technology.

7. Othmer, D.F. 1940. Correlating Vapor Pressure and Latent Heat Data, _Industrial Engineering Chemistry_, 32, 6: 841-856.

8. Roy, D. and Gidaspow, D. 1974. Nonlinear Coupled Heat and Mass Exchange in a Cross-Flow Regenerator, _Chemical Engineering Science_, 29: 2101-2114.

THERMAL CURING

Investigation of Thermal Performance of Concrete Pavement in Cairo, Egypt

G.B. HANNA and O.A. SALEH
Building Research Center
P.O. Box 1770
Cairo, Egypt

ABSTRACT

The thermal performance of a concrete pavement with sandy subgrade has been investigated theoretically in a climate and location of Cairo, Egypt. The finite difference method is used to estimate the temperature response in the concrete pavement and within the subgrade due to the influence of periodic changes of incident solar radiation, outdoor air temperature and wind speed. A computer program has been developed for any climatic conditions and location and used to predict the temperatures within the concrete and subgrade as well as amplitude and time-lag.

The effects of summer and winter weather conditions on the concrete pavement temperatures have been studied. The study shows that the time-lag in summer is much greater than that in winter for the same depth. The studies show that the temperature gradient has much affected during day hours and month of the year.

NOMENCLATURE

Symbols
A   dimensionless parameter
c   specific heat, ($J/Kg.^{\circ}C$)
h   film coefficient, ($W/m^2.^{\circ}C$)
I   total incident of solar radiation, (W/m^2)
k   thermal conductivity, ($W/m.^{\circ}C$)
L   thickness, (m)
n   an integer
N   reduced time, (non)
P   period, (s)
t   time, (s)
T   temperature, ($^{\circ}C$)
y   distance, (m)
Y   reduced distance, (non)

α absorptivity or emissivity, (non)
ρ density, (Kg/m^3)
θ reduced temperature, (non)

Suscripts
ao   outdoor air
b   bottom
c   convection

i   interface
max   maximum
min   minimum
r   radiation
s   surface

1. INTRODUCTION

     The thermal response of a concrete pavement is defined as the reaction of
the pavement construction to the periodic changes of incident solar radiation,
outdoor air temperature and wind speed. It is also depends on the thermal
properties of the concrete pavement, mass, thickness, colour and surrounding
conditions.

     The climatic conditions are one of the factors that affect the durability
of the concrete pavement. These climatic conditions create a thermal deformation
which may be in the form of deflection and strains. The deformation in the
concrete slab due to temperature variation differs from different locations in
the slab, from hour to hour and also from season to another.

     The extreme temperature changes throughout the day in the multilayered
concrete pavement causes a deformation effects [1] which is a major problem;
specially in hot arid climate such as Egypt. This deformation causes a significant warping stress and cracking deterioration leading to high maintenance
costs.

     No attempt had been made in Egypt so far to estimate and to measure, in
quantitative terms, what the temperature response of the pavement would be to
the different weather conditions. A field measurements has been carried out
for a concrete slab of 5.0 x 3.0 x 0.2 m with a sandy subgrade of 1.0 m
thickness. The surrounding weather conditions as well as the temperature performance and the resultant deformation have been recorded. A comparison between
the estimated and primary measured results have been made and shows a very good
agreement [2].

     Details specification of the concrete pavement and subgrade for these
studies are shown in Fig.1. Necessary climatic data such as , solar radiation
and outdoor shaded air temperature are plotted in Figs. 2 and 3. Average values
of wind speeds are also included. Main results of the studies are depicted
graphically in Figs 4 to 11.

2. PERFORMANCE EQUATIONS OF THE HEAT TRANSFER MODEL

     The temperature distribution within the concrete slab is governed by the
general equation of conduction in one dimension which has the following form:

$$k \frac{\partial^2 T}{\partial y^2} = \rho c \frac{\partial T}{\partial t} \tag{1}$$

where k is thermal conductivity of concrete slab and ρc is volumetric heat
capacity. Equation 1 is also valid within the subgrade with different values
of thermophysical properties. Here y is the distance from the top surface
and t is time; both are variables on which temperature, T, depends

$$T = T(y, t) \tag{2}$$

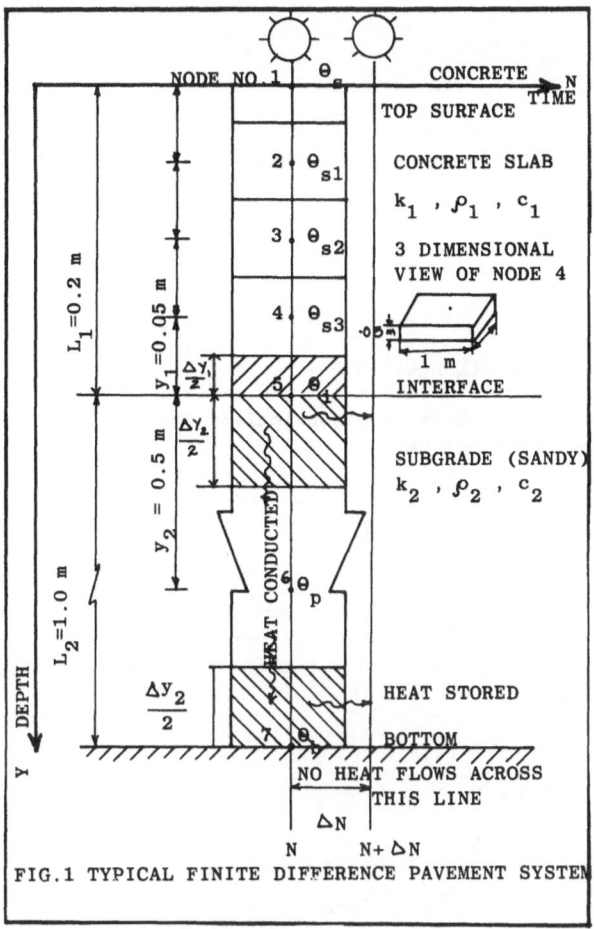

FIG.1 TYPICAL FINITE DIFFERENCE PAVEMENT SYSTEM

It is assumed that the temperature repeats itself periodically with a period P(=24 hour). This gives the condition of periodicity [3] as the first boundary condition for Eq.1. Mathematically, this is given by

$$T(y, t) = T (y , t + n P) \tag{3}$$

where n is an integer. The boundary conditions needed for Eq.1 are obtained by considering the heat balance at the top, interface and bottom of the subgrade [4]. This yields

$$- k_1 \left. \frac{\partial T}{\partial y} \right|_{y_1=0} = h_c (T_{ao} - T_s) + h_r (T_R - T_s) + \alpha I \tag{4}$$

$$- k_1 \left. \frac{\partial T}{\partial y} \right|_{y_1 = L_1} = - k_2 \left. \frac{\partial T}{\partial y} \right|_{y_2=0} \tag{5}$$

$$- k_2 \left. \frac{\partial T}{\partial y} \right|_{y_2 = L_2} = 0 \tag{6}$$

The outdoor air temperature, T_{ao}, and the total intensity of solar radiation, I, are functions of time t, and changes periodically for a certain period, Fig.2 and 3, i.e.

$$T_{ao} = T_{ao} (t) \tag{7}$$

$$I = I (t) \tag{8}$$

3. DIMENSIONLESS PERFORMANCE EQUATIONS

To set the present problem in dimensionless form, the following dependent and independent variables are used.

$$\Theta = (T - T_{min}) / (T_{max} - T_{min}) \tag{9}$$

$$Y = y / L \quad \text{and} \quad N = t / P \tag{10}$$

The following dimensionless parameters are also difined :

$$
\left.
\begin{aligned}
A_B &= h_o L / k \\
A_c &= h_r / h_c \\
A_F &= k P / (\rho c L^2) \\
A_h &= h_o P / (\rho c L) \\
A_I &= \propto I_{max} / h_o (T_{max} - T_{min}) \\
A_R &= I / I_{max} \\
A_y &= \rho c L y / (\sum_{1,2} \rho c L y)
\end{aligned}
\right\} \tag{11}
$$

Further, the weather conditions given by Eqs. 7 and 8 take the following dimensionless form

$$\Theta_{ao} = \Theta_{ao} (N) \quad \text{and} \quad A_R = A_R (N) \tag{12}$$

The following statements are obtained from substituting the dimensionless groups of Eqs. 9 and 10 into Eqs. 1 to 6 and using the dimensionless parameters of Eqs. 11 and 12. Therefore,

$$\frac{\partial \Theta}{\partial N} = A_F \frac{\partial^2 \Theta}{\partial Y^2} \tag{13}$$

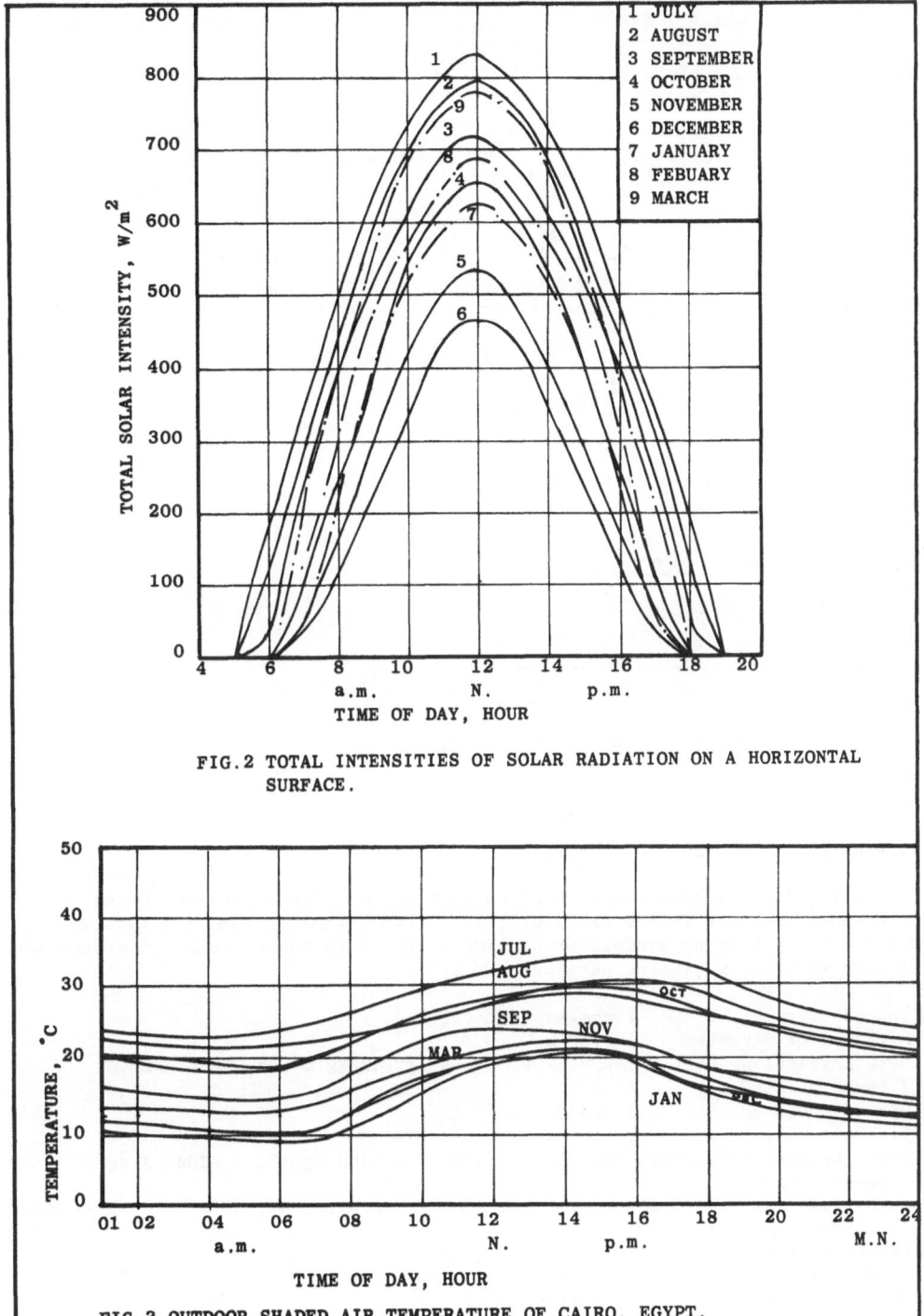

FIG.2 TOTAL INTENSITIES OF SOLAR RADIATION ON A HORIZONTAL
SURFACE.

FIG.3 OUTDOOR SHADED AIR TEMPERATURE OF CAIRO, EGYPT.

$$\Theta = \Theta \, (\, Y \, , \, N \,) \tag{14}$$

$$\Theta(\, Y, N)= \Theta \, (\, Y \, , \, N + n \,) \tag{15}$$

$$\left.\frac{\partial \Theta}{\partial N}\right|_{\Theta=\Theta_s} = (2/\Delta Y_1)(A_I \, A_h \, A_R + A_h \, A_c \, (\, \Theta_R - \Theta_s) + (A_h/A_B)\left.\frac{\partial \Theta}{\partial Y}\right|_{Y=Y_1} \tag{16}$$

$$\left.\frac{\partial \Theta}{\partial N}\right|_{\Theta=\Theta_i} = - 2 \, A_{F1} \, A_{Y1} \left.\frac{\partial \Theta}{\partial Y}\right|_{Y=Y_1} + 2 \, A_{F2} \, A_{Y2} \left.\frac{\partial \Theta}{\partial Y}\right|_{Y=Y_2} \tag{17}$$

$$\left.\frac{\partial \Theta}{\partial N}\right|_{\Theta=\Theta_b} = - 2 \, A_{F2}/ \Delta Y_2) \left.\frac{\partial \Theta}{\partial Y}\right|_{Y=Y_2} \tag{18}$$

To put the problem in a finite difference form, using forward difference in time interval, where

$$\frac{\partial \Theta}{\partial Y} \simeq \frac{\Theta_{Y+\Delta Y} - \Theta_{Y,N}}{\Delta Y} \tag{19}$$

$$\frac{\partial \Theta}{\partial N} \simeq \frac{\Theta_{Y,N+\Delta N} - \Theta_{Y,N}}{\Delta N} \tag{20}$$

$$\frac{\partial^2 \Theta}{\partial Y^2} \simeq \frac{\Theta_{Y+\Delta Y,N} - 2 \Theta_{Y,N} + \Theta_{Y-\Delta Y,N}}{2(\Delta Y)^2} \tag{21}$$

4. PARTICULARS OF THE STUDY CASE

A field measurements have been carried out [2] for a concrete pavement with sandy subgrade in the field of tests of the Building Research Center, Cairo. The following example is considered for both theoretical and experimental studies with the following particulars [1]:

Type of pavement   : cement concrete
Thickness of pavement   : 0.20 m
Thermophysical properties: k_1=1.2 W/m.C, ρ_1 =2242 Kg/m^3, c_1 =838 J/Kg.C
of pavement used
Type of subgrade   : sandy
Thickness of subgrade   : 1.00 m
Thermophysical properties: k_2 =0.13 W/m.C, ρ_2= 1650 Kg/m^3, c_2=2093 J/Kg.C
of sandy subgrade
Monthly average wind speed 2.0, 1.73, 1.89, 1.86, 2.0, 2.1, 2.7, 2.6, 2.71 m/s
(started July ended March
Average radiation   : 4.79 W/m^2.C
  coefficient (h_r)
Average surface   : 18.91 W/m^2.C
  coefficient (h_c)
Pavement absorptivity(α): 0.65

5. RESULTS AND DISCUSSION

The results obtained for the temperature variation in the concrete pavement and within the subgrade for nine months of a year, are shown graphically in Figs. 4 to 7.

Figures 4 to 7 show the temperature response curves at the top surface and within the concrete pavement structure due to the diurnal variations of outdoor air temperature, incident solar radiation and wind speed. During the early morning hours, in the period between 01.00 and 06.00 hours, the outdoor temperature decreases gradually. This decrease affects the top surface pavement temperature, as well as within the concrete structure and underlying subgrade, until it reaches its minimum value occuring after the minimum of the outdoor air. During the hot and sunrise period, between 06.00 and 17.00 hours, the top surface temperature increases and reaches its maximum value some interval after the maximum of incident solar radiation. During night hours and after sunset, the outdoor air temperature decreases gradually. Therefore, the top surface temperature is quickly cooled relative to other temperatures at deeper depths.

Table 1 gives the maximum and minimum temperatures and the time of their occurance for four months presents the different seasons of the year. It is clear from table 1 and from figures 4 to 7, that all maximum temperatures decreased, while the minimum increased. Therefore, the greater the depth of the pavement, the lower temperature amplitude. The massive concrete pavement, the more heat it can store and the less heat will be passed onto the subgrade during the day. Similarly at night, as the concrete slab contains a great deal of stored heat to lose to the outside environment, it will only have a limited effect on the underlying subgrade temperature. It can be seen that the temperature amplitude at the top surface is 13 °C and decreased to 9 °C at a depth of 0.05 m from the top free surface. At 0.70 m from the top surface, the amplitude is approximately zero, i.e. in the steady state region. The time-lag curves are shown in the figures where the maximum temperatures occur.

Figure 8 shows the monthly variation of concrete pavement temperature for 15.00 hour at different depths which indicates the time at which the maximum air temperature occurs. It is clear that the minimum appears in December while the maximum shown in July.

Figure 9 shows the diurnal temperature variation at the interface surface between the concrete and subgrade. It can be seen that the curves are nearly flatter, i.e. the amplitude is approximately zero.

Figure 10 illustrates the temperature gradient curves within the concrete pavement for different months occuring at 03.00, 07.00, 11.00 and 19.00 hours. These hours were choosen at the time where significant strains and deflection expected to be measured. The figures shows that at 03.00 hour, the temperature increases with increasing depth. This is to be expected since the concrete pavement structure contains a great deal of stored heat and has a limited effect to lose such heat to the outside environment. Also, it depends largely on the mass of the pavement structure. It is clear that the maximum occurs in July and the minimum occurs in December. It can be seen that December and November temperature curves give higher values than Janurary and March curves respectievely at a depth less than 0.10 m. For greater depth, the curves are reversed. This is due to the fact that the total radiation received in December is less than that received in Janurary. Also, the outdoor shaded air temperatures in December are greater than that of Janurary.

At 07.00 hour, for summer weather conditions, the temperatures remain steady

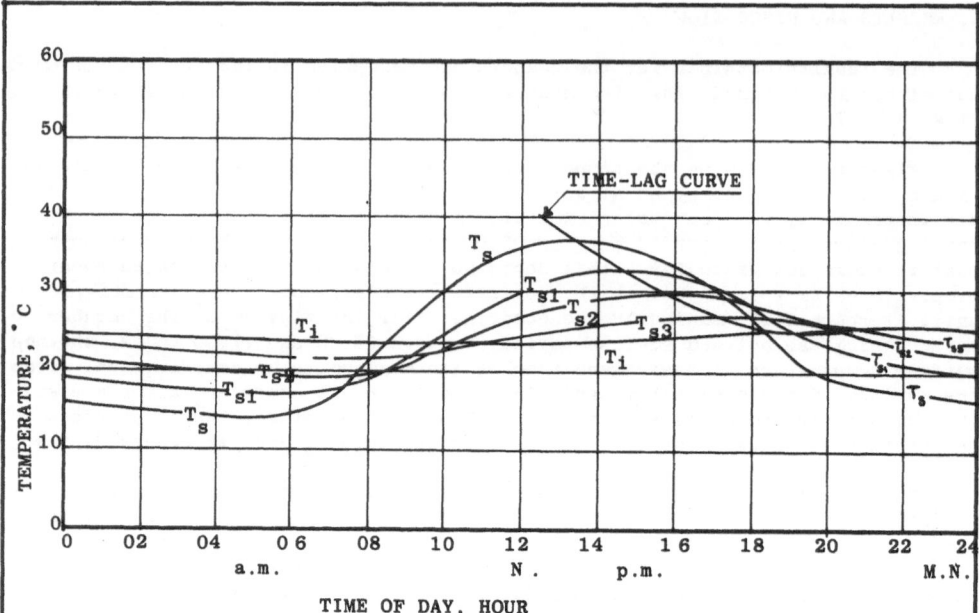

FIG. 4 TEMPERATURES OF A CONCRETE PAVEMENT AT DIFFERENT DEPTHS WITH SANDY
SUBGRADE IN MARCH.

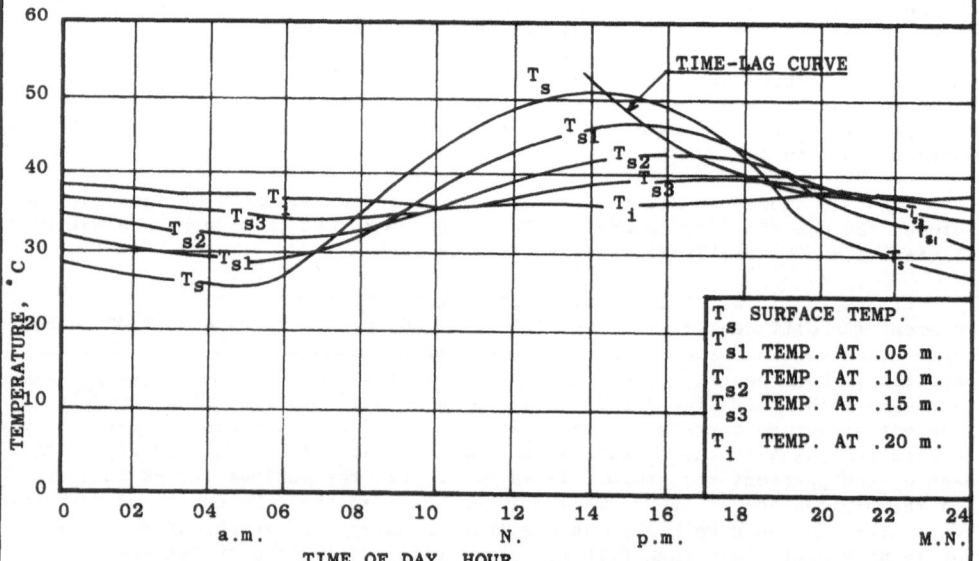

FIG. 5 TEMPERATURES OF A CONCRETE PAVEMENT AT DIFFERENT DEPTHS WITH SANDY
SUBGRADE IN JULY.

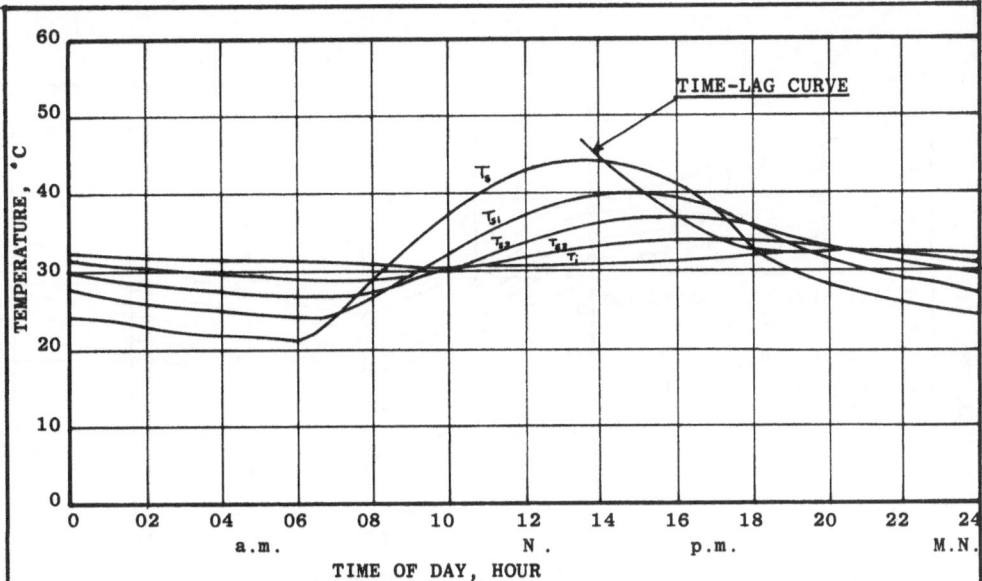

FIG.6 TEMPERATURES OF A CONCRETE PAVEMENT AT DIFFERENT DEPTHS WITH SANDY
SUBGRADE IN SEPTEMBER.

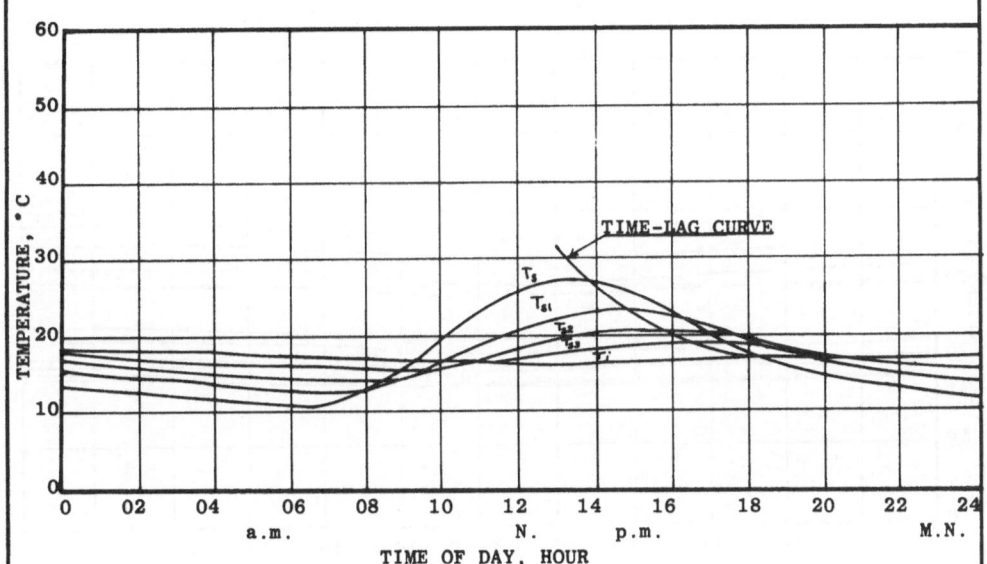

FIG.7 TEMPERATURES OF A CONCRETE PAVEMENT AT DIFFERENT DEPTHS WITH SANDY
SUBGRADE IN DECEMBER.

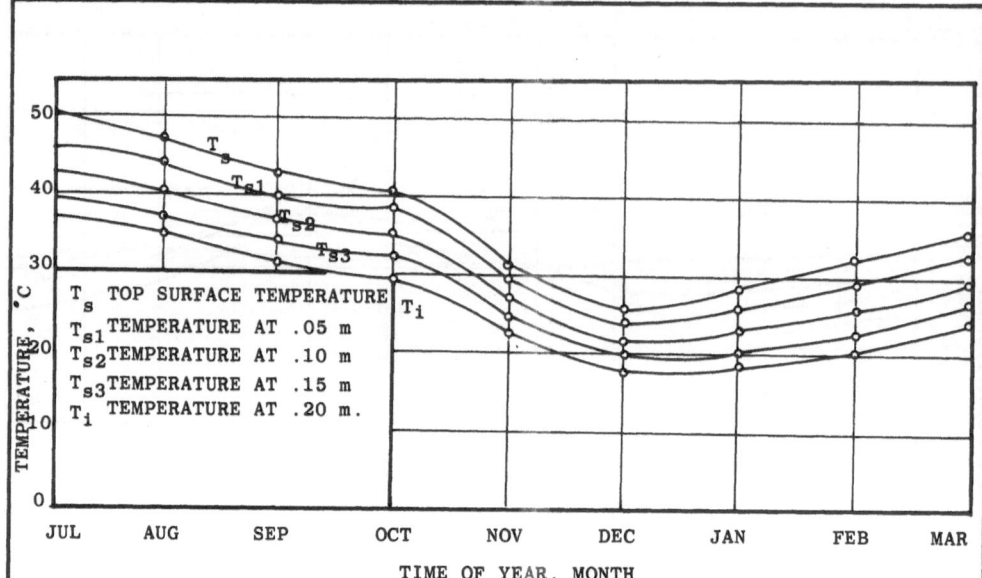

FIG.8 MONTHLY VARIATION OF CONCRETE TEMPERATURES OCCURING AT 1500 P.M.

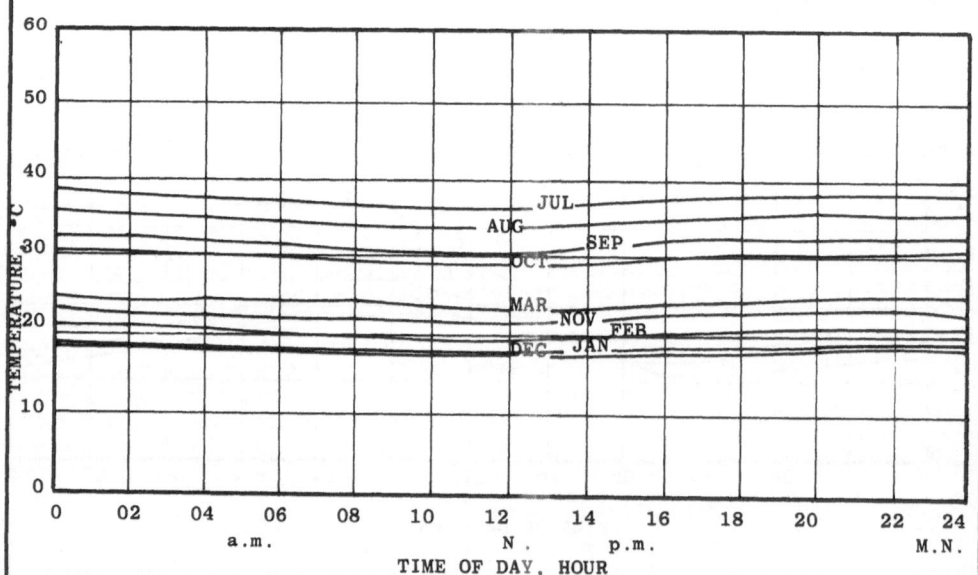

FIG.9 INTERFACE TEMPERATURES BETWEEN THE CONCRETE PAVEMENT & SANDY SUBGRADE.

TABLE 1 MAXIMUM AND MINIMUM TEMPERATURES WITH TIME FOR OUTDOOR AIR,
TOP SURFACE AND WITHIN THE CONCRETE PAVEMENT AT DIFFERENT DEPTHS.

Depth / TIME / OF YEAR	Outdoor air T_{ao} , °C		T_s , °C		T_{s1} , °C		T_{s2} , °C		T_{s3} , °C		T_i , °C	
	Max.	Min.	Max.	Min.	Max.	Min.	Max.	Min.	Max.	Min.	Max.	Min.
JUL	34.9* 1600	22.6 0500	51.4 1400	25.5 0500	46.9 1500	29.1 0500	43.5 1600	31.9 0600	40.5 1700	34.6 0700	38.4 2100	36.5 1100
SEP	30.3 1600	18.8 0500	44.0 1330	21.1 0600	39.9 1500	24.2 0600	36.9 1600	26.7 0700	34.2 1700	29.0 0800	32.4 2100	30.7 1100
DEC	20.9 1400	10.6 0600	28.0 1300	11.2 0600	24.3 1400	13.0 0700	21.8 1500	14.7 0800	19.8 1700	16.2 0800	18.5 2100	17.3 1100
MAR	22.6 1400	12.7 0500	37.6 1400	14.1 0600	33.2 1500	16.9 0600	29.8 1600	19.4 0700	27.0 1700	21.6 0800	25.1 2100	23.3 1100

*Time of occurance.

at a certain value then gradually increased with increasing depth. The shape
of the temperature curves differ from that shown at 0300 hour, specially for
summer weather conditions. This could be explained as the sun rises at 0600 hour
and the minimum outdoor air occurs at o500 hour. The temperatures nearer to the
top surface remains at equilbruim between that gain from lower elements and that
lost to the outdoor environment by forced convection and long-wave radiation.
For cold weather condition, the curves gradually increased with increasing depth,
since the temperatures at interface and bottom surfaces are much higher than the
surrounding environment. At this time, a considerable heat is lost to the outside environment and this could be shown in the curves from October to March.

During the hot period at 1100 hour, the top surface received and absorbed
a great deal of solar heat. As may be seen, the surface temperature in July
rapidly increased to 48°C, i.e. 20°C above ambient temperature. The rate of
decrease of temperature through the concrete pavement is much faster, reaching
to 36°C at 0.20 m depth, i.e. 12°C less than that of the top surface. It is
apparent that the temperature decreases gradually within increasing depth until
it reache the equilbrium nearer to the interface surface. The maximum occurs in
July and the minimum shown in December.

During night hours, at 1900 p.m., the top surface temperature is much warmer
than the other surroundings. The stored heat is lost to the outdoor air and
sky by forced convection and long-wave radiation. A part of the stored heat is
conducted to the inner element which is much nearer to the top surface and rises
its temperature above than that at the top surface. For greater depth, the
temperature is slowely decreased with increasing depth.

Figure 11 shows a relatioship between time-lag and depth for summer (July)
and winter (Janurary) weather conditions. It is clear from figure 11 that,
the time needed to reach the maximum temperature in summer is much greater
than that for winter. AT .o5 m depth from the top surface, the time-lag is
3.2 hours for July and is 2.7 hours for Janurary.

6. CONCLUSION

The thermal behaviour of a concrete pavement with a sandy subgrade is studied.

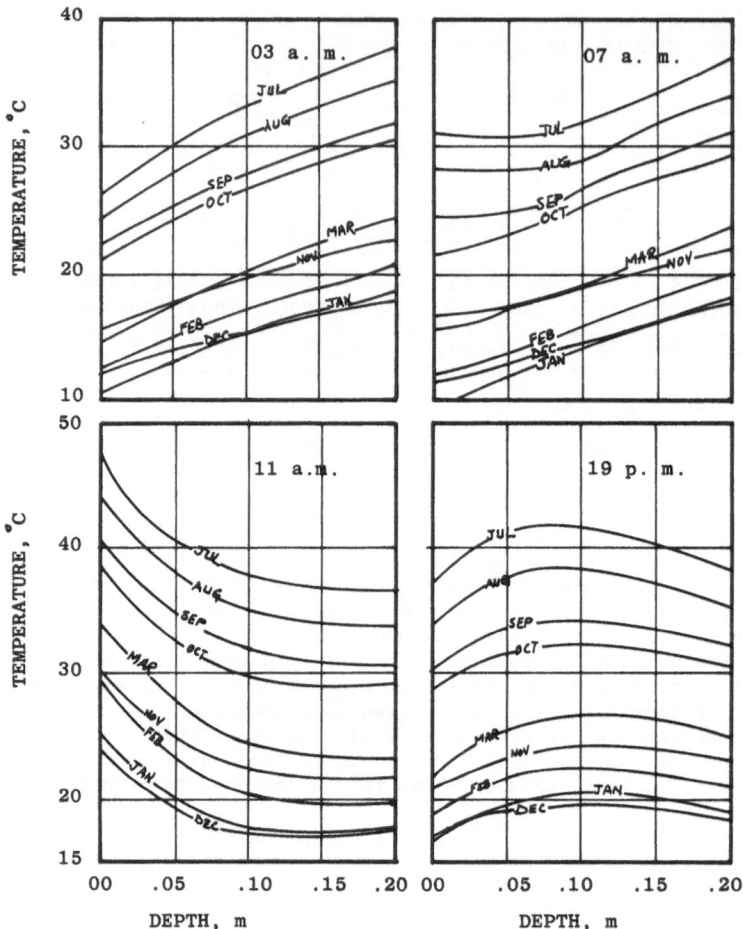

FIG.10 TEMPERATURE GRADIENT CURVES IN CONCRETE PAVEMENT
WITH SANDY SUBGRADE.

The particular case of the effect of weather conditions on the concrete pavement
temperatures are solved in a dimensionless form. The results are analysed and
shows that there are extreme temperature changes at the top and within the
pavement throughout the day and from season to another and could cause a deformation effect which is a major problem.

An experimental field investigation has been made for this particular study
case and the primary results show a very good agreement.

7. ACKNOWLEDGEMENT

This paper has been written as a part of a Ph.D. Thesis and the work was
made possible through the kind support of the Egyptian Building Reasearch Center.

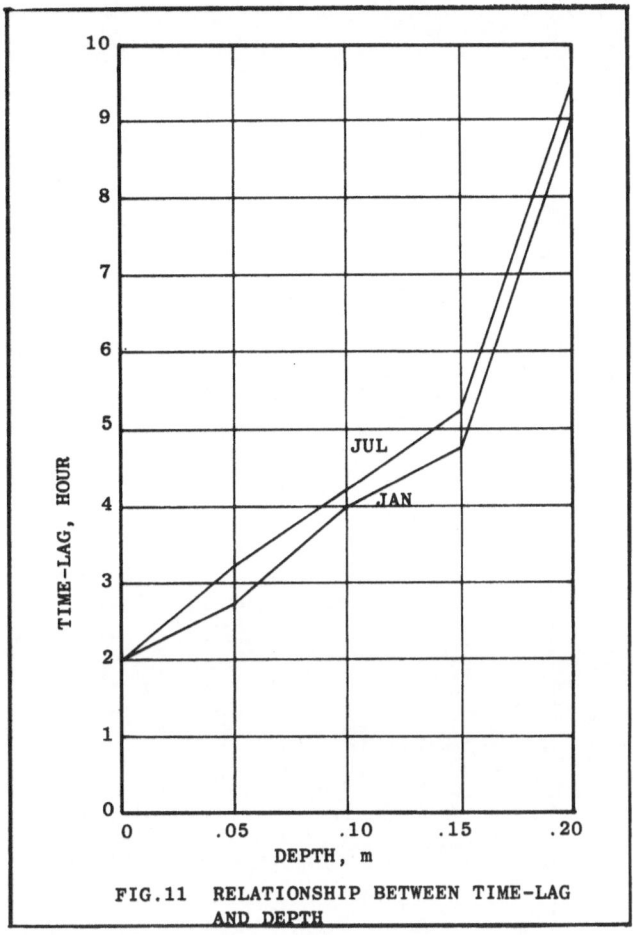

FIG.11   RELATIONSHIP BETWEEN TIME-LAG
         AND DEPTH

8. REFERENCES

1. Saleh O.A.S, " Durability of Concrete Pavement and its Resistance to Physical
   Thermal Effects", Ph.D. Thesis, Faculty of Engineering, Ain Shams University,
   (1980).
2. Saleh O.A.S, Atta A. and Hanna G.B, " Experimental and Theoretical Investigation of a Concrete Pavement", To be published.
3. Hanna G.B, " Computer Analysis of the Thermal Behaviour of Buildings and Walls",
   M.Sc. Thesis, Cairo University, (1969).
4. Hanna G.B, " Computer Analysis to Estimate the Temperature Response of an
   Enclosure by the Finite Difference Method", B.S.E. 45 (5), (1977).

Application of Thermal Energy in Curing Sodium Silicate Bonded Sand

YAW A. OWUSU
Wayne State University
Department of Industrial Engineering/Operations Research
Detroit, Michigan 48202, USA

ABSTRACT

The paper discusses a new alternative method, "thermal energy process," for curing (hardening) sodium silicate as a foundry sand binder. It also discusses the theory of sodium silicate polymerization and gelation and the subsequent hardening (curing) mechanism involved in the sodium silicate-thermal energy process. Some pertinent data have been provided to further elucidate the theoretical concepts involving the curing process.

The bond strength of bonded sand cured by the thermal energy process has been compared with those cured by the already known processes, namely, the carbon dioxide process (CO_2-process) and microwave energy process. The thermal energy process was found to provide a higher bond strength and a superior resistance to humidity than both CO_2 and microwave energy processes. The CO_2-process and thermal energy process were combined to produce a superior initial bond strength and a higher resistance to humidity than any one individual process. The thermal energy used in this investigation was hot (warm) convection air at 150C blown at 50 c.f.m. (cubic foot per minute).

1. INTRODUCTION

Sodium silicate, like clay, is a generic term. However, unlike clay, it is a manmade material. Sodium silicate solution is an opalescent, colliodal, Newtonian liquid until the viscosity rises sharply when the siloxane linkage (bonds) (Si-O-Si) are neutralized or overcome by either chemical reaction or by thermal energy. Sodium silicate also has a number of popular names such as sodium metasilicate, soluble glass, silicate of soda, liquid glass, or water glass. When dry, it is white or gray-white lumps or powder; it is soluble in water and alkalies, but insoluble in alcohol and acids. Often, it is available in the form of water solutions ranging from viscous to fluid liquids, depending on the ratio of soda to silica (Na_2O; SiO_2).

Presently, the United States of America alone produces over 2.3 million metric tons of sodium silicate per year, the majority of which are used for cleaning process, pulp and paper processes, and for the mining industry. Only a small fraction of the total production may be used in the foundry industry as an inorganic binder for molds and cores for the purpose of making castings.

Even though sodium silicates have been used to bond foundry sands into

molds and cores since the early 1920's [1,2], it was not until the early
1950's that the CO_2 gas curing process became established. Since the late
1970's, the microwave energy has been employed, though to a limited extent,
to cure sodium silicate bonded foundry sand cores. This current research [3]
was undertaken, in part, to find an alternative method of curing sodium silicate bonded sand molds and cores and also to use inorganic compounds as additives to improve humidity resistance and to promote easy sand collapsibility
after the casting is made. Previous articles have been published on humidity
resistance and collapsibility of sodium silicate bonded sand [4,5].

Improvement in sand technology has become very important to the foundry
industry over the past 15 years because of the urgency to reduce energy usage
and the need for environmental cleanliness. The increasing use of inorganic
sodium silicate binders compared with organic binders is both economically
feasible and ecologically advisable.

2.   THEORETICAL DISCUSSIONS ON SODIUM SILICATE POLYMERIZATION AND GELATION

Sodium silicate as a foundry binder is a complex colloidal system. The
bond consists of a precipitated gel, sodium silicate, and silicic acids.
These semi-solid substances bond the sand grains together into a continuous
three dimensional arrays of sand known as molds and cores. Figure 1 shows
a typical adhesive bonding in foundry sands. This process of forming semi-
solid substances is known to chemists as gel formation. The term "gel" describes the cohesive substance consisting of countless individual colloidal
particles.

Hurd [6] points out that gel formation is accounted for by the splitting
out of water molecules, accompanied by the dissipation of the charge and the
formation of an oxygen linkage between two groups, each containing one silicon ion and three OH⁻ groups (Figure 2). The reaction occuring during the
polymerization process which leads to the formation of a silica gel, and the
resulting equilibrium established thereby, can be expressed as:

$$\text{Sodium Silicate + Water} \rightleftharpoons \text{Silicic Acid} \overset{\nearrow}{\rightleftharpoons} \begin{array}{c} \text{Polysilicic Acid} \\ \updownarrow \\ \text{Gel} \end{array}$$

Thus, sodium metasilicate solutions (sodium silicate solutions) are in equilibrium with silicic acid and its ions. The silicic acid under proper pH
conditions or a change in thermal energy, polymerizes to form polysilicic
acid. The polysilicic acid, in turn, transforms into a gel, which bonds the
sand grains together into a mold or core.

Figure 3 represents schematically the chain polymerizations of sodium
silicate solutions or films and their final syneresis to silica hydrogel.
The term "syneresis" is used to mean the characteristic shrinking of a solid
with the squeezing out of a liquid phase. Figure 3-I illustrates an $Si(OH)_4$
molecule as it has formed from sodium silicate solution. In an aqueous medium such a molecule is hydrated; the water molecules, because of their di-
polar character, are attracted by the cation for better screening of its
positive field. Figure 3-II represents schematically such a molecule of
orthosilicic acid with two H_2O molecules. The number of H_2O molecules which
can come close to Si^{4+} is restricted by space requirements and by the dipole-
dipole repulsion between the oriented H_2O molecules. Because of their lower
polarizability, water molecules cannot screen the potential field of the Si^{4+}

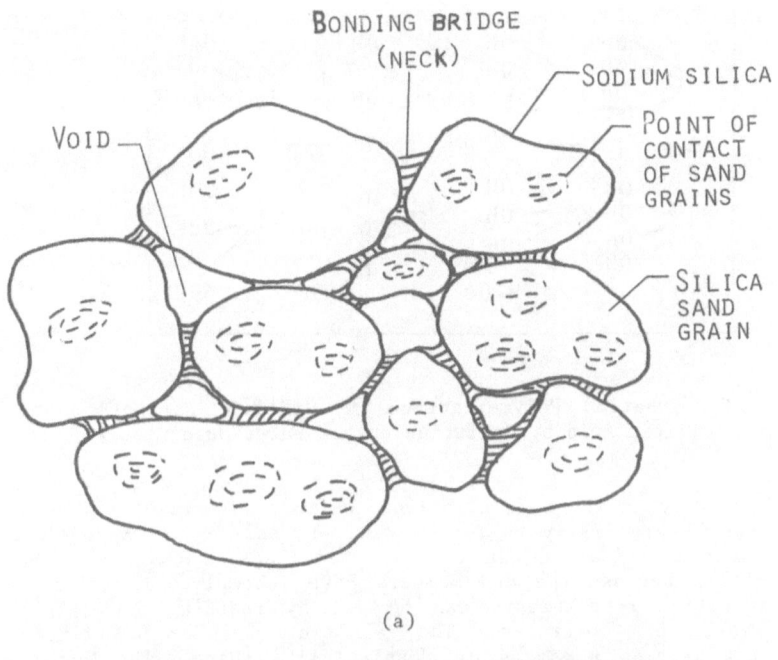

(a)

(b)

Fig. 1: Sodium Silicate Adhesive Bonding in Foundry Sand
         (a) Schematic diagram showing the nature of the bond
         (b) Scanning electron micrograph of a typical sodium
              silicate bond between two sand grains

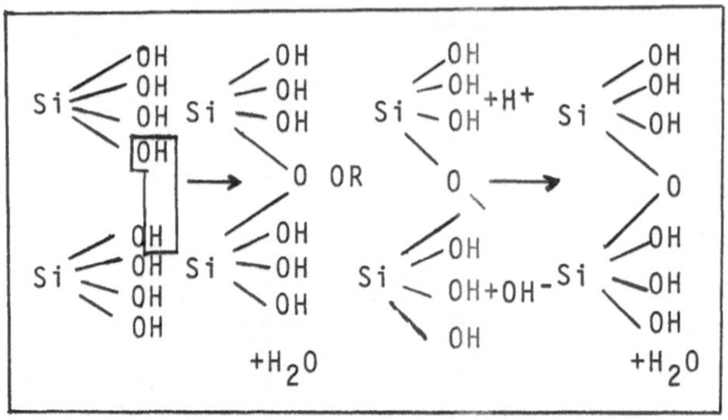

Fig. 2:   Condensation Polymerization of Silicate Particles into Poly-
          silicic Acid by Splitting out of Water Molecules [6]

ion as effectively as the same number of the more polarizable OH^- ions.   Two
units of Type II are likely to form a dimer by sharing two hydroxyl groups.
This type of polymerization can continue and leads to chain-like molecules
(Figure 3-III), because the end members have incomplete coordination, but
their coordination requirements can be met by reacting with the monomer
$Si(OH)_4$ to form new molecules.   The resulting chain is not likely to be
straight, but assumes a zigzag or a spiral-like shape.   The length of the
chains or the size of the macromolecules depends upon a number of factors,
such as the rate of polymerization, the pH of the solution, its concentration
of silica, temperature, and the presence of foreign molecules.

Formation of these chain-like molecules, especially in close proximity
to OH^- groups within the potential fields of two Si^{4+} ions, makes it possible
for a second reaction to occur, namely, a condensation reaction within the
silicic molecules.   Figure 3-IV illustrates two joining groups of one of the
polymer chains.   The clearing of water from the OH^- groups leads to Figure
3-V, that is, the grouping which now has the two Si^{4+} ions linked by an O^{2-}
ion.   This reaction lowers the energy of the system at least in part, because
two OH^- groups are replaced by one O^{2-} ion which has a higher polarizability.
This condensation process (final syneresis to a silica hydrogel) shortens the
Si^{4+} - Si^{4+} distances, as shown in Figure 3-V.   This phenomenon has little
consequence as long as the molecule is small and mobile.   However, once the
polymerization process has advanced to such an extent that the solution of
the silicic acid has formed long chains through the development of the gel,
the condensation process leads to the build-up of tension within the chain,
which consequently imparts rigidity to the gel.

2.1   Mechanism of Curing Sodium Silicate Bonded Sand by the Application of Thermal Energy

The thermal energy process of curing sodium silicate bonded sand is
based on the phenomenon of dehydration; and this takes place when there is
sufficient heat energy to evaporate the water content from the silicate solution.   Thus, dehydration takes place by the application of convection hot
(warm) air, microwave energy, or exothermic reaction resulting from organic

or inorganic compounds.

If sodium silicate solution or film is exposed to carbon dioxide gas, gelation starts 'in situ' because of an instant increase in viscosity resulting from a pH change.   A decrease in the pH value by introducing acidic species (carbon dioxide) causes an immediate precipitation of the silica hydrogel which bonds adjacent sand grains together to form a mold or core.

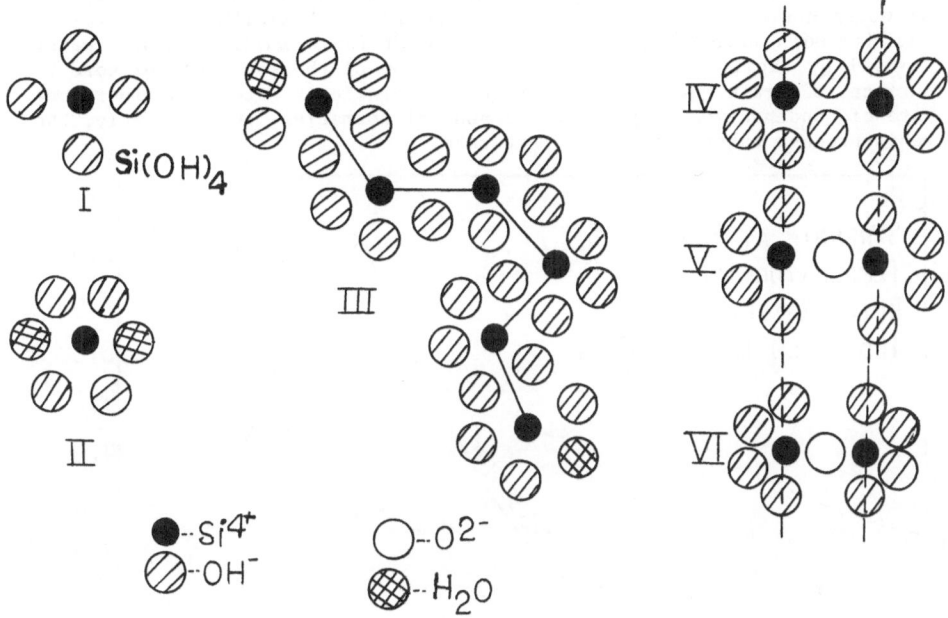

Fig. 3:   Condensation Polymerization of Orthosilicic Acid Molecules

          (I) $Si(OH)_4$ molecule;   (II) same plus two water molecules;
          (III) chain arrangement; (IV) neighboring chain groups;
          (V) condensation; and (VI) syneresis [Weyl and Hauser, 7]

The only difference between CO_2 process and the thermal energy dehydration process is that the end products of the former are Na_2CO_3 and/or $NaHCO_3$ and hydrated silica gel, whereas the latter process results in a dehydrated or semi-dehydrated silica gel, as the following two chemical reaction equations show:

$$Na_2O \cdot 2SiO_2 + CO_2 + 6H_2O \rightarrow Na_2CO_3 + 2Si(OH)_4 + H_2O \qquad (1)$$

$$Na_2SiO_3 + H_2O \; \underset{hydration}{\overset{dehydration}{\rightleftarrows}} \; Na_2SiO_3 \text{ (glassy phase)} - H_2O\uparrow \qquad (2)$$

The Si-O-Si linkage, which is called a "siloxane" linkage, is the only bond

which imparts rigidity to the bonded sand grains (Figure 1). This bond is
far stronger than either the Na_2CO_3 bond or the hydrogel bond developed when
cured with the CO_2 gas process. Therefore it would be expected that sand
cores and molds cured by the dehydration process would be stronger than those
cured by the CO_2 gas process.

Dehydration speeds up the condensation polymerization process by which
the silicic acid is formed from the alkaline silicate solution. Drying
occurs by evaporation at the surface; and the rate is constant and independent of water content, but is limited to temperature, humidity and air velocity, if forced (convection) air is used for drying. During curing, water
must travel by capillary force from the interior of the sand mold or core to
the surface at a rate equal to the rate of evaporation. Water loss increases
the rate of condensation polymerization and gel formation. Consequently, the
syneresis process takes place at a fast rate.

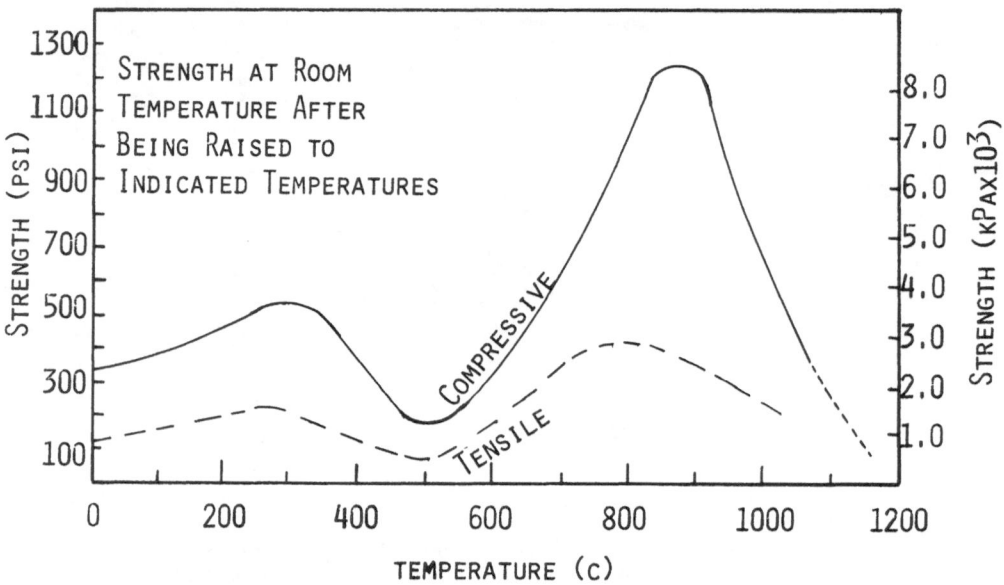

Fig. 4: Effect of Thermal Energy on Sodium silicate Bond
         Strength in Foundry Sands at High Temperatures

Figure 4 illustrates the effect of curing temperature on the bond
strength of sodium silicate when used as a binder in foundry sand. Normally,
the test samples are soaked at different temperatures for five minutes and
allowed to cool to room temperature before testing.   Many investigators
[9-11] have found similar results. Examination of the curves shows that the
initial strength is caused by the hydrated silica gel. The gel strength increases between 100C and 300C (212-570F) when all the free water has been
evaporated except the chemically bonded water (water of crystallization).
Following the first peak, there is a decrease in strength caused by the
release of chemically bonded water at about 315C (600F). In the area of
450-550C (842-1022F) the gel bond goes through a partially liquid state,
which results in the minimum strength. Beyond 550C (1022F) the strength

increases because of the formation of vitreous bond and at 800C (1472F) more glass is present. This glass has a high viscosity which produces greatest strength. At 982C (1800F) more glass is formed. According to Wales [9], with a three-hour soak at 982C (1800F), the hot tensile bar becomes very ductile and yields like copper.

For the purpose of making foundry molds and cores, the silica gel formed at the temperature range of 120-200C (248-392F) is necessary to obtain the required strength. Hence, 150C (302F) was chosen as the appropriate temperature for the thermal energy (hot air) curing process, in order to remove all the physically adsorbed water, leaving most of the chemically bonded hydroxyl group in the gel for an effective bonding. Eitel [12] observed that in silica hydrogels, the difference between physically adsorbed water (free water) and chemisorbed water (chemically bonded water) was most evident by the ease with which the first type of "moisture" was removed by heating in dry air at 120C (248F), and the firm binding of the hydroxyl groups in the silanol group Si-OH of the surface. Boer et al. [13] also observed small losses of chemically bonded water (OH hydroxyl group) at 150C (302F) and a continuous dehydroxylation above 380C (716F). In the same way that the hydroxyl groups are slowly disrupted and neighboring silanol groups undergo chemical reactions with the removal of water and the formation of condensed siloxane bindings, the ability of the gel to adsorb polar molecules like water is reduced. Thus, the more the hydroxyl groups (OH) are removed from the silica gel, the more the previously strong hydrophilic surface character becomes hydrophobic. The ternary phase diagram of sodium silicate solution at room temperature (Figure 5) can be used to illustrate gel formation from sodium silicate solution.

As an example, the sodium silicate solution, after mixing with a sand aggregate, is rammed or vibrated into a core. At this point, it is in the state marked by region an "ordinary commercial liquids" on the phase diagram. As this silicate reacts with CO_2, the Na_2O is consumed, and the gelled silicate shifts on the diagram to the region of unstable liquids and gels. This silica gel is high in SiO_2, as shown in the phase diagram, and it constitutes the bond strength of the core or mold aggregate. The phase diagram shows that the stable forms of sodium silicate exist in the region where the silica:soda weight ratio range is 2:1 to 3.75:1. This is the silicate ratio range used in the foundry industry. The silicate with high silica content (3.75:1) reacts faster with CO_2 than does that with low silica content, such as a silicate with a 2:1 silica:soda ratio.

If a forced draft of hot air is used, or if evaporation is accelerated by the application of thermal energy following the direction of the arrow in Figure 5, it can be seen that the liquid silicate changes first to a viscous liquid and then to a semi-solid and finally to a dehydrated liquid. On further evaporation, possibly by heating, hydrated glass and then conventional glass can be obtained, in principle. Silicates of higher $SiO_2:Na_2O$ ratios undergo these changes more rapidly. In the case of a conventional mold and core hardening process, however, the last changes may not occur, and the highest required strengths should be obtained within the region of the dehydrated liquid (Figure 5).

3.   EXPERIMENTAL RESULTS AND DISCUSSIONS

The results and discussion reported herein include: characteristics of sodium silicate solutions and dried silicates; comparison of different core and mold curing processes; and the effect of humidity on bond strength integrity of sodium silicate bonded sand cured by different processes. These data are presented here to further elucidate the theoretical concepts discussed previously in this paper.

3.1 Characteristics of Sodium Silicate Solutions and Dried Silicates

The data in Table 1 show the fundamental properties and characteristics of sodium silicate solutions used in this investigation. Thermal energy was used to evaporate the water, where necessary. The data indicate that the amount of chemically bonded water which remains in the silicate gel (dried silicate) is inversely proportional to silica:soda ratio of the solution. For example a 2:1 ratio has 18.80 percent chemically bonded water remaining the dried silicate, whereas a 3.75:1 silicate ratio has only 8.90 percent chemically bonded water, based on the weight of solid silicate.

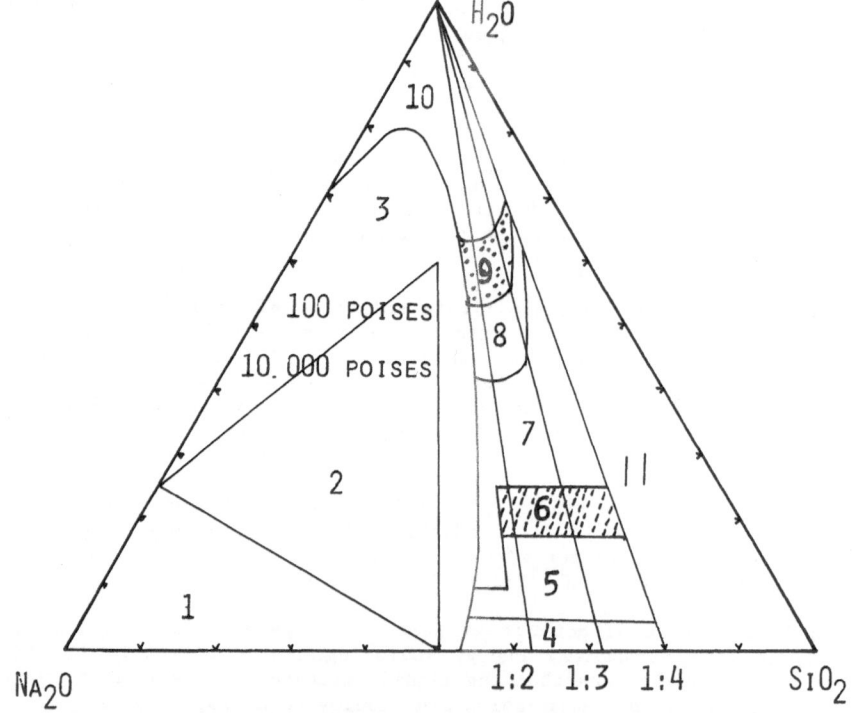

Fig. 5:   Ternary Phase Diagram Showing Phase Changes of
Sodium Silicate Depending on Water Content

(1) Anhydrous sodium orthosilicate and mixtures
with NaOH; (2) Crystalline alkaline silicates;
(3) Uneconomical partially crystallized mixtures;
(4) Glasses; (5) Uneconomical hydrated glasses;
(6) Dehydrated liquids; (7) Uneconomical semi-
solid; (8) Uneconomical viscous liquid; (9) Ordinary commercial liquids; (10) Dilute liquids;
(11) Usually unstable liquids and gels

Table 1. Fundamental Characteristics of Sodium Silicates
(Dry and Solution)

Silicate Weight Ratio SiO_2/Na_2O	Na_2O (%)	SiO_2 (%)	Total Solid Silicate in Solution (%)	Total Water in Solution (%)	Chemically Bonded Water of Dried Silicate (%)	Viscosity (C_p)	pH value
2.0:1	14.70	29.40	44.10	55.90	18.80	400	12.7
2.4:1	13.85	33.20	47.05	52.95	15.80	2100	12.0
2.58:1	12.45	32.10	44.55	55.45	14.83	780	11.8
3.22:1	9.15	29.50	38.65	61.35	12.40	420	11.3
3.75:1	6.75	25.30	32.30	67.95	8.90	220	11.1

The core samples were made in accordance with the American Foundrymen's Society (AFS) standard sand core specimen called "dog bones." These compacted sand cores are shaped like dog bones. The core tensile strength (or compressive strength) were measured on the Dietert No. 405 Universal Tensile Strength Machine. The initial core strengths were usually recorded one hour after curing to allow for complete syneresis [15].

Table 2 shows a comparative data for the three curing processes used in this investigation. The data indicate that cores cured by the thermal energy process (hot air) were slightly stronger than those cured by the microwave energy process because more water was removed from the core samples cured by the former process (150C) than the latter process (100C). The CO_2 gas cured cores had the weakest bond strength because the bond consisted of silica hydrogel, which still contained a considerable amount of unevaporated free water and all chemically bonded water. The combined process of CO_2 and hot air showed a higher bond strength than those cores cured by the CO_2 gassing only because of continuous dehydroxylation of the free water from the bond film.

Table 2. Effect of Core Curing Processes on Sodium
Silicate Bonding in Foundry Sand, No Additive

Type of Sodium Silicate SiO_2/Na_2O	Core Tensile Strength, 1-Hour After Curing			
	Microwave Process (6kW for 5 min) $kPax10^3$	Hot Air Process (150C for 10 min) $kPax10^3$	CO_2 Gas Process (10-15 sec gassing) $kPax10^3$	CO_2 Gas (2-5 sec) & Hot Air Process (150C for 5 min) $kPax10^3$
2.0:1	2.48	2.76	0.41	1.45
2.4:1	2.34	2.55	0.34	1.45
2.58:1	2.34	2.48	0.34	0.34
3.22:1	2.21	2.40	0.07	0.00
3.75:1	2.00	2.07	0.00	0.00

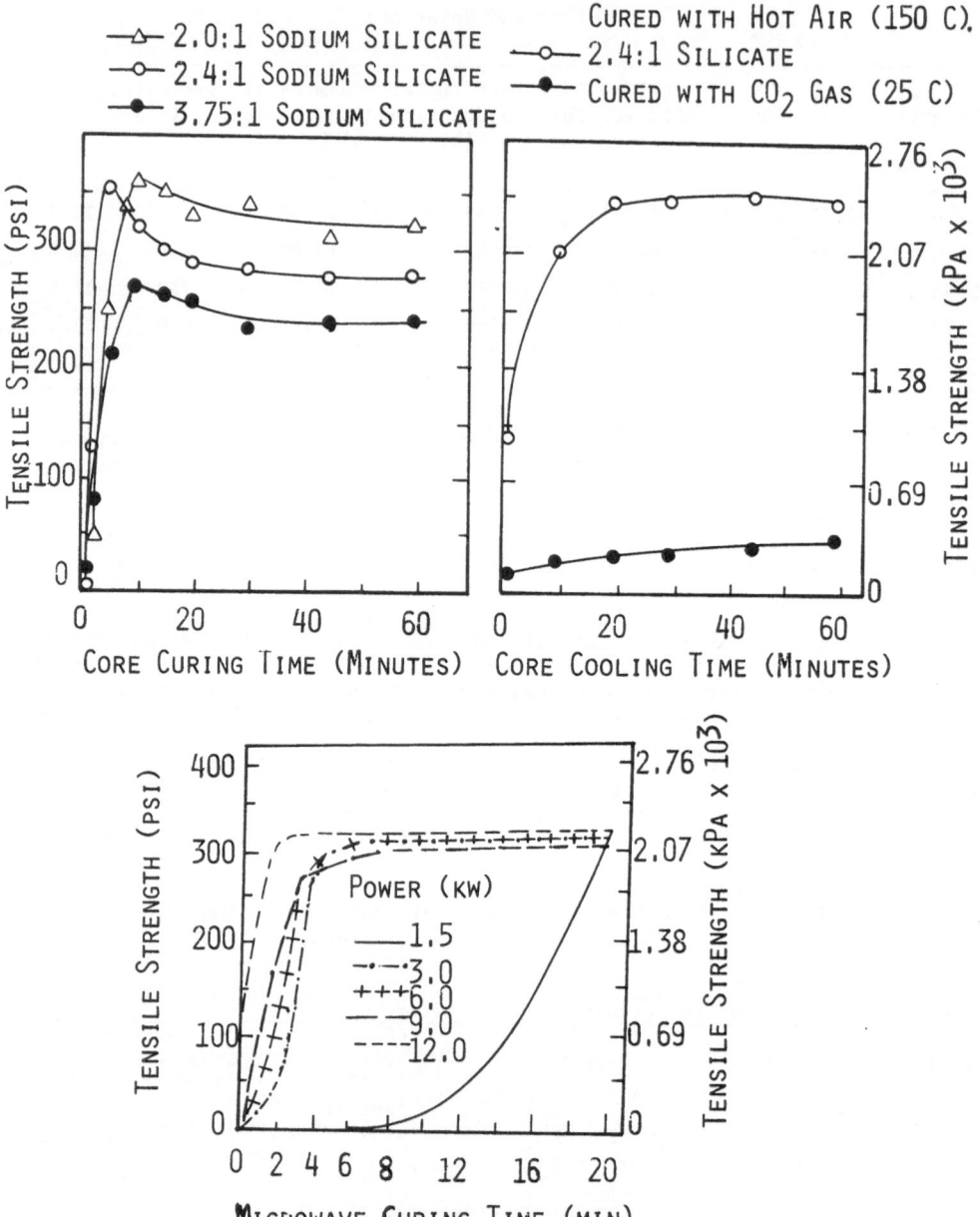

Figure 6. Core Curing Characteristics of Different Processes
          a) Thermal Energy Process (150C at 50 c.f.m.)
          b) Effect of the Phenomenon of Syneresis
          c) Microwave Energy Process

3.2   Comparison of Different Curing Processes on Sodium Silicate Bonded Sand

Three distinct curing processes used in this investigation were the CO_2-process, microwave energy process and the thermal energy process. All the CO_2 cured core samples were gassed for 15 seconds at a pressure of 0.14 x 10^3kPa (20psi). The cores cured by the microwave process were all processed for 5 minutes at 6 kilowatts. The cores cured by thermal energy process utilized convection air at 150C (302F) blown at 50 c.f.m. through an oven.

Figure 6 shows the curing characteristics of sodium silicate bonded sand. Figure 6(a) shows that thermal energy process achieves optimum bond strength within the first ten minutes; and Figure 6(a) shows the effect of the phenomenon of syneresis, where maximum bond strength is achievable 30 minutes after curing. While the curing time of CO_2 process is instantaneous (about 20 seconds), the optimum curing time for microwave energy is about 5 minutes.

Table 3 shows the superiority of thermal energy curing process over CO_2-process and microwave energy regarding bond strength even when different types of foundry sand were bonded with sodium silicate.

Table 3.   Sodium Silicate Bonding in Different Foundry Sand Using Different Curing Processes (1.5 Percent Solid Silicate of 2.4:1 Sodium Silicate, BOS)

| | Core Tensile Strength, 1-Hour After Curing, kPa($\times 10^3$)* | | | |
| | Type of Core Curing Process | | | |
Type of Sand	Microwave Process (6kW for 5 min) kPax10^3	Hot Air Process (150C for 10 min) kPax10^3	CO_2 Gas Process (10-15 sec gassing) kPax10^3	CO_2 Gas (2-5 sec) & Hot Air Process (150C for 5 min) kPax10^3
Wedron 5010 Silica Sand	2.34	2.55	0.34	1.45
Chromite Sand[†]	2.34	2.34	0.21	1.03
Lake Michigan Silica Sand	2.14	2.21	0.34	0.74
New Jersey Silica Sand	1.24	1.38	0.14	0.55
Olivine Sand[†]	1.03	1.56	0.14	0.41
Michigan Bank Sand	0.83	0.97	0.03	0.14

*Pounds/sq.in. x 6.895 = kilopascals (kPa) x 0.145 pounds/sq.in.
Pascal = Newton/sq. metre
[†]Unit volume equal to that of Wedron 5010 sand

3.3 Effect of Humidity on Bond Strength Integrity of Sodium Silicate Bonded
     Sand Cured by Different Process.

In order to monitor the effect of different humidity levels on sodium
silicate bond integrity, some cores were stored in humidity chambers for at
least 24 hours. The retained bond strength and the water absorbed by the
sample cores within the 24-hour period were measured concurrently by the
weight change of the core before and after exposure to humidity. Different
humidity levels were established by placing supersaturated salt solutions
of the following materials in sealed chambers (desiccators) as described in
ASTM E105-15 [16]: potassium sulphate for 97% RH, ammonium sulphate for 81%
RH, calcium nitrate for 52% RH and magnedsium chloride for 32% RH.

Table 4 shows that after exposure to humidity the bond strengths of all
core samples cured by microwave energy were zero, while those cured by the
thermal energy (hot air) and CO_2-process still retained considerable strength.

Table 4. Sodium Silicate Bond Integrity (Retained Bond Strength)
          After Exposure to 97% R.H. for 24 Hours. (Similar Core
          Samples as in Table 3.)

Type of Sand	Microwave Process $kPax10^3$	Hot Air Process $kPax10^3$	CO_2 Gas Process $kPax10^3$	CO_2 Gas (2-5 sec) & Hot Air Process $kPax10^3$
Wedron 5010 Silica Sand	0.0	0.14	0.21	0.41
Chromite Sand	0.0	0.10	0.14	0.34
Lake Michigan Silica Sand	0.0	0.10	0.14	0.28
New Jersey Silica Sand	0.0	0.10	0.10	0.14
Olivine Sand	0.0	0.05	0.07	0.14
Michigan Bank Sand	0.0	0.0	0.0	0.0

The cores cured by the combined process of CO_2 and thermal energy maintained
highest bond integrity among all the processes. These results are in perfect
agreement with the concepts discussed previously in theory. Microwave-cured
cores could not maintain any strength because of the hygroscopicity of the
bond. At 150C, the thermal energy process removed all the free water and
some of the hydroxyl group (chemically bonded water) and therefore rendering
the silicate bonds slightly hydrophobic. The CO_2 cured cores even though
contained a considerable amount of water after gassing, the presence of
Na_2CO_3 caused some degree of hydrophobicity. Of course, the combined process of CO_2 and hot air created the highest hydrophobic bond instead of
being hygroscopic; and therefore, the bond integrity was high after humidity
exposure. The disparity in bond strength existing among the different types
of sand is the result of the chemistry of the sand, surface area and impur-
ities in the sand.

SUMMARY AND CONCLUSIONS

Three different processes were used to cure sodium silicate bonded sand; namely CO_2 gas, microwave energy, and hot air. The Hot Air Process (HAP) (thermal energy) offers a realistic alternative process for curing sodium silicate bonded cores and molds because there is no hazard to either the environment or the operator. This investigation showed that the cores cured by the Hot Air Process were stronger and had a higher resistance to humidity than those cured by microwave process. The Hot Air Process combined with CO_2 Process produced strong cores, and had the highest resistance to humidity without the use of any additive, provided low silicate ratio (2.0:1 to 2.60:1) binder was used.

Treating sodium silicates in the light of colloidal chemistry helps one to understand their polymerization, gelation and bonding mechanism in foundry sands.

REFERENCES

1. Kirkham, A., "CO_2 and Why," The British Foundryman, Vol. 51, No. 7, 1958, pp. 323-330.

2. Sarkar, A.D., "Carbon Dioxide Process," Foundry Trade Journal, Nov. 1961, pp. 537-541.

3. Owusu, Y.A., Sodium Silicate Bonding in Foundry Sands, Ph.D. Dissertation, The Pennsylvania State University, Pennsylvania, USA, May 1980.

4. Owusu, Y.A. and A.B. Draper, "Inorganic Additives Improve the Humidity Resistance and Shakeout Properties of Sodium Silicate Bonded Sand," American Foundrymen's Society Transactions, Vol. 89, 1981, pp. 47-54.

5. Owusu, Y.A., "Processes of Improving Humidity Resistance and Collapsibility of Sodium Silicate Bonded Sand," University of Cairo 2nd Conference on Mechanical Design and Production Proceedings, Egypt, 1982.

6. Hurd, C.B., "Theories for the Mechanism of the Setting of Silicic Acid Gels," Chemical Reviews, Vol. 22, 1938, pp. 403-422.

7. Wehl, W.A. and E.A. Hauser, "Polymerization of Silicic Acid into Silica Gel," Kolloid Zeit, Vol. 124, 1961, pp. 72-76.

8. Nicholas, K.E.L., The CO_2-Silicate Process In Foundries, Published by British Cast Iron Research Association (BCIRA), Alvechurch, Birmingham, United Kingdom, 1972.

9. Wales, W.F., "Evaluation of Silicate Binders for Ceramic Mold Casting of Steel," American Foundrymen's Society Cast Metals Research Journal.

10. Vlasoc, A.F., "Methods of Improving the Knockout Properties of Sodium Silicate Bonded Sands," Russian Casting Production, 1964, pp. 232-234.

11. Skazhennik, V.A., A.A. Semenko, I.S. Sychev, and A.A. Limonova, "Easily Knocked-out Silicate Bonded Sands," Russian Casting Production, 1974, pp. 361-362.

12. Eitel, W., Silicate Science--Silicate Structures, Vol. 1, Academic Press, New York and London, 1964.

13. DeBoer, J.H., M.E.A. Hermans, and M. Vleesken, "Chemisorption and Physical Adsorption of Water on SiO_2 Surface," Koninkl. Ned. Akad. Proceedings, Vol. 60B, 1957, pp. 45-49 and 234-244.

14. Vail, J.G., Soluble Silicates, Vol. 1, Reinhold Publishing Corp., New York, 1952.

15. Owusu, Y.A., G.S. Cole, and R.M. Nowicki, "Characteristics of Sodium Silicate Bonded Microwave Cured Cores," American Foundrymen's Society Transactions, Vol. 87, 1979, pp. 605-612.

16. American Society for Testing Materials, "Maintaining Constant Relative Humidity by Means of Aqueous Solution," ASTM E-104-51 Part 41, 1977, pp. 93-96.

HYBRID ENERGY SYSTEMS

A Model for Assessing the Economic and Energy Savings Implications of Cogeneration with Steam Turbines in Citrus Engineering Plants

P.J. BISHOP, HAROLD CARPENTER, and A. MINARDI
Department of Mechanical Engineering
University of Central Florida
Orlando, Florida 32816, USA

ABSTRACT

A cogeneration system using a noncondensing steam turbine to simultaneously provide electricity and process steam to a citrus plant was modeled in order to assess the source energy savings and the economic implications with the employment of this type system under conditions of time varying plant energy demand.

Average monthly energy demand data from one citrus plant was analyzed. It was determined that the important parameter, in addition to a minimum demand level, for assessing economic acceptability is the demand thermal to electric ratio. One set of steam conditions will not necessarily provide the maximum source energy savings and at the same time be the most economically beneficial. The values of the economic criteria will remain relatively constant over a range of rated turbine capacities for each set of steam conditions.

NOMENCLATURE

A, B	constants dependent on steam temperature and pressure
E, E_i, E_o	electrical power load, from utility, sold to utility
E_r	rated load of turbine
f	fraction of turbine loss due to mechanical losses
H, H_b, H_p, H_t	thermal energy (discharge, from boiler, by plant, by turbine)
HPG	thermal discharge/electric power output of operating turbine
h	enthalpies
Δh_s	isentropic enthalpy change
m	fraction of total turbine steam flow rate at a state
$\dot{m}_o$, $\dot{m}_r$	steam rate (at zero turbine load, rated load)
$\dot{m}_t$	turbine theoretical steam rate
N	number of energy demand periods
PL	partial load fraction
PLCF	partial load correction factor
PF	plant electric power factor
PY	number of demand periods per year

Q_{cu}, cb	energy produced by (utility, boiler) during cogeneration
$Q_{u, b}$	energy produced by (utility, boiler) without cogeneration
Δt_j	time interval of jth demand, hours
α	half rate load steam correction factor
η_o, η_r	efficiency at operating load, rated load
η_g	generator efficiency
η_R	reduction gear efficiency
η_{th}	throttle efficiency
η_t	turbine efficiency from inlet to exhaust
η_b	steam generating system efficiency without cogeneration
η_{bc}	steam generating system efficiency with cogeneration
η_u	utility plant overall generating efficiency

1. INTRODUCTION

With the realization of the limitations on the size of our current known energy resources has also come an increasing awareness of the need for energy conservation. New as well as old conservation concepts are being studied to determine their utility in the current and forecasted future economic and social environment. In previous studies [1,2] it was found that citrus plants appeared to be good candidates for cogeneration in the State of Florida. But large variation with time of energy demands, both electrical and thermal, required an analytical model to determine the optimum cogeneration system and to assess the economic and energy savings implications of such a system.

The research effort involved development of the model and an assessment of the feasibility of cogeneration in terms of energy and cost savings using monthly energy demand data supplied by one citrus plant.

Most of the energy consumed in citrus processing plants is used in the production of citrus juices and concentrate while very small amounts are involved in packaging fresh fruit. The production of citrus concentrate involves first juicing the fruit, followed by evaporation of the water in the juice to produce concentrate. The pulp is pressed to remove moisture and fed to drier kilns. The pulp is then pressed into pellets and sold as cattle feed. Concentrate from various varieties of oranges is mixed to give proper flavor and sugar content.

Evaporators are the primary users of process steam. Freezing and cold storage are the largest users of electricity. Steam demand is dependent on the flow of citrus fruit to the plant. Electrical demand is dependent not only on this flow but also on the demand for juice products. As a result, the plant energy demands vary with time and the thermal demand and electrical demand will vary somewhat independently of each other. A histograph showing this particular citrus plant's average monthly electric and thermal demands over the approximately 8 months of production is presented in Fig. 1.

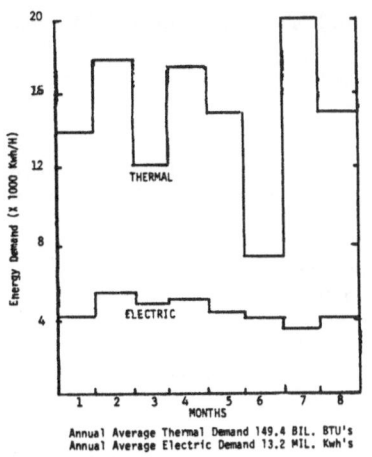

Fig. 1   Citrus Plant average
monthly electric and
thermal demands for
major processing
months

An energy survey of one citrus plant [3] reported that 80% of the plant thermal demand could be satisfied by steam at 5 psig for juice and molasses evaporator heating. Visits at other plants, including the one in this study, confirmed this percentage of use in evaporator heating.

2.   MODEL DESCRIPTION

A noncondensing turbine with an option to include a single extraction point will be used. Although additional electrical generation is possible with a condensing turbine, it requires a greater consumption of source energy than if obtained from a utility. The additional savings in utility electric costs possible with a condensing turbine must more than offset the additional investment and operating costs. With current fuel and electric utility prices this will not happen. Boiler fuel costs are projected to inflate at a greater rate than the cost of electricity [1,2] and therefore it is unlikely that condensing turbines will become more economically attractive in the future.

The choice of automatic extraction will depend primarily on economic advantage. About 80% of the steam demand at citrus plants requires pressures less than 20 psig which can be supplied from the turbine exhaust. The remaining 20% is at pressures between 90 and 250 psig. The additional costs for automatic extraction to supply the 20% of demand at the higher pressures must be offset by lower energy costs.

The model must permit identification of cogeneration system characteristics for optimum energy savings and economics and allow for evaluation of the interrelationship between energy savings and economics. This is accomplished by stepping through a range of rated turbogenerator capacities for a given set of steam operating conditions. The maximum rated capacity and increment is established by input data but the lowest rated capacity is set at 500 kw.

To compute energy outputs of the cogeneration system the plant energy demand must be matched to the turbogenerator energy output characteristics. Electrical demand (E in Kwh/h), thermal to electric demand ratio (HP), and time interval over which the demand occurred are important demand quantities handled by the model.

When it is substituted into Equation (5) the PLCF can be obtained as

$$PLCF = PL/[2\ PL + \alpha\ (1 - PL) - 1] \tag{8}$$

From [4] the value of α over the range of rated loads considered is relatively constant. Turbine efficiencies and half rated load factors were found from Ref. [4].

The enthalpies of steam discharged from the exhaust (State 2) and the extraction point (State 3) is determined by

$$h_2 = h_1 - \Delta h_{s1}\ (n_t + f) \tag{9}$$

$$h_3 = h_1 - \Delta h_{s2}\ (n_{th} + f) \tag{10}$$

where State 1 is defined at the throttle. If no extraction occurs h_3 is set equal to zero. Fig 2 shows the various states across the turbine.

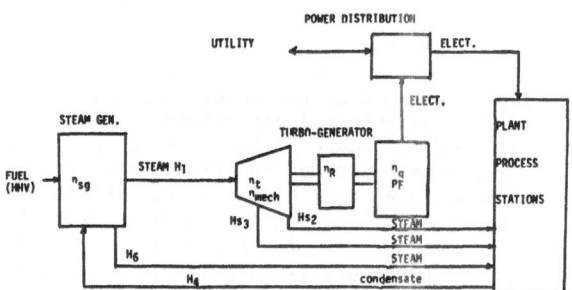

Fig. 2 Block schematic of cogeneration system

The thermal content of the steam discharged from the turbine per unit mass of steam (State 5) in terms of the energy extraction at the process stations (State 4) is then

$$h_5 = m_2(h_2 - h_4) + m_3(h_3 - h_4) \tag{11}$$

The thermal discharge to electric power output ratio of the turbine at operating load (HPG) is given as

$$HPG = h_5/[(m_2\ \Delta h_{s1}\ n_t + m_3\ \Delta h_{s2}\ n_{th})\ n_R\ n_g\ PF] \tag{12}$$

The program assumes only turbogenerators with rated electric loads of less than 2500 kw would be equipped with reduction gears.

2.1 Turbogenerator Energy Output

The turbine throttle to exhaust efficiency was computed using a curve-fitted equation to data [4] over a range of rated capacities from 1000 to 6000 BHP. The equation obtained was

$$n_t = (A + B \times \log_{10} E_r) \times PLCF \tag{1}$$

where

P_{gage}(psig)/T(F)	900/750	600/750	400/750	200/750
A	-0.018	0.255	0.305	0.447
B	+0.199	0.133	0.110	0.083

The relationship between turbine shaft output and steam flow rate is nearly linear and is known as a Williams line. This relationship is used to find PLCF, a correction factor for partial loading, as

$$PLCF = n_o/n_r \tag{2}$$

The turbine steam rate ($\dot{m}$) is equal to the theoretical steam rate ($\dot{m}_t$) divided by the efficiency at load

$$\dot{m} = \frac{\dot{m}_t \times E}{n_o} \tag{3}$$

and at rated load

$$\dot{m}_r = \frac{\dot{m}_t \times P_r}{n_r} \tag{4}$$

Therefore, the equation for the Williams line is

$$\dot{m} = PL \times (\dot{m}_r - \dot{m}_o) + \dot{m}_o \tag{5}$$

Substituting Equation (3), (4), and (5) into Equation (2) gives

$$PLCF = \dot{m}_r/[\ \dot{m}_r - \dot{m}_o + (\dot{m}_o/PL) \ ] \tag{6}$$

The half rate load steam correction factor is defined as

$$\alpha = \frac{(\text{steam rate at } 1/2 \text{ load})}{0.5 \ \dot{m}_r} \tag{7}$$

The electric power generated (E) and the thermal energy discharged (H) are finally obtained from

$$E = PL \times E_r \tag{13}$$

$$H = HPG \times E \tag{14}$$

2.2 Matching Demand Strategies

Matching the plant energy demand to the turbogenerator output is critical to the economic feasibility of a cogeneration system. Fig. 3 represents all possible plant demand points that were considered in the program, divided into six regions where the strategies for cogeneration would be different. A representative turbine energy output characteristic line (solid) is shown. The sloped dashed line is drawn from the origin to the turbine rated load point such that the slope is equal to the thermal to electric output ratio for the turbine at rated electric load.

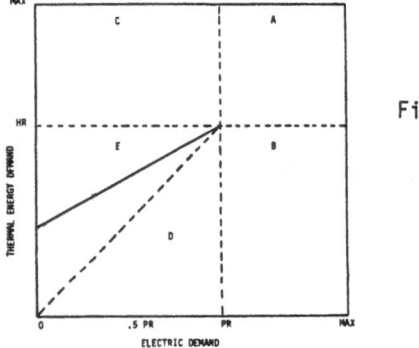

Fig. 3   Energy
demand
plot

Once the applicable region is chosen from the input demand information, the amount of energy supplied from each energy source to satisfy the demand is computed. The power generated, electric power received from or sold to the utility, the thermal energy supplied by the turbine and that direct from the boiler is determined.

In region A both thermal and electric demand exceeds the rated capability of the turbogenerator. In region B demand points exceed the rated electric load of the turbogenerator but are less than the rated thermal discharge. Since the turbogenerator loading cannot exceed that which would satisfy the plant thermal demand the turbine will operate at a partial load.

Thermal demand exceeds rated turbine thermal discharge but electric demand is less than the turbogenerator rated electric load in region C. If excess electricity is not to be sold, the turbogenerator is operated at partial load. Both the thermal and electric demand in region D is less than the rated turbine load and the turbine loading is computed similarly to region B.

Excess electricity can be generated in region E. If the excess is to be sold then the turbine output is computed similarly to region B; if excess electricity is not sold then the turbine energy output is calculated based upon a partial load. In region F the energy matching for demand points is similar to region D.

Demand points in regions B, D and F fall below the point of intersection of the turbine energy output characteristic line and thermal energy cannot be practically matched to turbine output since zero electric output would occur. Hence, partial loading is limited to 10% or greater.

2.3 Determination of Energy Savings

Energy savings is determined by comparing the fuel savings using cogeneration to the plant's conventional method of supplying energy. The steam flow rate at each jth demand point is calculated by

$$\dot{m}_j = \dot{m}_{1,j} + \dot{m}_{2,j} \tag{15}$$

where

$$\dot{m}_{1,j} = H_t / h_{5,j} \tag{16}$$

$$\dot{m}_{2,j} = H_b / h_6 - h_4 \tag{17}$$

The total energy from fuel used by the utility in providing electricity during the jth period with cogeneration is

$$Q_{cu,j} = (E_{i,j} - E_{o,j}) \, \Delta t_j / \eta_u \tag{18}$$

The total energy from boiler fuel used during the jth demand period with cogeneration is

$$Q_{cb,j} = [\dot{m}_{1,j} (h_1 - h_4) + \dot{m}_{2,j} (h_6 - h_4)] \, \Delta t_j / \eta_{bc} \tag{19}$$

The average annual values of fuel used by the utility and boiler using cogeneration are calculated by

$$Q_{cu} = \sum_{j=1}^{N} Q_{cu,j} / PY \tag{20}$$

$$Q_{cb} = \sum_{j=1}^{N} Q_{cb,j} / PY$$

The average annual values of fuel used by the utility and boiler without cogeneration are calculated by

$$Q_u = \sum_{j=1}^{N} [E_j \, \Delta t_j / \eta_u] / PY \tag{21}$$

$$Q_b = \sum_{j=1}^{N} [H_{p,j} \, \Delta t_j / \eta_b] / PY$$

The average annual energy savings (ES) is determined by subtracting the energy used with cogeneration from the energy used without cogeneration

$$ES = Q_u + Q_b - (Q_{cu} + Q_{cb}) \tag{22}$$

2.4 Economic Criteria

Regardless of energy savings resulting from cogeneration, industry will not adopt a system unless there is a reasonable return on investment for it compared to competing projects. The simple payback period is used most often as the initial criteria for evaluating the economic acceptability of a project. Simple payback period is defined as the length of time it takes to recover the initial investment from the net before tax cash flow savings without discounting for interest or inflation rates. It is calculated as

$$PB = I_c / (AS - CM) \tag{23}$$

where

I_c = additional investment costs for cogeneration

CM = maintenance and operating costs

AS = annual savings with cogeneration

The annual savings is determined by

$$AS = (E - E_i + E_o) \times C_E + (Q_b - Q_{bc}) \, C_F \tag{24}$$

where

C_E = cost of electricity

C_F = cost of fuel

If the project falls within the industry's range of acceptable payback values a more detailed economic analysis using the present worth approach can be accomplished. The rate of return would be obtained after the effects of inflation had been removed. It was assumed that the full investment tax credit of 10% of investment costs could be taken in the first year of operation. Further it was assumed that operation and maintenance costs remain constant over the life of the system except for inflation. Depreciation was computed by the Double Declining Balance Method. The equation for the return on investment analysis is

$$0.9\ I_c = \sum_{K=1}^{N} \left[\frac{\Delta E \times C_E \times (1 + EEE)^{K-1} + \Delta Q \times C_F \times (1 + EE)^{K-1}}{(1 + i)^K \times (1 + r)^K} \right.$$

$$\left. - \frac{CM}{(1 + r)^K} (1 - t) + \frac{D \times t}{(1 + i)^K \times (1 + r)^K} \right] \qquad (25)$$

where

$\Delta E = E - E_i + E_o$, $\Delta Q = Q_b - Q_{bc}$,

i   = inflation rate

r   = rate of return on investment

t   = tax rate

EE   = inflation rate of boiler fuel costs

EEE = inflation rate of electricity costs

The computer program iterates this equation with increasing values of r until the equality is satisfied at a value of n equal to the economic life of the equipment.

The cost of energy and energy inflation rates were determined from the projected prices of fuels and electricity in the South Atlantic Region 1980-95 as published in the Federal Register, Jan. 23, 1980. The general inflation rate was determined from the total price index change through 1990 as published in Chase Econometrics Associates, Inc., forecast dated November 1979.

Reference 5 has reported that steam generator costs vary with size raised to the 0.8 power and that heat exchanger costs vary with size raised to the 0.6 power. The relationships were used to calculate equipment costs as:

$$B_c = B_{cc} \times \text{Boiler Capacity}^{0.6} \qquad (26)$$

$$T_c = T_{cc} \times \text{Turbo-Generator Capacity}^{0.8}$$

where B_{cc}, T_{cc} are cost constants for the boiler and turbine, respectively and B_c and T_c are the system costs of the boiler and turbine, respectively. System costs include equipment costs and all related costs such as engineering, site construction, and installation.

Initial investment costs are calculated as

$$I_c = B_c + T_c - C_c \tag{27}$$

where C_c is the cost associated with the conventional system without cogeneration.

Annual operating and maintenance costs (CM) include those incremental costs due to cogeneration and is defined as

$$CM = L_c + I_c \times f_c \tag{28}$$

where L_c = additional labor costs

F_c = fraction by which investment cost is to be multiplied to obtain 0 + M costs other than labor. Investment and 0 + M costs were determined from information in References 5, 6, and 7.

2.5 Model Output

At each rated turbo-generator load the model computes the (1) turbo-generator capacity, (2) thermal and electrical energy required by the plant, (3) the electricity cogenerated, bought from utilities and sold to utilities, (4) the thermal energy supplied by the turbine and the boiler, (5) the maximum steam boiler rate, (6) annual energy savings with cogeneration, (7) investment, operating and maintenance costs, (8) monetary value of energy savings with cogeneration, (9) the payback period, and the (10) rate of return on investment.

3.   DISCUSSION OF RESULTS

Monthly energy use data over a three year period was reduced to an average monthly demand. These plant demands were matched to turbogenerators with electrical output capacities ranging from 500 kw to 5000 kw in increments of 500 kw. Four different steam conditions at the turbine throttle were analyzed.

(a) 900 psig, 750 F

(b) 600 psig, 750 F

(c) 400 psig, Saturated Temperature

(.d) 200 psig, Saturated Temperature

Automatic extraction at 250 psig was included for rated capacities equal to or greater than 2500 kw when operating at steam throttle pressures of 400 psig or greater, and the exhaust back pressure was set at 5 psig.

The total average annual energy use by the processing plant available for cogeneration was 140×10^9 Btu of thermal and 13.2×10^6 kwh of electricity.

Table 1 is the summation of the model results. With the exception of the 200 psig steam case, total plant thermal demand was supplied by the turbine over the range of rated loads, while only 50% or less of the electrical demand was cogenerated. This was caused by having the thermal to electric demand ratio generally lower than the turbine output ratio. For this particular plant the average thermal to electric demand ratio was about 3.5. Turbines operating with the exhaust and extraction steam conditions used in the analysis will have thermal to electric output ratios of between 5 and 8. The increasing values of maximum energy saved with increased values of steam conditions is also partially explained by this difference in energy ratios, since the higher the values of steam conditions at the throttle the lower the turbine thermal to electrical output ratio. The steam generation efficiencies will increase with increasing steam pressure and temperatures, further enhancing the energy savings at the higher steam pressures.

TABLE 1. SUMMATION OF ANALYSES RESULTS

	THROTTLE PRESSURES			
PARAMETER/RATED TURBO-GEN. LOAD, KW	200	400(MS)	600	900
MAXIMUM SOURCE ENERGY SAVED - 10^9 BTU	27.5 1500	44.4 2500	58.6 2500	73.6 4000-5000
MAXIMUM ELECTRICITY SUPPLIED FROM TURBO-GENERATOR -10^6 KWH	3.8 1500	5.9 2500	6.5 2500	6.9 3000
MAXIMUM THERMAL ENERGY SUPPLIED FROM TURBO-GENERATOR -10^9 BTU	119.4 2000	149.4 3500-4000	149.4 3500-4000	149.4 4000-5000
MAXIMUM STEAL RATE -10^3 lb/hr	77.7 2500	83 3500	71.9 3500-4000	73.6 4000-5000
MAXIMUM PAYBACK PERIOD, YEARS	4.5 500	3.1 500	7.2 2000	9.2 2000
RATE OF RETURN AT MINIMUM PAYBACK PERIOD, PERCENT	14	22.5	7.0	3.0

Average annual energy savings by cogeneration are shown in Figs. 4 and 5 for rated turbo-generator capacities and payback period. In Fig. 4, at rated loads less than the maximum point, the ability of the turbogenerator to satisfy plant demand is limited by its rated capacity. At the rated loads greater than at the maximum point, any increase in the ability of the turbine to satisfy plant electrical demand is more than offset by the decrease caused by an increasing turbine thermal to electric output ratio due to lower partial loading.

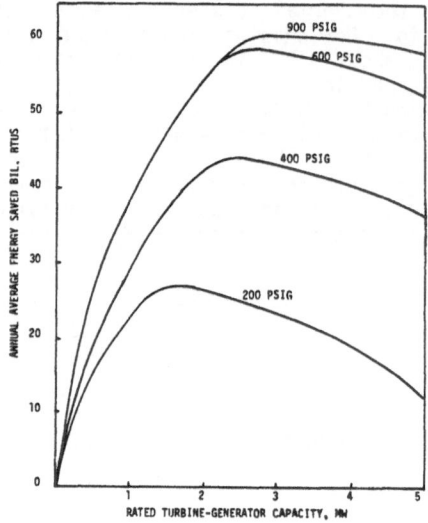

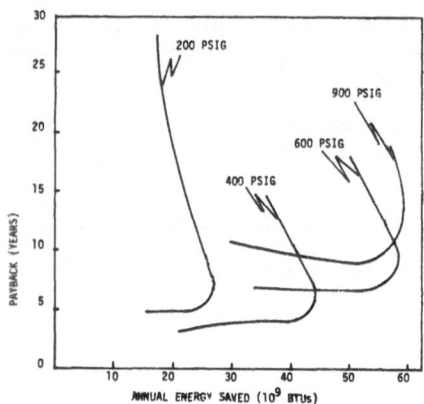

Fig. 5   Average annual energy
          saved versus payback
          period

Fig.4   Average annual energy
         saved versus rated
         turbogenerator capacity

Table 1 shows the total thermal demand is supplied by the 600 and 400 psig
systems at intermediate rated capacities but not at the higher ratings, indicating the thermal demands for some demand points is of such a low value as to
cause the turbines at the higher ratings to be operated at a partial load below
the 10% cut-off.

It is evident from Fig. 5 and Table 1 that the minimum payback period does
not coincide with maximum source energy savings. This is caused by increased
investment and operating costs at higher rated loads.

The 400 psig system gives the lowest payback period and is more economical
if a major replacement of the existing steam generation equipment is necessary.
When pressures and temperatures were used differing from those at the existing
installation, it was assumed that major replacements would be required. The
200 psig system reflects the economics of cogeneration utilizing the existing
steam generation equipment.

Sensitivity analyses were conducted on the 400 psig system and steam generation system efficiency was found to be the most sensitive of all operating characteristic parameters. Figs. 6 and 7 show the percent change in energy savings
and payback period with changes in this parameter. A 5% change in efficiency
results in a 30 to 50% change in energy savings, depending on the rated load
of the turbogenerator; also, a 5% reduction in efficiency increases payback
by 60 to 80% while an increase of 5% gives only a 25% decrease in payback period.

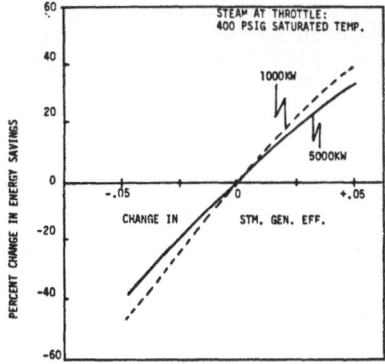

Fig. 6   Percent change of annual energy savings vs change in steam generating system efficiency

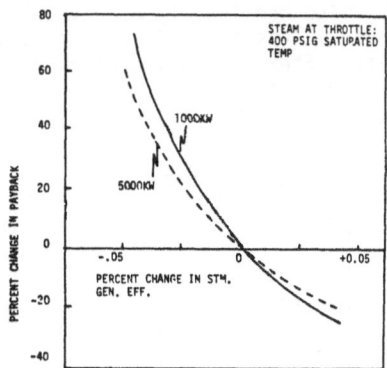

Fig. 7   Percent change in payback period with change in steam generating system efficiency

The effect of errors in turbine efficiencies, Figs. 8 and 9, is significant at the higher rated loads but not at lower ratings. Errors in reduction gear and generator mechanical efficiencies and the power plant factor will have comparable effects on the results. The effects vary considerably over the range of rated loads with small errors being introduced at a rated turbine load of 1000 kw. The relative insensitivity to turbine efficiency suggests the use of less efficient and less costly turbines at the lower rated loads where a single stage turbine could be used instead of a multistage turbine.

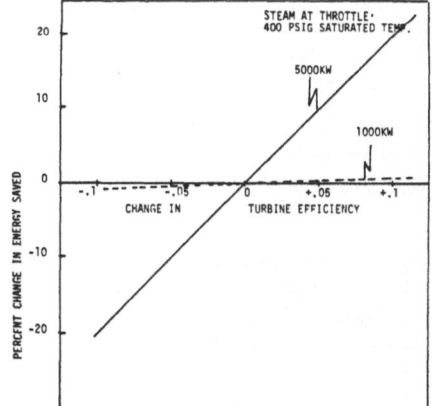

Fig. 8   Percent change in annual energy savings versus change in turbine efficiency

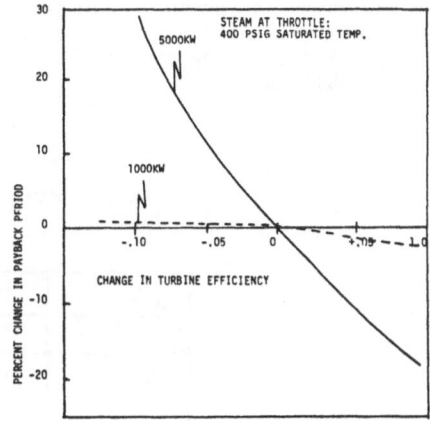

Fig. 9   Percent change in payback period with change in turbine efficiency

The sensitivity of payback period, Fig. 10 to the cost of boiler fuel (natural gas for this analysis) is small when compared to the cost of electricity. From Fig. 11, a doubling of the cost of electricity decreases the payback period by about 50% for a 1000 kw system while the doubling of boiler fuel costs increases payback by only about 2%.

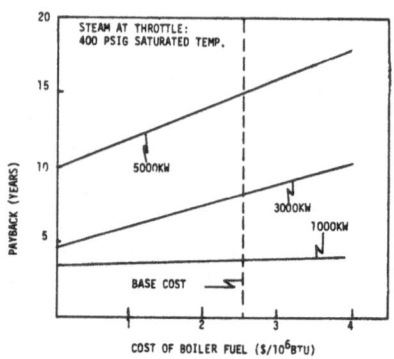

Fig. 10 Payback period versus cost of boiler fuel

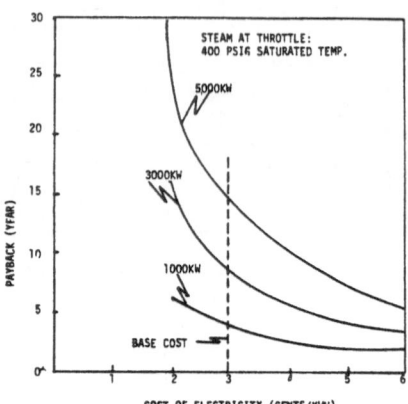

Fig. 11 Payback period versus cost of electricity

Time averaging of plant energy demand data introduces error into the analysis. The greater the dispersion of demand points about the mean the greater will be the error introduced. During months of near full production little error would be introduced since demand is relatively stable. However, during months of partial production the error could be significant. Table 2 shows the comparison of results attained using the averaged monthly data and an annual average yearly demand computed from average monthly data. Where marginal payback periods are encountered the smallest demand period available should be used for the analysis. A further study has shown more quantitative understanding of this using daily demand data and examining the effect of scattering of data points throughout the energy demand plot, Fig. 3. These results will be reported in a future publication.

TABLE 2

COMPARISON OF RESULTS USING AVERAGE MONTHLY AND

AVERAGE YEARLY DEMAND DATA WITH 400 PSIG COGENERATION SYSTEM

RATED CAPACITY, KW	ENERGY SAVED 10^9 BTUs		PAYBACK PERIOD YEARS	
	MONTHLY	YEARLY	MONTHLY	YEARLY
1000	28.5	28.5	3.7	3.2
2000	42.7	44.7	4.2	3.5
3000	43.4	43.5	8.0	7.1
4000	40.2	40.4	11.1	10.0
5000	36.9	37.0	14.7	13.5

The plant demand data used in this analysis did not include demand points in those regions of the demand plot, Fig. 3, where excess electricity could be generated. Thus, the option of selling excess electricity that is cogenerated was not evaluated for this citrus plant.

4. SUMMARY

An analytical model for evaluating the noncondensing steam turbine in co-generation with varying plant energy demands was developed. It was designed to permit plant managers to determine the feasibility of expending monetary resources to allow future energy and dollar savings. The utility of the model will depend upon the availability of plant demand data over sufficiently small time periods to give the desired accuracy of results.

The model further identifies the range of system parameters to be considered in a more detailed design study. The use of approximate relationships between pertinent system operating parameters plus the combining of many parameters essentially limits the use of the model for system design to the role of initial sizing.

The plant energy parameter which most influences the economic acceptability of a noncondensing steam turbine cogeneration system is the plant thermal to electric demand ratio. Plants with demand ratios of less than 4 to 5 will probably not be good candidates for this type of cogeneration because of marginal to unacceptable economics.

There will be an optimum steam condition and rated turbogenerator capacity for each set of plant demands that will be most economically advantageous. This will not necessarily be the system giving the maximum energy savings.

There will be a range of rated turbine sizes for a given set of steam conditions over which the payback period will remain relatively constant.

Further research is now being conducted using actual hourly demand data at a citrus plant to determine the effect of demand period on the feasibility of a cogeneration system.

REFERENCES

1. Porter, W. A. and Bishop, P. J. 1979. Analysis of energy use and recommendations for potential savings at a citrus concentrate plant in Florida. EIES Report, University of Central Florida.

2. Bishop, P. J., Doering, R. D. and Minardi, A. 1980. Potential for cogeneration and waste heat recovery in Florida. Final Report. Star # 79065. Governor's Energy Office.

3. Energy survey at a citrus processing plant - private communication.

4. Shepherd, D. G. 1956. Principle of turbomachinery. Macmillan Publishing Co. pp. 360. New York.

5. Gerlaugh, H. E., et. al. 1980. Cogeneration technology alternatives study (CTAS). General Electric final report. Vol I. Summary Report. NTIS DOE/NASA/0031-80/1. Springfield, Va.

6. United Technologies Corporation. Power Systems Division. 1979. Cogeneration technology alternatives study (CTAS). Vol. I. Summary Report.

7. TRW, Inc. 1980. Industrial cogeneration optimization program. Final report. September, 1979. NTIS/DOE/CS/4300-1. Springfield, Va.

A Thermo-electric System Using Concentrated Solar Energy with Photovoltaic Cells

KAU-FUI V. WONG and SCOTT DORNEY
Mechanical Engineering Department
University of Miami
Coral Gables, Florida 33124, USA

ABSTRACT

The work reported here is a preliminary study of a thermo-electric system using concentrated solar radiation with photovoltaic cells. The system incorporated a novel concentrating system with the designed purpose of utilizing those portions of the solar spectrum that the silicon solar cells were not responsive to. The experimental set-up consisted of a 3.8 cm diameter glass tube, 76.2cm long with the glass silicon cells positioned at equal intervals on its underside. Fresnel lenses were placed so that the incoming solar radiation was concentrated onto the silicon cells. The glass tube was filled with water which absorbed energy from the solar radiation before it reached the cells. The water had its temperature raised nine degrees Fahrenheit while also allowing the solar cells to produce approximately 0.05 watt. This results in an efficiency of around 35%. The results obtained are preliminary, and represent typical conditions in South Florida.

1. INTRODUCTION

As energy consumption increases with population growth and with the fact that people are trying to improve their living standards which are directly related to the amount of energy consumed, current energy sources like fossil fuels are rapidly being depleted. It is imperative therefore, that man should turn to longer term, renewable energy sources. Solar energy shows promise of becoming a dependable energy source, potentially capable of the world's needs. It does not require a highly developed technology and there appears to be no significant polluting effects from its use.

The photovoltaic cell or solar cell can be used to convert sunlight directly to electricity. This effect was first noticed by Becquerel in 1839 when light was directed onto one of the electrodes in an electrolyte solution. It was not until 1954 though that workers became interested in this effect as an energy source.[1] These early cells had efficiencies in the range of about 6%. Presently PV cells are being commercially produced with efficiencies in the range of 10 to 12% with high reliability, however the cost of these cells prohibit them from being economically competitive with current methods of electrical production except in isolated areas where power lines are not in place.

With the current energy problems that face this country the federal government has placed increased emphasis on the research being conducted. The government has set a goal of producing PV cells at a cost of $0.50/watt by 1985 compared with the current cost of about $6/watt.[2] There are currently two approaches that are being pursued to reach this goal. The first is to improve the manufacturing techniques being used to produce the cells. This approach emphasizes methodical cost reduction of the various process steps and lowering unit cost through large scale production. The second approach is to devise efficient methods of concentrating sunlight thereby reducing the area of cell required for a given electrical output. It is this approach that this paper concerns itself with.

There are various types of concentrating systems and devices that have been developed in the past several years. These can be generally classified into two categories, the refracting type, and the reflecting type. Those that fall into the first category employ some type of lens or group of lenses to achieve concentration. The best suited lens for solar applications seems to be the flat fresnel lens. These lenses can be inexpensively made with a high degree of optical accuracy. They are also very lightweight, generally being made of plastic, which is an important consideration since solar tracking devices are necessary. Those falling into the second category that have had wide use are the parabolic cylinder and the paraboloid. The parabolic cylinder is ideally suited to concentration onto a pipe that is located at its focal line and has been used to supply most of the electrical and hot water requirements of a hospital in Hawaii.[3] But due to the higher weight of such systems these do not seem to be the best solution to the concentrating problem.

The main problem however when considering concentration with PV cells is the heating of the cell. As is well known the power output of silicon cells decreases as the cell temperature increases. This increase in temperature is primarily due to the portions of the solar spectrum that is not converted to electricity by the cell. Silicon cells are only responsive to wavelengths between 0.42 and 1.2 microns.[4] All other wavelengths only cause the rise in temperature. Also the usable wavelengths are not entirely converted to electricity. This is due to several factors, the primary factor being recombination time. Recombination time is that period of time during which an electron that has been freed by a photon of light finds its way back into the lattice structure of the silicon. For this reason the grid pattern on the cell must be designed for the expected insolation level.[5] Methods have been proposed to keep the cell temperature down under high insolation. These include flux splitting,[6] external cooling,[7] and heat sinks.[5] Each has its own drawbacks but the external cooling method does allow for the use of that energy not used in the production of electricity. Other methods of temperature control proposed have been special coatings on the cell itself and filters.[4] These methods however do not aid in the total cost reduction of the system and again do not make use of the energy not used in electrical production. A system is proposed in the present paper that makes use of some of the methods already mentioned while introducing a novel approach to their application.

2   THE EXPERIMENTAL SYSTEM

Owing to the lack of available information on the properties of the silicon cells being used in this research a qualitative rather than quantitative approach was used. The primary problem was the method of keeping the cells from overheating. From the literature previously cited it was decided that a combination of the main proposals seemed necessary in order to make use of as much solar insolation as possible. This is to develop an inexpensive filtering device to divert those portions of the solar spectrum not used by the cells and at the same time actively cooling the cells in the event there was an excessive heat build-up.

A materials search was undertaken to determine those materials that had the necessary optical characteristics, were readily obtainable, and were inexpensive. Two such materials found were plexiglas and glass. Both materials are highly transmittant to those wavelengths that silicon cells are sensitive to (below 2 microns). They are also highly opaque to wavelengths above 3 microns. These qualities, as well as the fact that they are inexpensive and readily available, make them ideal candidates for filtering. Also, it is noted that water also has the necessary optical qualities needed for filtering the unusable portions of the solar spectrum.[8] It is also readily available and inexpensive. It was at this point that it was necessary to devise a means to bring these materials together to form a filter.

One such device is the water-filled transparent tube. This device will serve two important functions. The first is that of a filter for the previously mentioned wavelengths of light and the second is that this configuration renders a concentrator. It has been shown that a concentration of between 6.5 and 10 is possible.[9] The decision had to be made whether to use the plexiglass or the glass for the tube. Because predictions as to the temperature that would be reached on the exterior of the tube were not possible and because plexiglass can withstand only temperatures in the range of 150°F for any prolonged periods it was decided that the glass would be a better choice for experimentation. Because concentration levels of interest (200 and above) were not achievable with the water-filled tube an additional method of concentration was necessary. As previously mentioned fresnel lenses are well suited to solar applications and for this reason a 6" square fresnel lens was used for additional concentration. The fresnel lens would concentrate sunlight onto the water-filled glass tube where filtering and further concentration would take place.

A close inspection of the transmittance of glass[9] and the absorptivity of of liquid water,[11] will show that wavelengths above 1.4 microns will be filtered but wavelengths below 0.42 microns will still reach the cells. These wavelengths cannot be used by the cell for electrical production and will therefore show up as heat. To aid in keeping the cell cool the cells will be mounted on a copper tube through which cold water will be passed. With this approach the heat built up in the cell will be carried off through conduction. To insure good thermal contact the copper tube was flattened in those areas that the cells were located and a silicon compound was applied between the rear of the cell and the copper tube. This compound is a highly conductive material commonly used on electronic parts where heat build up is likely to occur. Finally the copper tube was connected to the water-filled tube to make the system complete. Sketches of the system are shown on Figures 1 and 2.

To prevent extensive amounts of energy being refracted out of the focal point of the water-filled tube highly polished aluminum reflectors are placed from the tube to the cells as shown in the sketches, Figures 1 and 2. The cells were cut over-sized from 3" diameter cells that were rated at 0.011 W/cm^2 @ 0.1 W/cm^2 insolation. The effective cell area was approximately 5 cm^2.

The glass tube needed some preparation for the purposes of experimentation. Holes were blown in the tube so that thermocouples could be placed in them for temperature measurements. They were blown at 6" intervals in between each illuminated portion of the tube. Measurements were taken for the incoming cold water at the copper tube entrance and before the water reached the first illuminated point. Temperature measurements were also taken after each illumination. Iron-constantan thermocouples were used for these measurements. Temperatures were recorded every fifteen minutes until a steady state condition was achieved. Voltage and amperage outputs from the cells were monitored continuously. The entire system was placed in a southward facing direction and was tilted at periodic intervals so as to act as a sun-tracking system. The

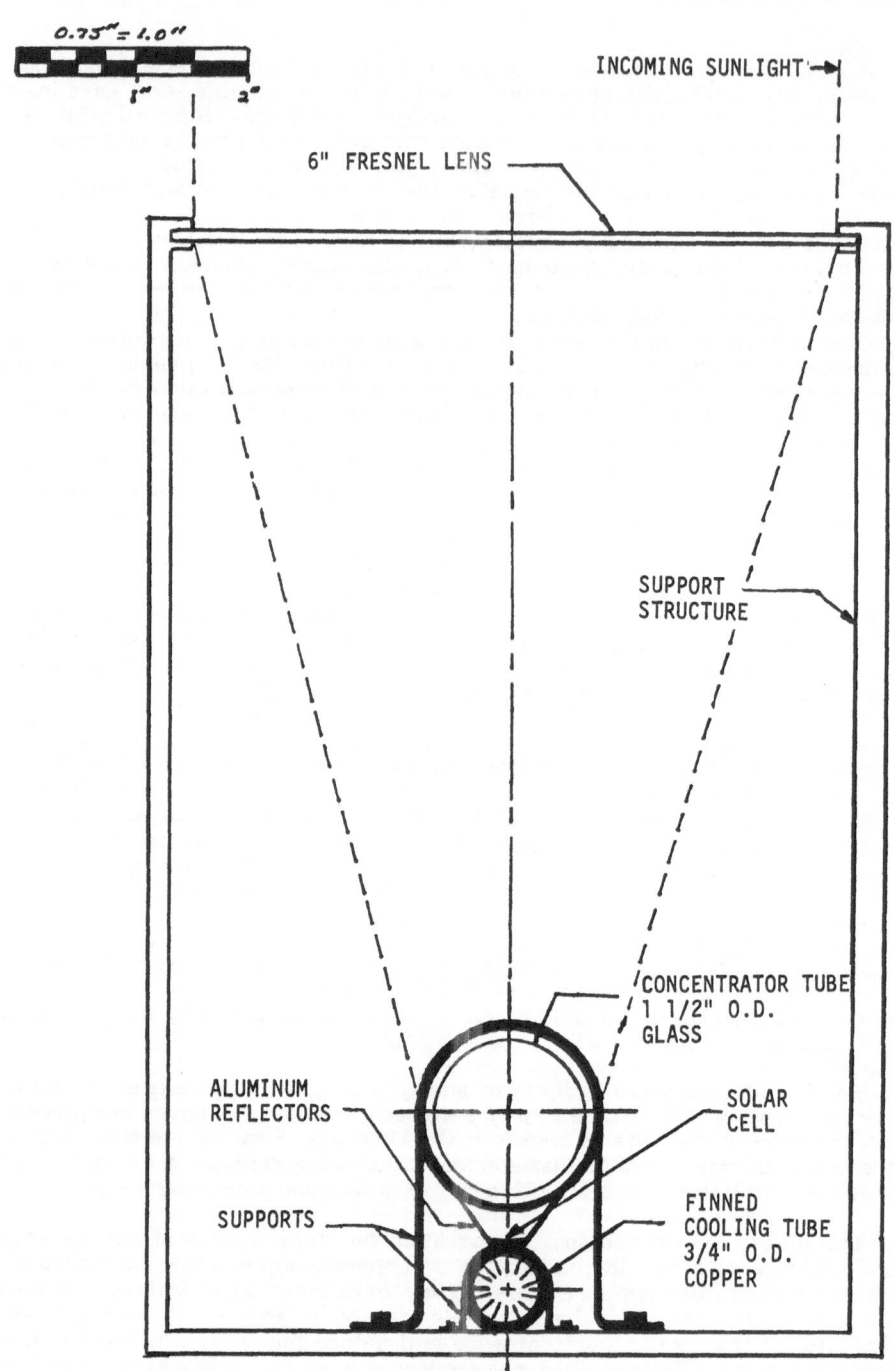

0.75" = 1.0"

1" 2"

INCOMING SUNLIGHT

6" FRESNEL LENS

SUPPORT
STRUCTURE

CONCENTRATOR TUBE
1 1/2" O.D.
GLASS

ALUMINUM
REFLECTORS

SOLAR
CELL

SUPPORTS

FINNED
COOLING TUBE
3/4" O.D.
COPPER

Fig. 1   THE EXPERIMENTAL SYSTEM

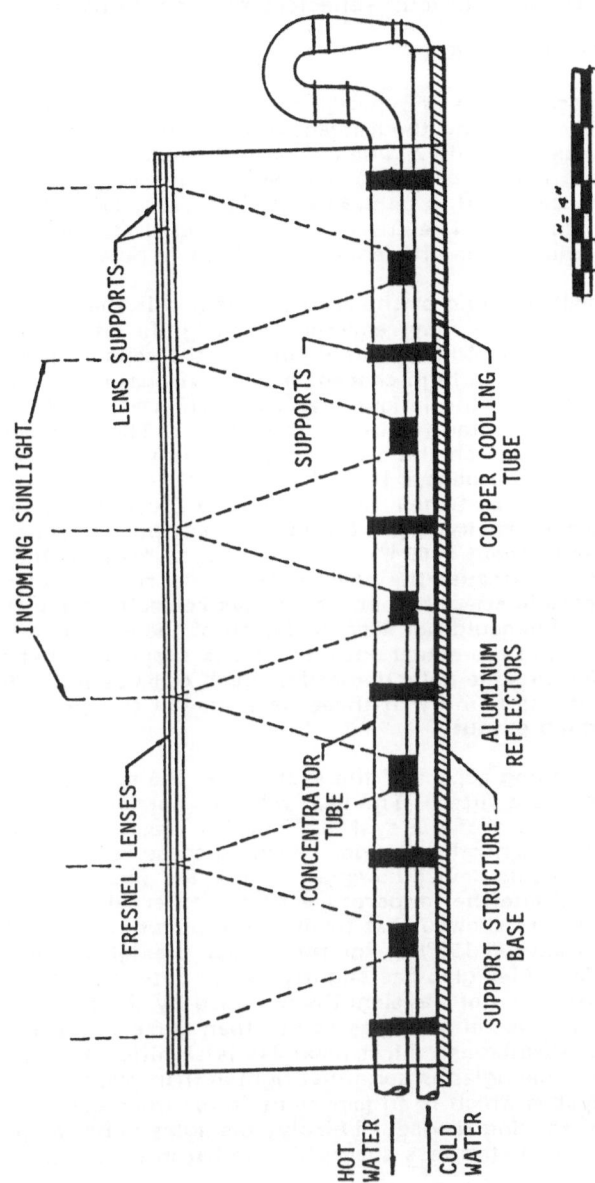

INCOMING SUNLIGHT

LENS SUPPORTS

SUPPORTS

FRESNEL LENSES

CONCENTRATOR TUBE

COPPER COOLING TUBE

ALUMINUM REFLECTORS

SUPPORT STRUCTURE BASE

HOT WATER

COLD WATER

1" = 4"

Fig. 2   THE EXPERIMENTAL SYSTEM

amperage output was used as a guide to determine the correct positioning of the system. It should be pointed out that for the results presented in the next section the polished aluminum reflectors were not in place.

3. RESULTS AND DISCUSSION

Figure 3 represents a set of typical experimental results obtained. The weather was partly cloudy and the ambient temperature was 67°F. The flow rate of the water was 12.9 L/hr. The experiment was conducted between 12 noon and 1:00 p.m., and three sets of temperature readings were taken. The energy output was calculated to be 3.93×10^{-2} Btu/sec. heat energy and 0.0378 W electrical energy. The energy input was estimated to be 0.112 Btu/sec. solar isolation. The efficiency worked out to be about 35%.

Perhaps the most puzzling of the results obtained is the cell's electrical output. It can be seen that they represent an insignificant amount to the overall efficiency. First it should be pointed out that the cells used were not in any way designed for use in high concentration environments. The cells were in fact intended for normal insolations levels (100 mW/cm^2), but because of their availability were selected for use in this study. The concentration level they were subjected to was on the order of 230X. As was mentioned earlier the grid pattern should be denser than the one on these cells. Still if these cells were to operate at their rated efficiency (10%) they would be expected to produce approximately 4.8 W of electrical power. The power produced was in fact only about 0.05 W. on the average. Some of the discrepancy could be accounted for because the polished aluminum reflectors were not in place and an appreciable amount of insolation was refracted away from the cell surface. However, this could not account for all of the difference between the expected output and the actual output. It was suspected that the remaining discrepancy was due to some defective cells. Each cell was individually checked for its output and it was found that three cells were in fact producing only about 4% of their rated output.

The most encouraging aspect of the system was the apparent efficiency of the water-filled tube as a filter. The heat transfer portion for practically all of the calculated efficiency of the system. This approach seems to have some merit as a filtering/concentrating device based on these results. Some improvements can be made on this device however. The glass was selected because it was difficult to estimate the temperature on the external walls. The experimental results showed that for a length of tubing equal to 2½ feet the temperature did not exceed 150°F. For this reason plexiglas could be used for the tube material. Plexiglas has slightly better optical characteristics for the purposes of absorption of wavelengths not used by the cells for electrical production.[4] Also the use of plexiglas rather than glass as the tube material has other inherent advantages. First, plexiglas is significantly less expensive than glass. Second, plexiglas is somewhat lighter than glass resulting in a more lightweight system which is an important factor when considering power requirements for a tracking device. Thirdly, plexiglas is not as prone to break as is glass making the system less susceptible to harsh environments.

4. CONCLUSION

The proposed system could be easily adapted to the structure of most buildings and houses and could be used to provide electrical power as well as hot water. This would aid in combating the current dependence that exists on fossil fuels for our energy requirements and utilizes a non-polluting source as the alternative. Further experimentation is required, with the above mentioned modifications, to gain additional data on system performance. This data can then be used to size the system to meet the specific needs of potential users.

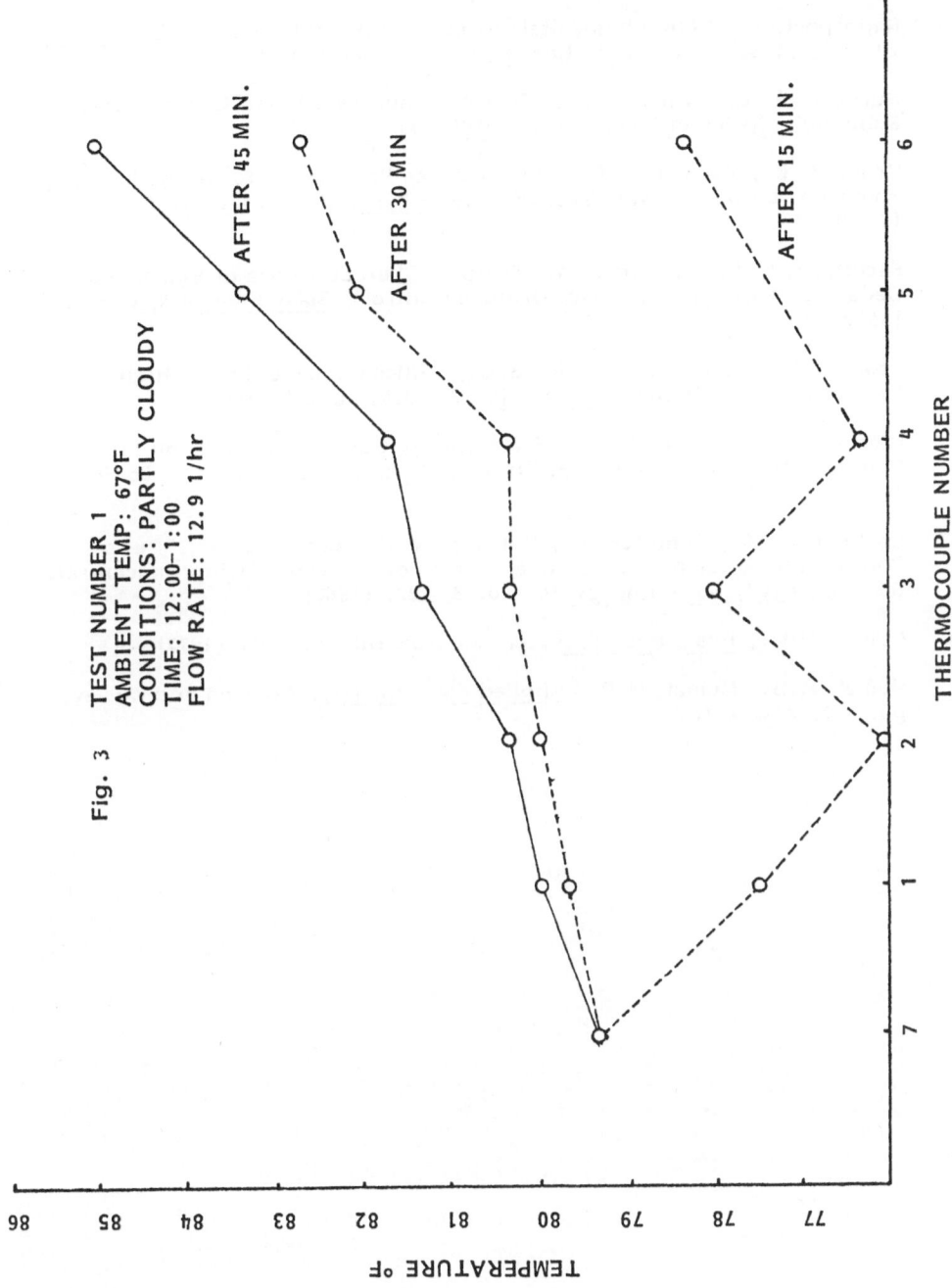

Fig. 3 TEST NUMBER 1
AMBIENT TEMP: 67°F
CONDITIONS: PARTLY CLOUDY
TIME: 12:00–1:00
FLOW RATE: 12.9 1/hr

AFTER 45 MIN.

AFTER 30 MIN

AFTER 15 MIN.

THERMOCOUPLE NUMBER

TEMPERATURE °F

REFERENCES

1. Rappaport, P., "The Photovoltaic Effect and its Utilization," RCA Review
    20, 373, (1959): also Solar Energy 3, December (1959)

2. Burgess, E. L., "Photovoltaic Energy Conversion Using Concentrated
    Sunlight", Optics in Solar energy Utilization II (1976)

3. Brink, D.F., Yasuda, A.K., "Performance Predictions for a Total Energy
    Photovoltaic Concentrator System", ASME Paper No. 80-C2/Sol-7, March
    (1980)

4. Escoffery, C.A., and Luft, W., "Optical Characteristics of Silicon Solar
    Cells and of Coatings for Temperature Control", Solar Energy 4, October
    (1960)

5. Dean, R.H., Napoli, L.S., Liu, S.G., "Silicon Solar Cells for Highly
    Concentrated Sunlight", RCA Review 36, 324, June (1975)

6. Blocker, W., "High Efficiency Solar Energy Conversion Through Flux
    Concentration and Spectrum Splitting", Proc. of the IEEE, Vol. 66 No. 1,
    January 1978)

7. Beckman, W.A., Schoffer, P., Hartman, W.R., Lof, G.O.G., "Design
    Considerations for a 50 Watt Photovoltaic Power System Using Concentrated
    Solar Energy", Solar Energy 10, No. 3, 132, (1966)

8. Ozisik, M.N., Basic Heat Transfer, McGraw Hill, pp. 348 (1977)

9. Meinel, A.B., Meinel, M.P., Applied Solar Energy, Addison and Wesley,
    pp. 143, 234, 267.

Index